Shell Structures for Architecture

Bringing together experts from research and practice, *Shell Structures for Architecture: Form Finding and Optimization* presents contemporary design methods for shell and gridshell structures, covering form-finding and structural optimization techniques. It introduces architecture and engineering practitioners and students to structural shells and provides computational techniques to develop complex curved structural surfaces, in the form of mathematics, computer algorithms, and design case studies.

- Part I introduces the topic of shells, tracing the ancient relationship between structural form and forces, the basics of shell behaviour, and the evolution of form-finding and structural optimization techniques.
- Part II familiarizes the reader with form-finding techniques to explore expressive structural geometries, covering the force density method, thrust network analysis, dynamic relaxation and particle-spring systems.
- Part III focuses on shell shape and topology optimization, and provides a deeper understanding of gradient-based methods and meta-heuristic techniques.
- Part IV contains precedent studies of realised shells and gridshells describing their innovative design and construction methods.

Sigrid Adriaenssens is a structural engineer and Assistant Professor at the Department of Civil and Environmental Engineering at Princeton University, USA, where she directs the Form Finding Lab. She holds a PhD in lightweight structures from the University of Bath, adapting the method of dynamic relaxation to strained gridshells. She worked as a project engineer for Jane Wernick Associates, London, and Ney + Partners, Brussels, on projects such as the Dutch National Maritime Museum in Amsterdam. At Princeton, she co-curated the exhibition 'German Shells: Efficiency in Form' which examined a number of landmark German shell projects.

Philippe Block is a structural engineer and architect and Assistant Professor at the Institute of Technology in Architecture, ETH Zurich, Switzerland, where he directs the BLOCK Research Group, and is founding partner of structural engineering consultancy Ochsendorf, DeJong & Block LLC. He studied at the VUB, Belgium, and MIT, USA, where he obtained his PhD. He has received the Hangai Prize and Tsuboi Award from the International Association of Shells and Spatial Structures (IASS) as well as the Edoardo Benvenuto Prize. He developed thrust network analysis for the analysis of historic vaulted masonry and design of new funicular shells.

Diederik Veenendaal is a civil engineer and a research assistant at the BLOCK Research Group, ETH Zurich, Switzerland. He received his Masters from TU Delft, Netherlands and started his career at Witteveen+Bos engineering consultants, working on groundfreezing analysis for the downtown stations of the Amsterdam North/South subway line and the structural design for the largest tensioned membrane roof in the Netherlands, the ice skating arena De Scheg. His current research involves the comparison of existing form-finding methods and development of new ones for flexibly formed shells and other structural systems.

Chris Williams is a structural engineer and a Senior Lecturer at the University of Bath, UK. He specializes in computational geometry and structural mechanics, in particular for lightweight structures and tall buildings, and his work has been applied by architects and engineers, including Foster + Partners, Rogers Stirk Harbour + Partners and Buro Happold. He worked at Ove Arup and Partners, where he was responsible for structural analysis of the Mannheim Multihalle. Since then, he has worked on such projects as the British Museum Great Court roof, Weald & Downland Museum gridshell, and the Savill Gardens gridshell.

Contents

Acknowledgements — vii

Forewords

    On architects and engineers — viii
    Jörg Schlaich

    Sharing the same spirit — xii
    Shigeru Ban

Introduction — 1

Part I Shells for architecture — 5

  1 Exploring shell forms — 7
       John Ochsendorf and Philippe Block

  2 Shaping forces — 15
       Laurent Ney and Sigrid Adriaenssens

  3 What is a shell? — 21
       Chris Williams

  4 Physical modelling and form finding — 33
       Bill Addis

  5 Computational form finding and optimization — 45
       Kai-Uwe Bletzinger and Ekkehard Ramm

Part II Form finding — 57

  6 Force density method: design of a timber shell — 59
       Klaus Linkwitz

  7 Thrust network analysis: design of a cut-stone masonry vault — 71
       Philippe Block, Lorenz Lachauer and Matthias Rippmann

  8 Dynamic relaxation: design of a strained timber gridshell — 89
       Sigrid Adriaenssens, Mike Barnes, Richard Harris and Chris Williams

  9 Particle-spring systems: design of a cantilevering concrete shell — 103
       Shajay Bhooshan, Diederik Veenendaal and Philippe Block

  10 Comparison of form-finding methods — 115
       Diederik Veenendaal and Philippe Block

  11 Steering of form — 131
       Axel Kilian

Part III Structural optimization — 141

12 Nonlinear force density method: constraints on force and geometry — 143
Klaus Linkwitz and Diederik Veenendaal

13 Best-fit thrust network analysis: rationalization of freeform meshes — 157
Tom Van Mele, Daniele Panozzo, Olga Sorkine-Hornung and Philippe Block

14 Discrete topology optimization: connectivity for gridshells — 171
James N. Richardson, Sigrid Adriaenssens, Rajan Filomeno Coelho and Philippe Bouillard

15 Multi-criteria gridshell optimization: structural lattices on freeform surfaces — 181
Peter Winslow

16 Eigenshells: structural patterns on modal forms — 195
Panagiotis Michalatos and Sawako Kaijima

17 Homogenization method: distribution of material densities — 211
Irmgard Lochner-Aldinger and Axel Schumacher

18 Computational morphogenesis: design of freeform surfaces — 225
Alberto Pugnale, Tomás Méndez Echenagucia and Mario Sassone

Part IV Precedents — 237

19 The Multihalle and the British Museum: a comparison of two gridshells — 239
Chris Williams

20 Félix Candela and Heinz Isler: a comparison of two structural artists — 247
Maria E. Moreyra Garlock and David P. Billington

21 Structural design of free-curved RC shells: an overview of built works — 259
Mutsuro Sasaki

Conclusion

The congeniality of architecture and engineering – the future potential and relevance of shell structures in architecture — 271
Patrik Schumacher

Appendices

Appendix A: The finite element method in a nutshell — 274
Chris Williams

Appendix B: Differential geometry and shell theory — 281
Chris Williams

Appendix C: Genetic algorithms for structural design — 290
Rajan Filomeno Coelho, Tomás Méndez Echenagucia, Alberto Pugnale and James N. Richardson

Appendix D: Subdivision surfaces — 295
Paul Shepherd

Bibliography — 299
List of contributors — 304
List of credits — 309
List of projects — 311
Index — 317

Acknowledgements

The editors would like to express their gratitude to the following people and institutions. Without their help it would not have been possible to realize this book in its current form.

First of all, we are grateful to our independent reviewers, Jack Bakker, Daniel Piker and Samar Malek, for their helpful comments during the final revisions of the manuscript. Specific parts of the book were also proofread, for which we would like to acknowledge Hannah Bands, Victor Charpentier, Allison Halpern, Matthew Horner, Alex Jordan, Lorenz Lachauer, Luca Nagy, Renato Perucchio, Daniel Reynolds, Landolf Rhode-Barbarigos, Edward Segal, Matthew Streeter, Peter Szerzo and Mariam Wahed.

Copyright permissions for specific figures were obtained with the help of Gianni Birindelli, Claudia Ernst, Ines Groschupp, Lothar Gründig, Helmut Hornik, Toni Kotnik, Gabriela Metzger, Matthias Rippmann and Hans-Jörg Schek. We thank Rafael Astudillo Pastor, René Motro, Sergio Pellegrino and John Abel of the International Association of Shell and Spatial Structures (IASS) for granting copyright permission for the use of several publications (Bletzinger, 2011; Ramm, 2004), which were partially reproduced in Chapter 5.

Several people assisted in the editorial process, whom we would like to thank: Masoud Akbarzadeh for his help in draughting the first manuscript; Madeleine Kindermann, Lucas Uhlmann and Ramon Weber for creating and adapting illustrations; Yuki Otsubo, Ryuichi Watanabe and Meghan Krupka for translating the chapter submitted by Mutsuro Sasaki; Shajay Bhooshan, Yoshiyuki Hiraiwa, Monika Jocher, Guy Nordenson, Junko Sakuta and Katrien Vandermarliere for supporting our correspondence and communication with particular authors; and Astrid Smitham for providing editorial assistance during the first months of the book project. We wish to acknowledge the supportive team at Routledge Architecture, Taylor & Francis: Francesca Ford, commissioning editor; Laura Williamson and Emma Gadsden, senior editorial assistants for architecture books; Alanna Donaldson, senior production editor; Janice Baiton, copy-editor; and Christine James, proofreader.

The matter of whether or not Antoni Gaudí used hanging models to design the Sagrada Família was settled with the help of Rainer Graefe, Santiago Huerta and Joseph Tomlow, who independently confirmed that this hypothesis is unsubstantiated and unlikely.

We are especially indebted to the authors for their excellent contributions, and their patience during the editorial process: Bill Addis, Shigeru Ban, Mike Barnes, Shajay Bhooshan, David P. Billington, Kai-Uwe Bletzinger, Philippe Bouillard, Rajan Filomeno Coelho, Maria E. Moreyra Garlock, Richard Harris, Sawako Kaijima, Axel Kilian, Lorenz Lachauer, Klaus Linkwitz, Irmgard Lochner-Aldinger, Tomás Méndez Echenagucia, Panagiotis Michalatos, Laurent Ney, John Ochsendorf, Daniele Panozzo, Alberto Pugnale, Ekkehard Ramm, James N. Richardson, Matthias Rippmann, Mutsuro Sasaki, Mario Sassone, Jörg Schlaich, Axel Schumacher, Patrik Schumacher, Paul Shepherd, Olga Sorkine-Hornung, Tom Van Mele and Peter Winslow. We would also thank their respective companies and institutions for allowing them the time to contribute to our publication and providing environments in which they were able to develop their invaluable knowledge and expertise.

Finally, we would like to thank our friends and families for supporting us during the writing and editing of this book, and in particular Felix, Hannah, Julia, Madeleine, Paul, Pieter and Regine.

FOREWORD

On architects and engineers

Jörg Schlaich

There is still a widespread misunderstanding concerning the role of architects and structural engineers: it is said that architects are the designers of a building from concept to detail, whereas the engineers (only) care for its stability. In fact, it is its function which clearly attributes a building to either an architect or an engineer only, or to both: to an architect only if it is multifunctional in a social context – typically a family house where no engineer is needed – and to an engineer only if it serves a singular structural purpose – typically built infrastructure such as a bridge where no architect is needed. A high-rise building typically needs both, an architect and an engineer.

The more the form of a building or structure develops from its flow of forces, the more it is under the responsibility of the engineer.

Especially due to the fact that most infrastructure, such as towers, power plants, long-span roofs and bridges, is large and long-lasting, a responsible engineer will seek the advice of an architect or a landscape designer when deciding on the material or the scale of their bridge or sports hall in an urban or natural environment.

> It is only culture that can convert our built environment into civilization.

Shells play a special, singular role for engineers. Their shape directly derives from their flow of forces, and defines their load-bearing behaviour and lightness, saving material by creating local employment, their social aspect. This is especially true for thin concrete shells with their characteristic curvatures: single curvature (cylindrical and conical), synclastic (dome-like), anticlastic (saddle-like) or free (experimental).

If well formed, there are no bending but membrane forces only (axial compression and tension) in a shell, permitting its thickness to be around 80mm for reinforced or prestressed concrete, even down to 12mm for fibre reinforced concrete (Fig. 0.1). Though these concrete shells initially do not leave much space for an architect (or even for the fantasy of an engineer), it is fortunately not unusual that the two collaborate fruitfully or that an engineer himself has the courage and imagination to go beyond strict logic.

So Pier Luigi Nervi's Palazzetto dello Sport in Rome would have fulfilled its purpose at considerably lower cost with a tensile ring on vertical supports instead of the inclined Y-shaped columns as built, but it would have looked like a boring and ugly tank (Fig. 0.2). Only a creative engineer could have made such a proposal as built!

In case of the large hypar roof for the Hamburg Alster-Schwimmhalle, the architect insisted that the edge beams should be free cantilevering beyond the facade. This caused vivid discussions, because according to the classical shell literature a hypar shell cannot transfer shear forces from edge beams but needs direct support as evident from the wonderful Candela shells (Fig. 0.3).

But the architect insisted and thus made us find a revolutionary though very simple solution by super-imposing the classical saddle surface for the surface loads with the straight line generators' surface for edge loads (Fig. 0.4). The result was a perfect structure thanks to the insisting architect.

During the last decades concrete shells lost more ground to: 1) cable nets; 2) textile membranes; and 3) steel grids. In terms of their load bearing, these

FOREWORD: ON ARCHITECTS AND ENGINEERS   IX

Figure 0.1 Bundesgartenschau Pavilion, with a 12mm fibre reinforced concrete shell, Stuttgart, 1977

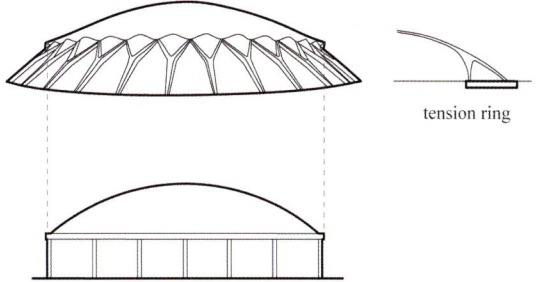

Figure 0.2 Palazzetto dello Sport by Pier Luigi Nervi, Rome, 1958

Figure 0.3 Hypar shell with direct edge beam supports for the Church of San José Obrero by Félix Candela, Monterey, 1959

can also be considered as shells. Let us discuss one example for each of them.

Cable nets

In 1967, the international competition for the sport fields of the 1972 Olympic Games in Munich was won by architects Behnisch+Partners from Stuttgart, even though they hardly fulfilled the requirements. In fact, they instead brought in an idea which was absolutely convincing, to bring together under one continuous and floating roof all sports facilities: the stadium, the sport hall, the swimming hall, and all transitions connecting them (Fig. 0.5). This is exactly what we engineers expect from our architect if structure plays a significant role: an idea, a concept, a proposal, but leaving the structural solution to us. Together with Frei Otto and Fritz Auer from Behnisch+Partners, we developed a

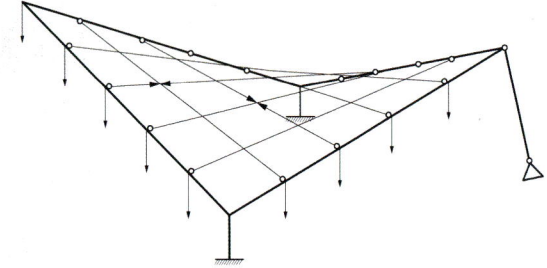

Figure 0.4 The cantilevering hypar of the Alster-Schwimmhalle, Hamburg, 1967

Figure 0.5 The cable-net structure of the Munich Olympic Roofs, 1972 (from left to right: Heinz Isler, Fritz Auer, Frei Otto, Jörg Schlaich, Fritz Leonhardt, Rudolf Bergermann and Knut Gabriel)

Steel gridshells

The courtyard of the Museum of Hamburg History, L-shaped in plan, was to be covered with a glass roof, as light and transparent as possible. The architect Volkwin Marg expressed with his sketch his wishes, his ideas which stimulated our adequate structural solution: a quadrangular mesh or grid, 1.2/1.2m from 60mm × 40mm steel members, diagonally stiffened by thin, prestressed cables (Fig. 0.6). By change of angles, the grid can easily adapt to a smooth doubly curved transition between the two cylindrical shells, which themselves are stiffened by radial spokes (Fig. 0.6). Thus, the fear of the client that his historic building would suffer from the roof could be relieved.

prestressed cable-net structure with 75cm quadrangular mesh-width, adaptable to any shape, subdivided by edge cables, supported by masts, held down by anchors and, finally, covered with Plexiglas. This huge but nevertheless light and floating roof has been very well accepted and is still very popular, thanks to an ideal cooperation between architects and engineers, each of them playing their role in a useful manner.

FOREWORD: ON ARCHITECTS AND ENGINEERS

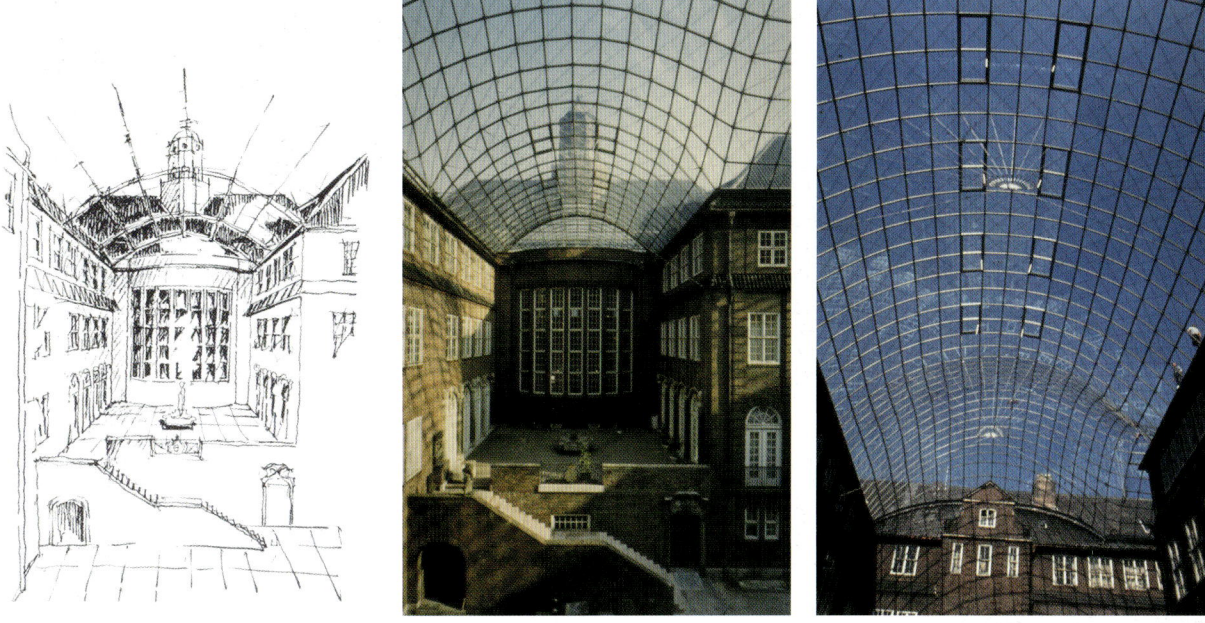

Figure 0.6 Sketched and completed steel gridshell for the Museum of Hamburg History, 1989

Figure 0.7 Roof for the ice skating rink of the Wolfgang Meyer Sports Centre Hamburg-Stellingen, Hamburg, 1994

Textile membranes

As we know from our clothes, we can produce doubly curved surfaces from plane textile with the help of a cutting pattern. If prestressed (by reducing the cutting patterns), they behave under loads as an ideal membrane shell and can even permit convertible roofs, such as those for the Wolfgang Meyer Sports Centre Hamburg-Stellingen (Fig. 0.7).

In a fruitful cooperation of architect and engineer neither of them will impose their opinion because it is only the result that counts! To build in untouched nature can only be justified by creating a responsible building culture.

FOREWORD
Sharing the same spirit

Shigeru Ban

Shell structures are but one of many different, interesting structural systems. If a gridshell structure happens to be suitable for a project, I use it, but otherwise I design another appropriate system. An architect shouldn't concentrate on one type of structure, and should take notice of all possible structural systems. Without an understanding of structures, we cannot design a building. If you would have some

Figure 0.8 Japan Pavilion at Expo 2000, Hannover

preference, it would be very difficult to adjust to different programmes.

Frei Otto is a notable exception. He has his own specialities: cable-net structures, membrane structures and lightweight structures. He does not design everything, yet he's also an architect. When I got the commission for the Japan Pavilion at the 2000 Expo in Hannover (Fig. 0.8), I immediately contacted him, and he agreed to collaborate with me. This had been a dream ever since I was a student. He does not like to make complicated connections, he always strives for minimum effort, minimum labour, minimum materials, and so on; ideals I really share with him. Like me, Frei Otto is not a form-making architect, meaning that we always design according to a combination of structural logics, architectural constraints and the programmes contained within the structure. That's why our collaboration went very well. We share the same spirit.

If I design just a shell, I always try to look for the most appropriate, most structurally efficient shape to reduce bending moments and the size of the

Figure 0.9 Centre Pompidou, Metz, 2010

Introduction

In this book, leading experts from academia and practice describe design and optimization methods for the generation of efficient shell forms and topologies. Some of these techniques are part of the long history of the theory of shells, others have been developed or first used in practice by the authors themselves. Recent years have seen a renaissance of shells. Through advances in computational techniques and power, as well as novel fabrication and construction methods, engineers and architects have been imagining and creating elegant thin-shell structures. For shells to be efficient, their shape should depend on the flow of forces and vice versa and therefore their design requires a process of form finding. The 'ideal' form of a shell may need to fulfil additional architectural, mechanical or technical aspects necessitating some form of optimization. This makes the design and engineering of shell structures a highly involved process.

Shell Structures for Architecture: Form Finding and Optimization offers a comprehensive overview and hands-on textbook for the form and topology generation of shells. Intended for students, researchers and professionals, both in architecture and in engineering, this book presents many new and established methods, and explains them through theory and through working design examples.

This book is written so that each chapter exists as a complete entity in its own right, and can be read as such, but also so that the chapters fit together as a whole. Some of the chapters contain mathematics and for some people this aids their understanding and insight. However, it is equally true that some very gifted shell designers have used physical and computer models in their work and have not relied on advanced mathematical knowledge. Thus, it is not necessary to fully master the mathematics in one chapter before moving on to a following chapter. However, it is recommended to have some understanding of linear algebra and basic calculus, or have some references concerning these topics on hand. The mathematics applied to shell structures comes from diverse fields – geometry, structural mechanics, linear algebra, mathematical optimization, computer science and reliability theory – each with its own notation, and nobody is knowledgeable in all these fields. Engineers may know about structural mechanics, but not be well versed in geometry; mathematicians may know about geometry but not be particularly familiar with structural mechanics. Each person assembles their knowledge in a hotchpotch fashion since it is not possible to know all there is to know about shell structures. Nonetheless, this book is intended as a good starting point and provides advice on further reading throughout, explaining, for different topics, which references will best expand on particular topics.

Shells for architecture

The main application of the methods described in this book are shell structures in the context of the built environment.

> *Shell structures* are constructed systems described by three-dimensional curved surfaces, in which one dimension is significantly smaller compared to the other two. They are form-passive and resist external loads predominantly through membrane stresses.

A 'form-passive' structural system does not significantly, actively change its shape under varying load conditions, unlike 'form-active' structural systems such as cable or membrane structures. A shell transfers external loads to its supports predominantly through forces acting in the plane of the shell surface, which are called membrane stresses and might be compression, or a combination of compression and tension. The word 'membrane' might suggest a film or fabric that can only carry tension, but the compressive stresses in a steel, concrete or masonry shell are still called membrane stresses. A 'thin' shell has to be sufficiently 'thick' to carry these compressive stresses without buckling. Shell structures can be constructed as a continuous surface or from discrete elements following that surface. In the latter case, we speak of lattice, reticulated or gridshells. These terms are synonymous; 'gridshell' will be used throughout this book.

This book distinguishes between three types of geometries for shell structures:

- *Freeform*, free-curved or sculptural shells are generated without taking into consideration structural performance. If they are shaped digitally, then they are often described by higher degree polynomials (e.g. patches of Non-Uniform Rational Basis Splines (NURBS)).
- *Mathematical*, geometrical or analytical shells are directly described by analytical functions. These functions are often chosen for their convenience in performing further analytical calculations and their ability to describe a shell's shape for fabrication purposes. These are often lower degree polynomials (hyperboloids, elipsoids and hyperbolic or elliptic paraboloids), or trigonometric or hyperbolic functions (the catenary).
- *Form-found* shells include natural, hanging shapes associated with the funicular structures of Antoni Gaudí, Frei Otto and Heinz Isler, but also 'strained' gridshells that feature bending stresses. If the shape is found digitally, it is initially parameterized by piecewise or higher degree polynomials. Their final shape is the result of attaining a state of static equilibrium.

The general topic of shell structures is introduced in Part I of the book, 'Shells for architecture'. Chapter 1 elaborates on the vast range of possible shapes for shells. Chapter 2 describes how we can conceive good structural forms. Chapter 3 asks what a shell is, and discusses the matter from a mathematical standpoint. A brief history of the use of physical models in the design of shells is provided in Chapter 4. Chapter 5 presents computational analogues for digital design and subsequent optimization, and focuses on different strategies for geometrical parameterization of curved surfaces. This chapter serves as an introduction to the numerical methods discussed in Part II and Part III of the book.

Form finding

The process of designing form-found shapes is called form finding or shape finding, where the former term is used in this book. Computational models for form finding may be a numerical simulation of the physical model involving hanging chains or cloth, or they might use imaginary properties that could not be simulated physically. In both physical and numerical hanging models, we note that form-active systems are exploited to find the shape of form-passive shell structures. We define the concept as:

> *Form finding* is a forward process in which parameters are explicitly/directly controlled to find an 'optimal' geometry of a structure which is in static equilibrium with a design loading.

For shells, the design loading is typically the dead load, most often being its self-weight. For masonry or concrete shells, or glass-clad steel gridshells this load is dominant. Timber is relatively light, but creeps under load and therefore the permanent dead load assumes a greater importance. The parameters that can be imposed to control the form-finding process are:

- boundary conditions, supports, external loads;
- topology of the model and;
- internal forces, and their relationship to the geometry.

The geometry is an unknown; form finding is the generation of geometry. However, the process may require some arbitrary starting geometry. Once the final shape is found, the numerical model is updated by assigning real physical material and member properties. This new updated model serves as a basis for structural analysis.

Methods developed for the form finding of shell structures are discussed in Part II, 'Form finding'. These methods solve the problem for static equilibrium, either without requiring material properties (force density method in Chapter 6 and thrust network analysis in Chapter 7), or by incorporating (fictitious) material or spring stiffness, and solving for dynamic equilibrium (dynamic relaxation in Chapter 8 and particle-spring simulation in Chapter 9). Part II is summarized through a comparison of these methods, applying them to a single chain in Chapter 10, before concluding with Chapter 11 on steering of forms.

Optimization

When additional objectives or constraints are introduced, methods of structural optimization are needed to solve for them. These methods can be applied to structural shapes resulting from form finding, but also to freeform or mathematical shapes. Different strategies for the solution of constrained optimization problems, in the context of the structural design of shells, are presented in Part III, 'Structural optimization'.

> *Structural optimization* is an inverse process in which parameters are implicitly/indirectly optimized to find the geometry of a structure such that an objective function or fitness criterion is minimized.

One can optimize for multiple objectives and does not have to be constrained to a single design loading. In this case, the objectives are evaluated based on their importance, either a priori using weightings or a posteriori by exploring the Pareto front. In multi-objective optimization, the objectives can include goals that are non-structural. Optimization often leads to a large set of feasible shapes, called the design space. In cases where such a design space is explored and we tend to more gradually develop a form, we may also encounter the term *computational morphogenesis*.

The 'optimal' shape in the design space is subject to a set of given requirements (the *constraints* such as allowable deformations) and is defined with respect to one (or several) *objectives*, such as:

- the minimization of material and weight in a structure, aiming for material economy;
- the minimization of deflections or dynamic vibrations of a structure, ensuring serviceability of a structure;
- the maximization of stiffness (i.e. minimization of structural compliance), aiming for efficient load-bearing structures.

When using mathematical optimization for finding optimal structures, one needs to formulate a general definition of the optimization problem:

$$\text{minimize } f(\mathbf{x})$$

$$\text{subject to } \begin{cases} g_i(\mathbf{x}) \geq 0, & i = 1,\ldots,m \\ h_i(\mathbf{x}) = 0, & i = 1,\ldots,p \\ x_i \in \mathbf{x}, & i = 1,\ldots,n \end{cases}$$

Here, an objective function $f(\mathbf{x})$ of the n design variables $\mathbf{x}$ subject to $m + p$ constraints is minimized. These constraints are also expressed as functions of the design variables and they may be either equalities $h(\mathbf{x})$ – for example, that certain lengths are fixed – or inequalities $g(\mathbf{x})$ – for example, that the height should not exceed some value.

Even though there is only one objective function, it can be a function of various quantities (e.g. cost, energy consumption), which are themselves functions of the design variables. Instead of just adding up these quantities, it is possible, for example, to minimize the sum of the squares of the cost and energy consumption, in which a weighting is assigned to give a monetary value to energy consumption. Alternatively, for multiple objective functions, a Pareto front is explored to assess the trade-offs between those, often competing, objectives.

Optimization problems are encountered in a vast number of fields, and methods to solve them can be roughly divided in two groups: the *local* and the *global methods*.

- Local methods reach a *local optimum*, and therefore offer no guarantee in finding the global one, unless the problem is convex. The most common local methods are based on the computation (or approximation) of derivatives. Therefore, they require the functions to be differentiable, and the variables to be continuous (specific methods have also been proposed in the literature for discrete parameters).
- The most popular global methods to find a *global optimum* are often based on an imitation of phenomena observed in nature. They belong to *metaheuristic* methods, which can be defined as general trial-and-error strategies guiding the search for optimal solutions in hard problems. In these methods, there is no guarantee either that the

optimal solution is systematically found, but they are designed for the purpose of finding a global optimum in complex problems and numerical experiments have demonstrated their efficiency in certain applications. These techniques include *simulated annealing* and *evolutionary algorithms*.

The chapters in Part III use both local and global methods. Chapters 12, 13 and 17 use local, gradient-based methods. The remaining chapters of Part III describe metaheuristic techniques. Chapter 16 uses simulated annealing, whereas Chapters 14, 15 and 18 apply *genetic algorithms*, which are a subset of evolutionary algorithms. These are discussed in Appendix C.

Structural optimization is traditionally classified in three categories depending on the nature of the variables involved:

- *Shape optimization* has variables acting on the geometry of the structure, without modifying the topology. Practically, in discrete structures, the node coordinates are often used directly as parameters to modify the geometry, but more advanced parameterizations are available (see Section 5.4 and Appendix D).
- *Topology optimization* deals with the situation in which the topology of the structure is not prescribed by the designer. The structure's topology is defined by the connectivity of the nodes in the structure, and the existence or absence of elements.
- *Sizing optimization* has variables representing cross-sectional dimensions or transversal thicknesses (the geometry and the topology remaining fixed). For instance, in discrete structures (trusses, rigid frames and gridshells), the areas of the cross sections usually are the design variables.

Chapters 12, 13, 16 and 18 each deal with shape optimization, but towards different objectives and for different structural typologies. The approaches and parameterizations vary as well. Chapter 16 further discusses how to derive a discrete topology, or 'structural pattern', from the resulting shape. Chapter 21 from Part IV explains shape optimization as it has been applied by Mutsuro Sasaki in his practice.

Chapters 14, 15 and 17 apply topology optimization. In Chapter 14, shape variables are also included, whereas Chapter 15 includes sizing optimization. Chapter 17 discusses continuum topology optimization, which aims to find the optimal layout of material in a given region, but applies this also to shape optimization.

Precedents

Throughout the book, examples are given of existing shell structures. Part IV, 'Precedents', provides more specific insights from built examples. Chapter 19 compares two seminal gridshell structures: the Mannheim Multihalle and the roof of the British Museum's Great Court. In contrast, Chapter 20 compares two types of continuous concrete shells: the geometrical shapes by Félix Candela and the form-found shapes by Heinz Isler. The final chapter, Chapter 21, gives an overview of Mutsuro Sasaki's concrete shell structures.

The book concludes with 'The congeniality of architecture and engineering', a short note on the future and potential of shell structures.

PART I
Shells for architecture

CHAPTER ONE
Exploring shell forms

John Ochsendorf and Philippe Block

Shell structures will always have a role for architecture and engineering. More so than any other structural system, shells have the ability to create eye-catching forms, to provide freedom for design exploration and to resist loads efficiently. These attributes call for sustained interest in the mechanics and design of shell structures.

For even the most highly constrained geometry, an infinite number of solutions are available to the shell designer. But each different shell geometry has advantages and disadvantages, and all shells are not equal. How then does the designer find shell forms that are inherently structural? Shell designers can invent forms by taking inspiration from nature, by innovating from precedent structures, or by exploring various form-finding possibilities. In all cases, designers must seek forms that offer multiple load paths for all expected applied loads, and whose formal possibilities are closely linked to the modes of construction. Master shell builders are deeply concerned with the final appearance of their shells as well as the construction processes to create them.

Designers can always learn more from studying historical structures and this is especially true of shells. Traditional masonry shells have a long history in architecture and construction, and the inherent limitations of masonry material require that such structures work primarily in compression. Recent buildings such as the Mapungubwe Interpretive Centre, with structural shells made from unreinforced earthen bricks, demonstrate the potential for contemporary projects to take inspiration from historical construction systems (see page 6).

1.1 Hooke's hanging chain

The shell designer seeks forms to carry the applied loads in axial compression with minimal bending forces. The earliest example of structural form finding for an arch was published by English engineer and scientist Robert Hooke (1635–1703). In 1676, Hooke published ten 'Inventions' in the form of anagrams of Latin phrases in order to protect his ideas. The third invention would later become known as Hooke's law of elasticity, for which he is most known.

The second (Fig. 1.1), describing 'the true Mathematical and Mechanichal form of all manner of arches for building' is given as:

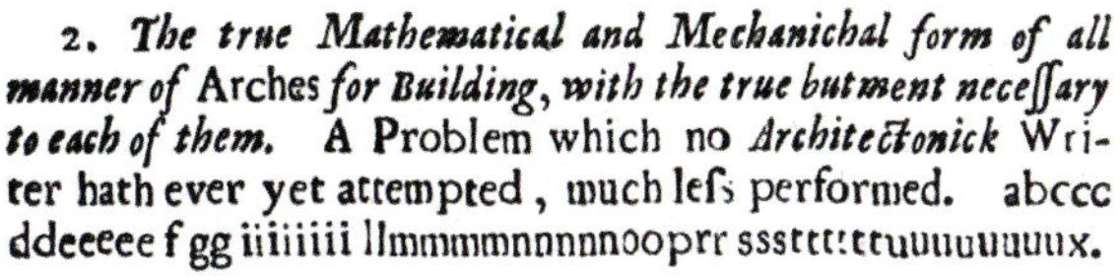

> 2. *The true Mathematical and Mechanichal form of all manner of* Arches *for Building, with the true butment neceſſary to each of them.* A Problem which no *Architeƈtonick* Writer hath ever yet attempted, much leſs performed. abccc ddeeeee f gg iiiiiiii llmmmmnnnnnooprr sssttttttuuuuuuuux.

Figure 1.1 Robert Hooke's anagram on the means to find the ideal compression-only geometry for a rigid arch (Hooke, 1676)

The solution to this architectonic riddle was posthumously published by the secretary of the Royal Society, Richard Waller (1705), and read:

> Ut pendet continuum flexile, sic stabit contiguum rigidum inversum.
> (As hangs the flexible line, so but inverted will stand the rigid arch.)

The idea is simple: invert the shape of the hanging chain, which by definition is in pure tension and free of bending, to obtain the equivalent arch that acts in pure compression.

The form of the ideal arch will depend on the applied loading. For a chain of constant weight per unit length, the shape of a hanging chain acting under self-weight is a catenary (Fig. 1.2). But if the load is uniformly distributed horizontally, the ideal arch would take the form of a parabola, and the chain would take different geometries according to the loading. In addition, the span/rise ratio (L/d) can vary widely, though most shell structures occur in the range of $2 < L/d < 10$. Thus, even a simple two-dimensional arch has infinite possible forms which would act in pure compression under self-weight, depending on the distribution of weight and the rise of the arch.

This principle, which will be referred to in this book as 'Hooke's law of inversion', can be extended beyond the single arch and considered for shell structures of various geometries. In the context of shell structures, the term *funicular* means 'tension-only' or 'compression-only' for a given loading, typically considered as the shape taken by a hanging chain for a given set of loads. Three-dimensional funicular systems are considerably more complex because of the multiple load paths that are possible. Unlike the case of a hanging cable with a single funicular form between its two supports, hanging membranes have multiple possible forms. And unlike the two-dimensional arch, the three-dimensional shell can carry a wide range of different loadings through membrane behaviour without introducing bending.

To continue the analogy with Hooke's hanging chain, a three-dimensional model of intersecting chains could be created. This hanging model could be used to design a discrete shell, in which elements are connected at nodes, or the model could be used to help define a continuous surface. If hanging from a continuous circular support, the model-builder could create a network of meridional chains and hoop chains. By adjusting the length of each chain, various tension-only solutions can be found when hanging under self-weight only. Once inverted, this geometry would represent a compression-only form. Such a model would quickly illustrate that many different shell geometries can function in compression due to self-weight. As an example, Figure 1.3 illustrates three shell structures supported on a circular base: a cone, a shallow spherical dome and a dome with an upturned oculus in the centre. All three forms can contain compression-only solutions due to self-weight according to classical membrane theory (see Chapter 3). Of course some solutions perform better under varying load conditions and the double

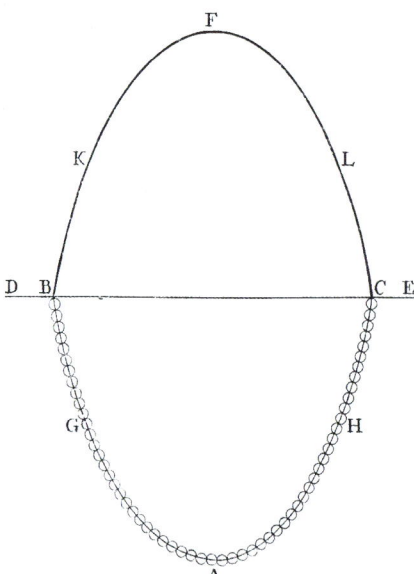

Figure 1.2 Hooke's hanging chain and the inverted rigid catenary arch, as depicted by Poleni (1748)

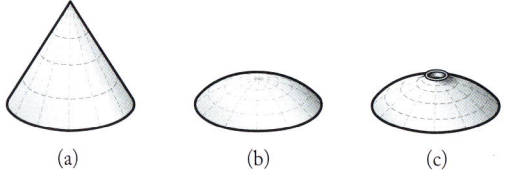

Figure 1.3 Examples of circular-plan shell structures which can act in pure compression under self-weight due to gravity: (a) conical shell; (b) shallow spherical dome; and (c) spherical dome with upturned oculus in centre

curvature of the dome is structurally superior to the single curvature of the cone in the event of asymmetrical live loading. The conclusion is clear: for networks of intersecting elements, there is no unique funicular solution.

1.2 Masonry shells in compression

Masonry shells have been built for centuries around the world in the form of arches, domes and vaults. Such shells exist primarily in compression, because masonry materials such as brick and stone are strong in compression and weak in tension. The challenge is to find geometries that can work entirely in compression under gravity loading. These geometries are not limited only to masonry, and will often provide efficient geometries for structures built of any material. However, for traditional masonry structures, the dominant loading is often due to the self-weight of the structure, and the applied live loadings due to wind or snow have a smaller effect.

As a practical demonstration of the multiple compressive solutions in three dimensions, consider the wide variety of shell geometries constructed in masonry tile commonly known as the Catalan vault or Guastavino method of construction. Compression-only tile vaults supported on a circular plan can take many possible geometries, ranging from conical to spherical as demonstrated by the structures in Figure 1.4. Even in brittle masonry structures, numerous openings can be made in shells and the resulting compressive solutions must therefore flow around the openings, as in the Bronx Zoo dome of 1909 (Fig. 1.4, middle). The supports must be capable of resisting the large reaction forces at the base of these structures, and in particular, shallow domes will create a large outward thrust at the base, which is commonly contained with a tension ring at the base of the dome to maintain equilibrium. Though traditional masonry shells were often constructed without structural calculations, historical vault builders developed a keen awareness of the multiple equilibrium states possible in doubly curved masonry shells.

These surfaces could be built in masonry because they can contain thrust surfaces in compression, similar to a grid of interconnected hanging chains. Thus, Robert Hooke's guiding principle for the

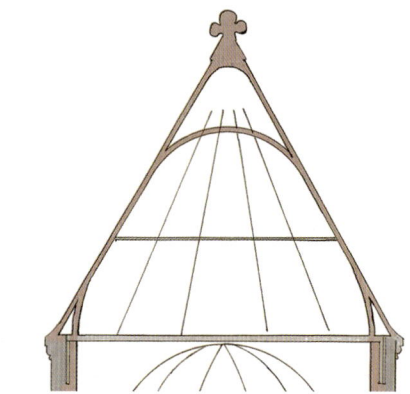

Figure 1.4 (top) Conical shell by Rafael Guastavino, Sr., Bristol County Courthouse, Taunton, MA, 1891, (middle) shallow spherical dome, Bronx Zoo Elephant House by Rafael Guastavino, Jr., New York City, NY, 1909, and (bottom) shallow spherical shell with upturned oculus, Pines Calyx by Helionix Designs, Dover, 2007.

ideal form of arches can also offer guidance on the form of shells. For the primary masonry dome of St Paul's Cathedral in London, Hooke suggested a cubico-parabolical conoid form to Christopher Wren (1632–1723), which is the ideal form of a compressive dome with zero hoop forces (Heyman, 1998) (see also

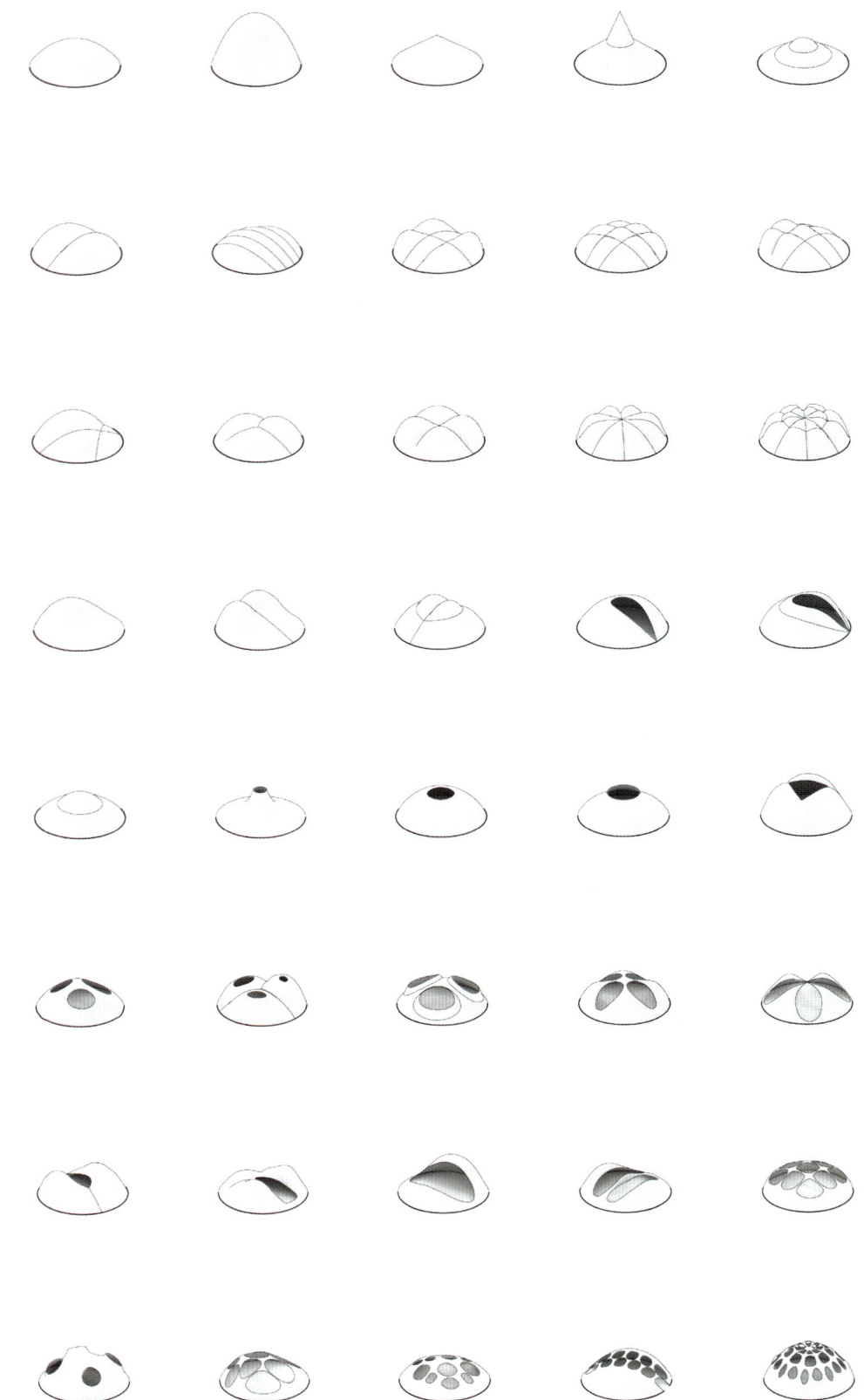

Figure 1.5 Compression-only shell geometries supported on a circular plan found using thrust network analysis

Section 4.2). However, many more forms are possible for shells in masonry.

1.3 An infinite number of structural forms

The challenge of shell design is to find the appropriate form for the given problem. And the joy of shell design is that an infinite number of structural forms are waiting to be discovered. Even highly constrained boundary conditions can still lead to a vastly rich landscape of forms to explore. For example, by continuing the challenge of finding compression-only solutions on a continuously supported circular base, one can discover an infinite array of options. Figure 1.5 demonstrates additional options found through thrust network analysis, which allows the shell designer to redistribute compressive forces within statically indeterminate networks (Chapter 7).

By exploring numerous boundary conditions, additional forms are possible. Historical spiral staircases in masonry are constrained within a cylinder, allowing the compression-only geometries to move beyond the planar support (Fig. 1.6). As with a dome or a cone in masonry, such spiral shells can stand in pure compression through a combination of meridional and hoop forces in equilibrium around a central oculus. It is important to note that the classical analysis methods applied to shells rely only on equations of equilibrium, and do not depend on invoking elasticity or other material properties in order to estimate the membrane forces in shells. The goal of the shell designer is to discover an array of forces in equilibrium with the applied loads.

While leading shell designers of the twentieth century demonstrated that numerous structural forms are waiting to be discovered, other projects demonstrated the pitfalls of shell design. The shell-like forms of the Sydney Opera House were created from sections of a sphere, which simplified the geometry (see also Section 4.3). But the chosen geometry did not have any particular structural advantages, and

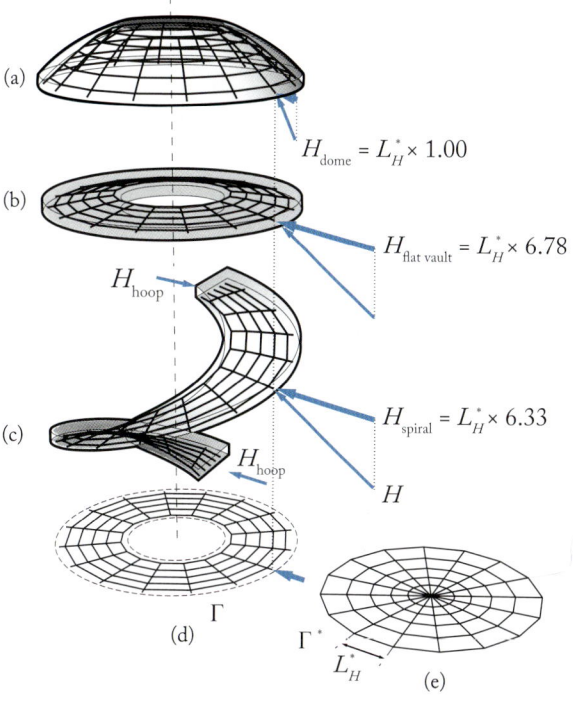

Figure 1.6 Guastavino thin-tile helicoidal stairway at the Union League Club, New York City, NY, 1901, and the force equilibria of (a) a dome with a circular oculus, (b) a circular flat vault with an oculus, and (c) a spiral stair can be explained with (d) the same force pattern topology, with radial arches and meridional hoops. The horizontal thrusts are balanced in the same manner, as visualized in (e) their shared reciprocal force diagram (see Chapter 7).

could not function effectively as a structural shell. The resulting ribcage of beam elements must resist significant bending stresses and was very costly to build. In fact, the project took three times longer to build than originally expected and the final cost was more than fourteen times the original estimate. Though the Sydney Opera House is a successful work of architecture, it is a poor structure and it serves as a cautionary tale to designers seeking to build shells without careful attention to the flow of forces in the conceptual design stage.

1.4 Conclusion

Shell designers can take inspiration from any number of sources. But Robert Hooke's powerful axiom provides a clear path forward: like the inverted chain, forces should flow in axial compression toward the supports with minimal bending. If Hooke's chain is extended into a three-dimensional network of elements and if multiple support conditions are considered, then an infinite number of forms exist for compression-only shells. By minimizing bending forces, designers can build more efficiently and can make better use of limited resources. And by understanding and exploring the infinite possibilities for even a highly constrained design problem, shell designers can continue to discover new forms for centuries to come. The builders of masonry vaults have discovered a remarkable variety of shell forms acting in compression, and the proof of their success can be found in the longevity of these vaults that future generations will continue to admire.

Further reading

- *Form and Forces: Designing Efficient, Expressive Structures*, Allen & Zalewski (2009). This book provides an exciting introduction to graphic statics for design, with a particular emphasis on form finding of funicular structures. Chapter 8 gives an overview of the key aspects to be considered in the design of an unreinforced masonry vaulted structure.
- *Equilibrium of Shell Structures*, Heyman (1977). This book offers a succinct and clear description of basic shell theory, and describes the application of membrane theory to historical masonry structures such as domes and vaults.
- *Structural Analysis: A Historical Approach*, Heyman (2007). This book describes the history of structural theory and introduces Robert Hooke's principle of the inverted hanging chain in its broader historical context.
- *Guastavino Vaulting: The Art of Structural Tile*, Ochsendorf (2010). This book describes the history and technology of the thin masonry vaults constructed by the Guastavino family in more than 1,000 buildings in the late nineteenth and early twentieth centuries.

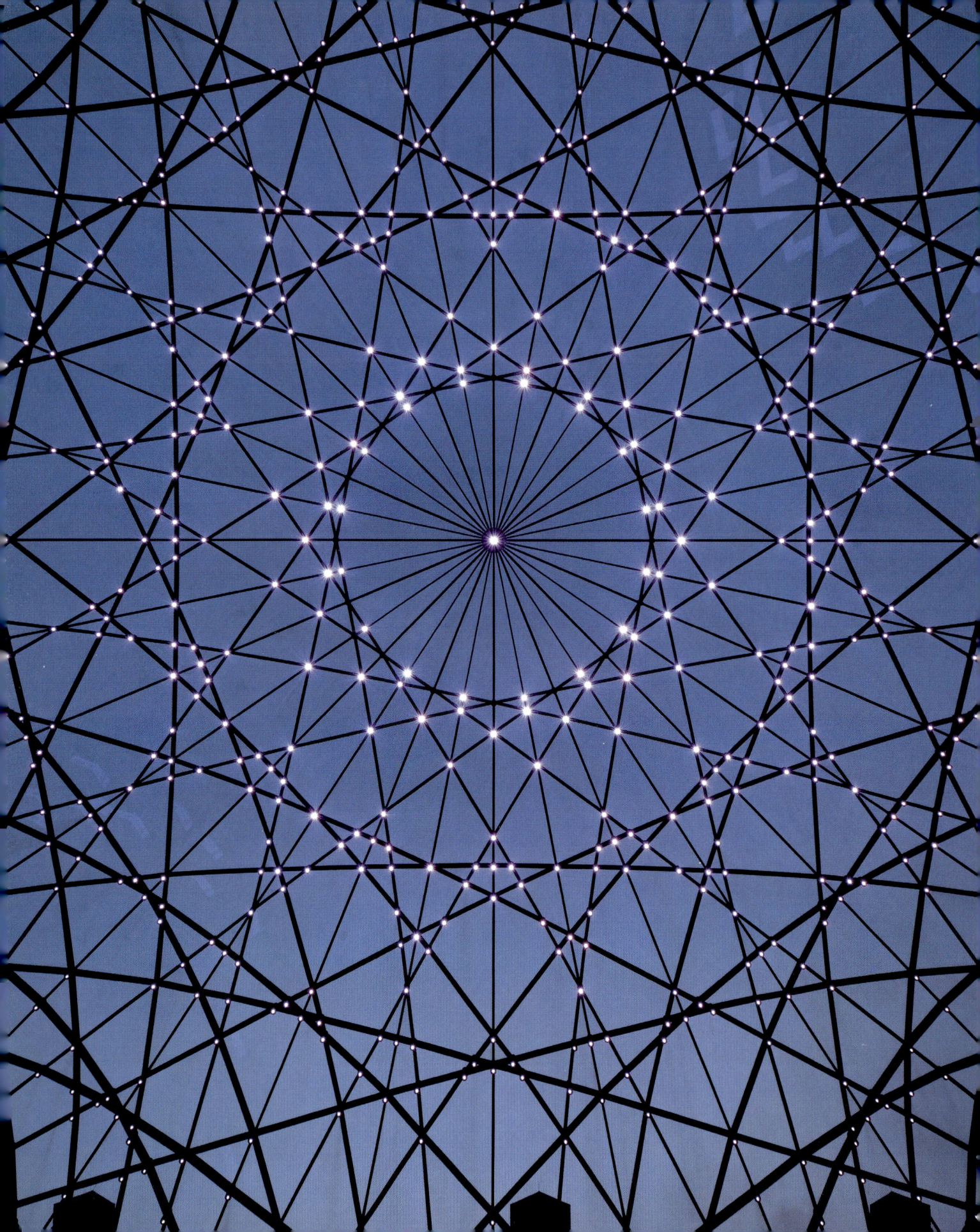

CHAPTER TWO
Shaping forces

Laurent Ney and Sigrid Adriaenssens

If you are reading this text, then you are a student, an engineer, an architect, or a person who is interested in the design of structures, specifically shells. You would like to learn how to design innovative structural forms by using context-sensitive parameters, exploring a variety of materials, creating aesthetical shapes and making your brainchild a reality. This is the creative mission, getting from the dream to the practical truth. But reality has its own constraints. In the world of structures, the dictates of science, statics and gravity, or in other words, the rules of nature itself, play the defining role. How does gravity influence a structure? Without gravity a dome would be *flat*. Gravity lies at the heart of our initial problem, but can also provide the beginning of the enjoyment in a creative design process. Where would be the excitement in designing a flat structure? The context of gravity is our playground. This design driver is a hard constraint but gives birth to a realm of intriguing complex spatial structural shapes.

No gravity, no fun.

Purely statically speaking, a structure is a device that channels loads to the ground. The funicular cable, for example, is a commonly misunderstood structural system (the word *funicular* comes from the Latin word *funiculus*, a diminutive form of *funis*, meaning a 'slender rope', or 'string'). The shape a cable takes under applied loads is the result of force equilibrium. The cable is a *form-active* system. With no flexural stiffness, the shape a cable assumes under applied loads mirrors the axial tensile forces acting in it, that is, the cable shape and the action line of forces coincide.

Well-designed shells, however, will mostly be loaded in compression. Their structural behaviour is different from form-active tensile systems in three ways:

- the shell cannot self-adjust its shape to varying loading conditions (it is *form-passive*);
- the deformations due to the elasticity of the material generate undesirable additional bending moments;
- the shell can be in unstable equilibrium – just like a ball on top of a hill (Fig. 2.1).

A small disturbance – for example, a small horizontal external force – will irreversibly disturb the equilibrium of the ball or shell system.

In summary, variations in applied loading on a tensile system (such as cables, cable nets, technical textiles) induce variations in shape. The new, resulting shape remains stable, the resulting stresses stay in the allowable range and the structure is safe. In other words, the sensitivity of a tensile system to load or shape variations is low, and the robustness of the structure is high. Our aim is to design a robust structure that remains safe under expected (or unexpected) load variations. The balloon, a tensile internally pressurized structure (Fig. 2.2), is, in this context, a good example when subjected to unforeseen wind gusts; it remains safe.

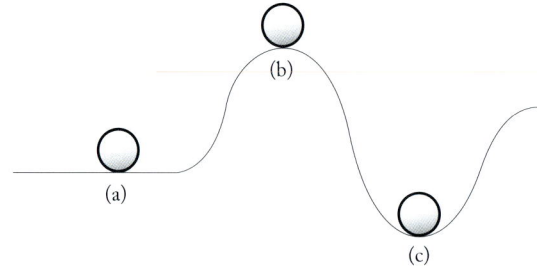

Figure 2.1 Ball in (a) neutral, (b) unstable and (c) stable equilibrium positions

Figure 2.2 A balloon in a wind gust, a robust structure under unexpected load variations

Physical, graphical or numerical techniques can exploit the form-active behaviour of tensile systems to generate forms. Interestingly enough, Antoni Gaudí (1852–1926), Heinz Isler (1926–2009) and Frei Otto (b. 1925) first exploited physical, gravity-loaded, inverted hanging models as form-finding tools for designing shell structures, based on ideas first introduced in Hooke's law of inversion (see Section 1.1): 'As hangs the flexible line, so but inverted will stand the rigid arch.'

Tensile shape inversion gives us forms for shells but no information about their stability or robustness. How safe and stable is our form-found shell? What is the influence of the material's elasticity, shrinkage and creep? How does load variation affect the shell's structural behaviour? These issues, negligible in a tensile structure, are crucial to the successful design of a shell. How to deal with these challenges in a form-finding process remains unclear to this day. Finite element methods can be a help but come with the caveat that they might lead to incorrect conclusions. Some general shell design guiding principles are known, such as a singly curved shell will be less robust than a doubly curved one, but these rules are not sufficient to design structurally efficient, safe shells.

Form-finding methods generate ideal shapes that are the result of stable force equilibrium. In the real world, the shell is the origin of force equilibrium and is itself not perfect! In the real world, we cannot construct exactly the desired ideal equilibrium shape. Due to the elasticity of the material, material imperfections, on-site construction tolerances and so forth, the built form will always slightly deviate from the perfect one. The shell should be able to manage this inherent difference as well as unpredicted load variations. In fact, we want to construct a structure that keeps the equilibrium of forces while being as close as possible to the form-found surface. If we achieve this objective, structural material is minimized and the resulting design is efficient, light and economic.

Like in the chicken and the egg quandary, we are trying to find a suitable, real structural design with an ideal force equilibrium situation. *Caveat lector* – the final form governs the channelling of forces rather than the forces determining the shape.

We have seen that gravity is a tough constraint in the design process. Do we conclude that gravity is a limitation of creativity? No, we believe that the possibilities of structures are infinite. We can call them *optimal* structures, whatever optimal means. The world of optimal forms is immense. We can compare it to a Pareto front, in that the optimal solution is never unique but rather an accumulation of ideal solutions. This being said, the world of bad solutions is also infinite. Chances of designing a terrible structure are of course higher. Compared to the universe, the optimal systems are the stars and the bad solutions are the rest of the universe. Pursuing this train of thought, galaxies are the different possible typologies of structures. Looking at the widespread construction of arbitrarily bulging, organic forms, or 'blobitecture', we see that it belongs to the emptiness of the universe and not to the stars. Strangely enough, starchitects are responsible for most of this contemporary design trend.

We see three reasons for this phenomenon:

- architectural dematerialization or virtualization;
- limitless finite element power;
- engineering design education and practice.

First, current architecture is shifting from the real to the virtual world. In the design process, a lot of thought and perspiration goes into the creation and rendering of a digital three-dimensional model. The virtual images show no difference when compared with pictures of a realized design. The crucial architectural marketing cycle can be reduced from sketch–plan–build–reality–picture–publish to sketch–render–publish. To be a celebrated architect, there is no longer a need to construct. A new generation of architects has built their fame in a virtual world. It is faster and easier. But this virtual world does not have the same rules and constraints as the real one: no gravity, no budget, no rain, no snow and no sun, nothing besides geometry. It stimulates *image* architecture. The actual realization of the image comes as an afterthought. Unfortunately, the structural solutions necessary to make these new shapes possible typically use an awkward and significant amount of material.

Second, the increasing power of structural design tools through Finite Element (FE) analysis gives us the feeling that everything is possible. Like Zarathustra, we master nature by building against it and not with it. This display of trendy demiurgic control leads to clumsy, silly, expensive and complex structures and occurs because of a lack of structural thinking in the preliminary design process. As engineers, we can always find solutions to difficulties in the post-design phase instead of trying to avoid these problems in the first place.

Third and last, any structural engineering 'design' course mostly focuses on FE modelling and code verification. Too few courses teach how to design an efficient structure, how to develop the right form. For most engineers this domain appears mysterious, complex and to be avoided. Due to the specificity of each project and time and cost restraints, form modification or optimization is difficult to integrate in a normal structural engineering workflow. In other related engineering disciplines (such as aerospace and automobile) or much older trades (such as ship building and armoury), optimization and modification are commonplace. As a consequence, structural engineers will say how much but not how. Form only seems to belong to the architectural and not the structural world. Geometry rests on an untouchable pedestal. But, we have seen that shape is the fundamental parameter to obtaining structural efficiency in a shell. We can thus state that the first thing to generate is the form. Form belongs to the structural playground too!

The question remains: 'How can we know if a shell is well conceived?' A more difficult question still is: 'How can we design a good structure "ex nihilo"?' This complex question cannot be answered in a few words; the aim of this book is to help us in this search. We can also learn from the past how to design for the future. We might ask ourselves why some historic structures remain interesting, appealing and intriguing.

The scalloped dome of the Hagia Sophia (537–present) – one of the most ambitious, advanced and largest monuments of late Antiquity – seems to float above its nave and, in doing so, has a mystical quality of light. Pendentives transition the dome elegantly into the square shape of its support piers. Although this masterpiece looks simple, its design has a complex and subtle quality to it. The central dome we see today, built in 558, has withstood nearly 1,500 years in a seismic region.

Today, most engineers fear shells and domes: they are mysterious and difficult to assess and master. Today, we are not able – or should we say, willing – to design these curved systems. Historic arches, shells and vaults are too easily seen as risks instead of opportunities. Under the motto of 'safety', these structures are destroyed and replaced by common beam-column systems. This is regrettable: their great historic and aesthetical value is lost forever. A better understanding and assessment of their structural behaviour holds the key to their conservation. It is not a sort of romanticism about historical structures that leads to these thoughts. Rather that shells and arches are highly efficient structures.

Vernacular vaulted structures can give us further clues to finding appropriate curved forms that evolve out of structural logic. The basis for their designs is informed by local reflections about urban and social planning, history, structure, stability, site, materials and local skills. In the highly dense urban fabric of the Cycladic islands, vaulted living units express the individual, yet emphasize the island's unity, community and permanence (Fig. 2.3). The island's hilly sites challenge the builder to explore views, natural lighting and ventilation in their vaults. The lack of forests and abundance of rock, stone and clay drive construction techniques towards

Figure 2.3 (top) One of many vernacular, curved structures in the Cycladic islands, and (bottom) a typical formwork for a vault

the use of masonry, a composite material excellent in compression. These are some of the constraints that form the base for the design process. The anonymous local master builder manipulates the curved vaults to optimize height, light or views. Although the form emerges out of a structural rationale, all the contextual constraints and considerations contribute to the final form, adding an extra richness to it. These hybrid designs might not offer the best solution but produce a form that offers a solution to all constraints.

In our projects we aim to design force-modelled forms, rooted in a structural, economic, cultural and technological logic. The two-dimensional geometry of the 2011 courtyard roof of the Scheepvaartmuseum, or Dutch National Maritime Museum, in Amsterdam reflects the symbolism and the building's history as a seventeenth-century gunpowder warehouse of the Dutch fleet (Figs. 2.4, 2.5 and 2.6, and page 14). Its grid is based on rosettes with loxodromes, figures on historical sea charts to mark out the courses of a ship. The dome's three-dimensional form is based on the numerical shape inversion of a hanging chain net using the dynamic relaxation technique presented in Chapter 8. The concentration of constraints in one definite structural context leads to a design with a poetic overtone. Its shape is unattainable to

Figure 2.4 Steel gridshell over the courtyard of the National Maritime Museum, Amsterdam, 2011

CHAPTER TWO: SHAPING FORCES   19

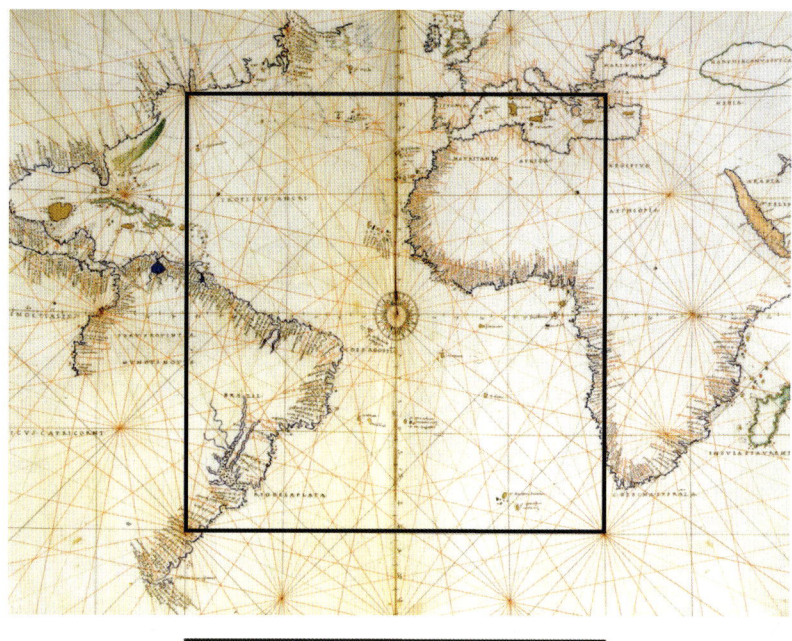

Figure 2.5 A geometric pattern, found on sixteenth-century sea charts, lies at the base of the gridshell mesh pattern of the courtyard of the Dutch National Maritime Museum, Amsterdam.

Figure 2.6 The Dutch National Maritime Museum as seen from outside revealing the facetted gridshell

anyone who attempts to generate it in an exclusively sculptural manner. This far-reaching combination of self-generated geometry, architecture and structure convinces us there is an infinite world of forms to be discovered. We see a new form of structural design emerging.

CHAPTER THREE

What is a shell?

Chris Williams

Structures can be classified in many ways according to their shape, their function and the materials from which they are made.

The most obvious definition of a shell might be through its geometry. A structure or structural element may be a fully three-dimensional solid object, or it might have some dimensions notably smaller than others. A beam is straight and it is relatively long in comparison to its cross section. Thus it is defined by a straight line. An arch is defined by a curved line and a plate by a plane.

A shell is a structure defined by a curved surface. It is thin in the direction perpendicular to the surface, but there is no absolute rule as to how thin it has to be. It might be curved in two directions, like a dome or a cooling tower, or it may be cylindrical and curve only in one direction.

This definition would clearly include birds' eggs and concrete shells, and nobody would argue with that. It would also include ships, monocoque car bodies (*coque* is one of the French words for shell) and aircraft fuselages, drinks cans, glasses cases (Fig. 3.1), all sorts of objects.

But this definition would also include tension structures such as sails, balloons and car tyres. If one wanted to exclude tension structures, one might stipulate that shells have to work in both tension and compression, but how about masonry vaults that can only work in compression? Most people would describe masonry vaults as a type of shell structure.

Figure 3.1 A glasses case is a shell

However, the word 'shell' has the implication of something relatively rigid, and this book is about such structures. We therefore need to have a separate category of tension structures to include sails and balloons as well as piano strings and fishing nets. Then we have six possible types of structure:

- straight-line elements (beams, columns);
- curved-line elements (arches, curved cables);
- plates (flat surface structures such as slabs, walls);
- tension structures (curved surface structures such as nets and fabric structures);
- shell structures (curved surface structures typically of timber, concrete, metal or masonry);
- fully three-dimensional lumps of material.

Figure 3.2 (a) A colander is a continuous shell, and (b) a sieve is a gridshell

A colander (Fig. 3.2a) is a curved surface structure. It contains holes for draining food, but these holes do not stop it being a shell. It is a continuous surface with a relatively small area removed. A sieve (Fig. 3.2b) is very similar, except that the surface is made from a large number of initially straight wires which are woven into a flat sheet and then bent into a hemisphere. It is also a shell, a gridshell.

Clearly there is some similarity between a sieve and a spider web – they are both lattice-like and are intended to catch things. The spider web is essentially flat and made up of straight elements and when the wind blows, it bows outwards like a sail and becomes curved. It therefore adjusts its shape to the loading (Fig. 3.3). We call this a 'form-active' structure. This feature is characteristic of tension structures. The sieve may be in tension, compression or a mixture of the two, but appears rigid. It does not significantly adjust its shape to the applied loading and, therefore, we call this a 'form-passive' structure. Where it is in compression, deflections lead to the structure becoming less able to carry the load, possibly leading to buckling. Columns carry loads via axial forces, but bending stiffness is required to stop buckling, and so it is with shells, although with shells buckling is resisted by a combination of bending and in-plane action.

Figure 3.3 A spiderweb is a form-active structure

The Temple of Mercury in Baiae, Italy (see page 20), once the largest dome in the world until the construction of the Pantheon in Rome, is an archetypal shell, is hemispherical, featuring an oculus at the top, and is included in our definition of the term 'shell'.

3.1 How do shells work?

Shells use all the modes of structural action available to beams, struts, arches, cables and plates, plus another mode that we might call 'shell action', which we will now try to pin down.

Structural elements that approximate to linear elements (i.e. with one dimension greater than the other two such as beams, arches and cables) or to surfaces (i.e. with one dimension smaller than the other two such as plates and shells) all share the same property: they are much easier to bend than to stretch. We use the word 'stretch' to mean change in length, possibly getting shorter, a 'negative stretch'.

Clearly a cable will stretch when we apply a tension to it. A column will undergo a negative stretch when we apply a compression to it. But if we apply more load, it will buckle and get shorter through bending, rather than axial strain.

A parabolic arch or cable can carry a uniform vertical load per unit horizontal length using only axial compression or tension. The component of load perpendicular to the cable is balanced by the axial force multiplied by the curvature. Thus load in kNm^{-1} is balanced by a force in kN multiplied by the curvature in m^{-1}. Note that curvature is defined as 1/radius of curvature.

Other loads will cause bending moments in the arch or deflection of the cable. The arch-bending moment is the product of the thrust and its eccentricity from the axis.

3.1.1 Flat plates and plane stress

In order to understand curved arches, we first learn about straight beams. Similarly, to understand shells, we first need to think about something simpler. We could start with arches and go from curved lines to curved surfaces. Or we could start with plates and go from flat surfaces to curved surfaces. Both approaches can be helpful, but let us start with plates.

A flat plate can be loaded by forces in its own plane (Fig. 3.4) or out of plane (Fig. 3.5). The term 'plane stress' is used for in-plane loading and it appears in all sorts of situations; for example, the bending of an I-beam. Clearly the beam is loaded perpendicular to its axis, but most of the stress in the web and flanges are in the plane of the steel plates. Out of plane loading of a plate or slab produces plate bending and, as we have already noted, it is much easier to bend a plate than to stretch it.

In Figure 3.4 we have introduced the components of membrane stress: normal stress in the x-direction σ_x, normal stress in the y-direction σ_y, shear stress perpendicular to the x-direction in the y-direction τ_{xy} and shear stress perpendicular to the y-direction in the x-direction τ_{yx}. In-plane membrane stress is a central concept in shell theory and corresponds to the axial stress in an arch – as opposed to the bending stress. Membrane stress is usually quoted as a force per unit length crossing an imaginary cut, rather than force per unit area. Equilibrium of moments about the normal tells us that $\tau_{xy} = \tau_{yx}$.

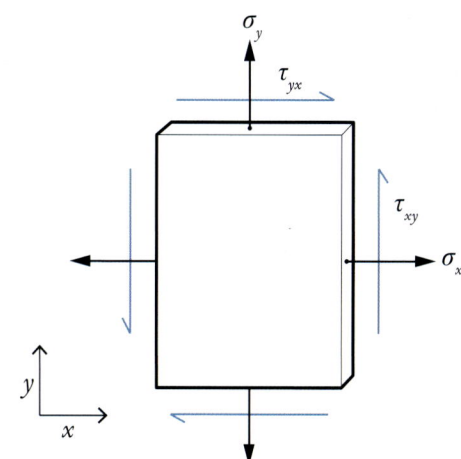

Figure 3.4 Plane stress

The shear stresses τ_{xy} and τ_{yx} are in the plane of the plate. We also get shear stress perpendicular to the plate due to plate bending. These are not labelled in Figure 3.5 because the notation for plate bending is rather confusing.

Thus for plane stress we have three unknown stresses, σ_x, σ_y and $\tau_{xy} = \tau_{yx}$. We have two equations of equilibrium

$$\frac{\partial \sigma_x}{\partial x} + \frac{\partial \sigma_{yx}}{\partial y} = q_x$$

$$\frac{\partial \tau_{xy}}{\partial x} + \frac{\partial \sigma_y}{\partial y} = q_y \qquad (3.1)$$

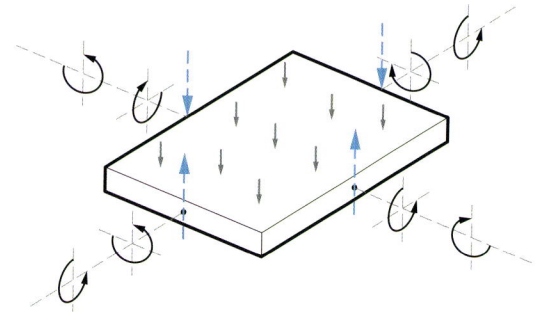

Figure 3.5 Plate bending

in the x and y directions respectively. The variables q_x and q_y are the loads per unit area applied to the plate, in its own plane, for example the self-weight of a wall. Thus we have three unknown stresses and only two equations of equilibrium so that plane stress is *statically indeterminate*.

If q_x and q_y are both zero, the stresses can be written in terms of the Airy stress function ϕ,

$$\sigma_x = \frac{\partial^2 \phi}{\partial y^2}$$
$$\sigma_y = \frac{\partial^2 \phi}{\partial x^2} \quad (3.2)$$
$$\tau_{xy} = \tau_{yx} = -\frac{\partial^2 \phi}{\partial x \partial y}$$

so that they *automatically* satisfy the equilibrium equations (3.1). Note that even if q_x and q_y are zero, the plate can still be loaded at its edges.

If the plate is elastic we can solve for ϕ and hence the stresses by using the stress-strain relationships,

$$\varepsilon_x = \frac{1}{E}(\sigma_x - \upsilon\sigma_y)$$
$$\varepsilon_y = \frac{1}{E}(\sigma_y - \upsilon\sigma_x) \quad (3.3)$$
$$\gamma_{xy} = \frac{2(1+\upsilon)\tau_{xy}}{E},$$

where E is Young's modulus and υ is Poisson's ratio, together with the compatibility equation,

$$\frac{\partial^2 \varepsilon_x}{\partial y^2} - \frac{\partial^2 \gamma_{xy}}{\partial x \partial y} + \frac{\partial^2 \varepsilon_y}{\partial x^2} = 0. \quad (3.4)$$

The compatibility equation comes from the fact that our three strains $\varepsilon_x = \partial u_x/\partial x$, $\varepsilon_y = \partial u_y/\partial y$ and $\gamma_{xy} = \partial u_y/\partial x + \partial u_x/\partial y$ are the result of only two components of displacement, u_x and u_y, which can be eliminated by differentiating the strains twice and subtracting. We finally end up with just one equation,

$$\nabla^4\phi = \frac{\partial^4 \phi}{\partial x^4} + 2\frac{\partial^4 \phi}{\partial x^2 \partial y^2} + \frac{\partial^4 \phi}{\partial y^4} = 0, \quad (3.5)$$

which is known as the biharmonic equation. Even though it looks complicated, it actually behaves very well and is not difficult to solve (Timoshenko & Goodier, 1970).

3.1.2 Membrane theory of shells

In the membrane theory of shells, we still have three components of membrane stress, exactly as in plane stress. But, we now have three equations of equilibrium. Two of them are in the directions tangent to the shell, exactly as in the case of plane stress. The third equation is perpendicular to the tangent to the shell surface. The load is balanced by the membrane stresses multiplied by the curvature. Here the load would be in kNm^{-2}, the membrane stress in kNm^{-1} and the curvature in m^{-1}.

Thus we have three unknown stresses and three equations of equilibrium so that shells should be statically determinate. Unfortunately, we have three partial differential equations of equilibrium in three unknown membrane stresses, and whether or not these equations have a solution depends upon the shape of the shell and the boundary conditions. This is a very difficult area of mathematics and it is often impossible to say whether a solution exists or not, let alone find one.

The simplest way to express this mathematically is using plane coordinates. The horizontal equilibrium equations (3.1) still apply if the stress components are redefined as the horizontal component of membrane stress per unit horizontal length. In particular, if a shell is only loaded in the vertical direction, the horizontal equilibrium equations are still satisfied by use of the Airy stress function. Then equilibrium in the vertical direction is simply

$$w = \frac{\partial^2 \phi}{\partial x^2}\frac{\partial^2 z}{\partial y^2} - 2\frac{\partial^2 \phi}{\partial x \partial y}\frac{\partial^2 z}{\partial x \partial y} + \frac{\partial^2 \phi}{\partial y^2}\frac{\partial^2 z}{\partial x^2}, \quad (3.6)$$

where z is the height of the shell and w is the load per unit plane area, both assumed to be known functions of x and y. This equation may not look any more complicated than the biharmonic equation (3.5), but depending upon the shape of the shell and the boundary conditions, it may be impossible to solve for ϕ. Equation 3.6 is 'exact' in that it does not assume that the slope of the shell is small.

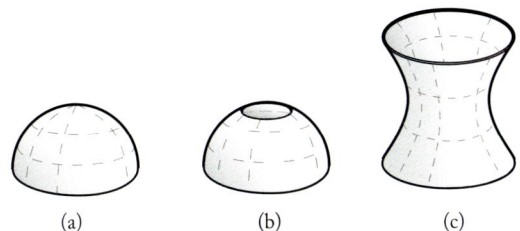

Figure 3.6 A dome, (a) without and (b) with an oculus, and (c) a cooling tower

If the shell is the wrong shape or it does not have enough boundary support, it may be a mechanism as far as the membrane theory is concerned and be able to undergo *inextensional deformation*, that is, deformation in which the shell is bent without stretching.

The dome and cooling tower in Figures 3.6a and 3.6c are statically determinate and cannot undergo inextensional deformation. The cooling tower has a big hole at the top, but putting a big hole in a dome produces a structure which is a mechanism, Figure 3.6b.

Figure 3.7 shows a detail of inextensional deformation in the region of a hole in a sphere. The deformation increases rapidly as the hole is approached, and the smaller the hole, the greater the deformation. However, for very small holes, the bending stiffness takes over and controls the deformation and this is what happens with the colander. The appendix to the paper *On bells* by Lord Rayleigh (1842–1919) contains the mathematics necessary to find the modes of inextensional deformation of any surface of revolution, with particular reference to the hyperboloid of one sheet – the cooling tower (Rayleigh, 1890). However, attaching the cooling tower to the ground prevents these modes.

The difference between the dome and the cooling tower is that the dome is synclastic and has positive Gaussian curvature whereas the cooling tower is anticlastic and has negative Gaussian curvature. Kelvin (1824–1907) wrote: 'We may divide curved surfaces into Anticlastic and Synclastic. A saddle gives a good example of the former class; a ball of the latter. The outer portion of an anchor-ring is synclastic, the inner anticlastic' (Kelvin & Tait, 1867). Here saddle refers to that on a horse and an anchor-ring is a torus.

The Gaussian curvature is the product of the two principal curvatures on a surface and they are of opposite sign on a cooling tower. Gauss's theorem (Theorema Egregium) tells us that the Gaussian curvature of a surface can be calculated by only measuring lengths on a surface, and therefore inextensional deformation *does not change the Gaussian curvature*. A *developable surface* is a surface with zero Gaussian curvature which can be laid out flat. Examples include cylinders and cones. Gauss's Theorem is derived in Appendix B.

The Cohen-Vossen theorem from differential geometry tells us that it is not possible to deform a closed convex surface such as an egg without changing lengths on the surface; in other words, inextensional deformation is impossible. However, part of an egg can be deformed inextensionally, explaining why it is so much more flexible. The stiffness of the part can be regained by glueing it to a support to form a dome on a foundation. This is a difficult concept because engineers are more used to thinking about forces rather than length changes. But a two-dimensional statically determinate pin-jointed truss can resist loads because it is not possible to deform it without changing the lengths of its members. The tension in a member is just the member trying to stop its length increasing.

3.1.3 Bending theory of shells and buckling

If a shell can undergo inextensional deformation, then it will have to rely on bending stiffness as well as membrane action in order to carry any load case,

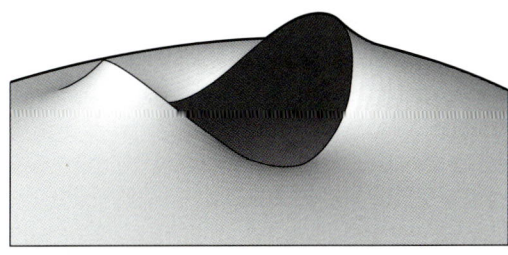

Figure 3.7 Inextensional deformation in the region of a hole in a sphere

including wind and snow. However, even if a shell has the correct shape and is properly supported, it must have bending stiffness to prevent buckling if there are any compressive membrane stresses. Thus, for efficiency, we want our shell to work primarily by membrane action, which is what shell action means, but we know that we must also have bending stiffness to resist buckling and inextensional deformation.

Shell buckling is particularly nasty because shell structures are so efficient. Almost no deflection occurs and then suddenly there is total collapse. Paradoxically, the less efficient the shell, in terms of shape, triangulation of the surface and boundary support, the better it behaves in buckling. This is because bending action of shells requires much more deflection than membrane action and therefore small irregularities in shell geometry and other initial imperfections have less effect.

Experiments show that a properly supported shell working primarily by membrane action can never support anything like the theoretical 'eigenvalue' buckling load or 'linear' buckling load even when the utmost care is taken to eliminate initial imperfections. The analysis of shell buckling by hand calculations is effectively impossible, even eigenvalue analysis of a spherical shell is very difficult and, as we have said, gives wildly optimistic answers. This means that there is no option but to use computer analysis, but this is quite an esoteric area, and even though many programs offer analysis for shell buckling, the results should be treated with a great deal of circumspection.

3.2 How much do we need to know to design a shell?

In the previous section, we tried to describe how shells work in a relatively qualitative manner. It should by now be clear that it is difficult to derive the equations (particularly for the bending theory) and usually impossible to solve them, except for the membrane theory for very simple shapes.

The theory of shell structures, as described in such classic works as Novozhilov (1959), Flügge (1960), Green & Zerna (1968), Billington (1982), Calladine (1983) is very mathematical. In Appendix B, there is an introduction to the differential geometry and the theory of shells, leaving a detailed reading to those with a strong constitution.

Thus in practice one has three possible approaches:

- simple hand calculations 'informed' by the classical theory of shells;
- numerical analysis using a computer;
- physical testing.

Numerical analysis almost invariably uses the finite element method (see also Appendix A) in the form of shell elements for continuous shells or beam elements for gridshells. The derivation of the finite element equations does not really depend very much on the theory of shells, except for being able to work in curvilinear coordinates.

The shape functions of the finite element method produce algebraic equations. These equations may be linear or nonlinear according to the material behaviour and whether one is concerned with buckling. The equations might be solved using an 'implicit' method involving inversion of a stiffness matrix or an 'explicit' method such as dynamic relaxation or Verlet integration.

However, the structural behaviour of shells can be so complicated that numerical predictions may be inaccurate and so there is still a place for physical testing. Physical testing of 'sketch models' can also give a qualitative insight that cannot be obtained from a numerical analysis.

3.3 Funicular shells

Figure 3.8 shows funicular polygons. The rope automatically adjusts its shape to carry the loads *without any bending moment*. The rope is a mechanism, which moves to carry one particular load case. We have seen that, if we are lucky, a shell can carry *any load* by membrane action only. However, if our shell has the wrong shape or is not properly supported, it will only be able to carry certain loads for which it is funicular. We might also have a shell which does not have sufficient bending stiffness to work in compression, like a fabric. Or we might have a masonry structure which only works in compression. Thus, in masonry, a funicular load is one that can be carried without any bending moment or tensile membrane stress.

CHAPTER THREE: WHAT IS A SHELL?   27

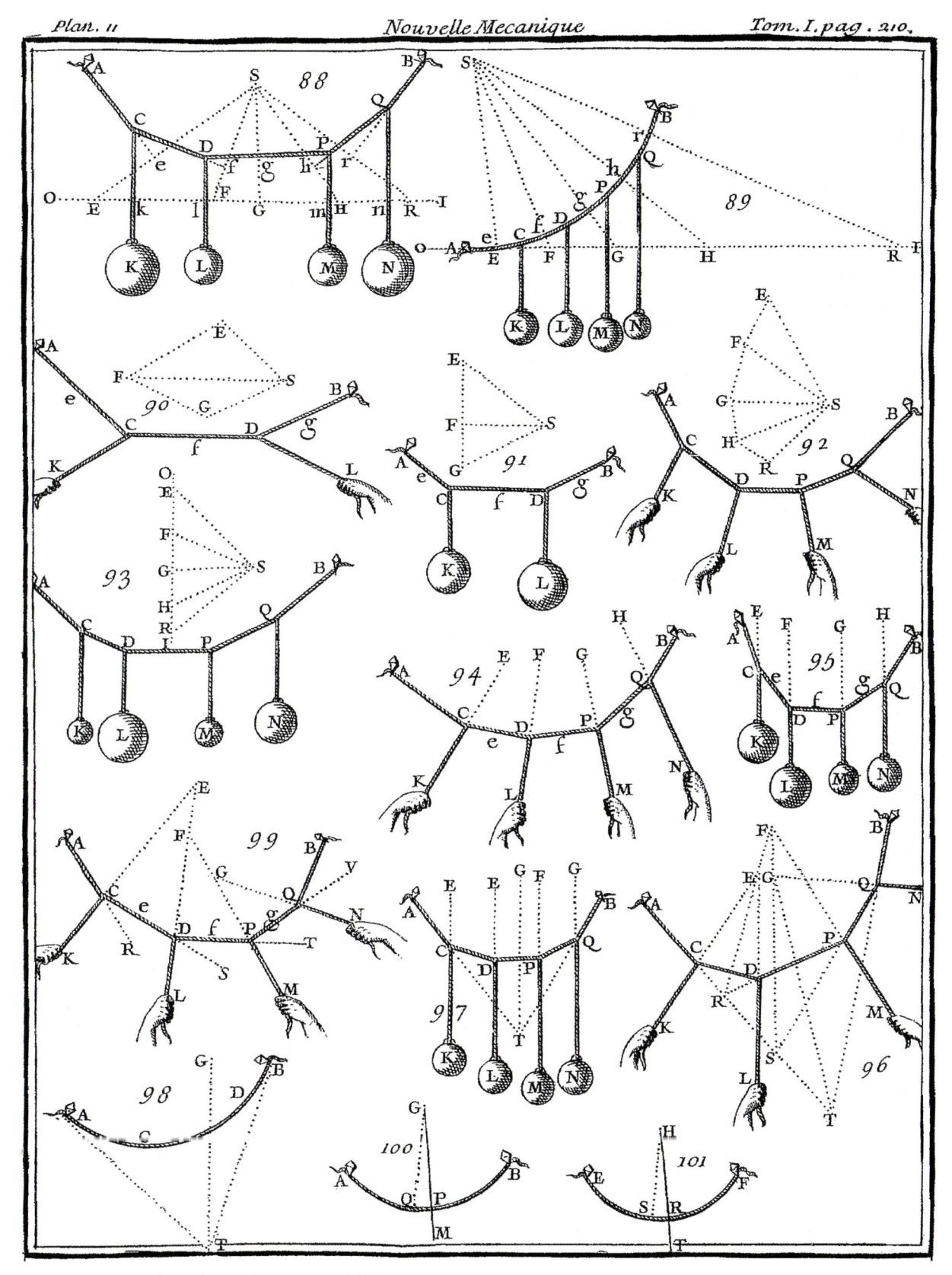

Figure 3.8 Funicular polygons by Varignon (1725)

Reversing the loads on a pure tension structure produces a pure compression structure, a fact used by Antoni Gaudí (1852–1926) for the Colònia Güell and Frei Otto (b. 1925) for the Multihalle.

3.3.1 Funicular arches and cables

A rope or chain hanging under its self-weight forms a catenary (from 'catena', the Latin word for chain). Thus, the catenary is a particular case of a funicular curve. The catenary is of some relevance to the design of shells, so it is worth deriving the mathematical form here.

If a cable is only carrying vertical loads, then the horizontal component H of tension in the cable,

$$H = T \cos \lambda = \text{constant}, \quad (3.7)$$

where T is the tension in the cable and λ is the slope between the cable and the horizontal. The vertical component V of tension

$$V = T \sin \lambda = H \tan \lambda = H \frac{dy}{dx}. \quad (3.8)$$

If the loading, w, is constant per unit arc length, s, then

$$w = \frac{dV}{ds} = \frac{dx}{ds}\frac{dV}{dx} = \cos \lambda \frac{dV}{dx}$$

$$= \frac{1}{\sqrt{1+\tan^2\lambda}}\frac{dV}{dx} = \frac{1}{\sqrt{1+\left(\frac{dy}{dx}\right)^2}} H \frac{d^2y}{dx^2}. \quad (3.9)$$

This can be integrated to give

$$\frac{dy}{dx} = \sinh\left(\frac{x}{c}\right), \quad (3.10)$$

in which $c = \frac{w}{H}$ and we have left out the constant of integration because it just moves the curve sideways. Integrating again,

$$\frac{y}{c} = \cosh\left(\frac{x}{c}\right) - 1, \quad (3.11)$$

in which the constant of integration is chosen so that the curve goes through the origin. This is the equation for the catenary, the solid curve in Figure 3.9, while the dashed curve is the parabola, the funicular curve for when the load is constant per unit horizontal length, as is the case for a suspension bridge.

It can be seen that the two curves are identical when their slope is low and they only peel apart when the load per unit horizontal length on the catenary increases with slope. The catenary is one of the few curves for which there is a simple relationship between x and y and the arc length along the curve, s, starting from the bottom,

$$s = \int_0^x \sqrt{1+\left(\frac{dy}{dx^2}\right)^2}\, dx = c \sinh\left(\frac{x}{c}\right). \quad (3.12)$$

It is relatively easy to find the funicular load for a given shape of cable or funicular shape for a given load, either by doing a simple physical experiment, mathematically, or graphically using graphic statics. Having found the shape, it can be inverted, or turned upside down, to find the best shape for the equivalent compression structure or arch.

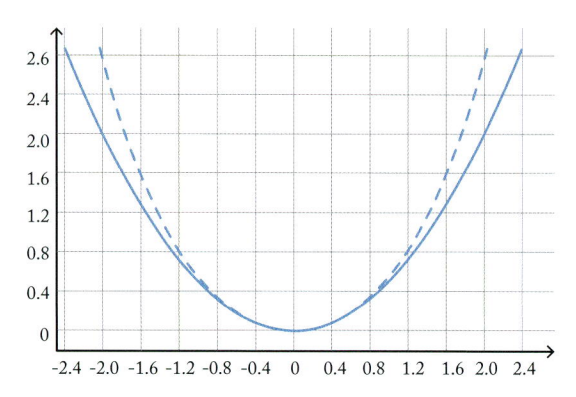

Figure 3.9 Catenary (solid curve) and parabola (dashed curve)

If an arch is carrying a funicular load, there will be no bending moment in it, which is equivalent to saying that the line of thrust is along the neutral axis of the arch. If a non-funicular load is added, it will produce bending moments and cause a deviation of the line of thrust.

The concept of funicular load applies particularly to structures that have to carry one dominant load case, perhaps their self-weight or some permanent load due to water or soil. Arch bridges such as the Gaoliang Bridge (Fig. 3.10) have to carry the extra weight of the masonry and fill over the support, together with the horizontal thrust from the fill. This means that more curvature is required towards the supports than would be the case for a catenary.

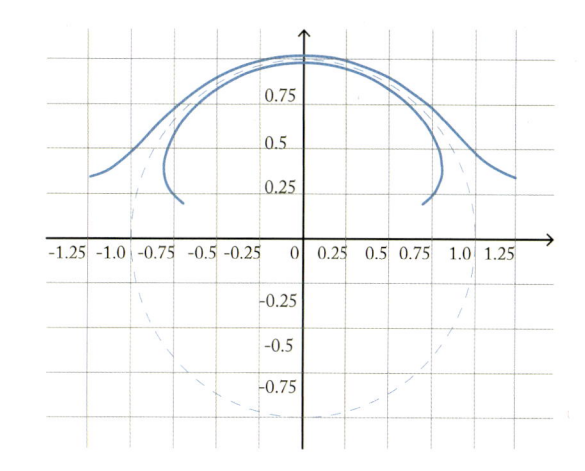

Figure 3.10 (left) The Gaoliang Bridge of the Summer Palace, Beijing, 1751–1764, resembles (right) a funicular circular arch of increasing thickness

Now let us suppose that we want to make a circular arch of varying thickness so its self-weight is funicular. If t is the thickness of the arch, R is its radius and ρg is its specific weight, or weight per unit volume,

$$\rho g t = -\frac{1}{R}\frac{dV}{d\lambda} = -\frac{H}{R}\frac{d}{d\lambda}(-\tan\lambda) = \frac{H}{R}\sec^2\lambda, \quad (3.13)$$

so that

$$t = \frac{H}{\rho g R \cos^2\lambda}, \quad (3.14)$$

in which $H/(\rho g R)$ is a constant with the units of length. Note that H and V are now forces per unit width and H is positive for a compression. Figure 3.10 shows how the shell gets thicker as it approaches the vertical. The stress in the arch

$$\sigma = \frac{H}{t\cos\lambda} = \rho g R \cos\lambda, \quad (3.15)$$

which reduces away from the top, the reverse of what happens with a catenary.

Another possibility is to say that the compressive stress σ should be constant. If that is the case,

$$\rho g t = \frac{dV}{ds} = H\frac{d}{ds}(-\tan\lambda) = -H\sec^2\lambda\frac{d\lambda}{ds} \quad (3.16)$$

and

$$H = \sigma t \cos\lambda. \quad (3.17)$$

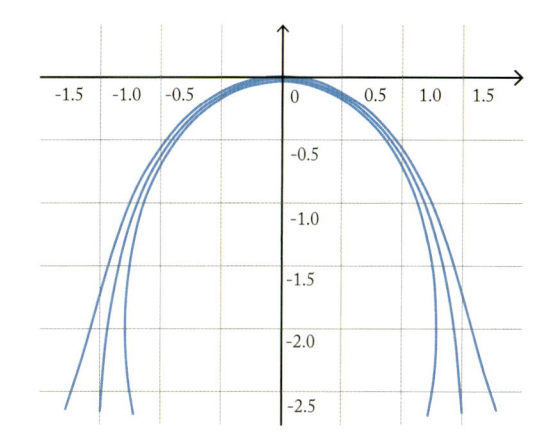

Figure 3.11 Constant stress arch

Thus

$$\frac{\rho g}{\sigma}\frac{ds}{d\lambda} = -\frac{1}{\cos\lambda} = \frac{\cos\lambda}{1-\sin\lambda} - \frac{\sin\lambda}{\cos\lambda},$$

$$\frac{\rho g}{\sigma}\frac{dx}{d\lambda} = -1,$$

$$\frac{\rho g}{\sigma}\frac{ds}{d\lambda} = \tan\lambda. \quad (3.18)$$

These equations can be integrated to give

$$\frac{\rho g}{\sigma}s = \log_e\left(\frac{\cos\lambda}{1-\sin\lambda}\right),$$

$$\frac{\rho g}{\sigma}x = -\lambda,$$

$$\frac{\rho g}{\sigma}y = \log_e(\cos\lambda),$$

$$t = \frac{t_0}{\cos\lambda} = t_0 e^{\left(\frac{\rho g y}{\sigma}\right)}, \quad (3.19)$$

where t_0 is the thickness at the top of the arch.

Figure 3.11 shows a constant stress arch. Let us now think about scale. If the arch is made from a weak masonry we might have $\rho g = 25 \times 10^3 \text{Nm}^{-3}$ and $\sigma = 0.5\text{MPa} = 0.5 \times 10^6 \text{Nm}^{-2}$. Therefore $\sigma/(\rho g) = 20\text{m}$ which means that if we decided to use the part of the arch in Figure 3.11 between -1.25 and +1.25 on the horizontal axis, the span would be 50m. If we used concrete at a stress of 25MPa, the corresponding span will be 1.25km for an arch under self-weight only.

Figure 3.12 shows a comparison of the catenary, circular and constant stress arches. They all have the same thickness and curvature (and therefore stress) at the top. However, the catenary will have stress increased by a factor of 2.5 at the supports. So the catenary is not a particularly good shape for an arch or a barrel, unless practical considerations mean that it has to have a constant thickness.

3.3.2 Uniform stress shell

The equivalent of the uniformly stressed arch is the uniformly stressed shell of revolution. Let us imagine a shell of variable thickness, t, which is only loaded by its self-weight, ρg per unit volume, where g is the acceleration due to gravity, and that there is a uniform compressive stress σ (force per unit area) in the material.

The equilibrium equation in the radial direction tangent to the surface is

$$\rho g r t \frac{dz}{dr} = -\sigma t + \frac{d}{dr}(rt\sigma) = -r\sigma \frac{dt}{dr} \quad (3.20)$$

and therefore

$$\frac{\rho g}{\sigma} = -\frac{1}{t}\frac{dt}{dz}, \quad (3.21)$$

which can be integrated to give

$$t = t_0 e^{-\frac{\rho g z}{\sigma}}. \quad (3.22)$$

This is exactly the same result that we obtained for the uniformly stressed arch, which is a special case of the uniformly stressed shell. In fact, this result applies for any plan shape of vertically loaded uniformly stressed shell. We can now use the equilibrium in the normal direction to find the shape of the shell,

$$\sigma \left[\frac{\frac{d^2z}{dr^2}}{\left(1 + \left(\frac{dz}{dr}\right)^2\right)^{\frac{3}{2}}} + \frac{\frac{dz}{dr}}{r\sqrt{1 + \left(\frac{dz}{dr}\right)^2}} \right] + \frac{\rho g}{\sqrt{1 + \left(\frac{dz}{dr}\right)^2}} = 0 \quad (3.23)$$

or

$$\frac{\frac{d^2z}{dr^2}}{1 + \left(\frac{dz}{dr}\right)^2} + \frac{1}{r}\frac{dz}{dr} + \frac{\rho g}{\sigma} = 0. \quad (3.24)$$

The quantity in the square brackets is the sum of the principal curvatures of the surface.

This equation probably cannot be solved analytically and Figure 3.13 shows a numerical solution. The smaller scale shell in Figure 3.13 is the two-dimensional cylindrical shell or arch from Figures 3.11 and 3.12. It can be seen that the shell of revolution will span roughly twice as far for the same stress, 100m for

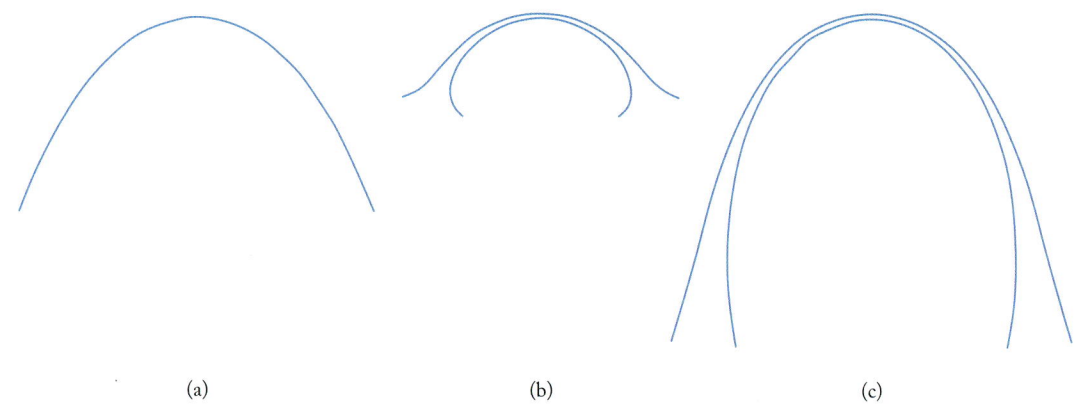

(a)   (b)   (c)

Figure 3.12 Three arches: (a) catenary, (b) circular, and (c) constant stress

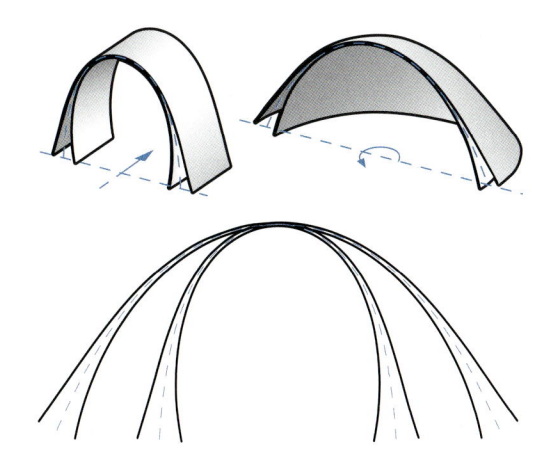

Figure 3.13 Constant stress shells – shell of revolution has larger span than cylindrical shell

the weak masonry and 2.5km for concrete – this is if the shell is only carrying its self-weight. The thickness does not come into the expression for maximum span, but if the shell is too thin, other loads will dominate the stresses and the shell may also buckle.

3.4 Conclusion

The uniform stress shell illustrated in the previous section describes some sort of 'optimum', at least for the case when the self-weight of the structure dominates. However, in practice, there will be all sorts of functional and aesthetic constraints which will mean that the shell will not be structurally optimum.

The Aichtal Outdoor Theatre, or Naturtheater Grötzingen (Fig. 3.14), by Michael Balz and Heinz Isler, is clearly not 'properly' supported all around its boundary. If it were, it would not fulfil the architectural and aesthetic constraints. However, the negative Gaussian curvature 'lip' at the free edges reduces the possibility of inextensional modes of deformation and therefore is an optimal design.

At the beginning we asked two questions: What is a shell? And how do they work? We can summarize our discussion as follows:

- Shell structures can be geometrically represented by surfaces.
- Shells are relatively rigid and this distinguishes them from tension structures such as balloons and sails.
- Shells work through a combination of membrane and bending action. Membrane action is more efficient but bending action is required to stop buckling and possible inextensible modes of deformation.
- The more efficient the shell, the more sudden the buckling collapse.
- Hand calculations for shells are very difficult or impossible. However, some understanding of shell theory will help with the choice of shell shape and interpreting computer and model test results.

Figure 3.14 Aichtal Outdoor Theatre with a 'lip' at the free edges, Germany, 1977

Further reading

- *Theory of Shell Structures*, Calladine (1983). This book is quite theoretical, but contains a clear explanation of the relative importance of membrane action and bending action in shell structures.
- *Theoretical Elasticity*, Green and Zerna (1968). If you only own one book on shell theory, it has to be this one. It unifies differential geometry and the theory of shell structures via the use of the tensor notation. You don't have to read the whole book, you can jump from the end of Section 1.13 on page 39 to Chapter 10 on page 373.

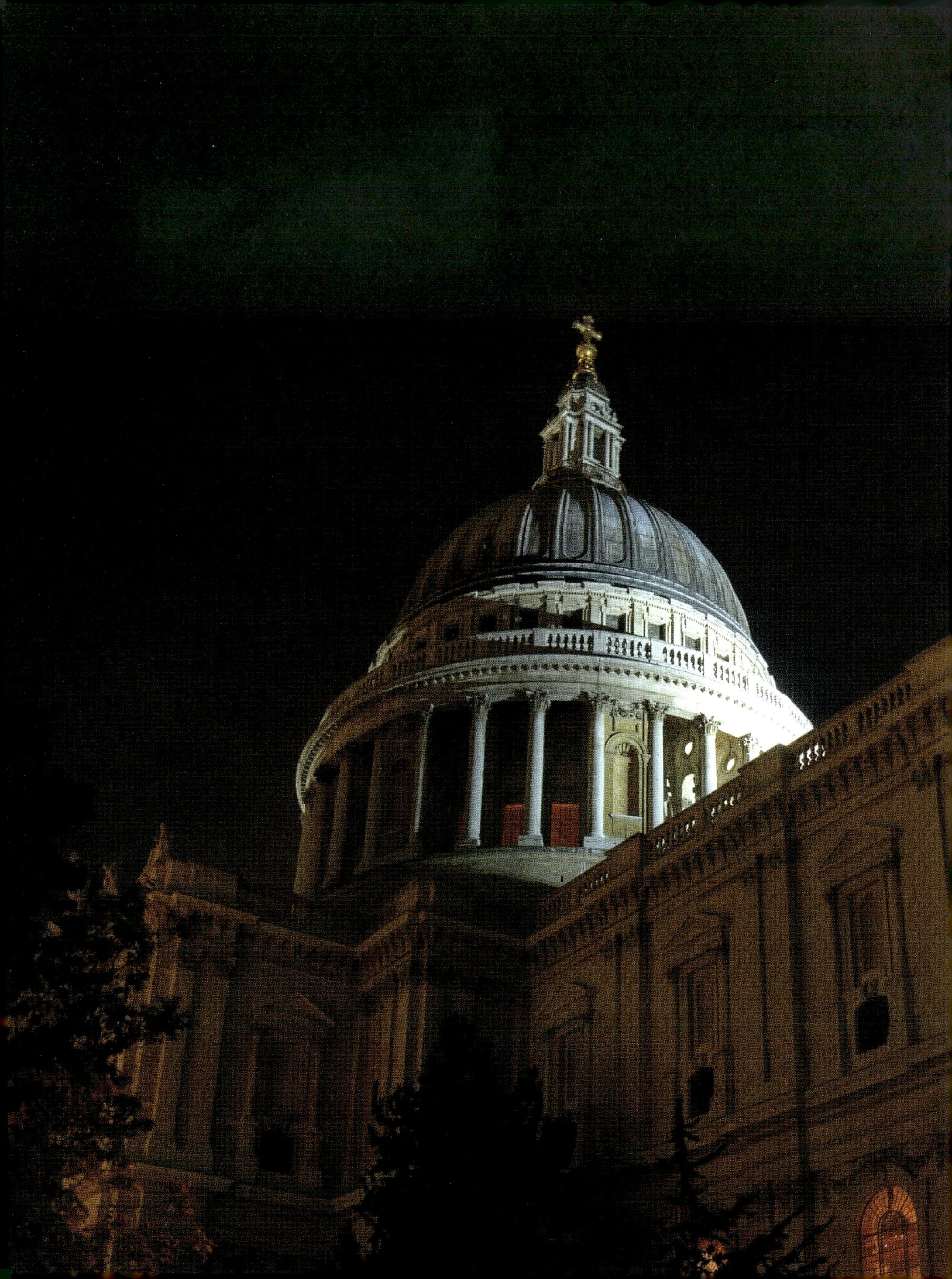

CHAPTER FOUR

Physical modelling and form finding

Bill Addis

The fundamental tasks of the structural engineer are:

- to provide sufficient information about the form of a structure, the dimensions and relative disposition of its components, as well as material specifications, to enable a contractor to begin construction and;
- to raise to a sufficient level the confidence of the engineer and builder that a proposed structure can be constructed and, once constructed, will work as intended.

In the context of designing a shell structure, this definition can be reduced to two questions:

- How can a three-dimensional geometric shape be defined, described or communicated from one person to another? Or, put another way, how can a three-dimensional form be generated in the mind of its creator?
- Which three-dimensional forms are most suitable for a shell structure in order that the designer gains sufficient confidence that the structure will work as intended?

Since ancient times, building designers have used both mathematical and physical models to answer both these questions. The concrete vaults and domes built by the Romans, which are the earliest ancestors of modern thin-shell structures, were cylindrical or hemispherical. In late medieval cathedrals, pointed arches and vaults, formed with two circular arcs, were widely used because they allowed the height and span of a vault to be varied independently. A large number of design procedures were developed using geometrical shapes (circles, squares and triangles) to help determine the shapes and dimensions of masonry arches and vaults as well as their abutments. In fact, it was widely believed that the resulting designs were suitable and efficient because they were based on these simple shapes – the same shapes that God had supposedly used to create the universe – and they were also used for designing the plans and elevations of entire cathedrals.

When designing a reinforced-concrete shell or timber gridshell today, many factors may affect the geometry of the shape; for example, the shape in plan, the rise, the loading, the desired degree of structural efficiency and, of course, the appearance and the nature of the space the architect wants to create. The type of shape chosen by an architect will depend on how it is created. It might be part of a mathematically definable, geometric shape such as a sphere, an ellipsoid, a cone, a toroid, or a hyperbolic paraboloid. Or it may be a natural, form-found shape created using a physical process, such as suspending a chain, net or membrane, creating a soap bubble, inflating an elastic membrane (a balloon), or even crumpling a sheet of paper.

In order to analyse the behaviour of a shell structure, it was (until the development of iterative computer programs) essential that the engineer could describe (in mathematical terms, usually as an equation) the geometry of the shape. This is one reason why many barrel vaults have circular or elliptical cross sections, and three-dimensional shells are parts of a sphere, cone, ellipsoid, toroid and so on. Such shapes are also easier to set out on site since the dimensions can be easily calculated.

As well as needing to guarantee the safety of a structure, structural engineers are often under pressure to optimize the structure they are designing. Apart from minimizing cost, this might mean designing for minimum weight, or minimizing a certain dimension such as rise or thickness, or making the structure easy to build; for example, by simplifying the geometry. And deep down in the soul of an engineer, there is the intellectual challenge to design for structural efficiency and elegance in ways that reflect the laws of statics and elasticity; for example, designing a beam whose shape reflects the bending moment diagram, or a compression structure such as a concrete shell with a shape that reflects the funicular form. In reality, however, a structure must be designed to carry many different load cases; thus, for example, an ideal funicular shape is not suitable for a real structure – there will always be a need for a shell structure to have some resistance to out-of-plane bending. Nevertheless, in the case of shell structures, determining the funicular shape – or rather the family of funicular shapes corresponding to the many load cases – is an important step in achieving a structurally efficient design.

4.1 Physical models and real structures

Designers of structures have used small-scale models when it is beneficial to do so, especially in order to raise the engineer's confidence in the design being proposed. This may have been for one of many reasons. For example:

- the available calculation methods were too complex or time-consuming;
- it would be too costly to build a full-size prototype;
- it was believed that normal structural analysis methods would not adequately model the structure;
- the geometry of the structure could not be defined using a mathematical equation;
- the type of structure was unprecedented;
- there were no other means available.

At an entirely intuitive level, many people believed (and still do) that a small model of a structure would provide some indication of how a larger version of the same structure would behave. For some types of structure, this is indeed the case. Masonry structures are one such category – an arch, a flying buttress or even a fan vault that works in a model will also work if scaled up twenty times to the size of a cathedral. However, there are many other types of structure that cannot be simply scaled up to give a similar structure that works at full size. There are two categories of structural behaviour: that which is independent of scale and that which is not.

4.1.1 Structural behaviour independent of scale

Some structural behaviour is independent of scale and can be scaled up linearly and used to predict full-size behaviour. For example:

- the statical equilibrium (stability) of compression structures, including masonry arches, vaults and domes;
- the shape of a hanging chain of weights or net; and, by Hooke's law of inversion (see Section 1.1), of funicular arches, vaults and domes.

For these types of structural behaviour, using models to assist with design is a reasonably straightforward process. A model arch, vault or dome made of cut stones can be a reliable predictor of the behaviour of a similar, full-sized structure. Although we have no convincing evidence, this characteristic of masonry structures explains how they were able to develop so spectacularly, long before any scientific or mathematical understanding of structures. The use of models of this type is discussed in Section 4.2.

4.1.2 Structural behaviour dependent on scale

Other structural behaviour is dependent on scale and cannot be scaled up linearly. For example:

- the strength and stiffness of a beam and;
- the buckling strength of a column or thin shell.

It must have been the case, although we have no historical evidence, that ancient builders knew that making models of some types of structure did work as a design process, while for others, it did not. However, they probably did not realize the significance of what they knew. The earliest mention of the nonlinearity of scaling was by Galileo Galilei (1564–1642) who noted what we now call the square-cube law – while area increases with the square of the scale factor, volume, and hence the mass, increase as the cube of the scale. He used this observation to explain that the finite strength of bone material means the size of animals cannot be scaled up indefinitely. Nor did ancient builders realize that what distinguishes the two categories of phenomena is that the first is independent of the structural properties of the materials involved, while the second is not. This was first understood only in the eighteenth century when scientists began to understand and use the concept of elasticity and stiffness.

4.1.3 Models for scale-dependent structures

For a structure that must be designed to resist bending, such as a real shell or gridshell, many factors need to be taken into consideration when relating the behaviour of the model and the full-size structure, not simply the geometric scale. These may include mechanical properties of the materials in the model and the full-size structure (e.g. density, stiffness, Poisson's ratio), non-mechanical material properties (e.g. the effects of temperature and moisture) and time-dependent material properties (e.g. creep, viscoelasticity) and, of course, the magnitude of the loads applied.

Often it may be necessary to make several scale models of a structure that represent the full-size structure and its behaviour in different ways, in which case the purpose of the model and the test must be carefully decided before making and testing the scale model.

Concrete shells, more than any other type of structure, and their designers have benefited from using model tests because the structural behaviour of shells is so complex. Not only were the available analytical methods often inadequate, but also the mathematics was complex and either beyond the ability of many designers or too time-consuming to perform within tight commercial constraints.

Finally, it should be noted that engineers have never relied wholly on the results of model tests. Test results complement the results of elastic and statical structural analysis, experience gained from previous full-size structures, and a designer's own feel for structural behaviour which, of course, is often developed using models that bring the forces and deflections into a range that can easily be seen and felt by a person. Each source of information contributes to raising the confidence of the designers to the point where they feel sufficiently confident that their design will deliver a structure that behaves as intended.

4.2 Scale-independent models

A funicular shape or structure is one formed by a hanging chain or a string loaded with any number of weights. In structural terms, the forces are pure tension and no portion of the structure is subjected to bending. The military and civil engineer Simon Stevin (1548–1620) was one of the first to develop the mathematical representation of forces as vectors. In an appendix (*Byvough*) to his book *De Beghinselen der Weeghconst* (The Principles of the Art of Weighing) published in 1586, he illustrated the parallelogram of forces and showed several examples of suspended weights creating funicular shapes in both two and three dimensions (Fig. 4.1).

The engineer and scientist Robert Hooke (1635–1703) worked with Christopher Wren (1632–1723) during the design of St Paul's Cathedral in London (see page 32) and used his understanding of the inverted catenary, Hooke's law of inversion (see Section 1.1), to help with the design of the dome. One of Wren's sketches for the 33m diameter dome shows the shape of a chain suspended over a cross section of the building (Fig. 4.2). This simple model test would have helped raise Hooke and Wren's confidence that the dome would work satisfactorily as a

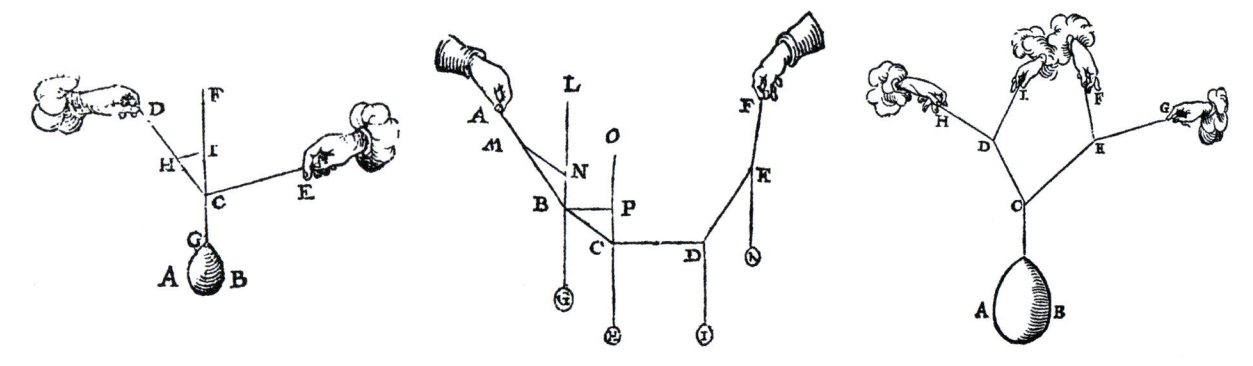

Figure 4.1 Diagrams of funicular shapes by Stevin (1586)

Figure 4.2 St Paul's Cathedral, London. Sketch of Hooke's catenary superimposed on a proposed section of the dome

compression structure and is the earliest known use of a physical model being used to help determine the form of a structure. At the very least it demonstrated that the catenary arch lay within the masonry which is a general condition for the stability of a dome. Nevertheless, it appears that a uniform chain was used, rather than a series of different weights representing the voussoirs of different sizes. Also, there is no evidence of a three-dimensional hanging model being used, although that would have represented a dome more faithfully.

Giovanni Poleni (1683–1761) used Hooke's principle in the 1740s to assess the safety of the hundred-year-old 41.9m diameter dome of St Peter's cathedral in Rome, which had developed a number of alarming radial cracks. The original design was for a solid and hemispherical dome, like the Pantheon in Rome. The artist, architect and engineer Michelangelo (1475–1564) decided the weight had to be reduced and proposed a double-skin dome, like that used by Filippo Brunelleschi (1377–1446) at the dome of the Florence Cathedral. Finally, in the 1580s, the design by Giambattista della Porta (1535–1615) increased the height of the dome in order to reduce the outward thrusts at its base and also introduced several iron chains to carry the tensile stresses in its lower part. Poleni was one of several engineers appointed to study the safety of the dome and propose remedial works if necessary. As well as analysing the stability of the dome using statics, he also used Hooke's hanging chain technique but, unlike Hooke, Poleni's model had different weights representing the voussoirs of different sizes (Fig. 4.3). Like Hooke, Poleni considered only a two-dimensional arch, but justified this approximation by arguing that the dome could be seen as a series of radial arches, in the form of semi-lunes (half-orange segments), acting independently of each other. Poleni used the model to determine the ideal shape of the masonry arch and, since the catenary lay safely within thickness of the arch, he concluded that each opposite pair of arches was stable, and hence, so was the whole dome. Nevertheless, he also recommended that additional ties were added to carry part of the hoop stresses.

A century later, the German engineer Heinrich Hübsch (1795–1863) also used Hooke's technique, making hanging-string models to determine the weights of voussoirs needed to achieve the desired shape of an arch or vault shape (Fig. 4.4). Hübsch's method was used in 1837 for the design of a (roughly) hemispherical dome covering a foundry

CHAPTER FOUR: PHYSICAL MODELLING AND FORM FINDING

(Gießhauss) in Kassel, Germany. The 16m diameter dome was made with hollow clay pots which were both lightweight and fireproof. Using Hübsch's method, the foundry owner who built the dome was able to reduce its thickness to just 175mm for the upper two-thirds of the dome.

During the second half of the nineteenth century, a number of books recommended the use of hanging models to establish the best geometry for arches and vaults. In the 1890 (revised) edition of Ungewitter's classic book on Gothic construction, Karl Mohrmann specifically recommended the use of three-dimensional hanging models. In the 1890s, Friedrich Gösling (1837–1899) also used both two- and three-dimensional models for some of his designs (Fig. 4.5).

Rather better known is the work of the Catalan architect Antoni Gaudí (1852–1926) who used both two- and three-dimensional hanging models made with strings and bags of sand to help establish the forms of arches and vaults for several of his masonry buildings (Fig. 4.6). The most well known of these was the crypt of Colònia Güell (Tomlow, 2011). Gaudí used the results of his model tests to complement his use of both statical calculations and graphical statical methods to determine the forms of the inclined columns and vaults.

Heinz Isler (1926–2009) was the last of the great concrete shell builders of the twentieth century, and, like many before him, brought his own unique approach to the challenge. His disarmingly simple idea was to take into three dimensions Hooke's technique by using a sheet of cloth (rather than a chain) to make hanging models, which he then scaled up to reproduce their funicular geometry at

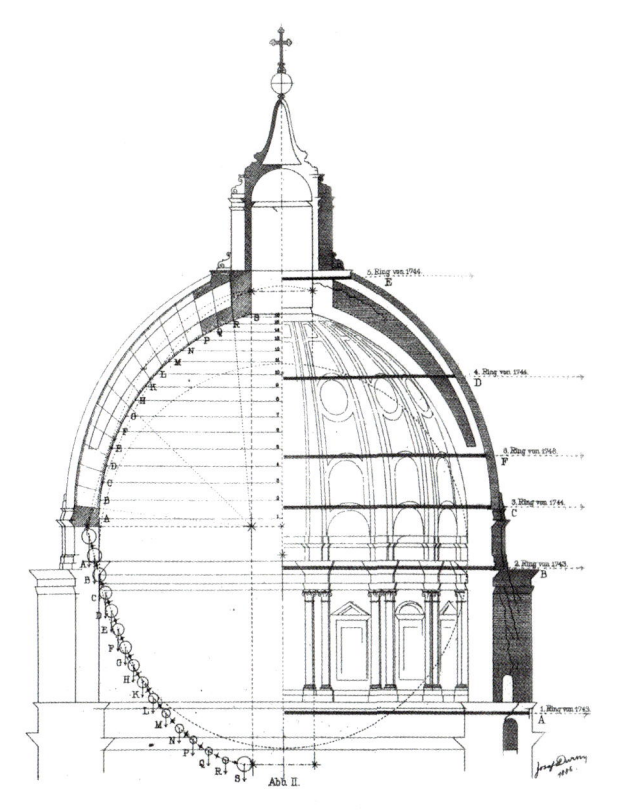

Figure 4.3 St Peter's Cathedral, Rome, 1506–1626. Sketch showing Poleni's hanging model superimposed on a section of the dome

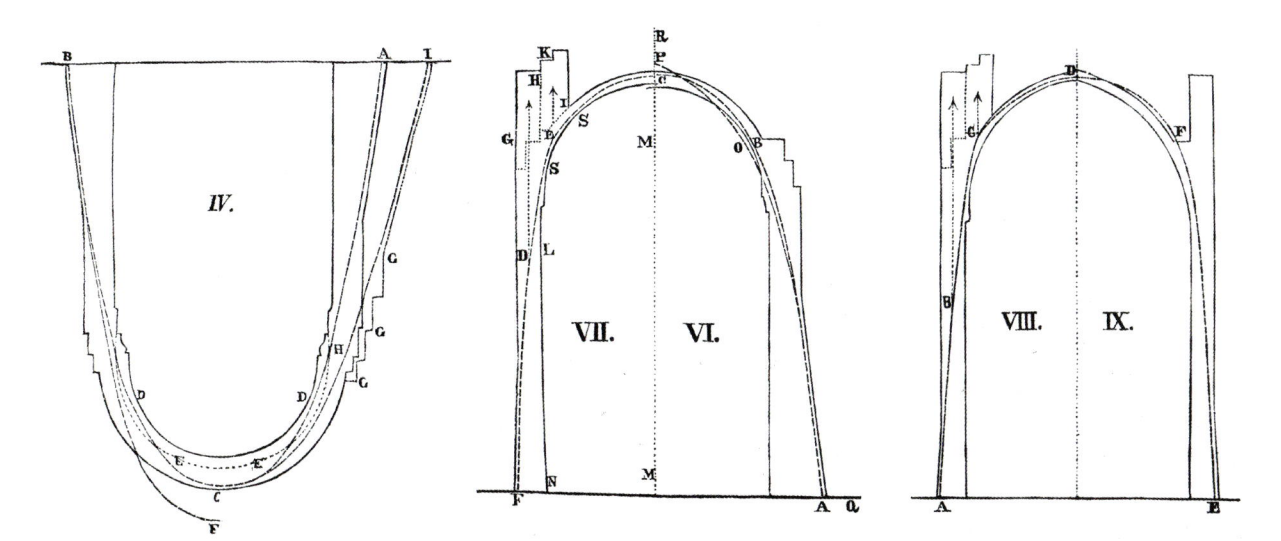

Figure 4.4 Investigation of the form and construction of various vaults using a hanging chain, by Heinrich Hübsch, c. 1835

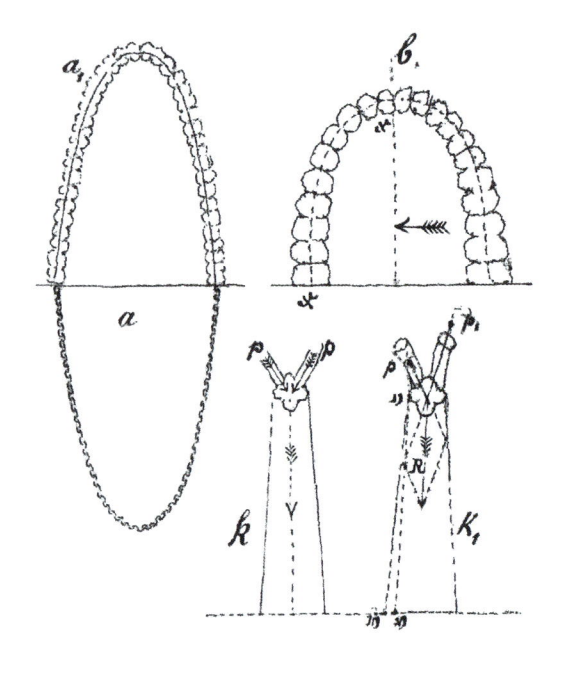

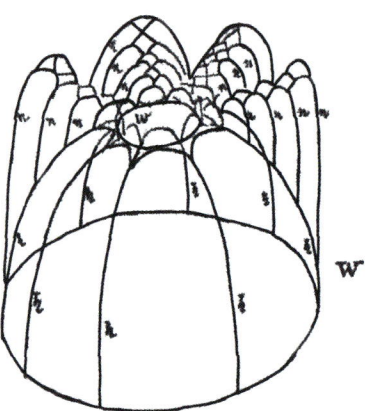

Figure 4.5 Form finding of arches and vaults using hanging chains, by Friedrich Gösling, c. 1890

Figure 4.6 Reproduction of Gaudí's hanging model for the crypt of Colònia Güell, Barcelona

Figure 4.7 Garden Centre Florelite, Plaisir, France, 1966 by Heinz Isler

full size (Figs. 5.2 and 20.13). The results were often breathtaking for their elegance and daring (Fig. 4.7). One technique he developed was to soak a piece of cloth in liquid plaster or resin and allow it to harden; another was to suspend a piece of wet cloth outdoors during in a Swiss winter night to freeze the shape (Fig. 4.8). Not only were these techniques able to generate the form of the main shell, but also the folds that provide stiffening to the free edges of the shells. Isler used other techniques to create both funicular and non-funicular structural forms, including using air pressure to inflate elastic membranes (Fig. 20.8). By these various means, Isler was able to generate forms whose geometry he could define and whose structural behaviour he could then analyse in detail, using a statical/elastic mathematical model, to determine the appropriate reinforcement needed to achieve the necessary strength, stiffness and resistance to buckling (Chilton, 2000) (see Section 5.3.1 and Chapter 20 for more on Heinz Isler's work).

Frei Otto (b. 1925) has been a great innovator in the use of models to determine the form of tension structures. He began using models in the 1950s as the only way of establishing the form of three-dimensional, membrane and cable-net structures whose final geometry could not, at the time (before computers),

Figure 4.8 Hanging model by Heinz Isler. A wet cloth frozen during a winter night to create the form of a reinforced-concrete shell roof

be determined using analytical methods. Since gravity loads played a minor part in establishing the form of the tensile structures, the models themselves were made of membranes or nets with different characteristics: soap bubbles which have a constant surface tension; elastic sheets whose surface tension depends on the strain; and nets whose surface tension arises partly from the elastic extension of fibres, and partly from shear deformations of the net (squares to rhombuses) (see also Section 5.3.1). Otto and his colleagues at the Institute for Lightweight Structures (IL) in Stuttgart developed a large range of modelling techniques, including making soap bubbles up to 1m wide (Fig. 4.9), as well as ingenious methods for measuring and surveying their complex forms which could not be defined using mathematical models. Having established the equilibrium geometry of the tensile structure, it was then possible to use analytical methods to determine in-plane stresses and forces at the boundary supports.

4.3 Scale-dependent models

Models used to study structural behaviour that is scale-dependent were of two main types: those used to study the elastic behaviour of shells loaded in ways which cause out-of-plane bending of the shell; and those used to study the buckling failure of shells due to high in-plane stresses.

For the competition for the Sydney Opera House, the architect Jørn Utzon (1918–2008) proposed a roof formed by two thin concrete shells, of undefined geometry, forming a shape that imitated the sails of yachts in the harbour. Various geometries for a thin-shell roof were considered and engineers Ove Arup & Partners commissioned structural tests to be undertaken on a 1:60 scale model of an early version (Fig. 4.10). These model tests led to the conclusion that this structural form could not provide sufficient stiffness and strength to carry the various loads that might act upon the roof. During six years of design development, many alternative structural solutions were considered which consisted either of substantial ribs supporting infill 'shells' or deeply corrugated shells (Fig. 4.11). A spherical surface was finally chosen to

Figure 4.9 Form-finding model by Frei Otto for a tensile membrane roof, using a soap bubble approximately 1m wide. Image inverted

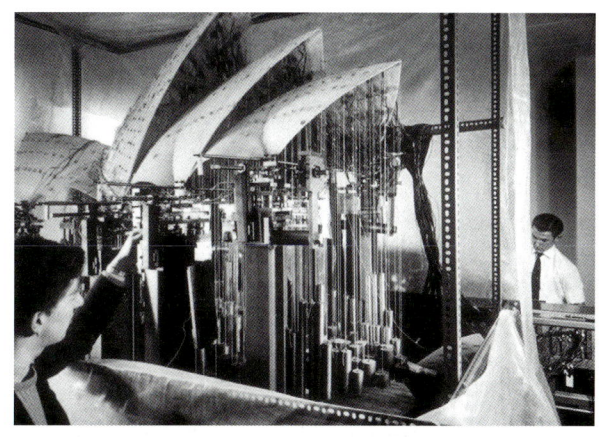

Figure 4.10 A 1:60 scale model of a thin-shell version of the Sydney Opera House roof being tested at the University of Southampton

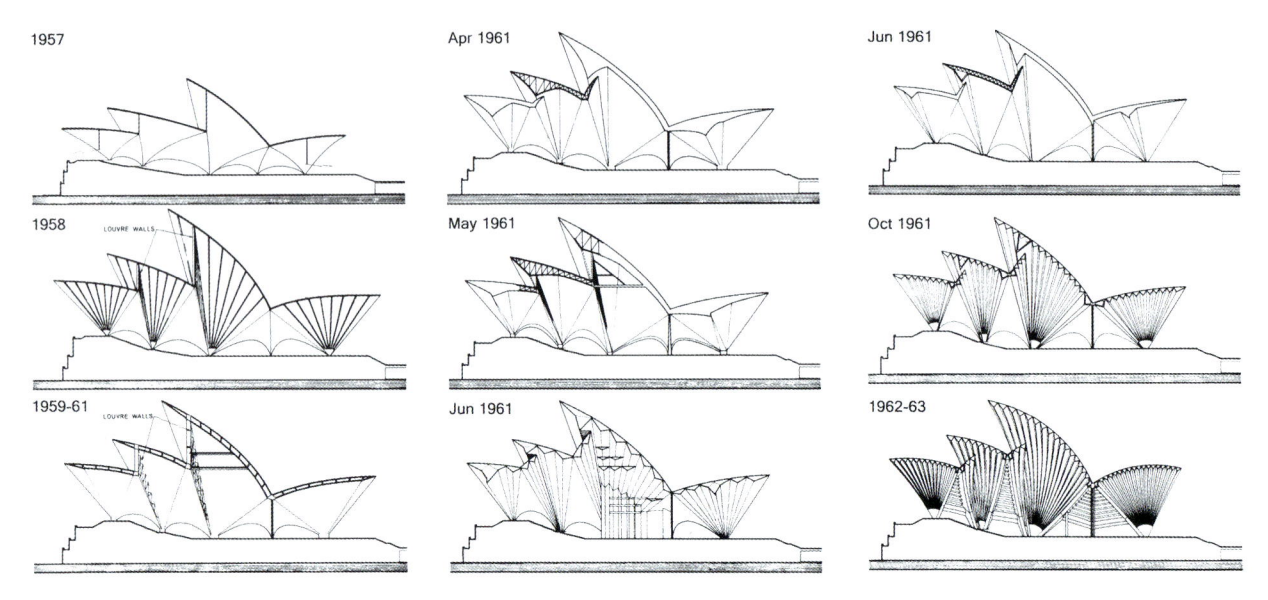

Figure 4.11 Variations of the designs considered for the roof of the Sydney Opera House

simplify the construction of the ribbed arches using pre-cast concrete elements, and fixing the ceramic-tile covering to the concrete structure.

Ove Arup & Partners were also the engineers for the Bundesgartenschau Multihalle in Mannheim, constructed in 1973–1974 (see also Chapters 12 and 19). The structure of the Multihalle is a form-found timber gridshell, on an irregular plan, covering around 7,400m². Being of unprecedented size and complexity, it tested the limits of structural analysis and a number of model tests were undertaken to help determine the optimal form for the grid and to supplement the manual and computer analyses.

The competition-winning design by the architect Carlfried Mutschler proposed a freeform timber gridshell, about 160m long with the largest shell spanning about 70m × 60m. Working with the architect Frei Otto, engineers Ove Arup & Partners used several different models to determine the overall form, the distribution of stresses in the grid and the best way to erect the structure. Close integration of the manual structural analysis, computer analysis and the results of testing the physical models was essential to meet the great challenges posed by this unprecedented structure (Happold & Liddell, 1975).

First, a wire-mesh model, about 1:300 scale, was made to establish the basic form of the structure, consisting of two main halls connected by a linking tunnel (Fig. 4.12). This enabled a more accurate 1:98.9 scale, hanging-chain model to be built to determine the geometry of the boundary supports that would give the best geometry for the roof. The mesh of hand-made wire links, each 15mm long, was hung from 80 supports whose positions were then finely adjusted to establish the boundary geometry that created the best funicular form, avoiding regions of low tension in the net (Figs. 4.13 and 4.14). The model was then surveyed using photogrammetry, by Professor Linkwitz and his Institut für Anwendungen der Geodäsie im Bauwesen, Stuttgart University, to determine the true geometry of the

Figure 4.12 A 1:300 scale wire-mesh model of the Multihalle to establish general form

CHAPTER FOUR: PHYSICAL MODELLING AND FORM FINDING

Figure 4.13 Making the 1:98.9 scale hanging-chain model to determine the precise funicular form of the Multihalle shell

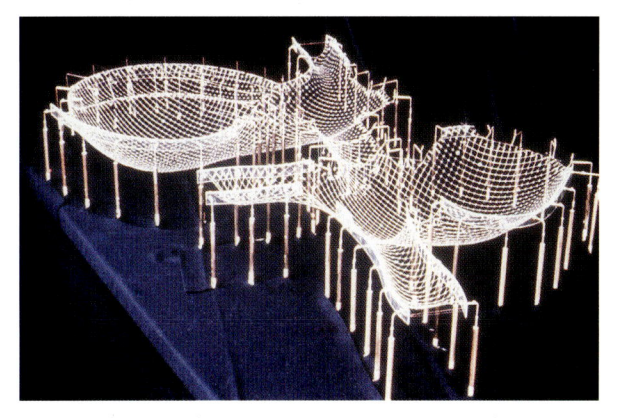

Figure 4.14 The finished hanging-chain model of the Multihalle surveyed using photogrammetry to determine its precise geometry

shell, and these results provided the coordinate data for manual and computer analytical models (see Chapter 12).

As there was no precedent for such a structure on this scale, it was considered essential to test an elastic structural model. However, there was also no precedent for making and testing such a model. It was decided to build and test a model of a similar, smaller gridshell dome that Frei Otto had previously constructed for the German Building Exhibition at Essen in 1962. This would give a better understanding of the behaviour of a gridshell dome, and would test the suitability of the model material and the proposed method of constructing the model, as well as determining the accuracy with which the model was able to replicate the behaviour of the full-size structure. A 1:16 scale model was constructed using laths of 3mm × 1.7mm, at 50mm centres, made of Perspex with a

modulus one-quarter that of timber. It was loaded with a Uniformly Distributed Load (UDL), on which was superimposed an increasing point load to initiate a local buckling failure. This was repeated with higher UDLs and thus the critical UDL was determined. Four versions of the model were tested – with pinned and rigid (glued) joints between the laths, and with and without diagonal ties.

Dimensional analysis (see also Section 8.3) was used to compare the results from the tests on the Essen model and the behaviour of the real Essen shell. This established that the relationship between the model collapse load and full-size collapse load was in proportion to the value of the dimensionless number

$$\frac{q_{critical}}{\left(\dfrac{EI_{xx}}{aS^3}\right)} \qquad (4.1)$$

for model and prototype, where $q_{critical}$ is the collapse load per unit area, E is the modulus of the material, I_{xx} the second moment of area, a the spacing between laths, and S the span of the dome. These results demonstrated the benefit of the diagonal cables in inhibiting overall buckling, and also indicated that a double-layer grid would be needed for the larger Multihalle shell.

Based on the experience of the Essen model, a 1:60 scale model of the main shell of the Multihalle was constructed in the same manner, using Perspex laths 1.4mm × 2.6mm, at 50mm spacing (Fig. 4.15). The 1m wide model was tested in the same manner as the Essen shell, applying the load using bundles of 100mm nails, each weighing 12.5g. Deflections of the grid were measured using dial gauges, and later

Figure 4.15 1:60 scale Perspex model of the Multihalle tested in the elastic range

refined by using wires and a lever arrangement to avoid the inevitable friction when a gauge touches the structure. The results were used to validate the results of the computer model and to provide a physical understanding of the nature of the buckling failures in the grid, and the location of critical areas in this complex three-dimensional structure.

Testimony to the success of the use of models during the design of this truly remarkable structure is that, despite being proposed as a 'temporary' building, it is still standing forty years later, after only one refurbishment in the early 1980s, and is a listed building.

4.4 Models and form finding in the twenty-first century

Physical models are as valid today as they always have been, as a means of creating potential geometries for shell and lattice structures. However, their role is no longer to predict structural behaviour in full-size structures as it was, for example, for the Sydney Opera House or Mannheim Multihalle; this can now be achieved, usually with a sufficient level of confidence, using computer models. Nevertheless, computer models are known to be fallible, due to programming errors or inappropriate assumptions, such as boundary conditions, in the building of the computer model. Physical models can still provide a valuable independent check on the validity of the output of computer programs. While similitude between full-size and model parameters may be slightly inappropriate, they, and the behaviour of the model shell itself, will always be approximately correct; unlike a programming error which may lead to errors of an order of magnitude or more.

At the very least, physical models can be used to generate forms that are structurally optimal and that cannot be generated using the normal vocabulary of geometric shapes such as spheres and ellipsoids. A physical model can serve to highlight the degree to which a geometric form deviates from the optimal shape and, hence, how much bending the shell will be required to endure – and the more bending must be resisted, the heavier the structural elements will be. The ease with which computers can generate three-dimensional forms can, however, lead architects to forget what many of them knew in the 1960s and 1970s – that optimal structural forms do exist, and that they are not parts of a sphere, ellipsoid or (as currently seems prevalent) a toroid. It is surely vital that educators of both engineers and architects continue to use models with the enthusiasm that Heinz Isler did to develop a basic understanding of shells and gridshell structures.

Further reading

- *Structural Engineering – the Nature of Theory and Design*, Addis (1990). This book contains useful discussion of the epistemology that underlies the relationship between the behaviour of models and full-size structures, and the role that models play in raising an engineer's confidence in a design for a proposed structure.
- *Building: 3000 Years of Design Engineering and Construction*, Addis (2007). This book contains much historical material regarding the development of masonry vaults and domes, concrete shells and lattice shells, as well as model testing.
- '"Toys that save millions" – a history of using physical models in structural design', Addis (2013). This paper is a review of model testing during the nineteenth and twentieth century, including many references to sources used for the present chapter.
- *Models in Architecture*, Cowan et al. (1968). This book provides a survey of the use of models by engineers and architects written in the heyday of model testing of structures, before the dawn of the computer age.
- 'Poleni's problem' is a 1988 paper reprinted in *Arches Vaults and Buttresses: Masonry Structures and their Engineering*, Heyman (1996). This article is an excellent review of what Poleni actually did when studying the stability of the dome of St Peter's cathedral, and how he interpreted the results of his analytical calculations and the hanging-chain model.
- *Model Analysis of Structures*, Hossdorf (1974). This is the best book on the subject.
- 'Generating shell shapes by physical experiments', Isler (1993). This paper is worth reading to understand Isler's philosophy in his own words, since he wrote very little.

- *Storia dei modelli dal tempio di Salomone alla realtà virtuale*, Piga (1996). This is a unique and excellent history of using models in building design.
- *Frei Otto – Spannweiten. Ideen und Versuche zum Leichtbau. Ein Werkstattbericht*, Roland (1965). This is a study of Otto's work at the time he was developing a large range of model testing techniques.
- 'Gaudí's reluctant attitude towards the inverted catenary', Tomlow (2011). This is a recent paper taking a critical look back at what Gaudí actually did and why.

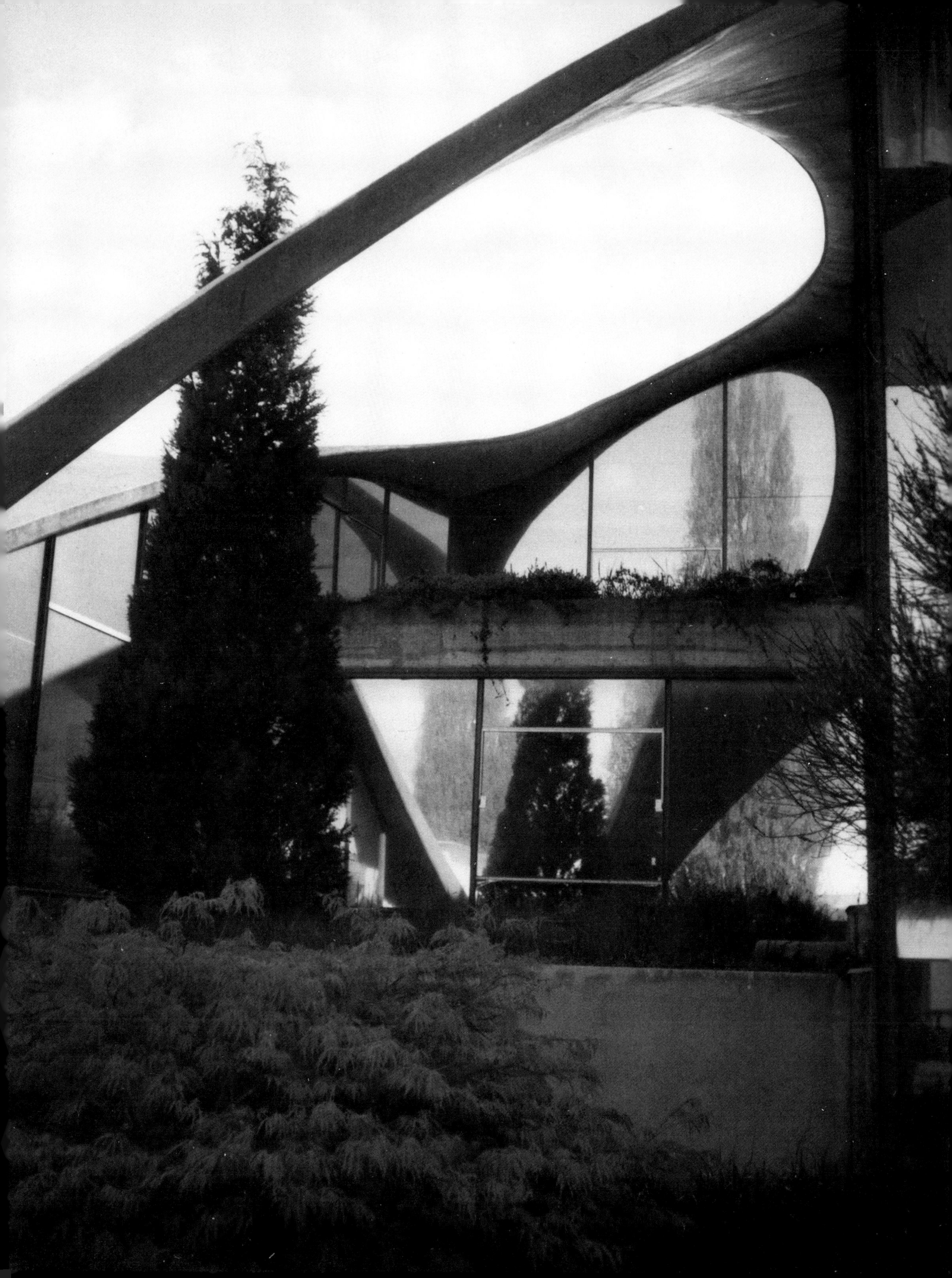

CHAPTER FIVE

Computational form finding and optimization

Kai-Uwe Bletzinger and Ekkehard Ramm

Shells are in general very thin structures. This *geometrical anisotropy* also leads to a *physical anisotropy*, expressed in load-carrying phenomena such as membrane action, bending and shear. Ideally, shells are optimized structures. They are used when heavy loads have to be carried or large distances have to be spanned. Their exceptionally favourable load-carrying capacity is due to their curvature that allows the shell to be in an almost pure membrane state. In other words, in contrast to bending, each fibre participates equally in the transfer of loads. And if the stress state tends to be homogeneous spatially, the situation is even more advantageous. Of course, these conditions have to reflect the construction material employed. A tension-oriented, thin fabric material has other requirements than a masonry or concrete structure, where compression is the favourable stress state.

This ideal situation is based on a number of prerequisites. It is well known that systems optimized with respect to selected parameters may be extremely sensitive to a change in these parameters or even other circumstances not considered. Optimization often is a generator of instabilities and imperfection sensitivities. This is particularly true for thin shells due to their extreme slenderness. Typical examples are buckling phenomena, or sudden failures due to excessive bending. A shell can thus be described as the 'prima donna' of all structures. The aforementioned physical anisotropy may lead to a physical sensitivity which almost automatically leads to a numerical sensitivity of simulation models. As a consequence, any sound design has to enable the shell to carry the loads primarily by membrane action to avoid bending, in particular inextensional deformations (see Section 3.1.2). At the same time, the shell needs to be as robust as possible, that is, imperfection insensitive. Most importantly, the design should guarantee these conditions for the entire lifetime of the structure.

This ideal situation depends of course on the usual design parameters, such as loads and their combinations, boundary conditions, and the material layout. It is obvious that the shape of a shell plays a dominant role in this design process. In this chapter, the geometry of the shell surface is synonymous either to *form*, usually used in the context of form finding, or to *shape*, as it is introduced in the community of optimization. It should be mentioned that in certain cases the design objective might just be the opposite, namely designing a shell for almost pure bending, inextensional deformations, and avoiding any stiffening membrane action. Examples are unfolding processes for space antenna or of plant leaves.

Shells can handle different loads by membrane action only if certain conditions are met, but this is of course not always the case. In the first half of the twentieth century, contrary to earlier centuries, roof shells were mainly based on pure, so-called geometrical shapes. These are shapes defined by elementary analytical formulae, such as spheres, cylinders, elliptic paraboloids and hyperboloic paraboloids. One reason of course was that geometrical shapes could be handled by analytical shell analysis. These classical shapes do not usually lead to a proper membrane-oriented response. Consequently, extra elements such as edge beams, stiffeners and prestressing had to be introduced. However, shape is a key parameter for the aesthetical appearance and can play a decisive role in the final design. This is particularly true in the context of shell roofs and meant that, contrary to the above-mentioned procedure where the response of a structure is determined for a prescribed shape, an inverse procedure had to be posed.

Given a few geometrical parameters, such as span and height, load, desired stress state and displacement limits, look for a natural and mechanically sound shape of the shell, at the same time satisfying the challenges of aesthetic and function as much as possible.

It has been shown that in many cases this concept renders more natural and elegant designs with free edges that avoid extra stiffeners such as heavy edge beams. In other words, following the laws of science in an optimal way often leads to an aesthetically pleasing structure. This brings us to the important issue of finding proper forms for thin shells.

5.1 Objectives for form finding

From these general statements we want to specify the general objective of an underlying form-finding concept:

Find the form and the thickness distribution of a shell for certain functional requirements, such that

- boundary conditions and all possible load cases are considered;
- material properties are taken into account (e.g. no or low tension for concrete);
- stresses and displacements are limited to certain values;
- an almost uniform membrane state results;
- buckling, excessive creep, and negative environmental effects are avoided;
- a reasonable lifetime is guaranteed; and hopefully
- manufacturing costs are justified;
- the design is aesthetically pleasing.

Here we follow the principle *form follows forces* – also discussed as *force follows form* – modifying Louis Sullivan's famous statement 'form follows function' (Sullivan, 1896). The aforementioned requirements and functional needs interact with each other and are in some cases even contradictory, such that a compromise has to be found. It is important to point out that, despite the many constraints governing a design, there is usually not one unique solution. Fortunately, the design space is not limited, allowing enough creativity and design freedom.

5.2 Infinity of design space and design noise

The principal challenge of form finding can be briefly explained by an illustrative example. The task is to design the stiffest structure made from a very thin piece of paper that is able to act as a bridge-carrying load. Because the flat piece of paper is able to only act unfavourably in bending, stiffeners have to be introduced by folding the paper. However, there exist an infinite number of solutions, all of which create stiff solutions of at least similar quality, which are far better than the quality of the initially flat piece of paper (Fig. 5.1). Surprisingly enough, even an arbitrary pattern of random folds appears to be a possible solution, although not being favourable from the manufacturing point of view.

The image of the randomly crumpled paper is an ideal paradigm for the infinity of the design space or, more ostensively, the *design noise*. This terminology indicates an extremely busy design. Together with the expressions 'filter' and 'mode', it follows the notions in signal processing for generally undesirable, random signals, e.g. white noise. As for the example, the crumpled paper can be understood as a weighted combination of all possible stiffening patterns and one can easily think of a procedure to derive any of the individual basic solutions of distinct stiffening patterns by applying suitable *filters* to the design noise.

Figure 5.1 Stiffened shell structures made from folded paper: (top left) without stiffening, or with a (top right) coarse, (bottom left) fine, or (bottom right) random stiffening pattern

In a general context, this means filtering out the essential shape parameters that mostly contribute to the objective. In a numerical context, a mathematical filter is chosen to condense the solution to a smaller set of geometrical mode shapes and to reduce the number of design degrees of freedom. This process is denoted as regularization or smoothing. It is clear that the kind of filter as well as the filter process can be freely chosen as an additional and most important design parameter. It is possible to define a procedure, as implied by the paper example, to first filter high-frequency design noise, and second apply geometrical filters. It is, however, also possible to apply *indirect* filters by pre-selecting and favouring certain classes of solutions in advance, such as a stiffener pattern of certain cross sections and spacing, chosen because of manufacturing or aesthetical reasons. There is no doubt that the latter approach is the more ingenious one, as a large set of other solutions, perhaps even better solutions, might otherwise be undetected. It is up to the insight and imagination of the designer how to define a procedure of pre-selection, regularization or pre-filtering. Most often, however, this task seems to be a somewhat mysterious one, or a vague one at best. From this point of view, form finding truly is an art.

5.3 Design methods for shell structures

In the following sections, two methods for form finding of shells are discussed that satisfy, at least approximately, the criteria mentioned in the Section 5.1.

In the first method, the shape of a membrane, deformed by a single, dominant load case, or design loading, is determined. An example is a hanging membrane, subject to gravity loads, that is inverted to form the final shape of a roof shell under dead load. This principle of an inverted hanging membrane can be investigated by two different procedures: experimentally, using a real, small-scale physical model, or numerically, using a computer simulation applying large displacement membrane analysis. Depending on the respective physical problem, other design load cases may be investigated; for example, uniform or hydrostatic pressure load.

The second method discussed is the most general procedure, namely shape optimization, which can be combined with simultaneous thickness and material optimization. Usually, the overall topology for shells is prescribed by the conceptual design. If, however, the design allows a change of the initial topology – for instance, including further openings or designing the optimal layout of a gridshell – topology optimization as a generalized shape optimization technique can be applied.

5.3.1 Physical and numerical form finding

The hanging model principle, or Hooke's law of inversion (see Section 1.1), is known to be one of the oldest form-finding methods for the design of arches and domes that are in pure compression by being free of bending. This principle allowed coping with the low- or no-tension material of masonry as the only available construction material at that time. By inverting the shape of the hanging chain, which by definition is free of bending, an equivalent arch that is in pure compression is obtained. As the structure is free of bending, the material is used in an optimal manner, which means that the structure can also be seen from the point of material minimization or stiffness maximization. For the exploration of the design space, the length of the chain is the one important control handle that defines the final structural height. If the inextensional chain is replaced by an elastic cord, the material properties serve as additional parameters which control the experimental result. The shape of the catenary is uniquely defined by the equilibrium of applied and reaction forces. Here the self-stabilizing effect of tensile forces in equilibrium serves as the aforementioned filter to detect the preferred result.

The situation becomes much more complicated if the hanging principle is applied to the design of domes and shells. The reason is that, due to the second dimension of the surface, there are always multiple possibilities for the forces to flow from the loading region to the supports. In the hanging model experiment, however, there is another very important complication when the one-dimensional chain is replaced by materials with a distinct two-dimensional stress state, which is the additional ability of a membrane to carry load by in-plane shear action. As a consequence, the final deformation of a hanging cloth is dominantly affected by its two-dimensional elastic properties, in particular its in-plane shear resistance. Wrinkles may develop when the structure buckles in compression, at which point the load carrying behaviour is locally reduced to one dimension.

The alternative of a woven, orthotropic material of negligible in-plane shear resistance, however, is characterized by the orientation of the fibres. All together, the result of a two-dimensional hanging model experiment for the design of domes and shells is inevitably driven by the material properties and the cutting pattern of the cloth which is used for the experiment. These facts make it difficult to compare results of hanging model experiments with different orientations or types of materials or to explain certain phenomena such as the shape of shells at free edges. Often there is very little, if any, information recorded about the details of the material properties for the hanging membrane, or, even more importantly, the cutting pattern and orientation of material anisotropy. This ambiguity of the material used in hanging models has also to be reflected to the properties of the material of the built structure. The masters of form finding, such as Heinz Isler (see Chapter 20), made superior use of experimentation with different materials and cutting patterns, resulting in seminal structures such as the 1969 Sicli Factory in Geneva (see page 44 and Fig. 20.16). Isler developed very finely tuned procedures with various materials from isotropic rubber to orthotropic textiles, which he carefully tailored for the applications at hand. Figure 5.2 shows examples of completely different shapes obtained by a hanging model experiment, using the same textile, but simply

CHAPTER FIVE: COMPUTATIONAL FORM FINDING AND OPTIMIZATION 49

modifying the angle of anisotropy. This experiment gives important hints about the secrets of this approach. Numerical simulations give similar results, obtained by large deformation finite element analyses with material anisotropy representing the warp and weft properties of a fabric (Fig. 5.2). The element geometry may, but does not have to be aligned to the fibre orientation. The experiments at Frei Otto's Institute of Lightweight Structures (IL), University of Stuttgart, nicely show the effect of shear-resistant materials (Fig. 5.3). Hanging forms made from shear-deformable cable nets give clear doubly curved surfaces whereas the shape generated by hanging cloths shows wrinkles and negative curvature at the free edges.

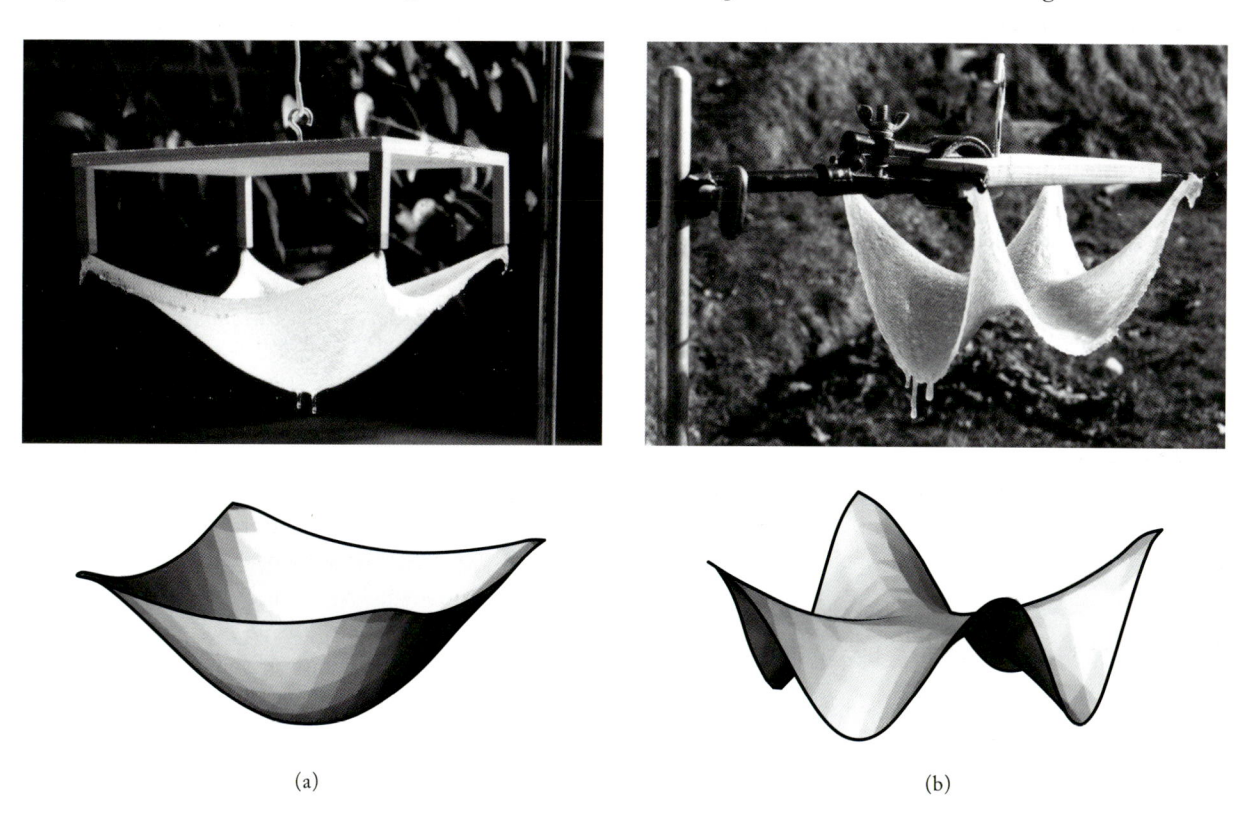

(a)   (b)

Figure 5.2 Ice and polyester experiments by Heinz Isler versus numerical hanging models. Negative and positive curvature by modifying the angle of material anisotropy, either (a) parallel, or (b) diagonally aligned to the edge

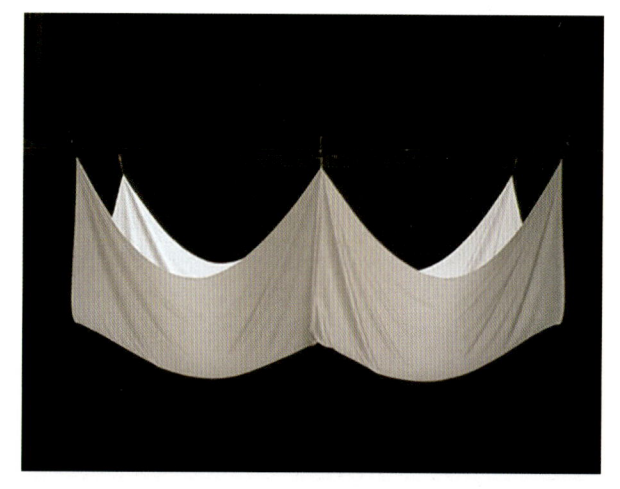

Figure 5.3 Hanging model experiments from IL

The physical hanging model experiment is an important tool in the conceptual design phase for whatever form-finding procedure will be finally chosen. It not only gives a lot of visual impression but also may provide substantial information for subsequent computer modelling. The numerical hanging models, as well as the optimization concept described in the next section, have the advantage that they automatically provide the geometrical data needed for subsequent structural analyses.

5.3.2 Structural shape optimization

The form finding described above – finding the final shape for a shell for desired or required constraints and objectives – is essentially an optimization process, usually followed in a trial-and-error manner, based on accumulated experience of the designer.

A logical progression is to cast this procedure into an automated approach of structural optimization (Fig. 5.4), demonstrated in Part II of this book. The key idea is that the three steps of design, namely the geometrical and material design, the analysis including the sensitivities, and the mathematical optimization, are iterated until the constraints of the problem are satisfied. In the terminology of optimization, these are:

- the objective(s);
- the equality and inequality constraints;
- the upper and lower bounds of the design parameters.

The objective, or in multi-criteria optimization the objectives, play the most important role. They may be classical, like minimizing weight or mass, or maximizing the ultimate load causing buckling and/or material failure, but may also be stress-levelling – getting a more or less homogeneous stress state (fully stressed design) – or minimizing strain energy, which is equivalent to maximizing the stiffness. This last parameter is a very promising objective for shells because it minimizes bending, yielding a membrane-oriented design. But further objectives may also be used, such as maximizing ductility and toughness, or avoiding certain frequency ranges. Often, conflicting situations exist, so a compromise solution has to be found that weighs the different objectives.

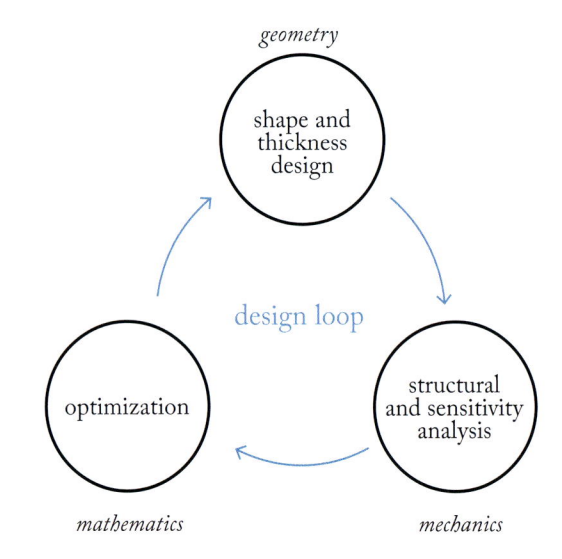

Figure 5.4 Three steps of shape optimization

The constraints are the usual design limits for stresses or displacements but may also be a prescribed construction mass/weight, or certain desired or undesired frequencies, just to mention a few. They may be defined as equality constraints (or as soft constraints) in the equilibrium equations.

The design parameters may be of different kinds. In the context of shape optimization, they are geometrical parameters controlling the shape and thicknesses of the shell. Thus, these basic features, shape and thickness, are parameterized by shape functions interpolating the initial and subsequently optimized geometry from selected nodal values.

5.4 Parameterization

The definition of geometry can be based on concepts with different parameterizations. This section briefly describes the three most common versions. Appendix D includes an explanation of subdivision surfaces.

5.4.1 CAGD-based

The industrial state of the art in structural optimization is characterized by the application of simulation techniques where Computer-Aided Geometric Design (CAGD) methods, Finite Element (FE) analysis, and Nonlinear Programming (NP) are combined for the design parameterization, the response analysis and the optimization, respectively. The idea is to define the

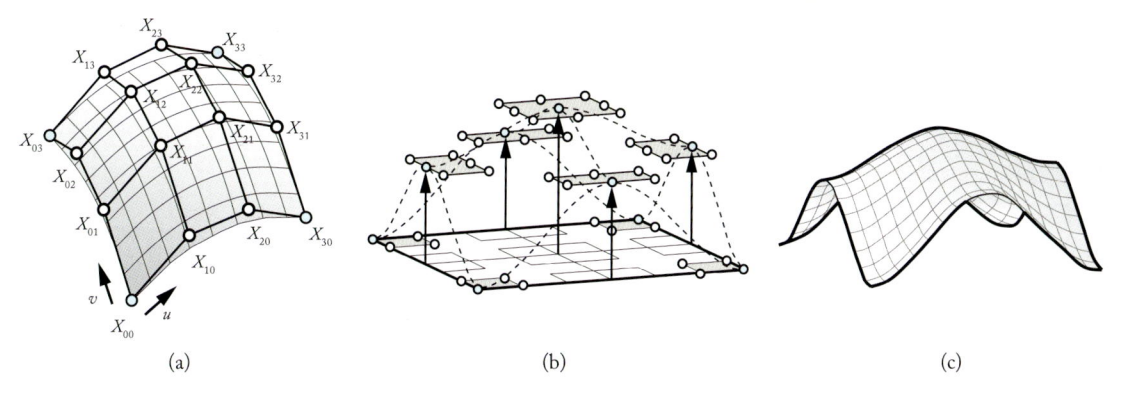

Figure 5.5 CAGD-based shape definition with design patches: (a) one NURBS patch, (b) four design patches with continuity patches and (c) generated shape

degrees of freedom for shape optimization and form finding by a few, but characteristic, control parameters of the CAGD model. Interestingly enough, the driving force behind the development of this approach was to find means to treat design and numerical noise as well as to avoid heavily distorted meshes. The solution is to use CAGD design patches and the related shape functions to prevent the high oscillations of numerical design noise. As a consequence, the design space is automatically reduced to the space of the chosen shape functions. The choice of a CAGD model is indeed identical to an implicit pre-selection of a design filter, which directly affects the result.

The design patches could be denoted as geometrical macro-elements, which in turn are subdivided by a finite element mesh. Typical representations use polynomial, Lagrangian interpolation, basis splines (B-splines), Bézier splines, or Coons patches as approximations, as shown in Figure 5.5.

Usually, C^1-continuity between patches has to be enforced. This means that the patches should not only be continuous, but also have a continuous derivative, making them continuously differentiable. For example, applying continuity patches, as shown in Figure 5.5b, generates a surface shape without sudden changes in angle. This, in turn, reduces the number of geometrical parameters due to the constraints of the nodal degrees of freedom of the patches.

To demonstrate shape optimization, we use the example of the Kresge Auditorium at the Massachusetts Institute of Technology (MIT) in Boston. This reinforced-concrete shell was designed by Eero Saarinen (1910–1961) in 1955 as a segment of a sphere on three supports. The structure, shown in Figure 5.6, which is certainly appealing from an architectural point of view, is far from optimal with respect to the requirements and objectives discussed earlier. As a result, it needed heavy edge beams and is still today subject to substantial bending. This

Figure 5.6 Kresge Auditorium, Cambridge, 1955, by Eero Saarinen, as of 2003, with detail of support

suggests that the structure is a candidate for finding a better shape, considering its structural behaviour. The original shape of the auditorium's roof structure (Fig. 5.7b) was subjected to a process that uses shape and thickness optimization with the objective of minimizing strain energy. Even without an edge beam, an almost pure membrane response in compression (see also Chapter 3) could be achieved. The quality of the modified design, shown in Figure 5.7a, was verified by geometrically and materially nonlinear analyses, and compared to that of the original design.

It has to be noted that the initial shell almost collapsed when the scaffolding was lowered; it had to be additionally supported by integrating a column every third mullion. The present analysis for the original shell did not take these extra supports into account. The concrete was modelled with an elasto-plastic material model. Geometrical imperfections were also taken into account. Figure 5.7 shows the dead load-displacement response for a representative point of the reinforced-concrete shell. It can be recognized that the original design shows a failure already for the dead load. The load-carrying capacity for the optimized shell without edge beams is substantially higher.

5.4.2 FE-based applying filters

An alternative to CAGD-based parameterization is the FE-based parameterization. Since in such a model all FE-nodes rather than only a few CAGD-control points can be activated for the shape definition, a huge design space is available. Now, explicit filters are applied for the selection of preferred optimal shapes. The procedure is such that the coordinates and the shape derivative at an FE-node are determined as the weighted mean of all the neighbouring nodes within the filter radius (Firl et al., 2012). Clearly, the filter decides which local optimum will be found from the design noise. As a matter of fact, this technique is based on sound theory and is strongly related to the method of subdivision surfaces (see Appendix D). The FE-based parameterization with filtering is most attractive for form finding and preliminary design because one can find all possible solutions with the same geometric model whereas alternative techniques need laborious reparameterization. Modifying filter functions and sizes is of no effort at all and appears to be most effective for exploring the design space. This is shown by an example in Figure 5.8. The shape of a shell, supported on three points and subjected to self-weight, is to be optimized for stiffness. The ground

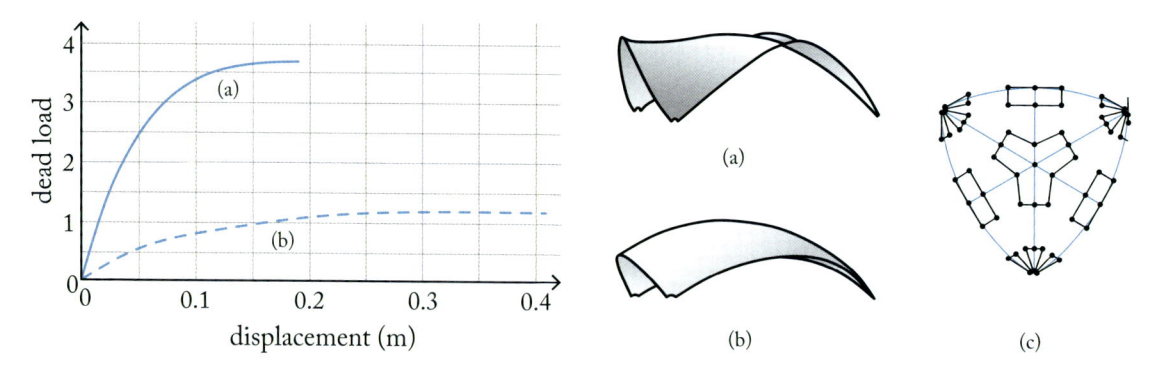

Figure 5.7 Nonlinear response of the Kresge Auditorium, for (a) the optimized shell without edge beams, and (b) the original shell with edge beams. The shape was parameterized with (c) six Bézier patches and control points

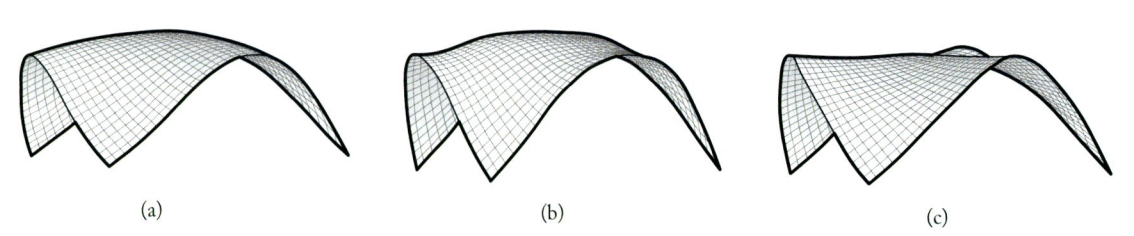

Figure 5.8 Optimal shell shapes by varying the filter radius size, with the ratio of interior to edge radii: (a) 11:3, (b) 7:3 and (c) 5:3

plan is fixed, and the thickness is held constant, particularly not allowing for thicker edge beams. Obviously, by simply playing with filters, a large variety of well-known shapes can be generated: (a) the positively curved shell similar to Otto's 'Segelschalen', (b) the negatively curved edges, well known from many of Isler's shells, and (c) the Candela-type, hypar-like solution. What is shown here was already obtained by the first few runs in the shortest possible time and serves as a basis for further optimization; for example, to improve the aesthetic quality. In turn, the example may give an impression about how design philosophies can be understood as tools to pre-filter the design space and how they affect the optimal result.

5.4.3 Isogeometric analysis

Usually, geometry is defined in a Computer-Aided Design (CAD) environment; for example, using Non-Uniform Rational Basis Splines (NURBS) as parameterization for the subsequent design. The geometric data are transferred afterwards to the analysts who transform the CAD model into an FE model. The FE parameterization is invoked for both the geometry and the mechanics. Typically, isoparametric finite elements with Lagrangian interpolation are applied. Isoparametric elements use so-called shape functions to represent both ('iso' means equal) the element geometry and the unknowns, typically the displacements (see also Appendix A).

More recently, so-called IsoGeometric Analysis (IGA) has been introduced (see for example Cottrell et al., 2009), where the analysis part also uses B-spline and NURBS parameterization. The advantage is, first of all, a consistent concept for both parts of the design. But, the NURBS representation also has a couple of advantages for the analysis part; for example, having a higher continuity between elements for the same order compared to the Lagrangian parameterization.

In the context of the design of shell structures, the IGA approach can be looked upon as a generalization of both concepts discussed above, using either coarse design patches or directly a fine finite element mesh for design. NURBS allow what is called *knot-insertion* to generate series of geometrically identical levels of refined meshes. Typically, the finest mesh is used as the analysis model whereas any of the meshes might be used as the geometrical model for shape optimization. It has to be noted though, that, by choosing one specific optimization model, the design space will be limited. A certain design filter will thus implicitly be applied.

The procedure is explained for a cylindrical shell under two opposite concentrated loads (Fig. 5.9). For the geometry of a cylinder with a circular cross section, a simple CAD model based on a NURBS representation with only a few parameters is sufficient, as shown on the left side of Figure 5.9. Based on this exact geometry, a NURBS refinement is necessary to discretize the shell for the mechanical

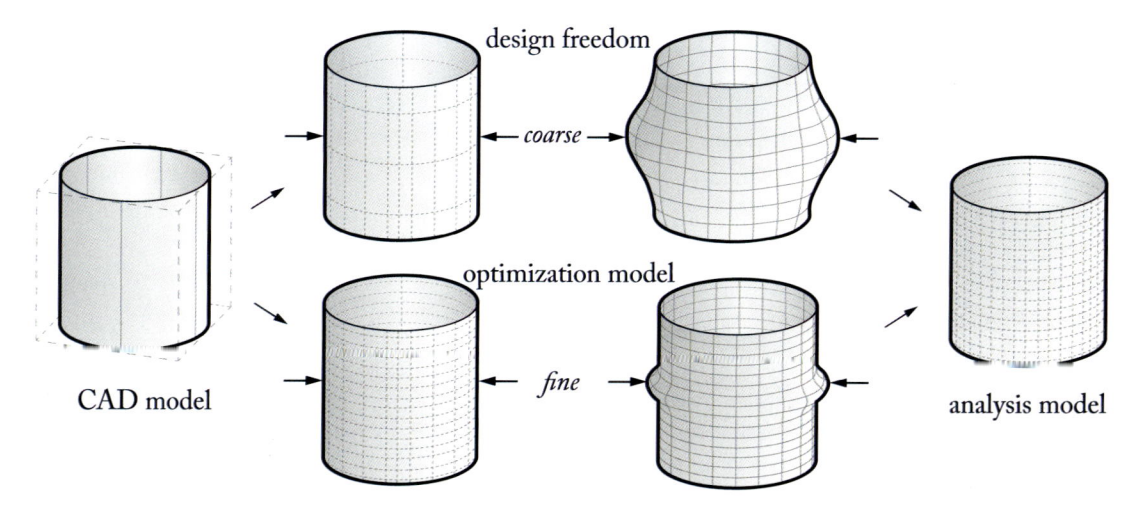

Figure 5.9 Isogeometric shape optimization for minimization of strain energy (Kiendl, 2011)

model, as indicated on the right side. Linear elastic material behaviour is assumed. In the following step, the shape of the structure is optimized with respect to minimizing the strain energy, which is equivalent to maximizing the stiffness. Now, the designer has an almost infinite design freedom, depending on how many parameters are inserted in the NURBS parameterized model to create the geometric model for shape optimization. Two options are indicated in the middle of Figure 5.9. The coarse optimization model in the upper part leads to a shape with two large bulges, increasing the curvature under the two loads. This case could be identified with the approach of applying design patches as described earlier. The optimization model in the lower part uses a very fine geometrical model, used for shape optimization, allowing a much more refined new shape, resulting in a thin ring stiffener. In the extreme case, the fine mesh of the analysis model can be used for the shape optimization as an FE-based parameterization, as mentioned above. In general, the density of the geometry may furthermore vary in different regions of the structure, which essentially replaces the filter described above by avoiding too much geometrical noise.

As has been mentioned already, optimization has the tendency to produce structures that are highly imperfection-sensitive. In order to avoid this critical situation, potential imperfections have to be included in the optimization process, that is, the imperfect shell has to be optimized; for example, by maximizing the failure load.

5.5 Conclusion

Ideal shells are optimized structures that can be extremely sensitive to imperfections, if not properly designed. This means that their shape has to be carefully adapted to the underlying design constraints. In other words, form finding is of utmost importance. This chapter discussed several methods, where the first one was based on the mechanical principle of inverting a hanging membrane. This can be verified either by a physical experiment or a corresponding numerical simulation based on large deformation analysis. A further, more general concept was shape optimization as a subset of overall structural optimization. Three different approaches to parameterize the geometry have been described. No matter which method is applied, it is important to have efficient means to control the design space. Ideally, they are implemented such that they are able to support the designer's intuition. The designer should always be aware that every detail of the design process affects the result as a design filter.

Experimental form-finding methods and, to a limited extent, also their computer simulation have the big advantage of being vivid and 'real'. They are ideally suited to preliminary design. Shape optimizations are much more general, but at the same time more abstract and sophisticated, and they need a basic design for the initial definition of the optimization problem. If the problem is properly defined, they have great potential.

This classification suggests a combination: hanging membranes as a means for the conceptual design stage, and shape optimization for its variation and refinement. However, despite a lot of progress in recent years, one must say that a magical toolbox does not exist. There is still enough room for design freedom and creativity.

Further reading

- 'Shape finding of concrete shell roofs', Ramm (2004). This paper discusses physical modelling in more detail and has been partially reproduced as part of this chapter with kind permission of the IASS.
- 'Form finding and morphogenesis', Bletzinger (2011). This article was published in the book *Fifty Years of Progress for Shell and Spatial Structures*, and includes discussions on minimal surfaces, tension structures and topology optimization. It has also been partially reproduced as part of this text with kind permission of the IASS.
- 'Heinz Isler Shells – the priority of form', Ramm (2011). This paper discusses the three form-finding methods employed by Heinz Isler, also in the light of computational models.
- 'Regularization of shape optimization problems using FE-based parametrization', Firl et al. (2012). This journal paper explains a fully stabilized formulation for shape optimization problems, featuring

several examples and proposing the filter radius as a means to control design.
- *Isogeometric Analysis: Toward integration of CAD and FEA*, Cottrell et al. (2009). This seminal book on IGA explains core concepts of this method as it has been applied to many types of problems.
- 'Isogeometric analysis and shape optimal design of shell structures', Kiendl (2011). This doctoral dissertation demonstrates the first comprehensive application of IGA to the problem of shell structures.

PART II
Form finding

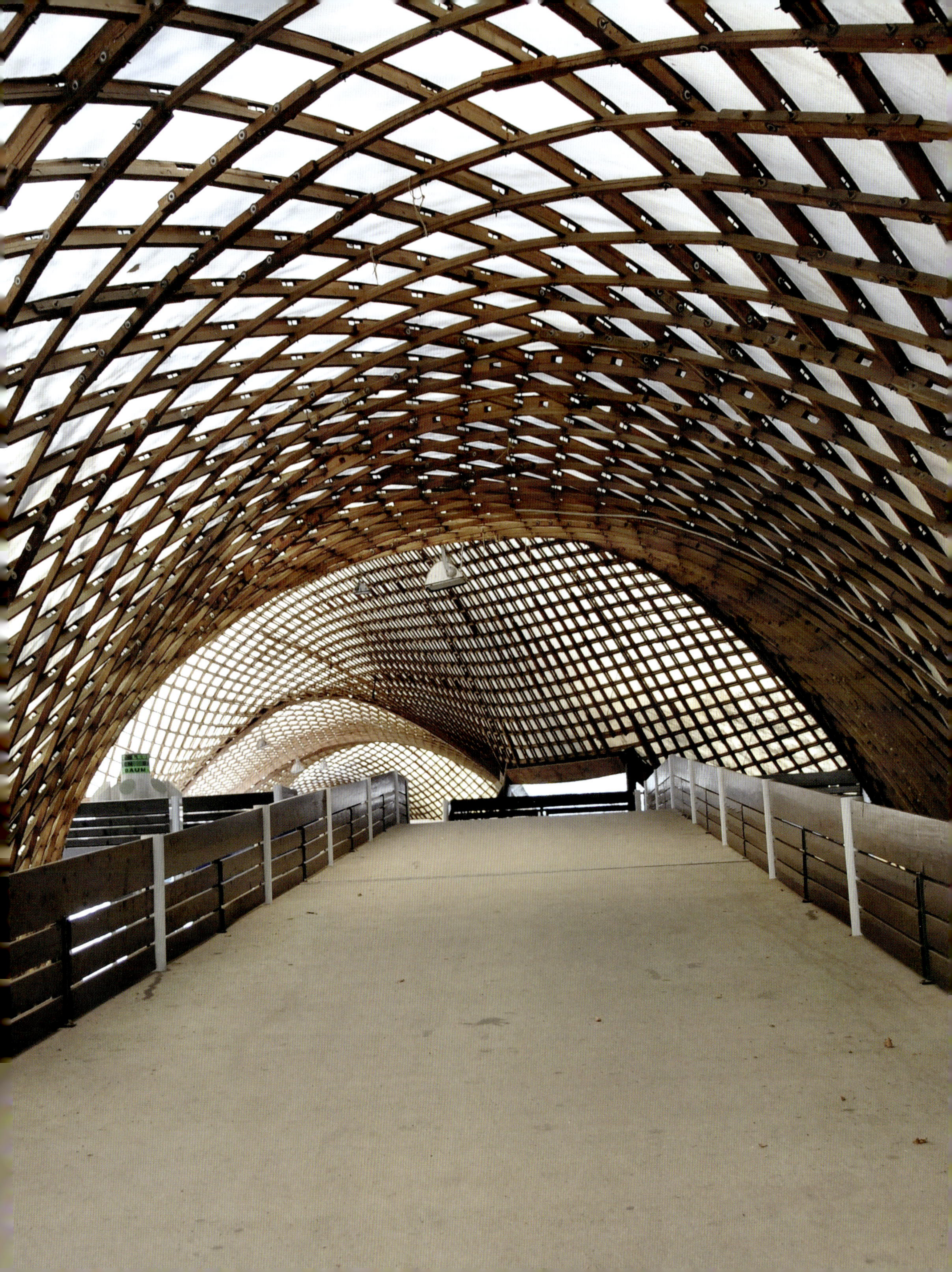

CHAPTER SIX

Force density method

Design of a timber shell

Klaus Linkwitz

LEARNING OBJECTIVES

- Derive the static equilibrium equations of a single node, with four bar forces and a load applied to it.
- Explain the consequences of introducing force densities into those equations.
- Apply matrix algebra and branch-node matrices to generalize this single-node formulation to arbitrary networks.
- Use the force density method to generate the shape for a shell based on such a network.

The use of 'force densities' presents an approach for the rapid generation of feasible shapes for prestressed and (inverted) hanging structures. This method allows, especially in the early stages of a new project, the instant exploration of large numbers of alternative, feasible solutions.

This chapter explains the basic premise and application of the Force Density Method (FDM), also known as the '(Stuttgart) direct approach'. It has been applied to the design of many built structures, particularly to tensioned roofs, but also to the timber shell roofs of the 1974 Mannheim Multihalle (see page 58) and the 1987 Solemar Therme in Bad Dürrheim, discussed in further detail in Chapter 12.

The brief

To expand its sports offerings, the municipality of Stuttgart is developing a new sports complex, which includes a swimming pool and an ice rink (Fig. 6.1). Each has to be covered independently to maintain a hot and a cold climate. Both structures are adjacent in order to exchange heat between their heating and cooling installations. For convenient access, both facilities share a central entrance. The Olympic-size swimming pool (25m × 50m) will be naturally ventilated. For this reason, a high-point roof is envisioned,

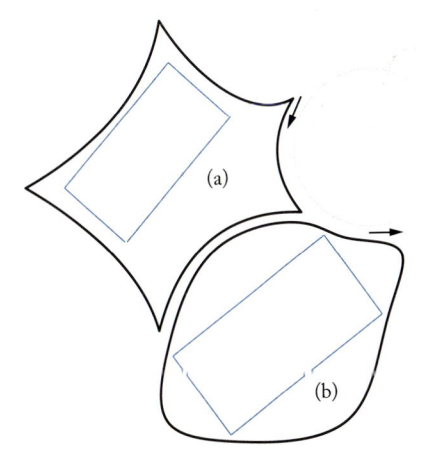

Figure 6.1 Outline for sports complex with (a) an Olympic-size swimming pool (25m × 50m) and (b) a standard-size hockey rink (30m × 60m)

creating a stack effect. The controlled climate for the standard-size hockey rink (30m × 60m) is maintained by a domed enclosure. The client wishes to use locally sourced timber for the main structure, so a timber gridshell is proposed as the structural system.

6.1 Equilibrium shapes

As part of the conceptual design of structures, especially domes, shells and membrane structures, generating an adequate structural shape is crucial to the load-bearing behaviour and aesthetic expression of the design. Their shapes cannot be freely chosen and conceived directly, due to the intrinsic interaction between form and forces. For such a problem one needs form finding. Typical structural systems that require form finding include:

- soap films within a given boundary;
- prestressed, or hanging fabric membranes;
- prestressed, or hanging cable nets;
- structures generated by pressure (e.g. air, water).

Membranes or cable nets can be used for the design of shell structures such as thin gridshells, but are only partially valid as the constituting elements are not necessarily free from bending.

For these types of structures, the force density method has proven an invaluable approach to generate equilibrium solutions, and thus feasible shapes for potential designs. Solutions are generated from simple linear systems of equations.

Another advantage of using force densities is that they do not require any information about the material for the later realization of the design. As we are dealing with non-materialized equilibrium shapes, no limitations with respect to material laws exist. The materialization follows in a second step. When introducing material, we may choose (independently for each bar in the net) the material, without changing the shape created with force densities.

6.2 A thought experiment

Looking at any prestressed, lightweight surface structure, we observe that the continuous surface is doubly curved at each point. In other words, when

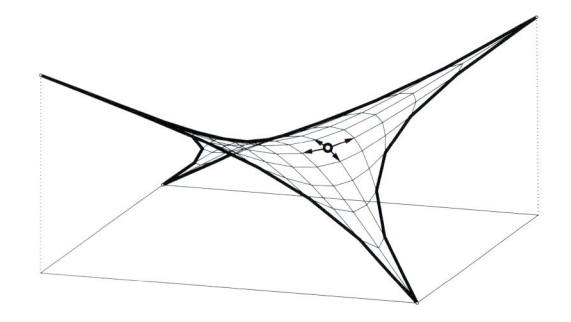

Figure 6.2 A prestressed surface, when discretized shows at each interior point, two 'hanging' bars curved upwards, and two 'standing' downwards

considering its discretization as a pin-jointed net, at each interior intersection point, two bars are curved downwards, or 'standing', and two curved upwards, or 'hanging' (Fig. 6.2). This opposite curvature is called 'anticlastic' or 'negative' curvature.

This characteristic doubly curved shape immediately becomes understandable in the following thought experiment. We put up four elastic rubber bands in a box. They will sag under their self-weight and assume the shape of catenaries (Fig. 6.3a). To stabilize these four hanging rubber bands, we place four new ones perpendicularly over them, connecting them at their intersections and attaching them to the bottom of the box (Fig. 6.3b). We now stabilize the net further by applying tension to the bands, by prestressing them. We pull the hanging bands by their ends, located at the sides of the box. Their initial lengths decrease, and as they are lifted up, they pull the net upwards (Fig. 6.3c). We continue by alternately pulling both sets of four elastic bands, as the geometry of the net changes less and less (Fig. 6.3d). Meanwhile, the net itself becomes increasingly stiff, and anticlastic curvature results at every point. Precisely this principle is applied everywhere in a prestressed, structural net.

6.2.1 A single node in equilibrium

The thought experiment in Section 6.2 can be further simplified. The stationary, prestressed net is characterized by the fact that at every node equilibrium must exist between the four cable-forces, induced by initial prestress, and any load acting on that node.

Let us consider such a single node P_0 in equilibrium (Fig. 6.4). The node P_0 is connected to fixed points

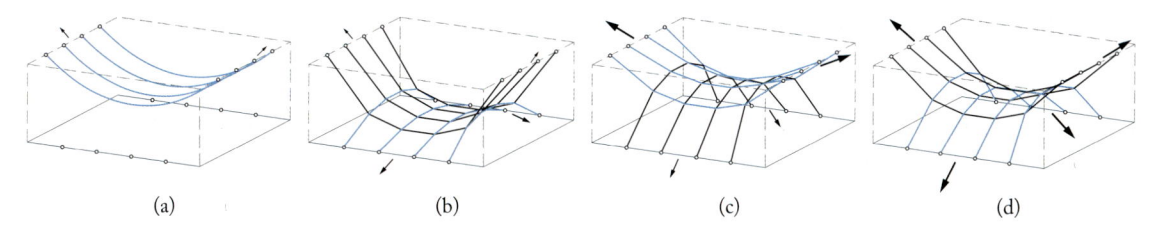

Figure 6.3 A thought experiment resulting from (a) four hanging, (b) four standing rubber bands and alternately (c, d) tensioning them

P_1, P_2, P_3, P_4 in three-dimensional space in an 'anticlastic' configuration, meaning that of the opposite pairs of points, P_1, P_3 must be 'high' and P_2, P_4 must be 'low' points, or vice versa.

The four elastic bars a, b, c, d between these points, are connected as pin-joints. In their slack state the four bars are too short to be connected at P_0. Consequently, tension forces F_a, F_b, F_c, F_d are generated when they are connected at P_0. A fifth force, representing self-weight, is applied as an external load P_z at node P_0.

6.2.2 Translation to equations

Now that we have modelled the prestressed net as a spatial, four-bar, pin-jointed network, the question is how to find a state of equilibrium and the resulting geometry. To this end, we first formulate three basic relationships:

1. Every individual bar is increased in length due to the tension force acting in it. The difference between the non-stretched and elastically stretched length of the bar results from material behaviour.
2. In the prestressed state of the net, every length of an elastically extended bar has to be equal to the distance of the nodes to which it is connected. This describes the compatibility between the elongation of the bars and the geometry of the net in the final, prestressed state.
3. The tension forces and self-weight applied to the unsupported node must be in equilibrium.

If we translate these three basic facts to mathematical formulae, we obtain the following relationships, for a single node and its four neighbours.

First, Hooke's law of elasticity applies without loss of generality to the changes in length. The tension force F_i, with $i = a, b, c, d$, in each of the four bars is

$$F_i = \left[\frac{EA}{l_0} \cdot e\right]_i. \quad (6.1)$$

where EA is the (axial) stiffness of the material, l_0 is the non-stretched length, and e the elastic elongation of the bar.

Second, the length l_i of every elastically elongated bar must be exactly equal to the spatial distance between the nodes at its ends

$$l_i = \sqrt{(x_k - x_0)^2 + (y_k - y_0)^2 + (z_k - z_0)^2}, \quad (6.2)$$

where x, y, z are the coordinates of the nodes, and $k = 1, 2, 3, 4$.

The elastic elongation e is the difference of stretched and non-stretched lengths l and l_0, so by substituting l with equation (6.2), the elongation of each of the four bars

$$e_i = l_i - l_{0,i}. \quad (6.3)$$

Third and last, there must be equilibrium in every node. This must also hold in each of the three dimensions, x, y and z. We decompose the force F_i in each

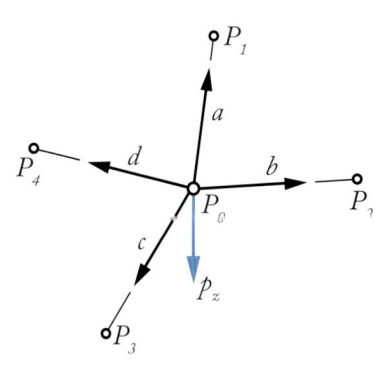

Figure 6.4 A single node with four forces and a load

bar into three components, which for example in x-direction, gives

$$F_{i,x} = F_i \cdot \cos\alpha_i \qquad (6.4)$$

where $\cos\alpha_i$ is the direction cosine, and α_i is the angle between bar i and the x-direction. The equilibrium of the four forces in point P_0 in the x-direction, with a load p_x acting on the node, is

$$F_a \cdot \cos\alpha_a + \ldots + F_d \cdot \cos\alpha_d + p_x = 0. \qquad (6.5)$$

Writing the direction cosine between points P_k and P_0 with reference to the three coordinate axes results in the expression

$$\text{direction cos} = \frac{\text{coordinate difference}}{\text{distance in space}}$$

and substituted as an equation in (6.5) gives

$$\frac{x_1 - x_0}{l_a} \cdot F_a + \ldots + \frac{x_4 - x_0}{l_d} \cdot F_d + p_x = 0. \qquad (6.6)$$

In structural mechanics, the following relationships apply:

- Static equilibrium requires the formulation of *equilibrium equations*, which relate external loads and internal forces. In our case, they relate load p to forces F_i in equation (6.6).
- Material behaviour is determined by *constitutive equations* relating the internal forces to deformations (more generally, strain) such as elongations and/or curvatures. In this case, Hooke's law of elasticity (without loss of generality) describes the material relationship (6.1) between forces F_i and elongations e_i.
- Geometry is governed by the *compatibility* or *kinematic equations*, relating the deformations to translations.

In order to find a solvable system of equations from (6.1–6.6), we carry out a number of substitutions.

First, by substituting equation (6.3) into (6.1), we can write

$$F_i = \left[\frac{EA}{l_0}(l - l_0)\right]_i. \qquad (6.7)$$

We insert these expressions for the forces into the equilibrium equations (6.6), and get

$$\frac{x_1 - x_0}{l_a} \cdot \frac{EA_a}{l_{0,a}}(l_a - l_{0,a}) + \ldots$$

$$+ \frac{x_4 - x_0}{l_d} \cdot \frac{EA_d}{l_{0,d}}(l_d - l_{0,d}) + p_x = 0. \qquad (6.8)$$

If the coordinates of the fixed points x_k, y_k, z_k and the unstressed lengths of the elastic bars $l_{0,a},\ldots,l_{0,d}$ are given, we are now able to determine the unknown coordinates x_0, y_0, z_0 of point P_0 and thus its position in three-dimensional space. We have to solve the system of the preceding equations for the unknown coordinates x_0, y_0, z_0. However, this is by no means trivial. The system to be solved is nonlinear as the unknown coordinates x_0, y_0, z_0 are also contained in the lengths $l_a,\ldots,l_d$, in equation (6.2).

Thus, we have to linearize the system, observing that the system is nonlinear with respect to geometry and material.

6.2.3 Force densities

To deal with the nonlinearity of the problem, we introduce force densities, also known as 'tension coefficients'. These are defined as

$$\text{force density} = \frac{\text{force in a bar}}{\text{stressed length of the bar}}.$$

Quantities of this type can already be recognized in our previous equations, (6.6) and (6.8). To find the spatial coordinates of P_0, we take the following approach to overcome the nonlinearity problem. First, we rewrite equation (6.6) to

$$(x_1 - x_0) \cdot \frac{F_a}{l_a} + \ldots + (x_4 - x_0) \cdot \frac{F_d}{l_d} + p_x = 0. \qquad (6.9)$$

The quotients F/l are declared as new variables q, called force densities, defined as

$$q_i := \frac{F_i}{l_i} \qquad (6.10)$$

and equation (6.9) thus becomes

$$(x_1 - x_0) \cdot q_a + \ldots + (x_4 - x_0) \cdot q_d + p_x = 0. \qquad (6.11)$$

Equation (6.11) is reordered in such a way, that the terms with the unknowns (i.e. the coordinates of P_0) are on the left-hand side and the terms with the constant factors (i.e. the given values of the coordinates of the fixed points and the force densities) are on the right-hand side of the equation. The resulting system of equations

$$-(q_a + q_b + q_c + q_d) \cdot x_0 =$$
$$-p_x - (x_1 \cdot q_a + x_2 \cdot q_b + x_3 \cdot q_c + x_4 \cdot q_d) \quad (6.12)$$

has the solution, now given in three dimensions,

$$x_0 = \frac{p_x + x_1 \cdot q_a + x_2 \cdot q_b + x_3 \cdot q_c + x_4 \cdot q_d}{q_a + q_b + q_c + q_d},$$
$$y_0 = \frac{p_y + y_1 \cdot q_a + y_2 \cdot q_b + y_3 \cdot q_c + y_4 \cdot q_d}{q_a + q_b + q_c + q_d}, \quad (6.13)$$
$$z_0 = \frac{p_z + z_1 \cdot q_a + z_2 \cdot q_b + z_3 \cdot q_c + z_4 \cdot q_d}{q_a + q_b + q_c + q_d}.$$

For each chosen set of four force densities we get a unique solution of the unknown point $P_0(x_0, y_0, z_0)$ from the linear system of equations (6.13). These unique solutions are equivalent and identical with the solutions of the nonlinear equations (6.6) and (6.8). Notice the equivalence of the equations (6.6) and (6.12), where the former is the nonlinear description, and the latter is the linear description, of the very same equilibrium solution.

6.3 Matrix formulations

So far, we have found a solution if we only have one unknown three-dimensional point in space. Practically, the solution for the single node is by no means sufficient. We are dealing with nets with arbitrary topology and numbers of given fixed and unknown free points. To find solutions for arbitrarily large nets, we have to extend our mathematical tools, and introduce matrix formulations combined with graph theory. In the following sections we discuss some conventions in our notation, then introduce two specific concepts: the branch-node matrix and the Jacobian. Using these, we rewrite the single-node problem in matrix form, before generalizing to arbitrary networks.

6.4 Notation

A vector is interpreted as a one-column matrix and written in bold lower case, and a general matrix is written as a bold capital letter. The same symbols are used for the components but they have an additional index i, j or k. The m-dimensional vector $\mathbf{a}$ – called the m-vector $\mathbf{a}$ – has therefore a_j as j-th component. The transpose of a vector is a one-row matrix,

$$\mathbf{a} = \begin{bmatrix} 1 \\ 2 \\ 3 \end{bmatrix} = [1 \; 2 \; 3]^T, \mathbf{a}^T = [1 \; 2 \; 3].$$

Further, we often need the diagonal matrix $\mathbf{A}$ belonging to any vector $\mathbf{a}$: $\mathbf{A}$ is simply defined to have $\mathbf{a}$ as diagonal, for example

$$\mathbf{a} = \begin{bmatrix} 1 \\ 2 \\ 3 \end{bmatrix}, \mathbf{A} = \operatorname{diag}(\mathbf{a}) = \begin{bmatrix} 1 & 0 & 0 \\ 0 & 2 & 0 \\ 0 & 0 & 3 \end{bmatrix}.$$

6.4.1 Branch-node matrix

Before proceeding to generalize our equations for a single node to those for an arbitrary network, we discuss some fundamental concepts of graph theory that can be used to describe net-like entities and are therefore useful for such a general formulation. In graph theory, a net-like entity consists of an aggregation of n nodes (also called points) and an aggregation of m branches (also called edges). Each branch connects two nodes.

The topological relationships between nodes and branches can be described in graph theory by a branch-node matrix $\mathbf{C}$ (or incidence matrix $\mathbf{C}^T$), consisting of the elements +1, -1 or 0 in each row, so

$$C_{ij} = \begin{cases} +1 & \text{if branch } j \text{ ends in node } i, \\ -1 & \text{if branch } j \text{ begins in node } i, \\ 0 & \text{otherwise.} \end{cases}$$

A few remarks characterizing the branch-node matrix $\mathbf{C}$:

- $\mathbf{C}$ does not contain 'geometry', only topology, that is, there are no metric relationships;
- there is precisely one element +1 and one element -1 in each row;
- the matrix is not necessarily regular with respect to

its columns, that is, it can have a different number of elements in each column;
- in the case of a contiguous net, **C** has the rank $m - 1$.

Now we are able to treat our 'single-node problem' using the corresponding branch-node matrix. The branch-node matrix **C** of the point P_0 with its neighbours $P_1,\ldots,P_4$ is

$$\mathbf{C} = \begin{bmatrix} P_0 & P_1 & P_2 & P_3 & P_4 \\ +1 & -1 & 0 & 0 & 0 \\ +1 & 0 & -1 & 0 & 0 \\ +1 & 0 & 0 & -1 & 0 \\ +1 & 0 & 0 & 0 & -1 \end{bmatrix} \begin{matrix} a \\ b \\ c \\ d \end{matrix}. \quad (6.14)$$

The usefulness of **C** can be demonstrated by noting that the coordinate differences,

$$\mathbf{u} = [x_1 - x_0 \;\; x_2 - x_0 \;\; x_3 - x_0 \;\; x_4 - x_0]^\mathrm{T}, \quad (6.15)$$

are obtained through the multiplication of the coordinate vectors

$$\mathbf{x} = [x_0 \;\; x_1 \;\; x_2 \;\; x_3 \;\; x_4] \quad (6.16)$$

with the branch-node matrix **C**, such that

$$\mathbf{u} = \mathbf{Cx}. \quad (6.17)$$

We subdivide the branch-node matrix **C** into the part $\mathbf{C}_\mathrm{N}$ containing the new, unknown points, and the part $\mathbf{C}_\mathrm{F}$ containing the fixed points, so

$$\mathbf{C} = [\mathbf{C}_\mathrm{N} \;\; \mathbf{C}_\mathrm{F}] \quad (6.18)$$

and in a similar manner we subdivide the vector of coordinates **x** into new, unknown points and fixed points, so

$$\mathbf{x} = [\mathbf{x}_\mathrm{N} \;\; \mathbf{x}_\mathrm{F}] \quad (6.19)$$

Substituting equations (6.18) and (6.19) into (6.16), and similarly for the y- and z-direction, we get

$$\begin{aligned} \mathbf{u} &= \mathbf{C}_\mathrm{N}\mathbf{x}_\mathrm{N} + \mathbf{C}_\mathrm{F}\mathbf{x}_\mathrm{F}, \\ \mathbf{v} &= \mathbf{C}_\mathrm{N}\mathbf{y}_\mathrm{N} + \mathbf{C}_\mathrm{F}\mathbf{y}_\mathrm{F}, \\ \mathbf{w} &= \mathbf{C}_\mathrm{N}\mathbf{z}_\mathrm{N} + \mathbf{C}_\mathrm{F}\mathbf{z}_\mathrm{F}, \end{aligned} \quad (6.20)$$

and using their diagonal matrices U, V, and W the corresponding bar lengths

$$\mathbf{L} = (\mathbf{U}^2 + \mathbf{V}^2 + \mathbf{W}^2)^{\frac{1}{2}}. \quad (6.21)$$

Declaring force densities and bar lengths as vectors **q** and **l**, or as diagonal matrices **Q** and **L**, we have everything at our disposal that allows us to solve the 'single-node problem' automatically.

6.4.2 Jacobian

To write the equations (6.6) and (6.8) in matrix notation, we also determine the gradient ∇ in Euclidean space, or the Jacobian $\partial \mathbf{f}(x_0)/\partial x_0$, of the function

$$\mathbf{f}(x_0) = \begin{bmatrix} f_a(x_0) \\ f_b(x_0) \\ f_c(x_0) \\ f_d(x_0) \end{bmatrix} = \begin{bmatrix} l_a \\ l_b \\ l_c \\ l_d \end{bmatrix} = \mathbf{l}. \quad (6.22)$$

As a result, we get for the transposed Jacobian

$$\left(\frac{\partial \mathbf{f}(x_0)}{\partial x_0}\right)^\mathrm{T} = \begin{bmatrix} \frac{\partial l_a}{\partial x_0} & \frac{\partial l_b}{\partial x_0} & \frac{\partial l_c}{\partial x_0} & \frac{\partial l_d}{\partial x_0} \end{bmatrix} = \begin{bmatrix} \frac{-(x_1-x_0)}{l_a} \\ \frac{-(x_2-x_0)}{l_b} \\ \frac{-(x_3-x_0)}{l_c} \\ \frac{-(x_4-x_0)}{l_d} \end{bmatrix}^\mathrm{T}. \quad (6.23)$$

Using the branch-node matrix, we can write the Jacobian

$$\left(\frac{\partial \mathbf{f}(\mathbf{x})}{\partial \mathbf{x}}\right)^\mathrm{T} = \mathbf{C}_\mathrm{N}^\mathrm{T}\mathbf{U}\mathbf{L}^{-1}. \quad (6.24)$$

The Jacobian corresponds exactly to the direction cosines of equation (6.6).

6.4.3 Solution in matrix form

Introducing the vector of forces **f** and the vector of load components **p**, equation (6.6) is then equivalent to

$$-\left(\frac{\partial \mathbf{f}(x_0)}{\partial x_0}\right)^{\mathrm{T}} \begin{bmatrix} F_a \\ F_b \\ F_c \\ F_d \end{bmatrix} + p_x = 0. \quad (6.25)$$

This equation can be written as

$$\mathbf{C}_{\mathrm{N}}^{\mathrm{T}} \mathbf{U} \mathbf{L}^{-1} \mathbf{f} + \mathbf{p} = \mathbf{0}. \quad (6.26)$$

With the definition for the force densities known already from equation (6.10),

$$\mathbf{q} = \mathbf{L}^{-1}\mathbf{f}, \quad (6.27)$$

we receive, by substitution, the system of equations

$$\mathbf{C}_{\mathrm{N}}^{\mathrm{T}} \mathbf{U} \mathbf{q} + \mathbf{p} = \mathbf{0}. \quad (6.28)$$

We want to find the linear system of equations for the determination of the solution. Given that $\mathbf{Uq} = \mathbf{Qu} = \mathbf{QCx}$, we rewrite equation (6.28) to equations of the form

$$\mathbf{C}_{\mathrm{N}}^{\mathrm{T}} \mathbf{QCx} + \mathbf{p} = \mathbf{0} \quad (6.29)$$

or

$$\mathbf{C}_{\mathrm{N}}^{\mathrm{T}} \mathbf{QC}_{\mathrm{N}} \mathbf{x}_{\mathrm{N}} + \mathbf{C}_{\mathrm{N}}^{\mathrm{T}} \mathbf{QC}_{\mathrm{F}} \mathbf{x}_{\mathrm{F}} + \mathbf{p} = \mathbf{0}. \quad (6.30)$$

We observe the independence of the equations for the respective coordinate components. For simplicity, we set $\mathbf{D}_{\mathrm{N}} = \mathbf{C}_{\mathrm{N}}^{\mathrm{T}} \mathbf{QC}_{\mathrm{N}}$ and $\mathbf{D}_{\mathrm{F}} = \mathbf{C}_{\mathrm{N}}^{\mathrm{T}} \mathbf{QC}_{\mathrm{F}}$, and obtain the system of equations of equilibrium in the form

$$\mathbf{D}_{\mathrm{N}} \mathbf{x}_{\mathrm{N}} = \mathbf{p} - \mathbf{D}_{\mathrm{F}} \mathbf{x}_{\mathrm{F}}, \quad (6.31)$$

which is a system of linear equations of the standard form $\mathbf{Ax} = \mathbf{b}$. This equation linearly defines the free node coordinates $\mathbf{x}_{\mathrm{N}}$. This system can be solved efficiently by using, for example, Cholesky decomposition (see Section 13.5.1). With a given load and a given position of fixed points, we get for each set of prescribed force densities exactly one equilibrium state with the shape, now given in three dimensions,

$$\begin{aligned} \mathbf{x}_{\mathrm{N}} &= \mathbf{D}_{\mathrm{N}}^{-1}(\mathbf{p}_x - \mathbf{D}_{\mathrm{F}} \mathbf{x}_{\mathrm{F}}), \\ \mathbf{y}_{\mathrm{N}} &= \mathbf{D}_{\mathrm{N}}^{-1}(\mathbf{p}_y - \mathbf{D}_{\mathrm{F}} \mathbf{y}_{\mathrm{F}}), \\ \mathbf{z}_{\mathrm{N}} &= \mathbf{D}_{\mathrm{N}}^{-1}(\mathbf{p}_z - \mathbf{D}_{\mathrm{F}} \mathbf{z}_{\mathrm{F}}), \end{aligned} \quad (6.32)$$

and the branch forces

$$\mathbf{f} = \mathbf{Lq}. \quad (6.33)$$

6.4.4 Generalization

We can generalize the solution in equation (6.32) for arbitrarily large nets of m branches and $n = n_{\mathrm{N}} + n_{\mathrm{F}}$ nodes, consisting of n_{N} unknown, new nodes, and n_{F} fixed nodes. Previously, we had only $m = 4$ branches and $n = n_{\mathrm{N}} + n_{\mathrm{F}} = 1 + 4$ nodes. So long as the $m \times n$ branch-node matrix $\mathbf{C}$ correctly describes the topology of our network, and consistent indexes are used for all the matrices and vectors describing properties of branches and nodes, equation (6.32) holds for any problem. We apply the same equation to another topology, belonging to that of the net drawn in Figure 6.5.

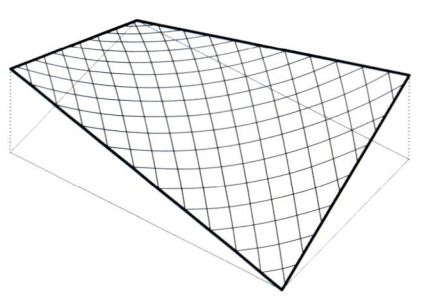

Figure 6.5 An arbitrary net with a large number of nodes and branches

The example, without external loads, for force densities varying in the edges and in the interior of the net, and the boundary conditions shown in Figure 6.5, leads to the solutions in Figure 6.6. Here the ratio of force densities in the edge to the interior branches is varied from 5:1, 2:1, 1:1 to 1:2, suggesting possible shapes for anticlastic surfaces. When vertical loads are introduced to Figure 6.7a in the range $\mathbf{p}_z = 0.1, 0.2, 0.5$, and the boundary edges are either free or fixed, it yields different, synclastic shapes shown in Figure 6.7b–d.

6.5 Materialization

A net has been determined with 'pure' force densities, without any information about the material used for its realization. Subsequently, any materialization of each and every individual bar is possible.

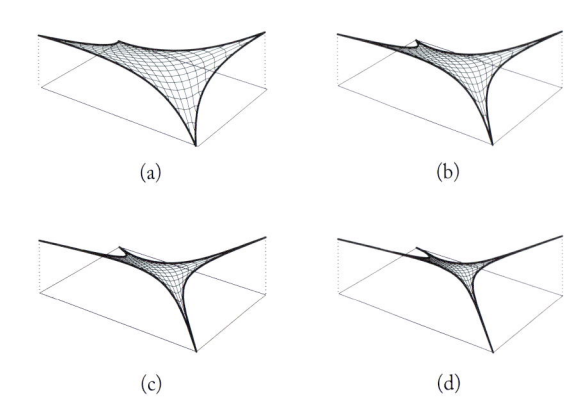

Figure 6.6 Figures of equilibrium for varying proportions of edge to interior force densities, (a) 5:1, (b) 2:1, (c) 1:1 and (d) 1:2

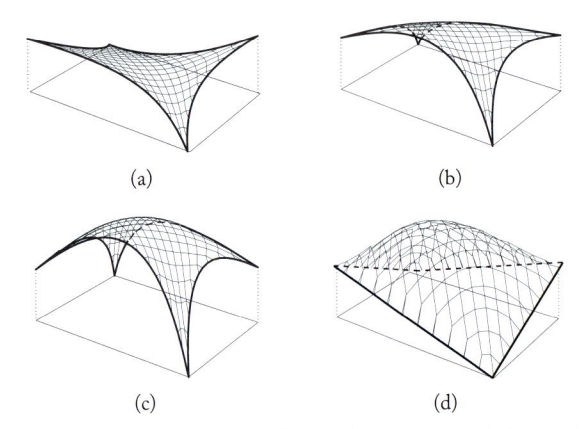

Figure 6.7 Figures of equilibrium for varying loads (a) $\mathbf{p}_z = 0$, (b) $\mathbf{p}_z = 0.1$, (c) $\mathbf{p}_z = 0.2$ and (d) $\mathbf{p}_z = 0.2$ with fixed, straight edges

We know the forces from equation (6.33), where the force densities $\mathbf{q}$ are given, and the lengths are calculated for each bar from equation (6.3) using the appropriate coordinates.

Then, selecting a diagonal matrix of axial stiffnesses $\mathbf{EA}$, we can calculate the corresponding vectors of elastic elongations $\mathbf{e}$ and initial lengths $\mathbf{l}_0$. The principle here is to select the initial lengths $\mathbf{l}_0$ in such a manner that the elongations $\mathbf{e} = \mathbf{l} - \mathbf{l}_0$, that is, the elongation which is necessary to generate the solution. According to Hooke's law of elasticity, rewritten and generalized from equation (6.1),

$$\mathbf{e} = \mathbf{l} - \mathbf{l}_0 = \mathbf{L}_0(\mathbf{EA})^{-1}\mathbf{f}$$
$$= \mathbf{L}_0(\mathbf{EA})^{-1}\mathbf{L}\mathbf{q}, \qquad (6.34)$$

$$\mathbf{q} = \mathbf{e}\mathbf{EA}\mathbf{L}_0^{-1}\mathbf{L}^{-1}. \qquad (6.35)$$

Continuing to rewrite equation (6.34)

$$\begin{aligned}\mathbf{l} &= \mathbf{l}_0 + \mathbf{L}_0(\mathbf{EA})^{-1}\mathbf{f} \\ &= (\mathbf{I} + (\mathbf{EA})^{-1}\mathbf{F})\mathbf{l}_0,\end{aligned} \qquad (6.36)$$

$$\begin{aligned}\mathbf{l}_0 &= (\mathbf{I} + (\mathbf{EA})^{-1}\mathbf{F})^{-1}\mathbf{l} \\ &= (\mathbf{I} + (\mathbf{EA})^{-1}\mathbf{QL})^{-1}\mathbf{l},\end{aligned} \qquad (6.37)$$

where $\mathbf{I}$ is an identity matrix of size m.

As long as each force F is larger than zero, the denominator of the fraction is always >1 and therefore we have tension in the bar if $l_0 < l$ and, vice versa, compression. Furthermore, the initial length l_0 of the bar, or cable segment, necessary to realize a given equilibrium shape, is only dependent on the chosen, individual axial stiffness EA.

6.6 Procedure

The steps in the force density method (Fig. 6.8) are simply constructing the right-hand side of equation (6.32), which consists of the boundary conditions, or fixed points $\mathbf{x}_F$, the topology described by $\mathbf{C}$, the force densities $\mathbf{q}$ and external loads $\mathbf{p}$. If the resulting solution, described by the coordinates $\mathbf{x} = [\mathbf{x}_N \; \mathbf{x}_F]$, is unsatisfactory to the designer, each of these four quantities can be changed to generate a new, unique solution.

The loads could also be calculated from the surface area or bar lengths surrounding each node, to approximate self-weight of the structure. For discrete structures, this is done using equation (10.28), while for continuous surfaces, approaches such as those adopted in Chapters 7, 13 and 14 can be used.

The effect of particular sets of force densities $\mathbf{q}$ on the resulting equilibrium shapes may be difficult to anticipate. Their value can also be determined indirectly: either by the user controlling the horizontal thrust components as in thrust network analysis (Chapter 7), or by adding constraints to form a least-squares problem (Chapters 12 and 13).

6.7 Design development

The roof of the ice rink is designed as a synclastic surface structure by applying vertical loads $\mathbf{p}$. Starting from a quadrilateral topology, Figure 6.9 shows some design possibilities by varying the loads **p** and the force densities **q**. Because both these parameters are given by the designer, and because their relation (6.28) is linear, one can obtain the same geometry, by scaling both parameters equally. For example, the solution in Figure 6.9a with **q** = 2 and **p** = 0.5 also results by scaling the loads and force densities by a factor 2, so with **q** = 4 and **p** = 1. In other words, once the geometry is found, one can calculate the real load afterwards and simply scale the force densities accordingly to obtain actual forces in the structure.

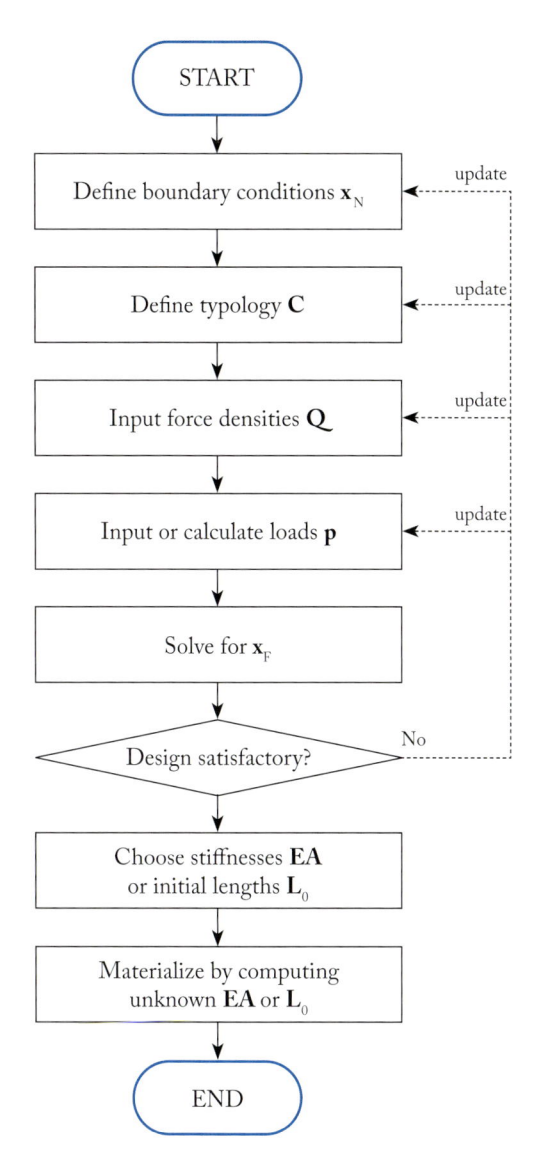

Figure 6.10 shows a few variations for the swimming pool roof, obtained by changing the boundary conditions (the height and size of the opening in the middle), and by adding point loads to the nodes in Figure 6.10c. The initial topology is radial, with its origin in the centre of the high point. Without providing a load, the resulting forms are anticlastic. The design team settles on the geometries shown in Figures 6.9 and 6.10b for the initial design of the sports complex, shown in Figure 6.11. This design can be materialized and then tested for other load combinations. However, the project may have additional architectural or structural constraints, ultimately expressed in the form of specific positions of nodes, target lengths of branches or values of forces. These constraints lead to a nonlinear FDM, as explained further in Chapter 12.

Figure 6.8 Flowchart for FDM

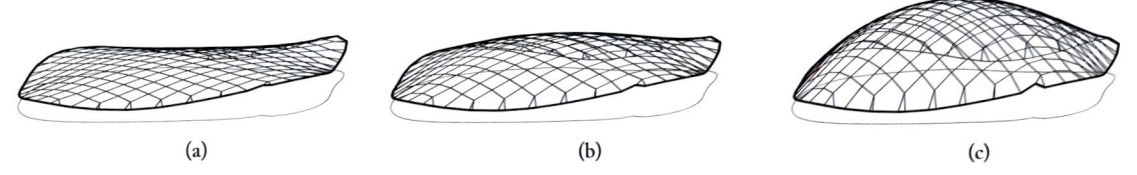

Figure 6.9 Variations for the ice rink roof with (a) **q** = 2, **p** = 0.5, (b) **q** = 2, **p** = 1, and (c) **q** = 1, **p** = 1

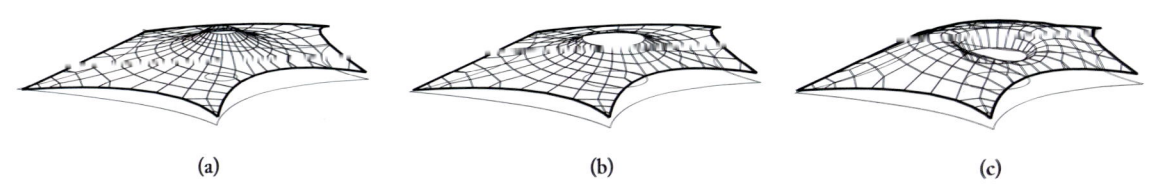

Figure 6.10 Variations for the swimming pool roof with **q** = 1 for (a) high point h = 12m, small opening, (b) h = 10m, large opening and (c) h = 6m, large opening and **p** = 1

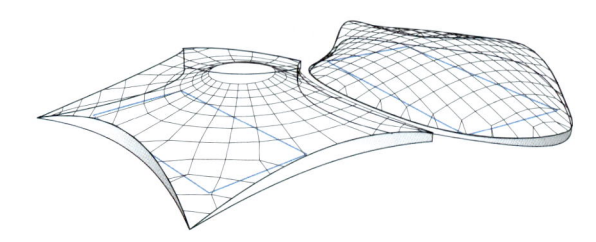

Figure 6.11 Preliminary design for the Stuttgart sports complex

6.8 Conclusion

The force density method is able to generate solutions of discrete networks, through linear systems of equations, that are in an exact state of equilibrium, without needing iterations or some kind of convergence criterion. Applied to the design brief, it enabled us to quickly generate solutions for a given topology, by varying the force densities and external loads. The method, originally developed for cable nets, is, to this day, very common in the design practice of tensioned membrane roofs. By introducing loads, it also allows the form finding of synclastic structures, highly suitable for efficient shell structures. Because the method is entirely independent of material properties, two interesting opportunities arise. First, resulting designs can be materialized arbitrarily, giving the initial lengths of the network in undeformed state, without affecting the final shape. Second, one can simply multiply the loads to any realistic value, and then calculate the internal force distribution, again without changing the geometry.

Key concepts and terms

The **force density** is the ratio of force over (stressed) length in a bar or cable segment. It is also known as the 'tension coefficient'.

A **branch-node matrix** is a matrix that shows the (topological) relationship between m branches and n nodes. The matrix has m rows and n columns. The entry in a certain row and column is 1 or -1 if the corresponding branch and node are related, and 0 if they are not. The sign depends on the direction of the branch. The transpose of the branch-node matrix is known as the incidence matrix.

The **Jacobian** is the gradient for functions in Euclidean space. It is the matrix of all first-order partial derivatives of a vector- or scalar-valued function with respect to another vector. In other words, the variation in space of any quantity can be represented by a slope. The gradient represents the steepness and direction of that slope.

Further reading

- 'Einige bemerkungen zur Berechnung von vorgespannten Seilnetzkonstruktionen or Some remarks on the calculation of prestressed cable-net structures', Linkwitz and Schek (1971). This German journal publication laid all the groundwork for what later would be called force densities and the force density method.
- 'The force density method for form finding and computation of general networks', Schek (1974). This seminal paper by Hans-Jörg Schek concisely describes the force density method and explains how constraints can be introduced.
- 'Formfinding by the "direct approach" and pertinent strategies for the conceptual design of prestressed and hanging structures', Linkwitz (1999). This journal paper explains the force density method and also linearizes the nonlinear, materialized equations for static analysis. This shows, indirectly, how the force density method relates to static analysis.

Exercises

- Four points, $P_1(0,0,0)$, $P_2(5,0,3)$, $P_3(0,7,3)$ and $P_4(7,5,0)$ are connected to a central node P_0 through links a, b, c and d. Each link has a force of 1kN. When a gravity load $p=5$kN is applied, determine the position of node P_0. Calculate the sum of forces in node P_0. What do you observe?
- Now, determine the position of node P_0 once more, except by imposing force densities $q=1$ in the four bars, instead of forces. What is the sum of forces in node P_0? What happens if the force density $q=2$ in bars b and d?
- Compose a branch-node matrix for the standard example grid in Figure 6.12 and assemble the vectors of force densities **q** and coordinates **x,y,z**.

- Change the boundary conditions (vertical position of anchor points z_F), force densities q and external loads p_z, to shape a shell structure.

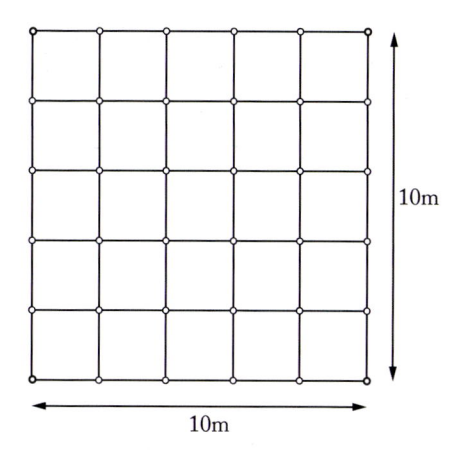

Figure 6.12 Standard grid

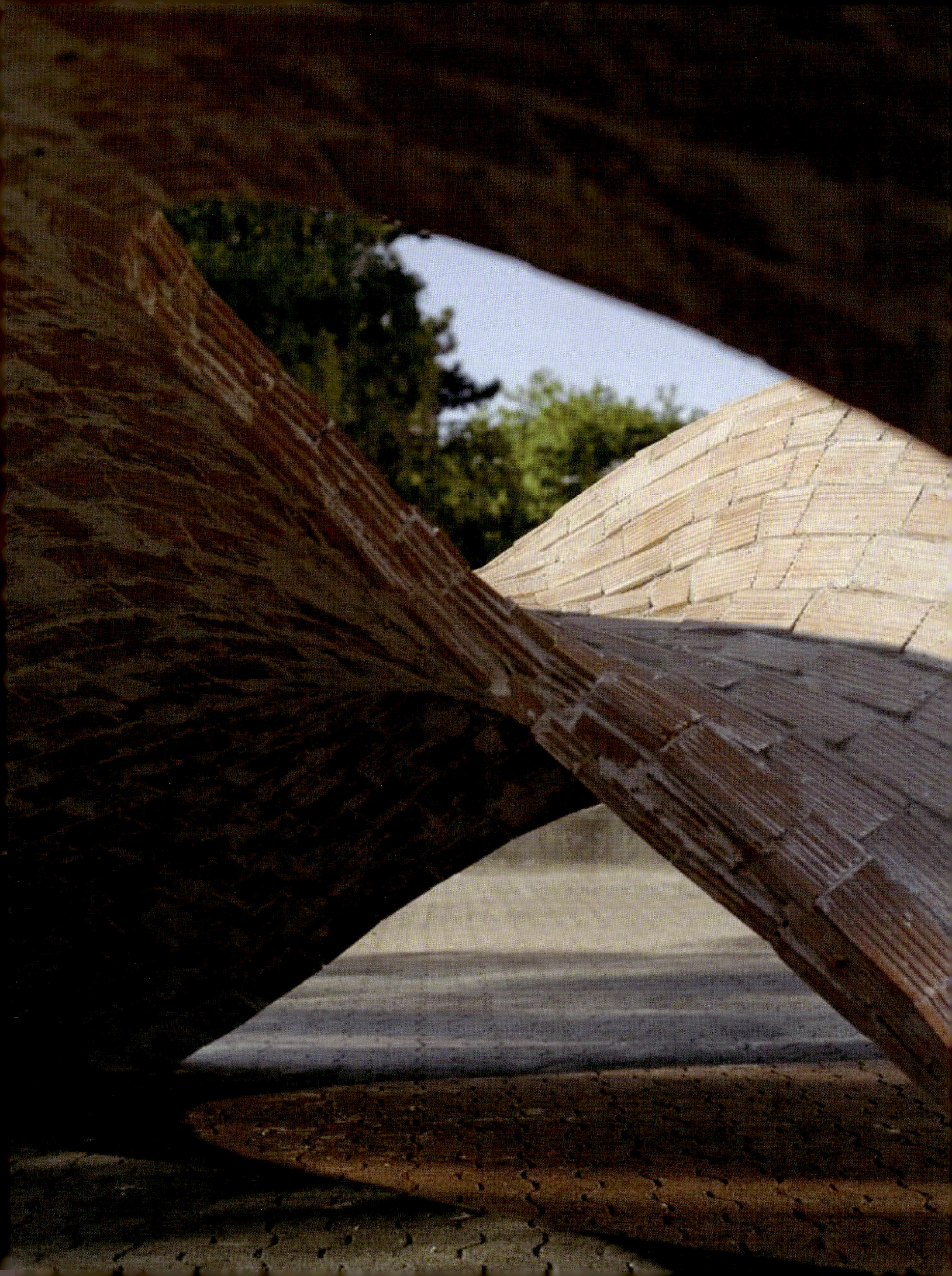

CHAPTER SEVEN

Thrust network analysis

Design of a cut-stone masonry vault

Philippe Block, Lorenz Lachauer and Matthias Rippmann

LEARNING OBJECTIVES

- Discuss the basic principles of equilibrium analysis for masonry.
- Relate the key concepts of thrust network analysis to graphic statics and reciprocal diagrams.
- Implement a simple thrust network analysis solver using linear algebra.
- Generate funicular shells through explicit control of form and internal force distribution.

PREREQUISITES

- Chapter 6 on the force density method.

The method presented in this chapter, Thrust Network Analysis (TNA), is appropriate for the form finding of compressive funicular shells, thus particularly for any type of vaulted system in unreinforced masonry. For example, based on TNA but using tile vaulting instead of cut-stone, a full-scale prototype of a 'freeform' funicular shell has been built on the ETH campus in 2011 (see page 70 and Figure 7.1). Figure 7.2 shows the form and force diagrams and thrust network of the structure. The colours represent the magnitude of thrust under self-weight in the network's branches.

The brief

We have been asked to design a pavilion for a park in Austin, TX, USA, that will cover the stage and seating of a performance area of 20m × 15m, providing shade to audience and performers.

To design a lasting landmark, we proposed an unreinforced, cut-stone vaulted structure. Using the locally quarried 'Texas Cream', a soft limestone, the pavilion would blend into its surroundings. Many important buildings in Austin, such as the State Capitol, have been built with this beige-coloured stone. The pavilion structure stands out from the other buildings though because of its structural use of stone instead of mere cladding.

Because Austin, at the heart of Texas, has a very low chance of earthquakes, we can convincingly propose the safe design of an unreinforced stone structure, and a 500-year design life can be guaranteed for the structure, due to the omittance of reinforcement steel.

7.1 Funicular structures and masonry vaults

This section introduces the main concepts, methods, and terminology related to the structural design and analysis of masonry structures.

Figure 7.1 Tile vault prototype at ETH Zurich, Switzerland, 2011

7.1.1 Hanging models

Before the existence of structural theory, the ancient master builders used design methods for masonry structures based on rules of proportion and geometry. The first scientific understanding of the stability of unreinforced masonry structures was formulated in 1676 by the English scientist Robert Hooke, in the form of the inverted hanging chain: Hooke's law of inversion (see Section 1.1). Note that this is of course only considering static equilibrium, and not, for example, instability, such as buckling.

Weights, proportional to the self-weight of each stone piece (voussoir), of an arch (Fig. 7.3a) are applied on the vertical lines of action through their centroids, to a hanging string (Fig. 7.3b). When inverted, it produces a thrust line that fits within the arch's geometry (Fig. 7.3a). This compression funicular can be used to show a possible compression-only equilibrium of the arch.

7.1.2 Graphic statics

For two-dimensional problems, graphics statics can be used instead of a hanging model. It allows finding the form of possible funicular shapes for given loads, but at the same time also the magnitude of the forces in them. The geometry of the structure, represented here by the funicular polygon, is named the form diagram (Fig. 7.3a). The magnitude of force in each element of the form diagram is simply known by measuring the length of the corresponding, parallel element in the force diagram, which is drawn to scale (Fig. 7.3c).

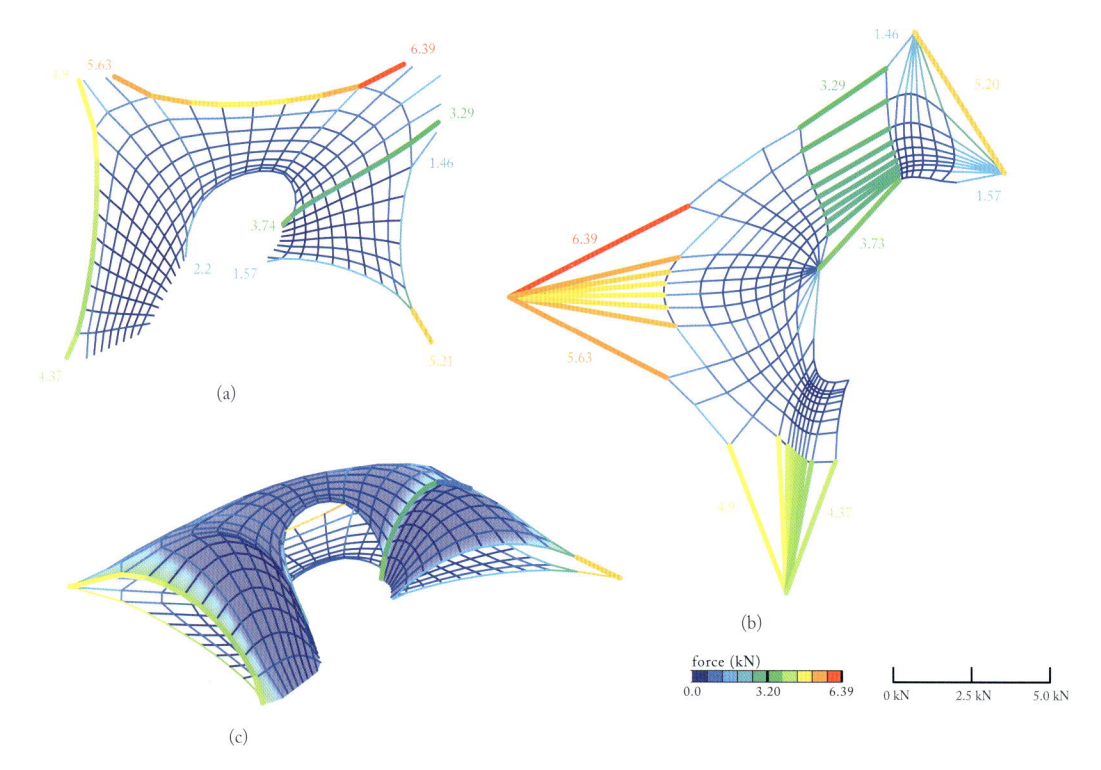

Figure 7.2 (a) Force network in plan, (b) corresponding reciprocal force diagram and (c) compression-only thrust network

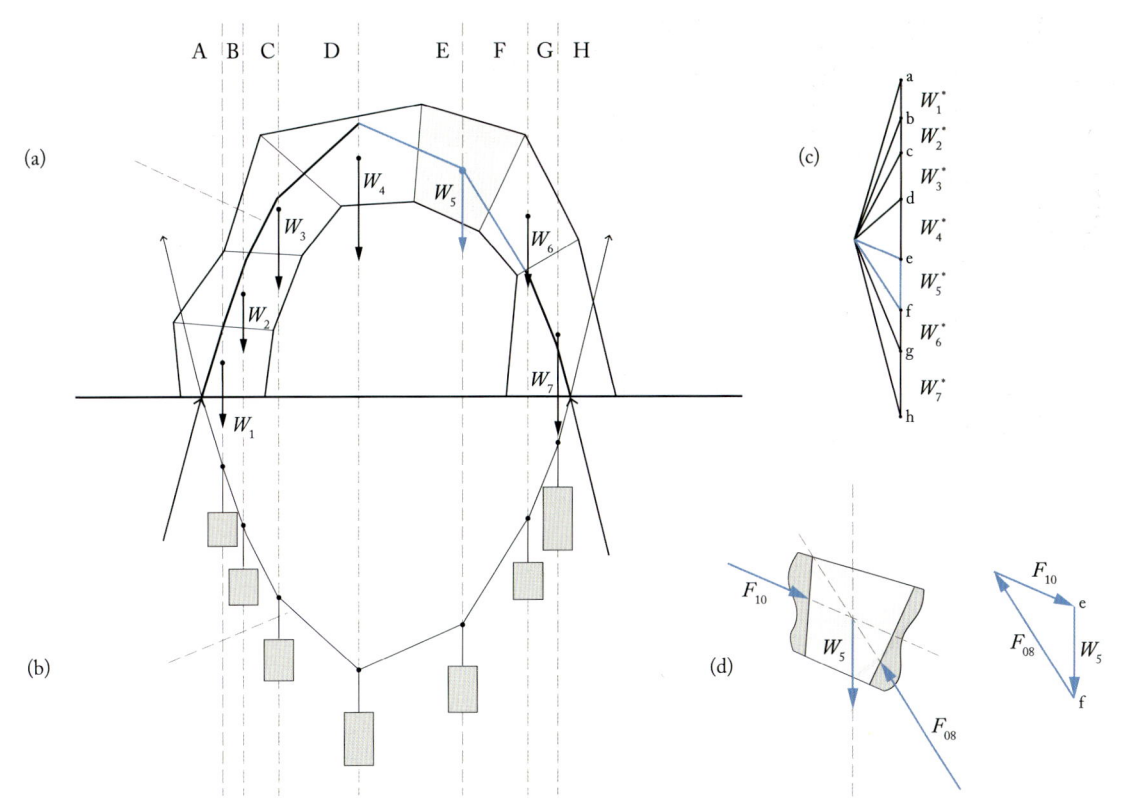

Figure 7.3 A masonry arch of arbitrary geometry with (a) thrust line and (b) corresponding hanging string, (c) the force diagram, showing (d) the equilibrium of one stone block

The geometrical and topological relationship between form and force diagram is called reciprocal (Section 7.2.2). An important additional aspect of the force diagram is that it shows both global and local force vector equilibrium of the force in the funicular line (Fig. 7.3c–d).

A particular quality of graphic statics is its bi-directional nature, meaning that the form can drive the forces and vice versa. So, if constraints are imposed on the form, such as maximum depth of the structure or given points for the structure to go through, then the forces have to follow; if constraints are imposed on the forces such as upper bounds on the thrust values, then the resulting form will emerge. Controlling either form or force diagram thus allows an informed design exploration. Unfortunately, graphic statics is practically limited to two-dimensional problems. In contrast, physical hanging models allow the 'analogue computing' of equilibrium shapes of fully three-dimensional networks, although force information of the resulting networks needs to be obtained separately.

7.1.3 Thrust line analysis and the safe theorem

Graphic statics can be used to generate *thrust lines*, which, when fitted within the masonry structure, visualize possible compressive 'flow of forces' through the structure.

Jacques Heyman formulated the lower-bound (or safe) theorem for masonry, which states that an unreinforced masonry structure is safe for a specific loading case as long as one compressive solution can be found that equilibrates those loads and fits within the structure's cross section. Assumed is that the interfaces between the voussoirs (i.e. masonry blocks) provide enough friction or interlocking to avoid sliding failure, and that crushing does not occur (typical historic masonry has stress levels two orders of magnitude smaller than the compressive strength of stone). Most historic masonry structures rely on their thickness to resist live loads, hence combining two stabilizing effects: first, a high self-weight reduces the influence of asymmetric live loading on the resulting thrust line; and second, a large structural depth allows the thrust lines to fit, changing under non-funicular load cases, within the structure's geometry. An important observation is that a range of possible and admissible thrust lines can be found within the section of a masonry structure; these are all possible equilibrium states in which the vaulted structure could stand.

Ideally, thrust lines should be kept within the kern (i.e. the middle-third zone) of the masonry section: as the thrust line visualizes the resultant of the compressive stresses throughout the structure, this means that the entire breadth b of the cross section is then effectively in compression (Fig. 7.4). Theoretically, a resultant force in the middle of the section represents uniform compressive stresses over the entire breadth of the section (Fig. 7.4a), while a resultant force applied at the middle third of the section corresponds to a triangular distribution of compressive stresses, still engaging the entire section (Fig. 7.4b). Going outside the middle third would result in 'tensile' stresses, which cannot exist as the masonry structure is unreinforced; the effective breadth b' of the section is therefore reduced (Fig. 7.4c). When designing a new masonry construction, the central 'axis' of the structure thus wants to follow the thrust line under its dominant loading condition, the dead load, typically its self-weight, as close as possible.

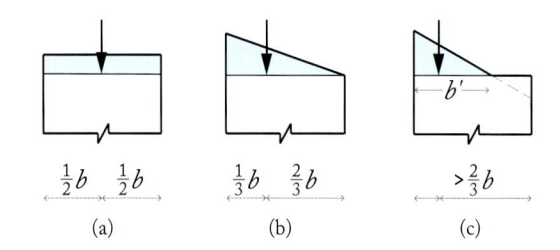

Figure 7.4 The middle-third rule with compression in the section: (a) uniformly distributed, (b) triangularly distributed and (c) reduced to the effective breadth b'

7.2 Method

The three-dimensional version of a thrust line is a *thrust network*. TNA extends discretized thrust line analysis to spatial networks for the specific case of gravity loading, using techniques derived from graphic statics. Analogously to the two-dimensional case of a masonry arch, the resulting thrust network is not necessarily a rigid structure by itself, but rather a representation of one possible static equilibrium in compression under a given set of loads. Based on the safe theorem, it is sufficient to show that one state of equilibrium exists that fits within the geometry of the masonry vault to guarantee the vault's stability for that loading case. Buckling, deflection, sliding or other (asymmetric) loading combinations have to be checked separately after the form-finding process.

TNA allows for the intuitive design of funicular networks with a high level of control due to the following key concepts:

- vertical loads constraint (Section 7.2.1);
- reciprocal diagrams (Section 7.2.2);
- statically indeterminate networks (Section 7.2.3).

7.2.1 Vertical loads constraint

Since only vertical loads are considered in TNA, the equilibrium of the horizontal force components (thrusts) in the thrust network can be computed independently of the chosen external loading. This allows splitting the form-finding process in two steps: solving for an equilibrium of the horizontal thrusts first, and then solving for the heights of the nodes of the thrust network, based on the external vertical loads, the given boundary conditions, and the obtained horizontal equilibrium.

Figure 7.5a shows the relationship between the form diagram Γ, which is the horizontal projection of the funicular equilibrium solution, the thrust network **G**, and the force diagram Γ^*, which is the reciprocal diagram of Γ. When referring to elements or properties of the reciprocal, an asterisk symbol (*) will be used.

7.2.2 Reciprocal diagrams

The in-plane equilibrium of Γ, and thus also the horizontal equilibrium of **G**, can be computed explicitly using its reciprocal force diagram Γ^*.

Since the diagram Γ is planar, methods from graphic statics allow for the finding of an equilibrium state. Considering Γ as form diagram, each force distribution is represented by a force diagram Γ^*, up to a given scale. Form and force diagrams are related by a reciprocal relationship. This means that Γ and Γ^* are *parallel dual graphs*: branches which come together in a node in one of the diagrams, form a closed space in the other, and vice versa, and corresponding branches in both diagrams are parallel (Fig. 7.5b). Structurally, this means that the equilibrium of a node in one graph is guaranteed by a closed polygon of force vectors in the other, and vice versa. When the closed polygons of the force diagram Γ^*, representing the equilibrium of the nodes of the form diagram Γ, are all formed clockwise, then the projected form diagram Γ, and as a result also the thrust network **G**, will be entirely in compression. The force diagram is furthermore drawn to scale such that the magnitude of the axial forces in the form diagram, and hence the horizontal components of the axial bar forces in the thrust network, can be found directly by measuring lengths in the force diagram.

7.2.3 Statically indeterminate networks

The static indeterminacy of networks with fixed horizontal projection and subjected to vertical loads can be explained with a geometrical analogy: for a given form diagram, there exist several reciprocal diagrams, that is, dual graphs that satisfy the constraints that corresponding branches are parallel. Because the force diagram represents the horizontal equilibrium of the network, these different, geometrically possible solutions represent different admissible equilibrium states for that form diagram, and consequently the

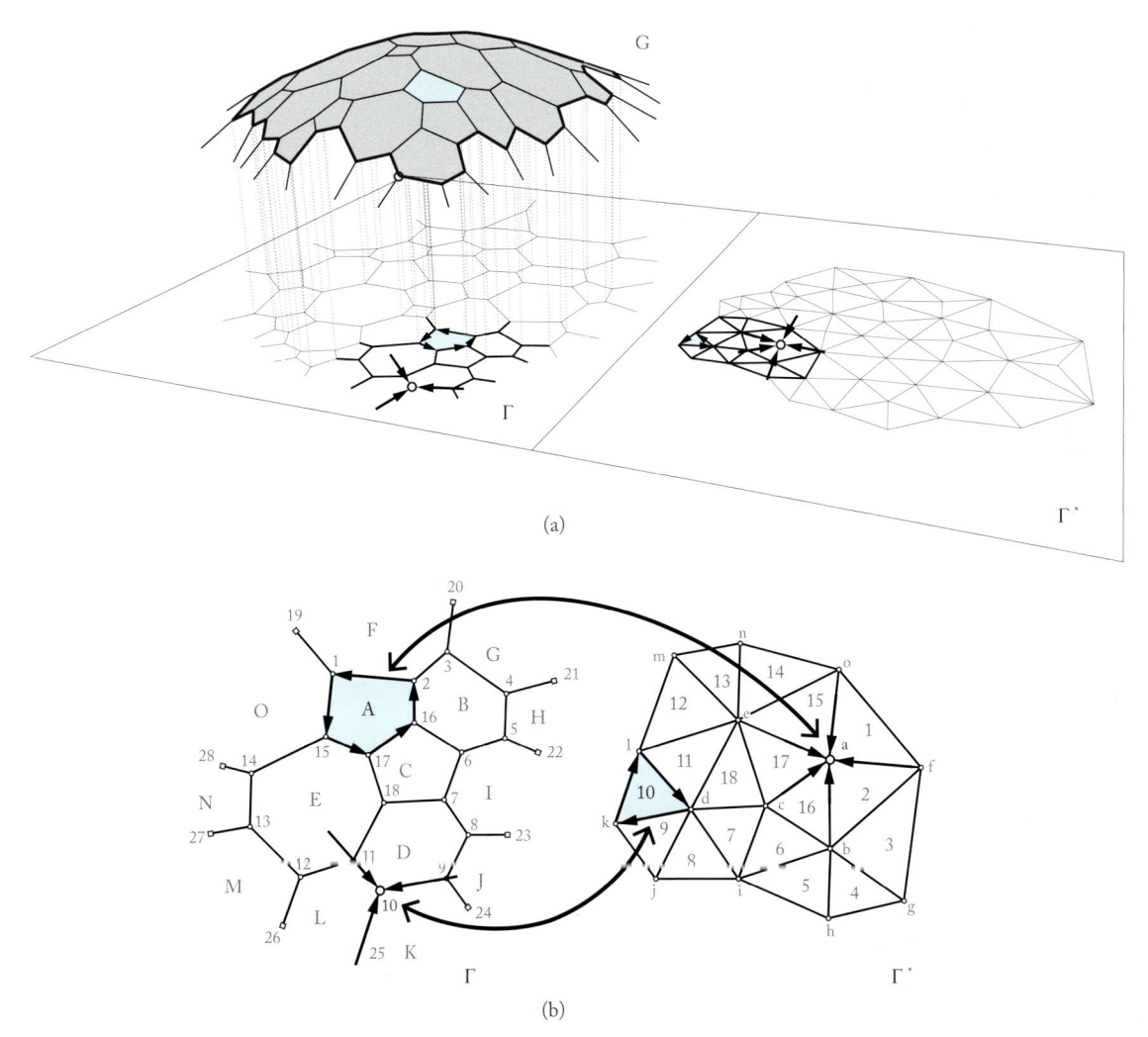

Figure 7.5 (a) Relationship between the thrust network **G**, its planar projection, the form diagram Γ and the reciprocal force diagram Γ^* and (b) the reciprocal relation between Γ and Γ^* using Bow's notation to label corresponding elements

three-dimensional equilibrium solution for given loading.

A three-valent form diagram, such as the one shown in Figure 7.5, is structurally determinate, which means that it has a unique internal distribution of forces, again up to a scale factor, which is clear from its triangulated reciprocal force diagram. Such networks thus only have one degree of freedom: the scale of their force diagram, which is, for the same loading, inversely proportional to the depth of the equilibrium solution. It is thus not possible to redistribute forces in such networks, when their horizontal projection is considered to be fixed.

For nodes in the form diagram with a valency higher than three, the network is structurally indeterminate, which means that the internal forces can be redistributed in the structure, resulting in different thrust networks for the given form diagram, but for each given form diagram Γ, force diagram Γ^*, and vertical loading P, a unique thrust network $\mathbf{G}$ exists (Fig. 7.6).

The key strategy in TNA is to give the designer direct control over the distribution of the thrusts in the system. The designer can choose these horizontal forces within the geometric constraints of the reciprocal relationship between form and force diagram. As in graphic statics, both form and force can be manipulated to determine the equilibrium shape. The intuitive force diagrams allow the designer to visually and explicitly distribute internal forces that define the three-dimensional equilibrium shape (Fig. 7.7). Boundary conditions and a solution space can be imposed on the equilibrium shape, which in turn controls the internal forces.

7.3 Computational set-up

This section introduces the equilibrium equations of the thrust network (Section 7.3.1), the branch-node data structure (Section 7.3.2) that allows for an efficient matrix notation (Section 7.3.3), for use in an implementation of TNA (Section 7.3.4).

7.3.1 Equilibrium equations

The equilibrium of a typical internal node i in $\mathbf{G}$ (Fig. 7.8) can be written as

$$F_{H,ji} + F_{H,ki} + F_{H,li} = 0, \qquad (7.1a)$$

$$F_{V,ji} + F_{V,ki} + F_{V,li} = P_i, \qquad (7.1b)$$

where $F_{H,ji}$ and $F_{V,ji}$ are respectively the horizontal components, combining x- and y-components as force vectors, and vertical components of the branch forces coming together in node i, and P_i the vertical load applied at the node.

Because the form (Γ) and force (Γ^*) diagrams are reciprocal, the branch forces in Γ, hence the horizontal components $F_{H,ji}$ of the axial forces F_{ij} of the thrust network $\mathbf{G}$, are equal to the corresponding branch lengths $l^*_{H,ji}$ in Γ^*, multiplied with the scale factor $1/r$ of the reciprocal diagram

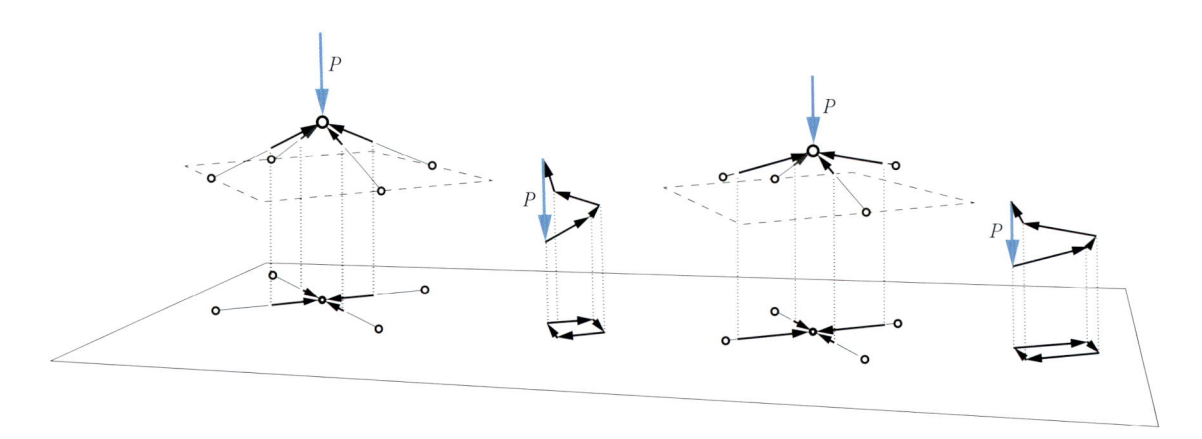

Figure 7.6 Indeterminacy of a four-bar node: for the same load P: left – an equal distribution of horizontal forces results in a symmetric network; and right – attracting more thrust in one direction results in a shallower network in that direction

CHAPTER SEVEN: THRUST NETWORK ANALYSIS   **77**

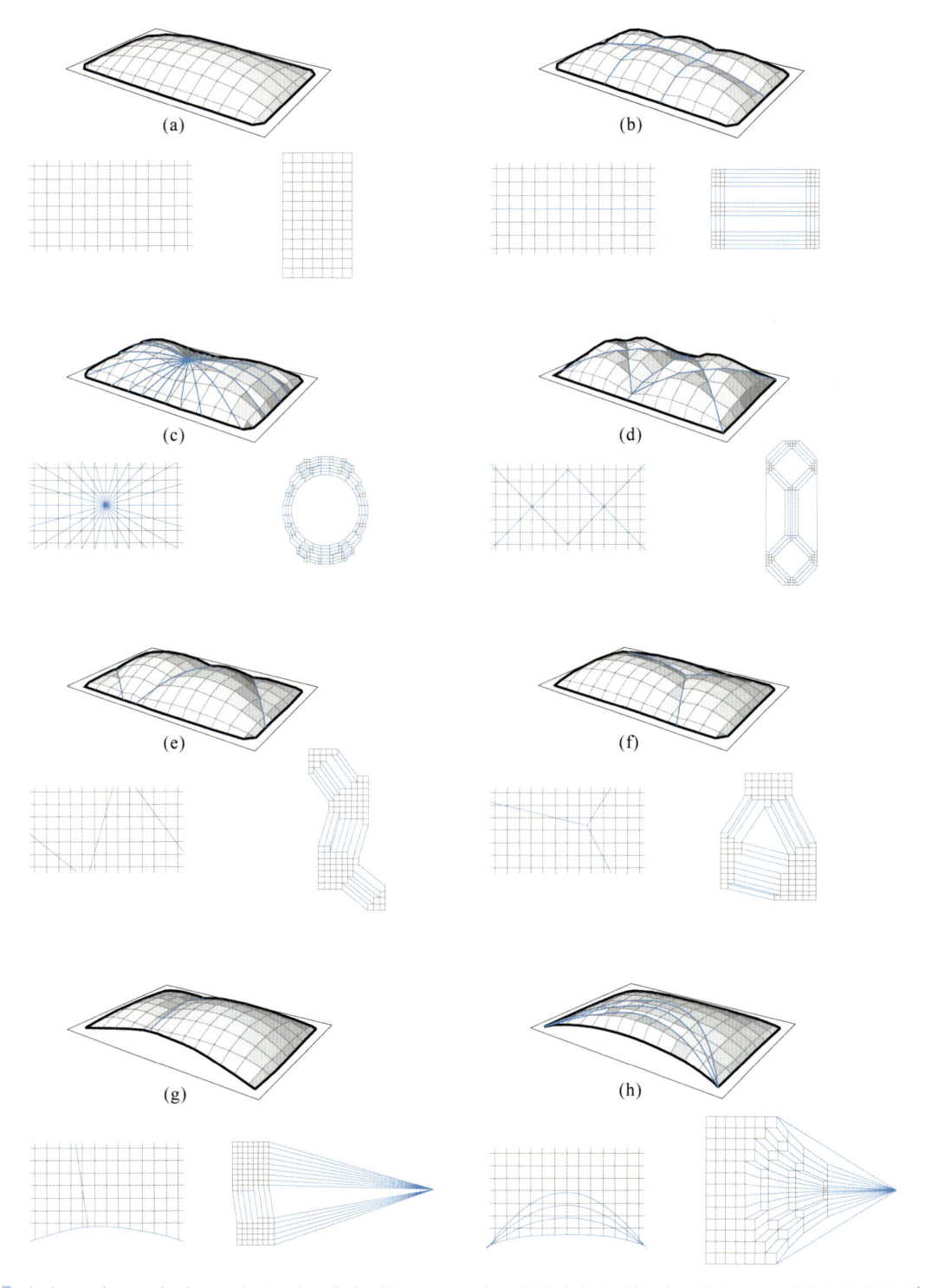

Figure 7.7 Indeterminacy of a four-valent network: for the same uniformly distributed loading, (a) an equal distribution of horizontal forces results in a thrust network with the typical 'pillow' shape and (b–h) the attraction of higher force in certain regions results in creases in the equilibrium solution

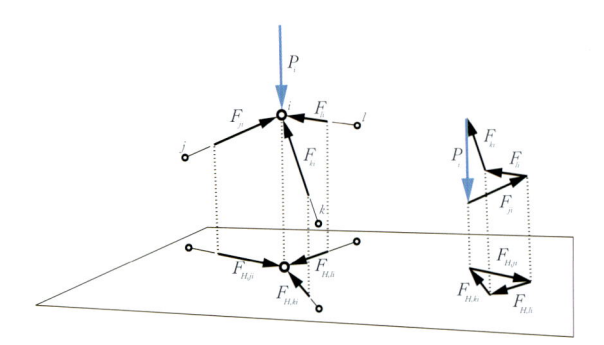

Figure 7.8 Static equilibrium of a single node i, with corresponding: left – form diagram Γ; and right – force diagram Γ^*

$$F_{H,ji} = \frac{1}{r} \cdot l^*_{H,ji}, \quad (7.2)$$

where the reciprocal branch lengths

$$l^*_{H,ji} = \sqrt{(x^*_i - x^*_j)^2 + (y^*_i - y^*_j)^2},$$

are defined as a function of the reciprocal node coordinates.

It is thus sufficient only to describe the vertical equilibrium of the nodes of **G** since their x- and y-coordinates are defined by the choice of a form diagram, Γ, and a horizontal equilibrium of the thrust network is guaranteed to be in equilibrium by the chosen 'closed' reciprocal force diagram Γ^*. There are thus only n_N equilibrium equations needed, one for each free (non-supported) node of the thrust network **G**.

The (vertical) equilibrium equations (7.1b) can be written as a function of the $F_{H,ji}$, and the geometry of the network **G**:

$$F_{H,ji} \cdot \frac{(z_i - z_j)}{l_{H,ji}} + F_{H,ki} \cdot \frac{(z_i - z_k)}{l_{H,ki}} + F_{H,li} \cdot \frac{(z_i - z_l)}{l_{H,li}} = P_i, \quad (7.3)$$

with

$$l_{H,ji} = \sqrt{(x_i - x_j)^2 + (y_i - y_j)^2}$$

the lengths of branches ij of the form diagram Γ.

Using equation (7.2), and plugging it into the nodal equilibrium equations (7.1b), after multiplying both sides by r, gives

$$l^*_{H,ji} \cdot \frac{z_i - z_j}{l_{H,ji}} + l^*_{H,ki} \cdot \frac{z_i - z_k}{l_{H,ki}} + l^*_{H,li} \cdot \frac{z_i - z_l}{l_{H,li}} = P_i \cdot r, \quad (7.4)$$

or after rearranging,

$$\left(\frac{l^*_{H,ji}}{l_{H,ji}} + \frac{l^*_{H,ki}}{l_{H,ki}} + \frac{l^*_{H,li}}{l_{H,li}} \right) \cdot z_i - \frac{l^*_{H,ji}}{l_{H,ji}} \cdot z_j$$

$$- \frac{l^*_{H,ki}}{l_{H,ki}} \cdot z_k - \frac{l^*_{H,li}}{l_{H,li}} \cdot z_l - P_i \cdot r = 0, \quad (7.5)$$

which is written as a linear combination of z_i, the unknown nodal heights of **G**, and the inverse of the scale of Γ^*, r, and by substituting with constants d_i, which are a function of the known branch lengths of Γ and Γ^*,

$$d_i \cdot z_i - d_j \cdot z_j - d_k \cdot z_k - d_l \cdot z_l - P_i \cdot r = 0. \quad (7.6)$$

7.3.2 Data structure

As in FDM (see Chapter 6), the topology of the thrust network **G**, and in TNA thus also Γ, can effectively be captured using an $m \times n$ branch-node matrix $\mathbf{C} = [\mathbf{C}_N | \mathbf{C}_F]$. The $m \times n^*$ (i.e. $m \times f$, with f the number of spaces in Γ) dual branch-node matrix $\mathbf{C}^*$ contains the connectivity information of the reciprocal force diagram Γ^*.

The branch-node matrix $\mathbf{C}^*$ of the reciprocal force diagram Γ^* can easily be constructed from observation of $\mathbf{C}$. For each j-th column of $\mathbf{C}^*$, which corresponds to the j-th space in the form diagram Γ or j-th node of the reciprocal force diagram Γ^*, the component c^*_{ij} is 1 if edge i is adjacent to the j-th space and is oriented in the same direction as a counter-clockwise cycle around that face in Γ, -1 if opposite, and 0 if the edge is not adjacent to that face. This is shown for a simple network in Section 6.4.1. For the network in Figure 7.9a, the **C**-matrix becomes

$$\mathbf{C} = \begin{bmatrix}
1 & -1 & . & . & . & . & | & . & . & . \\
1 & . & -1 & . & . & . & | & . & . & . \\
1 & . & . & -1 & . & . & | & . & . & . \\
1 & . & . & . & -1 & . & | & . & . & . \\
. & 1 & -1 & . & . & . & | & . & . & . \\
. & . & 1 & -1 & . & . & | & . & . & . \\
. & . & . & 1 & -1 & . & | & . & . & . \\
. & 1 & . & . & -1 & . & | & . & . & . \\
. & 1 & . & . & . & -1 & | & . & . & . \\
. & . & 1 & . & . & . & | & -1 & . & . \\
. & . & . & 1 & . & . & | & . & -1 & . \\
. & . & . & . & 1 & . & | & . & . & -1 \\
\end{bmatrix} \begin{matrix} I \\ II \\ III \\ IV \\ V \\ VI \\ VII \\ VIII \\ IX \\ X \\ XI \\ XII \end{matrix} \quad (7.7)$$

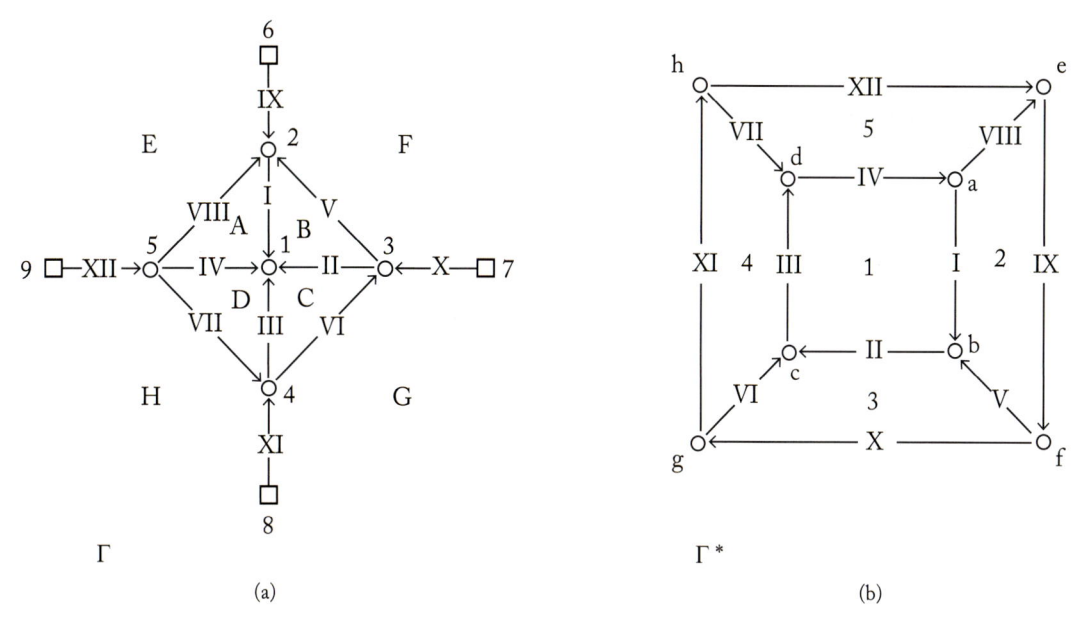

Figure 7.9 Directed (a) form and (b) force diagram. Nodes in Γ and corresponding reciprocal spaces are labelled using numbers, faces in Γ and reciprocal nodes using letters, and branches in Γ and Γ* using roman numbers

The dual branch-node matrix $\mathbf{C}^*$ can be constructed by inspection of Γ, and becomes,

$$\mathbf{C}^* = \begin{bmatrix} -1 & 1 & . & . & . & . & . & . \\ . & -1 & 1 & . & . & . & . & . \\ . & . & -1 & 1 & . & . & . & . \\ 1 & . & . & -1 & . & . & . & . \\ . & 1 & . & . & -1 & . & . & . \\ . & . & 1 & . & . & -1 & . & . \\ . & . & . & 1 & . & . & . & -1 \\ -1 & . & . & . & 1 & . & . & . \\ . & . & . & . & -1 & 1 & . & . \\ . & . & . & . & . & -1 & 1 & . \\ . & . & . & . & . & . & -1 & 1 \\ . & . & . & . & 1 & . & . & -1 \end{bmatrix} \begin{matrix} I \\ II \\ III \\ IV \\ V \\ VI \\ VII \\ VIII \\ IX \\ X \\ XI \\ XII \end{matrix} \quad (7.8)$$

with columns labelled a b c d e f g h.

The directed graph, shown in Figure 7.9b, was created by assigning directions to all branches of the reciprocal diagram Γ*. The reciprocal diagram was constructed from Γ following the clockwise convention necessary to guarantee a compression-only solution. It can be seen that when following these conventions, all corresponding directed branches in Γ and Γ* are not only parallel but also have the same orientation. This property is a requirement for a compression-only reciprocal.

There are several ways that the reciprocal diagram Γ* can be constructed from Γ:

- drawn manually, or equivalently constructed procedurally, which is the approach used for the implementation provided with this chapter (note that this approach only works for small and/or simple networks);
- automatically generated using an optimization problem (Block, 2009);
- computed directly by identifying the independent force densities $\mathbf{q}_{indep}$ and using algebraic methods (Block and Lachauer, 2013);
- obtained iteratively by enforcing the reciprocal constraints explicitly on the geometry of Γ and Γ* (Rippman et al., 2012) (see Chapter 13).

7.3.3 Matrix formulation

The equilibrium equations (7.5) can be written in matrix form as

$$\mathbf{C}_N^T(\mathbf{L}_H^{-1}\mathbf{L}_H^*)\mathbf{C}\mathbf{z} - r\mathbf{p} = \mathbf{C}_N^T(\mathbf{T})\mathbf{C}\mathbf{z} - r\mathbf{p} = \mathbf{0}. \quad (7.9)$$

When comparing equation (7.9) with (6.29) in the force density method, it is clear that the force densities $\mathbf{q}$ have to be the parameters $\mathbf{t}$ relating the lengths of corresponding branches of Γ and Γ^*, divided by the scale factor r,

$$\mathbf{q} = \frac{1}{r} \cdot \mathbf{L}_H^{-1} \mathbf{l}_H^* = \frac{1}{r} \cdot \mathbf{t}. \tag{7.10}$$

An important difference with FDM is that in TNA not all force densities $\mathbf{q}$ can be chosen freely. Only specific sets of $\mathbf{q}$ result in equilibrium solutions for $\mathbf{G}$ that have their fixed horizontal projection equal to the form diagram Γ. These possible sets of $\mathbf{q}$ correspond to the geometrically allowed reciprocal diagrams, that is, those who respect the parallelity constraints, for Γ. It is possible to identify the independent $\mathbf{q}_{indep}$ from the total set of $\mathbf{q}$ which can be chosen freely (Block and Lachauer, 2013).

By introducing the $n_N \times n$ matrix $\mathbf{D} = \mathbf{C}_N^T \mathbf{T} \mathbf{C}$, the $n_N \times n_N$ matrix $\mathbf{D}_N = \mathbf{C}_N^T \mathbf{T} \mathbf{C}_N$ and $n_N \times n_F$ matrix $\mathbf{D}_F = \mathbf{C}_N^T \mathbf{T} \mathbf{C}_F$, equation (7.9) can be written as

$$\mathbf{D}\mathbf{z} - r\mathbf{p} = \mathbf{D}_N \mathbf{z}_N - \mathbf{D}_F \mathbf{z}_F - r\mathbf{p} = 0, \tag{7.11}$$

separating the n_N free (non-supported) nodes from the n_F fixed (supported) nodes at the boundaries.

For a given scale of the reciprocal diagram $1/r$ and boundary heights $\mathbf{z}_F$, one immediately finds, for the chosen Γ and Γ^*, the inside geometry $\mathbf{z}_N$ of the equilibrium network $\mathbf{G}$

$$\mathbf{z}_N = \mathbf{D}_N^{-1}(\mathbf{p}r - \mathbf{D}_F \mathbf{z}_F). \tag{7.12}$$

It is also possible to find a solution within given boundaries, $\mathbf{z}^{LB}$ and $\mathbf{z}^{UB}$, by formulating equation (7.9) as a linear optimization problem, with as variables all the z-coordinates and the scale factor r

$$\min_{\mathbf{z},r} -r \quad \text{such that} \quad \begin{cases} \mathbf{D}\mathbf{z} - r\mathbf{p} = 0 \\ \mathbf{z}^{LB} \le \mathbf{z} \le \mathbf{z}^{UB} \\ 0 \le r \le +\infty \end{cases}. \tag{7.13}$$

which renders, if a solution exists, the deepest compression-only thrust network within the structural depth of the vault, for the chosen form and (proportional) force diagram, within the given boundaries.

The direct approaches in equations (7.11) and (7.12) can be used for problems in which the loading is known a priori. When form finding a shell, that is, the shape of the shell is not known in advance, then the loads due to self-weight are not known a priori.

Instead, they need to be found through an iterative procedure, taking at each step the weights proportional to the tributary area of the nodes. Because both equations (7.11) and (7.12) solve fast, this is not a problem.

After a thrust network $\mathbf{G}$ is found, the axial branch forces $\mathbf{s}$ are then obtained directly as

$$\mathbf{s} = \frac{1}{r}\mathbf{L}\mathbf{t}, \tag{7.14}$$

where $\mathbf{L}$ are the branch lengths in three dimensions.

7.3.4 Process

In Figure 7.10, the general computational design process for forward TNA is illustrated. The form-finding procedure begins with an initial form diagram Γ as user input. Based on the connectivity of the form diagram, the matrices $\mathbf{C}$ and $\mathbf{C}^*$ are generated. In the next step, a reciprocal force diagram Γ^* is generated, representing one possible horizontal equilibrium state for $\mathbf{G}$. For simple cases, that is, small networks, this can be done by manually using methods from graphic statics. For more complex networks, this is done using one of the strategies given in Section 7.3.2. Based on Γ^*, a feasible set of force densities $\mathbf{q}$ is calculated using equation (7.10). Subsequently, the user chooses the heights of the support nodes $\mathbf{z}_F$ and the scale r. The tributary load for each node is estimated – for example, by using the cells of a Voronoi diagram as tributary load areas – resulting in the vector $\mathbf{p}$ (see Section 14.2.3). Based on $\mathbf{p}$, r, $\mathbf{z}_F$, $\mathbf{q}$ and $\mathbf{C}$, the heights of the unsupported nodes $\mathbf{z}_N$ are solved; for example, using the simple forward TNA method, from equation (7.11).

Based on the topological information provided by $\mathbf{C}$, and the nodal heights z_F and z_N, the thrust network $\mathbf{G}$ can be visualized in three dimensions. This representation enables the designer to examine if the spatial and formal requirements are met. If the result is not satisfying, the form diagram Γ, the internal force distribution, represented by Γ^*, the overall scale factor r, or the heights of the supports z_F have to be changed iteratively.

CHAPTER SEVEN: THRUST NETWORK ANALYSIS   81

tool used for this example ensures that the reciprocal relationship between the form and force diagram is enforced during the design exploration and solves for the thrust network. For details about the implementation and algorithms involved, we refer to Rippmann et al. (2012).

The design process starts with a simple four-valent grid as a form diagram. Figure 7.11 shows the different design choices step by step, always showing the form (left) and force (middle) diagrams, and the resulting thrust network (right). The grid is generated based on the local (u,v)-coordinate system of a quadrilateral NURBS patch that is drawn by the user.

Here, the force diagram is initially generated such that its branches are about equal in length, that is, the horizontal forces in the thrust network are about equally distributed. This choice results in a thrust network with the shape of a 'pillow' (Fig. 7.11a).

In the subsequent step, a second network patch is added to the form diagram (Fig. 7.11b). The 'stitching' of the two patches results in two identifiable parts in the force diagram. The effect of the slight change in direction between the branches of the two subgrids of the form diagram is clear in the force diagram with the separation of the two patches. Longer lengths in the force diagram are equivalent to higher thrusts in the thrust network, which result in a gentle undulation in the equilibrium shape, as forces are being attracted along this line.

In the next step (Fig. 7.11c), this effect is exaggerated by attracting more force along the inner edge between the patches. This is done by locally 'stretching' the force diagram along that line, without changing the form diagram. As the manipulation does not necessarily preserve parallelity, a diagram is found that satisfies the reciprocal constraints closest to the new configuration. The resulting equilibrium shape now clearly features a crease line.

Two new effects are introduced in the following step (Fig. 7.11d). First, the form diagram is deformed such that the angle between the branches at the edge of the two patches increase towards the bottom middle support. The new, enforced 'flow of forces', represented by this change in the form diagram results in an accumulation of forces along the edge between the two patches, causing an increasing crease in the thrust network. Second, the top edges of the two

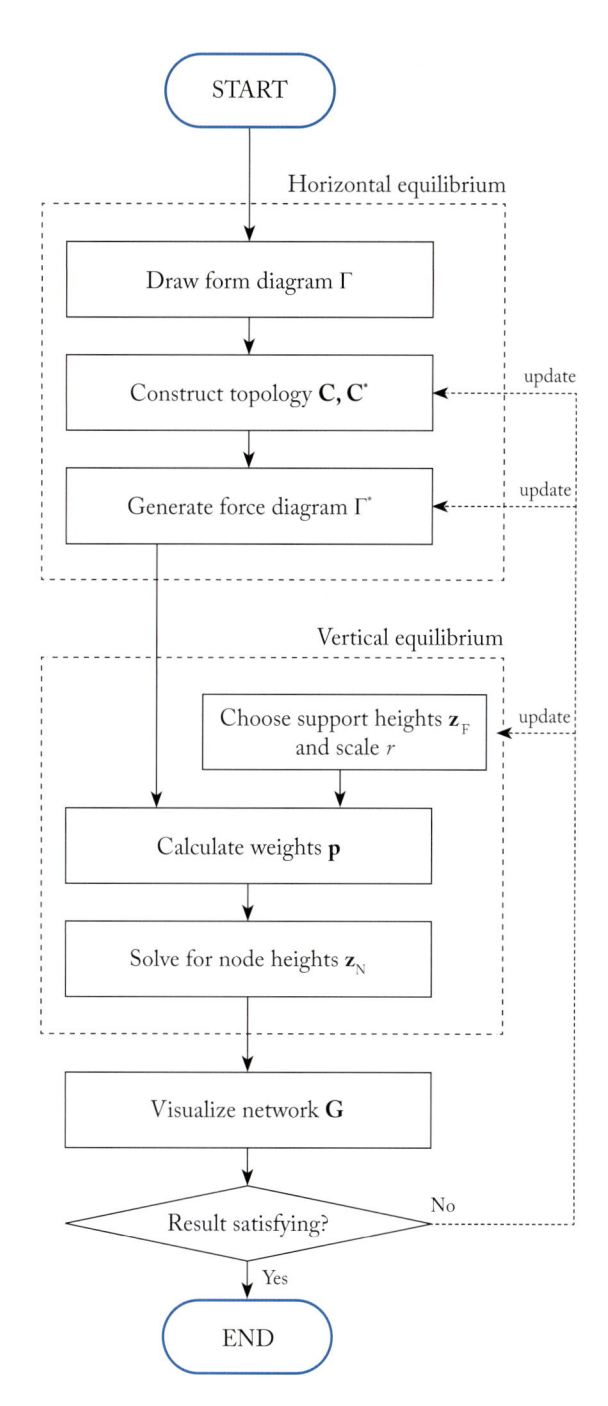

Figure 7.10 Overview of the TNA process

7.4 Interactive design exploration

This section will go through the design process of the pavilion vault, demonstrating how TNA allows the full control of three-dimensional equilibrium, and thus the ability to *steer* the form of the compression-only vault in a very intuitive and flexible manner. The

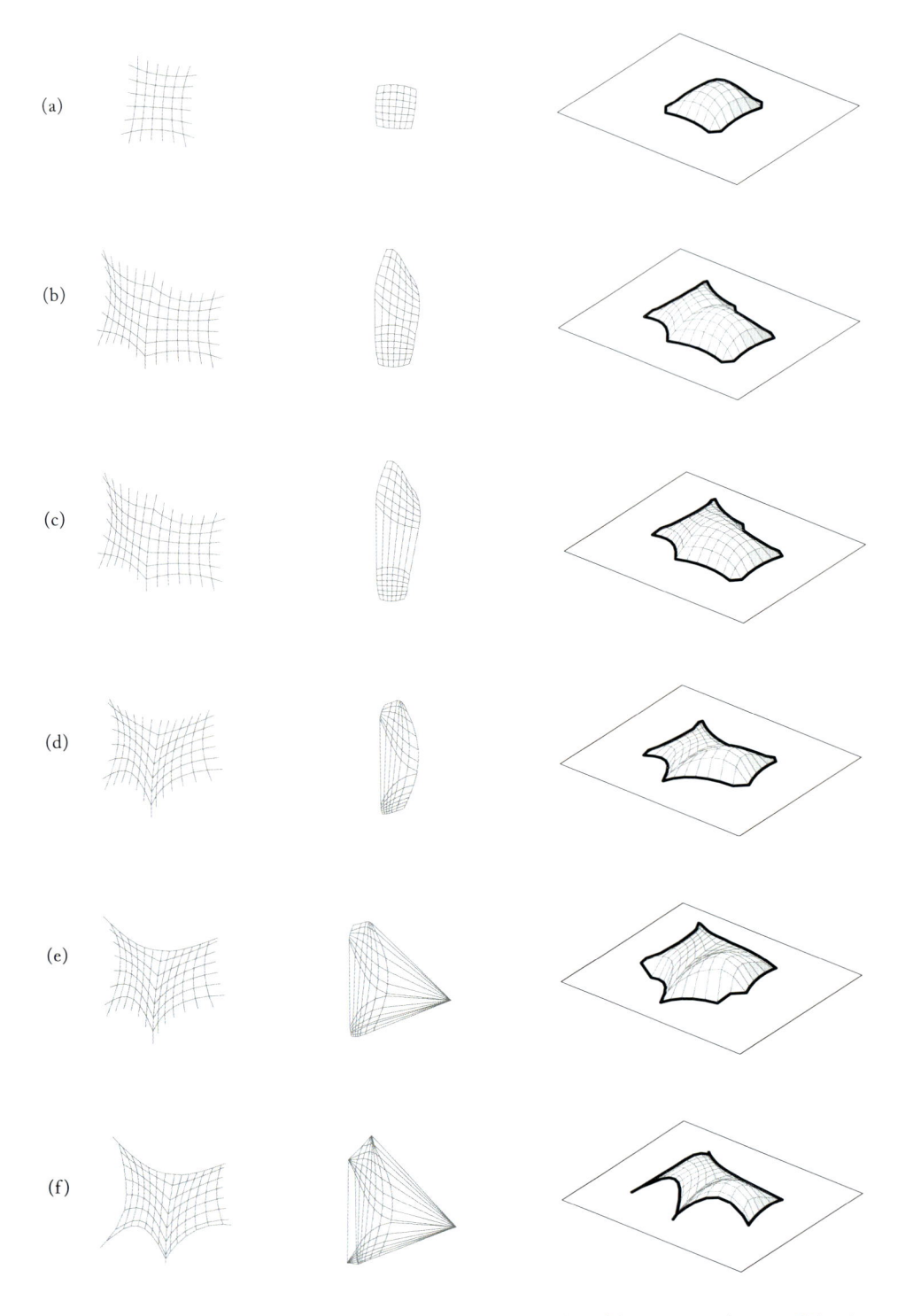

Figure 7.11 (left) Form and (middle) force diagrams, and (right) thrust networks, of the sequential steps of the design exploration

patches are aligned to generate a continuous edge condition.

Until now, all edges were considered fully supported. By adding branches to the edge spanning both patches on the top, an 'edge arch' is formed (Fig. 7.11e). This results in the fan-like part in the force diagram. This triangulated part of the force diagram exactly corresponds to the force diagram of a simple funicular arch in graphic statics (Fig. 7.3c). The geometry of this edge arch in plan is constrained by this part of the force diagram. In addition, the bottom supports are lowered, with the idea of trying to achieve an even more accentuated crease. This action is rejected in the next step by the designer.

Shown in the final step (Fig. 7.11f), all other supported edges are replaced by edge arches except one, added in the same manner as discussed above. The resulting thrust network represents the final design (Fig. 7.12).

This design example shows that with TNA an expressive funicular shell can be designed. This is achieved thanks to a controlled exploration of the degrees of freedom of the highly indeterminate, three-dimensional funicular structures. The control parameters were:

- the form diagram, defining the discretization and the choice of the flow of forces;
- the reciprocal force diagram, allowing (re-)distribution of the horizontal forces;
- the scale of the force diagram, controlling the overall depth of the equilibrium solution;
- the height of the boundary supports;
- the edge condition, that is, closed, which means fully or partially supported, versus open, which results in a three-dimensional edge arch.

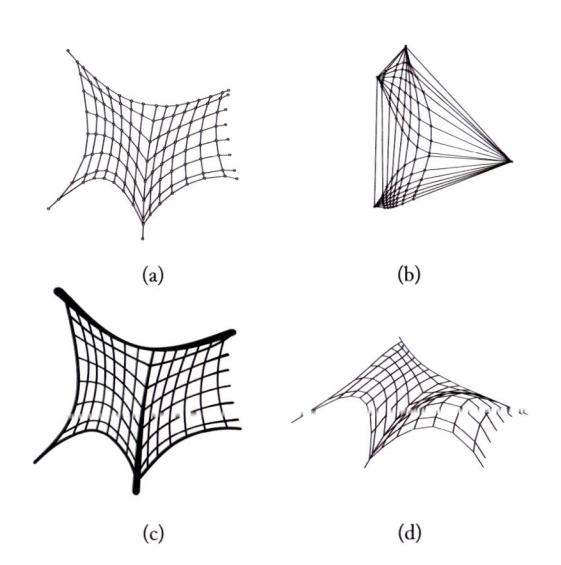

Figure 7.12 The final design: (a) form diagram, (b) force diagram, (c) force distribution and (d) the thrust network

7.5 Materials, details and construction challenges

The materialization process of discrete stone vaults begins with the planning and generation of the tessellation, which defines the cut pattern of the structure. Based on this tessellation, the geometry of the individual voussoirs is generated. The subsequent materialization and construction phases include the fabrication of individual elements and the erection of the structure on full in-situ falsework. The overall process is informed by structural and fabrication-related requirements to ensure structural stability and to minimize material usage, energy consumption and fabrication time.

7.5.1 Tessellation

The tessellation is based on the overall funicular shape of the vault generated with TNA. The resulting thrust network can be seen as an approximation of a continuous compression surface, representing the centre geometry of the vault, excluding the thickness of the structure. The digital representation of this surface is either a NURBS-surface or a very dense mesh geometry, which serves as a target surface for the tessellation. A user-assisted optimization process allows the generation of a feasible pattern on the irregular doubly curved thrust surface, incorporating various, correlated design criteria (Rippmann and Block, 2013).

Besides architectural and tectonic considerations, these criteria include structural stability as well as fabrication and construction feasibility. The tectonic expression of the vault is dominated by the topology of the tessellation, which is defined by drawing lines onto the thrust surface or by using assisting tiling strategies. A parallel, automated process guarantees that all lines are as perpendicular or parallel as possible to the local force flow, as the orientation of the tessellation needs to be aligned to the local force vector field to prevent sliding failure between the voussoirs. The force flow is sometimes equated to the direction of steepest descent, the so-called rainflow analogy. The force flow can also be deduced from the results of the TNA form-finding process, by taking the sum of forces in each node, with all the forces oriented

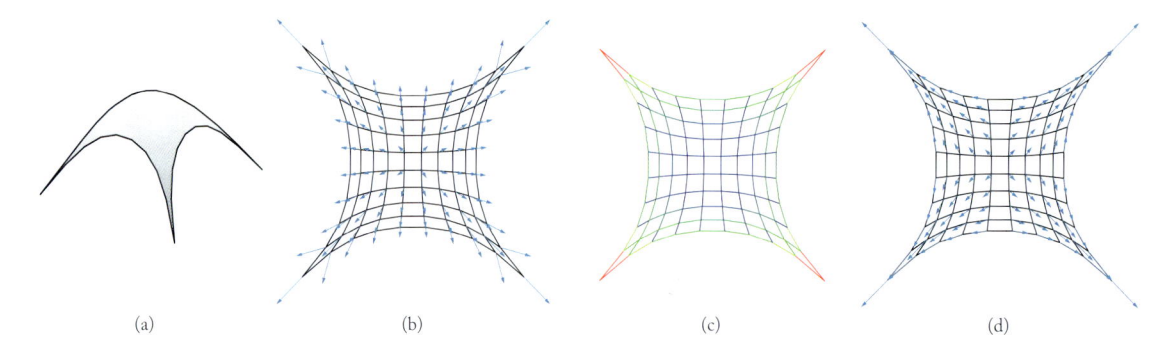

Figure 7.13 (a) A simple vault, with (b) vectors of steepest descent, that is, the rainflow analogy, (c) internal forces and (d) vectors from forces

downwards (multiplied by the sign of the vertical coordinate differences). The difference between both approaches is particularly evident near open edges of a vault (Fig. 7.13). The nodal force flow vectors are

$$\mathbf{f}_x = -\frac{\mathbf{w}}{|\mathbf{w}|} \cdot \mathbf{Dx},$$
$$\mathbf{f}_y = -\frac{\mathbf{w}}{|\mathbf{w}|} \cdot \mathbf{Dy}, \quad (7.15)$$
$$\mathbf{f}_z = -\frac{\mathbf{w}}{|\mathbf{w}|} \cdot \mathbf{Dz},$$

with $\mathbf{w} = \mathbf{Cz}$.

Furthermore, the prevention of sliding failure and full three-dimensional structural action is achieved using tessellation bonds in which neighbouring discrete pieces interlock. Fabrication and material-related parameters are taken into account by constraining the length of the edges of the tessellation pattern to a specific value, or within a given range, informed by machining limitations and maximum block dimensions. In this assisted design process, the best solution of a tessellation for a given surface, driven by the force flow, edge length restrictions and the topology, is found using an iterative solving algorithm based on a relaxation approach.

7.5.2 Voussoir geometry

In a next step, individual voussoirs are generated from the tessellation, the thrust surface and data regarding the local thickness of the structure, which is calculated based on the non-funicular live load cases (see Chapter 13). Each contact face is described by lofting through a set of lines normal to the thrust surface, resulting in an alignment normal to the force flow. The resulting contact faces are twisted ruled surfaces (Fig. 7.14).

The load-transmitting contact faces should have a flush alignment, and thus also a high geometric accuracy in the fabrication process, in contrast to the upper and lower surfaces of the voussoirs. Circular blade stone cutting fulfils these precision requirements and is at the same time one of the most efficient stone-machining processes (Fig. 7.15). However, to use this technology requires planar cuts. An iterative procedure was thus developed to planarize most contact faces. Computer numerical controlled (CNC) machines with five or more axes are used to process the voussoirs. In contrast to the planar contact faces, the upper and lower surfaces can only be approximated by progressively using parallel cuts tracing the doubly curved geometry.

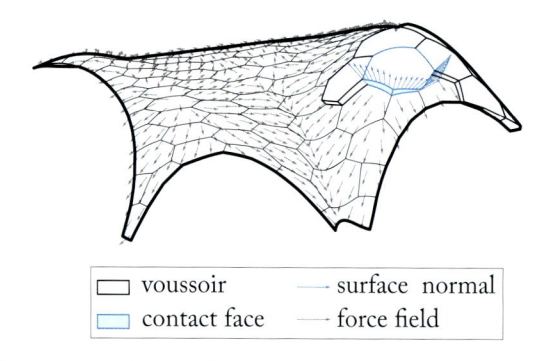

Figure 7.14 The tessellated vault geometry, showing voussoirs, their contact faces, surface normals (blue) and force field (grey)

CHAPTER SEVEN: THRUST NETWORK ANALYSIS   **85**

Figure 7.15 A 5-axis circular blade saw

7.5.3 Structural scale model

Once the tessellation is designed and the voussoir geometry constructed, the structural behaviour of the vault can be tested using a 3D-printed structural scale model. A qualitative understanding of the stability of the vault can be achieved by just manually applying 'point loads' to the model (Fig. 7.16). Discrete structural scale models provide insight into the structural behaviour of an unreinforced masonry structure – and compressive funicular shells in general. Because of the scalability of compression-only masonry structures, which is due to their very low stresses and stability based on their geometry (Heyman, 1995), the models enable a reliable prediction of the stability of a real-scale stone structure for corresponding load assumptions. These unglued, 3D-printed, 'masonry' scale models thus serve as convincing validation of the TNA results. By applying point loads and observing the partial and incremental collapse of the scale model, the force distribution assumptions made for the design can be checked directly and possible collapse mechanisms under extreme loading cases (see Section 13.7) can be determined.

7.5.4 Installation

After fabrication, all individual voussoirs are assembled in situ on falsework, starting at the foundations, which

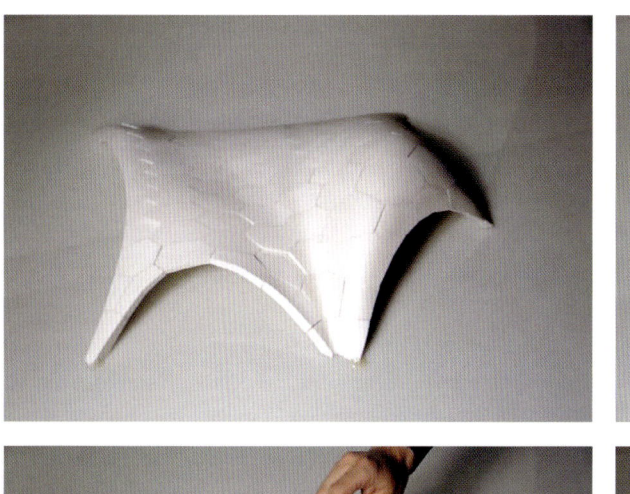

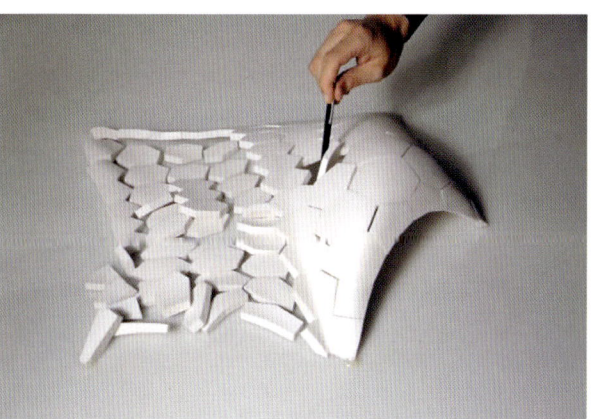

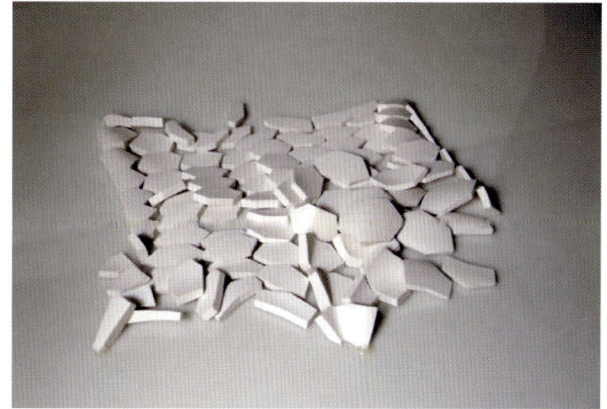

Figure 7.16 Gradual collapse of a 3D-printed, unglued structural model of the vault due to manually induced point loads

are connected by tension ties to avoid spreading due to the horizontal thrusts at the supports. This process demands accurate measurements through construction site survey to minimize the accumulation of errors due to fabrication and assembly tolerances, and consequently guarantee the exact position of all voussoirs during installation. The symmetry of traditional vault shapes helps to identify and build up stable sections, meaning that construction sequences are in static equilibrium and falsework can be removed, and reused during the erection process. In contrast, stable construction sequences for the installation of 'freeform' vaults are less obvious to identify and often limited to small patches of the structure. Thus, the falsework needs to support the entire structure during construction until the last voussoir is in position. The subsequent decentring of the falsework needs to be done uniformly in order to prevent asymmetric loading caused by partial, decentred falsework. Conventional, modular falsework systems for concrete structures can be used to carry the loads of the voussoirs during construction.

7.6 Conclusion

This chapter has shown how to generate geometries for vaulted masonry structures using thrust network analysis. The equilibrium of a simple arch was explained using thrust lines and graphic statics. This concept was then generalized to the three-dimensional and statically indeterminate thrust networks by introducing the reciprocal form and force diagrams. The reader was introduced to the equilibrium equations of the nodes in thrust networks and the construction of the data structure, which allowed the generation of the thrust networks using simple linear algebra.

The step-by-step design explanation for the Texas vault has been explained in depth, followed by a discussion on the materialization and construction of the cut-stone vault. This chapter is inspired by a real design case for an unreinforced, cut-stone vault, with a model for the real project shown in Figure 7.17.

Key concepts and terms

Graphic statics is a structural design and analysis method developed in the nineteenth century. It uses

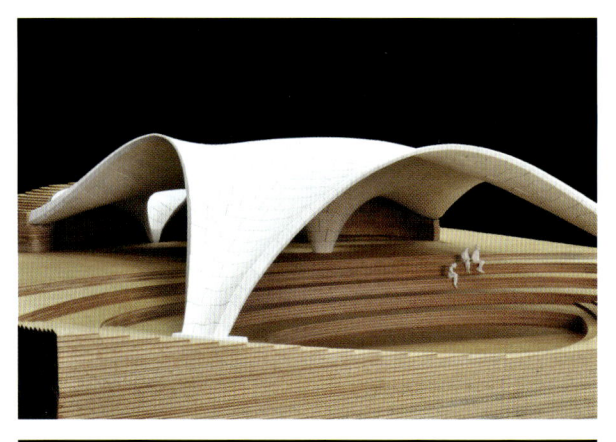

Figure 7.17 Final model of the real Texas pavilion project

form and force diagrams to calculate the equilibrium of pin-jointed structures graphically.

A **form diagram** is used in graphic statics as representation of the geometry of a pin-jointed structure.

A **force diagram** is used in graphic statics as representation of the (equilibrium of the) inner forces in a pin-jointed structure. Form and force diagrams are reciprocal diagrams.

Reciprocal diagrams are two planar diagrams that are topologically dual, with parallel corresponding edges.

A **funicular** represents the shape of a weighted hanging string and corresponds to an admissible equilibrium state of a planar arch for the same loads.

A **thrust network** is the generalization of the funicular or thrust line, representing the spatial equilibrium state of a vault as pin-jointed system.

The **middle third rule** states that a masonry arch is safe, and without tension, as long as the resultant force at each interface between blocks stays within the middle third of the cross section.

Exercises

- The thickness of the 21.5m hemispherical dome of the Temple of Mercury (see page 20) varies from 1.6m at the bearing to 0.6m at the crown. Can you explain why historic cut-stone domes have such a structural depth? Why is it possible to open up the middle with an oculus (3.65m)?
- Consider the standard grid (Fig. 6.12) as a simple four-valent form diagram. Construct its reciprocal force diagram (it helps to label the nodes and spaces of both diagrams using Bow's notation). Implement a routine that generates the **C** any **C*** matrices for the same standard grid, and check them with the labelled diagram you just produced.
- Implement a forward TNA solver using the previously implemented routine for the construction of **C** and **C*** matrices. Use this solver to generate a thrust network **G** for the previously drawn form and force diagrams, Γ and Γ^*, considering uniform loading (i.e. the same vertical load applied at each node). We want to explore a number of stable cut-stone shell forms as design options. Modify the internal force distribution by changing the reciprocal force diagram to generate, using the simple solver above, different thrust networks. Note that allowed reciprocal form and force diagrams have parallel corresponding edges.

Further reading

- *The Stone Skeleton: Structural Engineering of Masonry Architecture*, Heyman (1995). This book is the best reference to learn about how and why masonry structures work.
- *Form and Forces: Designing Efficient, Expressive Structures*, Allen & Zalewski (2009). This book, also recommended in Chapter 1, provides an exciting introduction to graphic statics for design. Chapter 8 gives an overview of the key aspects to be considered in the design of an unreinforced masonry (vaulted) structure.
- 'Thrust network analysis: exploring three-dimensional equilibrium', Block (2009). This PhD dissertation from MIT gives the most detailed introduction of the TNA method.
- 'Interactive vault design', Rippmann et al. (2012). This journal paper explains the algorithms behind the plug-in for the CAD-program Rhinoceros for funicular shell form finding, RhinoVAULT.
- 'Rethinking structural masonry: Unreinforced, stone-cut shells', Rippmann and Block (2013). This journal publication describes the details of the structurally informed, fabrication-optimized digital design chain developed to design and fabricate the cut-stone voussoirs of the MLK Jr Park Vault in Austin, TX, USA.

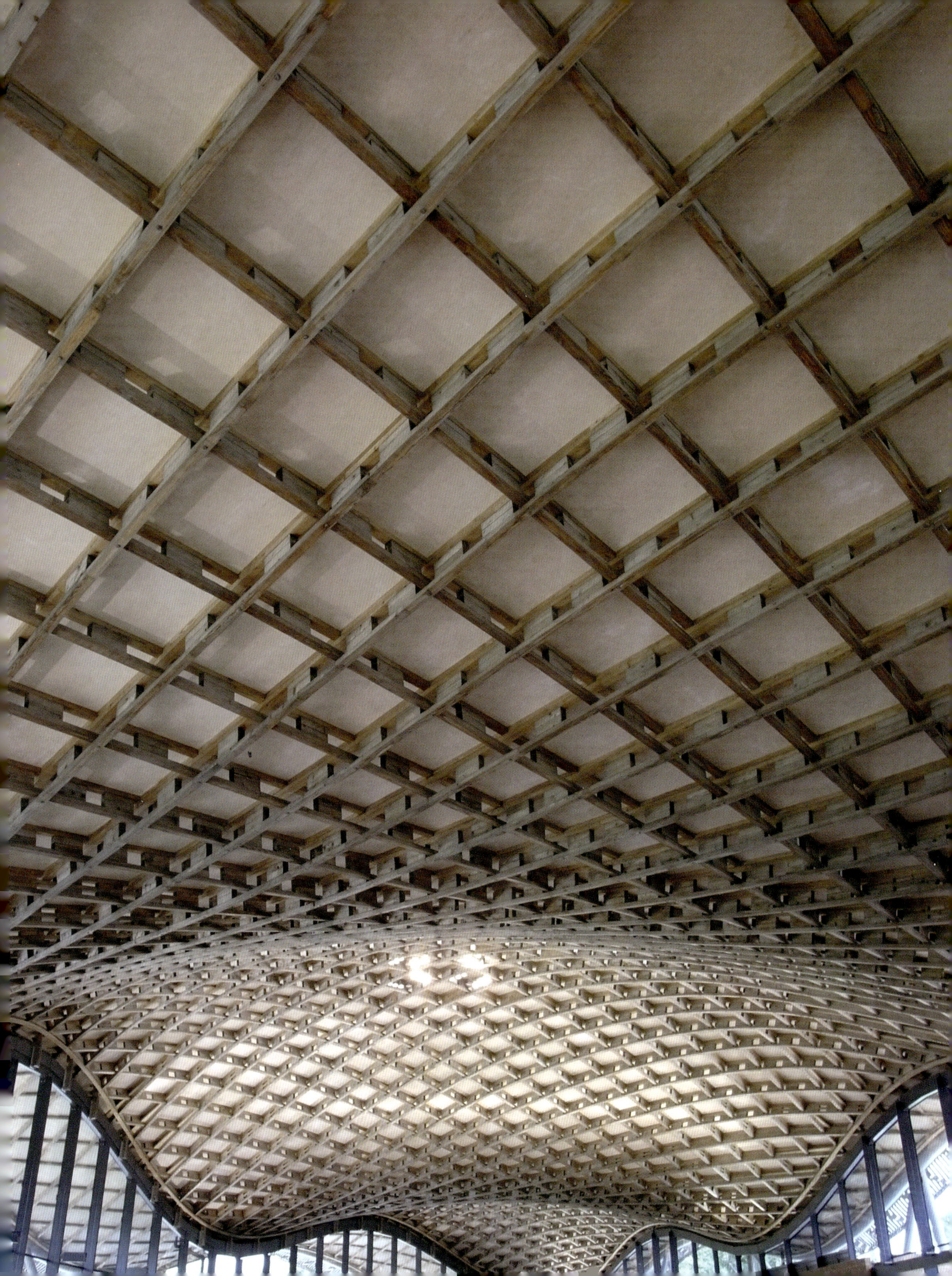

CHAPTER EIGHT

Dynamic relaxation

Design of a strained timber gridshell

Sigrid Adriaenssens, Mike Barnes, Richard Harris and Chris Williams

LEARNING OBJECTIVES

- Explain the difference between a strained and unstrained gridshell.
- Describe material constraints, connection details and construction techniques for gridshells.
- Explain the dynamic relaxation form-finding technique including the bending effect of continuous splines.
- Use the dynamic relaxation method to find the form of a gridshell.
- Perform the form finding of a strained gridshell.

A gridshell, such as the one shown in Figure 8.1 and page 88, is essentially a shell with its structure concentrated into individual members in a relatively fine grid compared to the overall dimensions of the structure. The members may be short in length and only pass from node to node, or they may be continuous, crossing each other at the nodes. The grid may have more than one layer, but the overall thickness of the shell is small compared to the overall span.

This chapter discusses gridshells that are either made from initially straight elements and or prefabricated from curved members. The method of dynamic relaxation can be used for the form finding of either.

The brief

Barbados, an island in the Caribbean Sea, is famous for its yearly Crop Over, a harvest festival held in the months of July and August. For the entire two months, life for many islanders is one big party. For this event, the capital Bridgetown needs a large-span pavilion with a 50m × 50m footprint to hold the crafts market and shelter the visitors from sun, wind and occasional rain. The pavilion will provide shelter to 150 stalls and two restaurants, which totals 2,500m^2 of covered area. The client wants this pavilion to harmonize with the local crafts and to promote sustainability at the same time. We, the designers, select a gridshell made of local wood. Barbados cultivates wood that can be cut with mobile sawmills.

8.1 The strained gridshell

Our proposal for the Bridgetown Pavilion uses long, slender, continuous wood laths arranged in a grid. These laths, or structural splines, are both flexible and strong. A spline is an initially straight member that is bent into a spatial, continuous curve. The word 'spline' originally denoted the flexible wooden or metal strip draughtsmen and women used to draw smooth ship lines and railway curves. The most simple and familiar structural spline examples are pole vaults and slender

Figure 8.1 The gridshell roof of the Savill Building at the Windsor Great Park, UK, 2006

battens that prestress umbrellas, camping tents and sails. When combined in a grid and bent, the splines form a structural, complex, curved surface. This initial bending action strains the shell, hence the term 'strained' gridshell. In contrast, the 'unstrained' gridshell is a curved system that in its initial state is stress-free (apart from stresses due to self-weight). Initially, the grid for the Bridgewater Pavilion is formed from continuous, straight wooden splines bolted together at uniform spacing in two directions. When flat, the grid with its scissor-pinned connections is a mechanism with one degree of freedom. If the grid members were totally rigid and connected with frictionless joints, the movement of one member parallel to another would cause a sympathetic movement in the entire grid. As a result, all squares would become parallelograms, and the diagonal length between the joints would change as shown in Figure 8.2. This grid distortion feature, combined with the grid's flexibility, is crucial to the erection method which moulds the initially flat grid into a three-dimensional structural surface (see Section 8.7). This construction method contrasts with that of the unstrained gridshells; these shells are assembled from prefabricated curved subframes.

In addition to the distortion of the squares to parallelograms, the layers of a multi-layer, strained gridshell have to slide over each other during erection. Sometimes slotted bolt holes are used to allow this sliding movement. The blocking pieces between the upper and lower layers, shown on page 88 and Figure 8.3c, can be added after erection to ensure composite bending action between the separate layers.

If the geometry of a gridshell is derived from a hanging model, and it is loaded with its own self-weight, then it experiences no bending as a result of

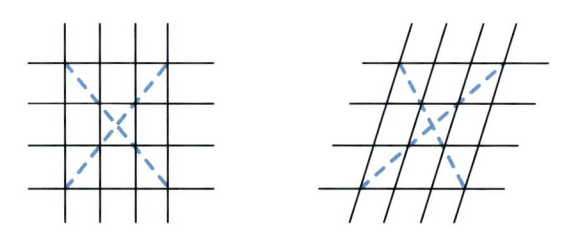

Figure 8.2 Frictionless pinned connections allow members to rotate and propagate the distortion throughout the grid

gravity loading. In reality, live loads are larger than self-weight, for these light and efficient structures, and unevenly distributed over the nodes. These live loads, especially the point loads, cause bending of the laths resulting in large shell displacements and changes in angle between the laths. To reduce this shell movement, diagonal stiffness can be introduced in different ways, through:

- joint rigidification;
- the addition of diagonal ties with a cross-sectional area less than the laths;
- addition of cross-bracing of equal area to the laths.

In the latter case, the diagonally braced shell will behave very much like a continuous shell.

Figure 8.3a shows in-plane membrane stresses acting upon a square element, cut out of a continuous shell surface, made of an isotropic material (see also Section 3.1.1). In a well-designed structure, this will be the prime mode of structural action. In addition, there will be out-of-plane bending and twisting moments and the associated shear forces acting normal to the surface. The element's orientation has no influence on the force-displacement characteristics of the shell. A similar element in a gridshell, however, resists membrane forces only in the direction of the laths. Out-of-plane bending can be resisted, but diagonal forces between parallel laths cannot be transmitted in the initial pinned situation. Once the grid is moulded and fixed into its final form, the additional diagonal cable layer or the joint rigidification gives the shell its in-plane shear stiffness.

Figure 8.4 shows the diagonal bracing network positioned outside the gridshell in the 2007 Chiddingstone Orangery. This in-plane shear stiffness, combined with the blocking pieces between the layers and the connections to the boundary supports, locks the flexible grid into shape. These locking methods together with the shell's spatial curvature must ensure that the thin, lightweight gridshell is stable under all loading combinations. Live loads cause bending and deflections that change the initial gridshell shape.

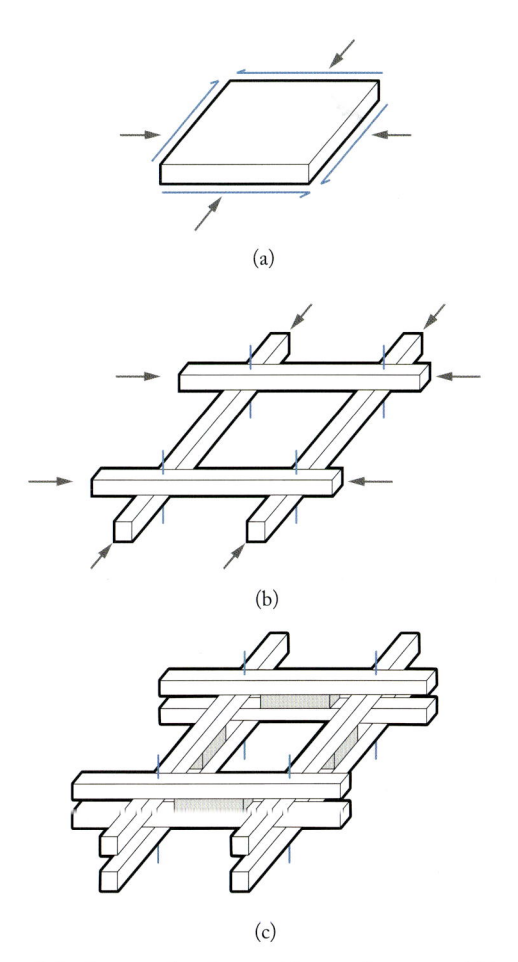

Figure 8.3 Structural action on (a) a continuous and (b) a gridshell element, and (c) a multi-layer gridshell element with blocking pieces

With the line of force then positioned eccentrically from the initially defined shell surface shape, these direct forces produce bending moments. As the loads increase, the stiffness decreases. At a certain critical value of load, the gridshell no longer resists an increase in load and collapses. This characteristic is typical of compression systems.

Since the axial stiffness of the laths is relatively high, the main techniques for increasing the collapse load are the addition of in-plane diagonal stiffness and increasing the out-of-plane bending stiffness. The cross-sectional dimensions of the laths are limited by the need to bend them into shape. To increase the second moment of area of the shell (and thus the out-of-plane bending stiffness) more layers can be added. Blocking pieces between the layers are put in to ensure composite action over the entire depth of the shell.

Gridshells may collapse due to buckling while all the members are still elastic. Alternatively, some part may break, leading to collapse before elastic buckling occurs. Because of the diagonal flexibility, and often lack of rigid boundary support, gridshells tend to deflect a relatively large amount before collapse, especially compared to conventional shell structures. This feature makes them less efficient, but it does mean that initial imperfections in their shape are relatively unimportant (see Chapter 3).

8.2 The unstrained gridshell

An unstrained gridshell differs from our proposed strained system in that it is curved and unstrained in its initial state and is made from an assembly of relatively short straight or pre-bent members. The curvature can be induced in two ways. The first method uses pre-bent steel or aluminium members or curved laminated timber. Alternatively, the members may be straight and the change in member direction is achieved at the nodes. The nodal connections have to be moment resisting to prevent buckling, or the shell has to consist of more than one layer, producing a curved-space frame. Unstrained subframes can be fabricated in the controlled environment of a workshop and assembled on site on falsework tailored to the form of the complete shell surface.

The shape development of recent unstrained gridshells is often driven by a combination of aesthetic,

Figure 8.4 The double-layer gridshell with external bracing of the Chiddingstone Orangery in Kent, UK, 2007

geometrical, physical and constructional considerations. The recent emergence of 'freeforms' illustrates a design approach with sculptural or aesthetic design intent as the major shape driver. With computer-aided modelling tools at hand, more designers base their freeform work on aesthetic considerations that achieve scenographic effects, in and around the shell, but pay no particular attention to structural efficiency of the form. The organically twisted, merging gridshells of the Murinsel in Graz, shown in Figure 8.5, exemplify this approach. These sculpturally merging steel and glass gridshells form a connection between the banks of the river Mur.

These contemporary freeforms or 'blobs' contrast sharply with smart shells based on 'simple' geometries. Since antiquity, analytically defined geometries have been favoured for their constructive and structural qualities. The hyperbolic steel shell of the Shukhov Tower demonstrates how surfaces of revolution lend themselves to shell action and discretization into straight elements. Translational surfaces, such as the Hippo House in Figure 8.6, have the additional advantage of discretization into planar meshes. From an economic cladding perspective, mesh planarity is desirable especially for sheet materials such as glass and steel.

The importance of the shell's geometry cannot be overemphasized. The form decides whether the thin shell will be stable, safe and sufficiently stiff. Finding the 'right' geometry under the chosen loading (usually gravity) means that under this design load any bending is eliminated and only advantageous membrane action results. The structural challenge lies in the determination of a three-dimensional surface within which the shell can be described. Both architects and engineers have developed physical and numerical methods to generate structurally and constructionally efficient three-dimensional gridshell shapes other than the 'simple' geometries. This distinction of form between freeform, mathematical and form-found shells is also discussed in the Introduction.

CHAPTER EIGHT: DYNAMIC RELAXATION

Figure 8.5 Freeform, Murinsel gridshell, Graz, 2003

Figure 8.6 Hippo House quadrilateral gridshell, Berlin Zoo, 1996

8.3 Dimensional analysis

The physical modelling of shells, including gridshells, is discussed in Chapter 4. Here we discuss the case of gridshells in more detail. The structural action of gridshell structures is so complex that even today with powerful and affordable computers, there is still a place for physical model testing. The most rudimentary physical model can give more accurate predictions of deflections and buckling load than hand calculations. Because of the complex interaction between membrane and bending action, the prediction of buckling loads by hand calculations is effectively impossible. However, the deflections and buckling load from a physical model can be scaled using dimensional analysis (see also Section 4.3). A sieve-structure is a good example; its behaviour is dominated by the bending stiffness of the members, whether they are wires or laths. The quantity

$$\frac{wS^3}{\left(\frac{EI}{a}\right)}, \qquad (8.1)$$

where w is the load per unit area, S the span, EI the bending stiffness of a member, and a the member spacing.

This quantity is dimensionless – it has no units. Therefore, it must have the same value for the small-scale model and the full-scale gridshell. The effect of the diagonal members can be represented by the dimensionless group

$$\frac{\left(\frac{(EA)_{dia}}{b}\right)S^2}{\left(\frac{EI}{a}\right)}, \qquad (8.2)$$

where $(EA)_{dia}$ is the axial stiffness of diagonals and b the diagonal member spacing. This dimensionless group would have the same value for the model and the actual full-scale gridshell.

Note that the model does not have to be made from the same material as the full-scale structure, provided that the correct Young's moduli are used. It may be convenient to use a wire mesh model to investigate the behaviour of a full-scale, timber gridshell.

8.4 Implementation

The numerical form-finding method we explore for the design of the Bridgetown Pavilion uses the Dynamic Relaxation (DR) method and includes the bending stiffness of the splines. DR was invented by Alistair Day in 1965 and is a numerical procedure that solves a set of nonlinear equations. Summarized, the technique traces the motion of the structure through time under applied load. The technique is effectively the same as the leapfrog and Verlet methods, which are also used to integrate Newton's second law through time.

The basis of the method is to trace step by step for small time increments, Δt, the motion of each interconnected node of the grid until the structure comes to rest in static equilibrium. For the form finding of the Bridgewater Pavilion, we start from a flat 50m × 50m square grid with a mesh size of 2m × 2m. This mesh is a relatively coarse grid and the actual choice of spacing depends on a number of factors including the loading on the shell and the type of cladding system. The grid is supported at its four corners. This condition means that forces are concentrated at the corners and gridshells are often reinforced at their boundaries by steel or laminated timber arches. All grid splines are assigned values for their axial and bending stiffness, *EA* and *EI*, respectively where *E* is the Young's modulus, *A* the cross-sectional area and *I* the second moment of area. The shell edge members are assigned higher stiffness values to model the boundary arches.

We cause the motion of the grid by applying a fictitious, negative gravity load at all the grid nodes. The upwards load avoids having to turn the structure upside down to get the hanging tension form. We also adjust the position of the four corners to give the correct amount of rise to the structure. During the form-finding process, the values of all numerical quantities (*EA*, *EI* and load) are arbitrary since it is only their ratios that affect the shape. The axial stiffness *EA* needs to be sufficiently high to give a low axial strain and the ratio of bending stiffness *EI* to the weight controls the relative effect of bending stiffness and (upwards) weight upon the form. If the bending stiffness is set to zero, the resulting form is 'optimal' but modelling is difficult because the structure tends to develop wrinkles in areas of low tension in the hanging state. This problem applies equally to both physical and computer models.

The real values of stiffness need to be used during the structural analysis of the completed structure under dead and live loads. If the structure is 'locked' into shape by blocking pieces, the unstrained shape of the structure is now curved. This locking can be included in the analysis by treating the bending stiffness and initial curvature as being quantities which depend upon the orientation relative to the direction normal to the grid surface. However, this introduces complexities beyond the scope of this chapter.

The DR formulation for this project uses Newton's second law governing the motion of any node *i* in the *x*-direction at time *t*. The residual force at node *i* in the *x*-direction at time *t* is

$$R_{ix}^t = M_i \dot{v}_{ix}^t, \quad (8.3)$$

where $\dot{v}_{ix}$ is the acceleration at node *i* in direction *x* at time *t* (the dot indicates that it is the time derivative of the velocity). It is the sum of all the forces acting on a node from the members connected to it and the applied loading. The mass M_i is the lumped, fictitious mass at node *i*, and for a bar-node system,

$$M_i = \frac{\Delta t^2}{2} S_i, \quad (8.4)$$

where S_i is the greatest direct stiffness that occurs at node *i*,

$$S_i = \sum_{m=1}^{k} \left(\frac{EA^s}{L_0} + G \frac{T^s}{L^s} \right). \quad (8.5)$$

This expression covers the worst possible case of all links at the node becoming aligned in one single direction, where L_0 is the initial length of the link, *G* the factor that allows for the increase of initial geometric stiffness due to possible shortening of link lengths in the form-finding process. Superscript *s* refers to parameters (initially) specified by the designer, *m* to each member of all *k* members meeting at node *i*.

Expressing the acceleration term in equation (8.3) in finite difference form and rearranging the equation gives the recurrence equation for updating the velocity components

$$v_{ix}^{t+\Delta t/2} = v_{ix}^{t-\Delta t/2} + \frac{\Delta t}{M_i} R_{ix}^t \quad (8.6)$$

and hence the updated geometry projected to time

$$x_i^{t+\Delta t} = x_i^t + \Delta t \cdot v_{ix}^{t+\Delta t/2}. \quad (8.7)$$

Equations (8.2) and (8.3) apply for all unconstrained nodes of the structure in each coordinate direction. These equations are nodally decoupled in the sense that the updated velocity components are dependent only on previous velocity and residual force components at a node. They are not directly influenced by the current $(t + \Delta t/2)$ updates at other nodes. Having obtained the complete, updated geometry, the new member forces can be determined and resolved together with the applied gravity load components P_{ix} to give the updated residuals

$$R_{ix}^{t+\Delta t} = P_{ix} + \sum_{m=1}^{k} \left[\left(\frac{F}{L} \right)(x_j - x_i) \right]^{t+\Delta t} \quad (8.8)$$

for all *k* elements connecting to *i*, and where F_m is the axial force in member *m* connecting node *i* to an adjacent node *j*, and L_m the current length of member *m* (calculated using Pythagorean theorem in three dimensions).

8.4.1 Viscous and kinetic damping

This process is continued, through each iteration, to trace the motion of the unbalanced structure. But, thus far, we have not introduced any damping, which means that the structure goes past static equilibrium and continues to oscillate. This phenomenon can be prevented by introducing a 'viscous damping' force in the opposite direction to the velocity. Alternatively, we can use 'kinetic damping' in which there is no viscous damping but instead all the nodal velocities are set to zero when a kinetic energy peak is detected (Figs. 8.7 and 8.8).

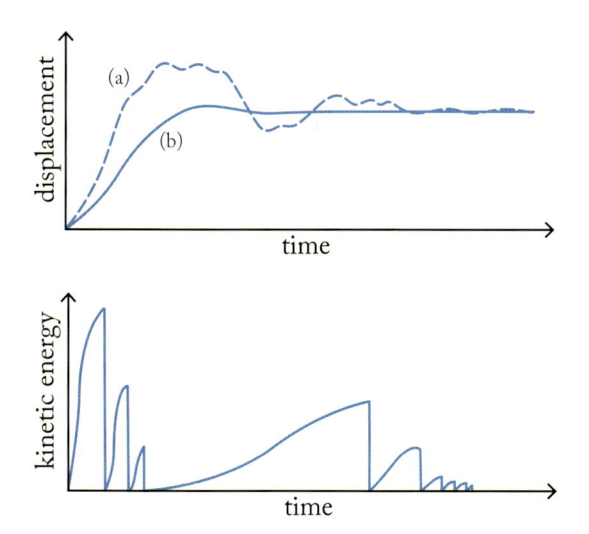

Figure 8.7 Effect of (top) viscous damping for (a) underdamped oscillations and (b) critically damped oscillations, and of (bottom) kinetic damping

Whichever technique is used, the process will never truly converge, but once the residual forces are below a certain tolerance, convergence has occurred for all practical purposes. At that point, we achieve a shape that is in 'static equilibrium' and have found the 'correct' spatial surface.

With viscous damping the recurrence equations for the velocities (8.6), is rearranged so that

$$v_{ix}^{t+\Delta t/2} = A \cdot v_{ix}^{t-\Delta t/2} + B \cdot \frac{\Delta t}{M_i} R_{ix}^t, \qquad (8.9)$$

where $A = (1 - C/2)/(1 + C/2)$, $B = (1 + A)/2$ and C is a constant for the complete structure. In cases where only kinetic damping is used, $A = 1$.

On starting or restarting the kinetic damping process, following a kinetic energy peak, velocities are set to zero. Thus, for the first iteration and after each energy peak, or re-initialization,

$$v_{ix}^{\Delta t/2} = \frac{\Delta t}{2M_i} R_{ix}^0, \qquad (8.10)$$

to give effectively $v_{ix}^{\Delta t/2} = -v_{ix}^{\Delta t/2}$, or $v_{ix} = 0$ at time zero. After detecting an energy peak, coordinates will have been projected to time $t + \Delta t$. But the 'true' kinetic energy peak will have occurred at some earlier time t^*. To determine the coordinates at time t^*, a quadratic can be fitted through the current (F) and two previous total kinetic energy values (D and E) in Figure 8.8.

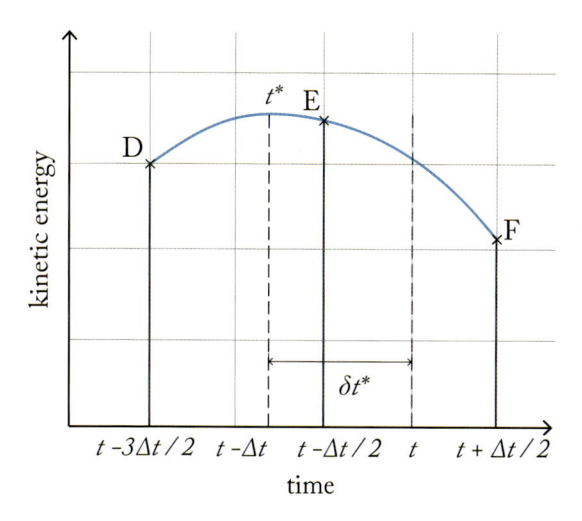

Figure 8.8 Kinetic energy peak at t^*

It is convenient for computation to keep records of the difference between the previous and current kinetic energies G and H. We define the elapsed time t^* since the energy peak in terms of these differences,

$$\delta t^* = \Delta t \frac{H}{H-G} = \Delta t \cdot q, \qquad (8.11)$$

where $H = E - F$ and $G = D - E$.

Since coordinates have been updated using average velocities (at midpoints of time intervals), they should be reset according to the same scheme. Thus,

$$x_i^{t^*} = x_i^{t+\Delta t} - \Delta t \cdot v_{ix}^{t+\Delta t/2} + \delta t^* \cdot v_{ix}^{t-\Delta t/2}. \qquad (8.12)$$

Hence, using equations (8.6), (8.7) and (8.11)

$$x_i^{t^*} = x_i^{t+\Delta t} - \Delta t(1+q) \cdot v_{ix}^{t+\Delta t/2} + \frac{\Delta t^2}{2} q \frac{R_{ix}^t}{M_i}. \qquad (8.13)$$

An alternative is to assume that the peak occurs at $t - \frac{\Delta t}{2}$ and hence $q = \frac{1}{2}$ in equation (8.13).

For a real dynamic analysis, the lumped mass would be the actual mass associated with a node. However, to perform form finding and obtain a static state in equilibrium, we can choose fictitious masses to get the quickest convergence. This phenomenon is less of an issue with modern fast computers, but essentially the number of iterations needed to get acceptable convergence depends upon the value of Δt, which in turn depends upon the highest natural frequency of the structure. Thus, if there is a particularly stiff part of a structure, like a steel edge beam on a timber grid, the fictitious masses should be increased in that area.

8.4.2 Spline elements

The form finding of the Bridgewater Pavilion needs to incorporate the effect of bending moments and shear forces, caused by bending the flexible wood splines from their initially straight state. The spline treatment described next, takes into account the resulting straining action. The technique adopted requires only three translational degrees of freedom per grid node and the usual rotational degrees of freedom used to accompany bending effects are not required. Often, the coupling of these rotational degrees of freedom with axial stiffnesses and translational degrees of freedom causes conditioning problems in an explicit numerical method such as DR. The scheme adopted is, in effect, a finite difference modelling of a continuous spline element with at least two segments. Figure 8.9 shows consecutive nodes along an initially straight spline, as well as two adjacent deformed segments, a and b, viewed normal to the plane of nodes ijk which are assumed to lie on a circular arc of radius R. The spacing of the nodes along the spline must be sufficiently close to model curvature as it varies along the spline, but the segment lengths need not be equal.

From the geometry of Figure 8.9, the radius of curvature through ijk,

$$R = \frac{L_c}{2 \sin \alpha} \quad (8.14)$$

and the consequent moment,

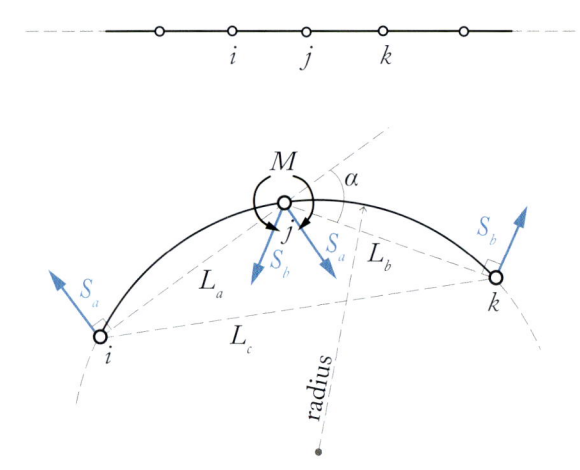

$$M = \frac{EI}{R}. \quad (8.15)$$

Figure 8.9 Consecutive nodes along an initially straight spline (above) and two adjacent deformed segments, viewed normal to the plane of nodes ijk (below)

Note that for a given R and L_c the value of α is independent of the position of the point j along the arc. This fact is a consequence of the inscribed angle being constant. The bending stiffness EI is assumed to be constant along the spline. The free-body shear forces, S, of links a and b are:

$$S_a = \frac{2EI \sin \alpha}{L_a L_c}, \; S_b = \frac{2EI \sin \alpha}{L_b L_c}. \quad (8.16)$$

These shear forces act on nodes i, j and k, as shown in Figure 8.9, and must be taken as acting normal to the links and in the local plane of ijk. To account for the stiffness component due to bending at node j, a term is added to equation 8.5, so that

$$S_i = \sum_{m=1}^{k} \left(\frac{EA^s}{L^s} + G \frac{T^s}{L^s} \right) + \sum_{a,b} \left(\frac{4EI}{L_s^3} \right). \quad (8.17)$$

The calculations and transformations required in a DR scheme are thus very simple. With sets of three consecutive nodes being considered sequentially along the entire spline, each set lies in a different plane when modelling a spatial curve. If the spline is pin-ended, as is normally the case for splines in gridshells, no special numerical treatment for end conditions is required. If the spline is a closed loop, then overlapping end links are required (a similar finite difference type of modelling would be required for fixed-ended splines using extended end segments). If the stiffnesses used when setting nodal masses in the DR process are unfactored, the minimum length of any traverse segment should not be less than the radius of gyration of the cross section. In practice, this limit is not likely to be approached. If this limit is approached, appropriate factoring of the bending stiffness must be applied when setting mass components in order to allow for coupling of the axial displacements with bending stiffnesses. Incorporating the effect of out-of-plane bending and twisting of the splines makes the form-finding process more complicated. For our design project, we do not consider these structural behaviours as we assume that all laths will be bent in plane.

8.4.3 Result

Figure 8.10 shows a form-found shape exploration for the Bridgetown Pavilion. When starting with a flat square grid with four corner supports, DR generates a tall, symmetric, curved shape (Fig. 8.10a). The splines in the curved surface have a large radius of curvature. This means they will only experience small bending action during the erection process from a flat to curved surface. The global shape lets rainwater run towards the boundary edges. Due to the height of these edges, the shell might not provide sufficient shading to the stalls located in the boundary areas. To create smaller spatial pockets that could have different programmes (restaurant versus stalls), and ensure better shading, we introduce an additional central support in the grid. The resulting, form-found shape is lower (Fig. 8.10b). As a result, the splines are bent to a tighter radius and need to be checked for overstraining. The created central funnel allows for central evacuation of drainage water. Figure 8.10c shows yet another shape variation using the same initial square grid layout with different point supports. In the resulting forms, we would need to perform spline curvature, shading and water run-off analysis to guarantee the overall feasibility of the shell's shape.

8.5 Procedure

The process for DR with kinetic damping can be captured in a flowchart, shown in Figure 8.11. The procedure iterates until satisfactory convergence has occurred. The factor of 0.999 in the calculation of velocity is in fact equivalent to viscous damping. If one were to use only kinetic damping, this would be replaced by 1.0. In practice, it is not always clear whether it is best to use viscous or kinetic damping, or a combination of the two.

8.6 Materials and details

Material selection for the gridshell differs depending upon whether the shell is strained or unstrained. Gridshells have been made from aluminium, concrete, steel, wood, bamboo and composite materials. Each material has advantages and disadvantages regarding strength, ductility, stiffness, cost, weight and durability. Connection and cladding design also inform the material choice.

For an unstrained shell, the material should allow for curved fabrication and shape retention. In this case, fabrication issues mostly drive material selection and element shape. The elements are produced with curvature about both transverse axes and twist along

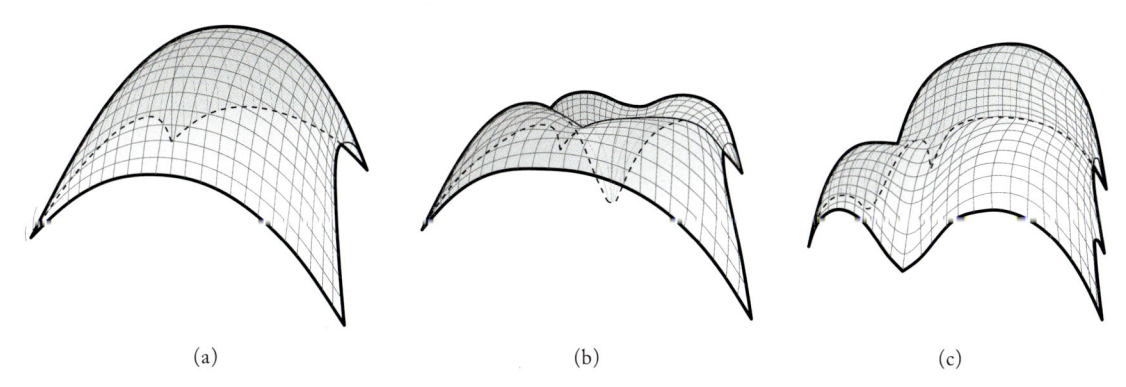

(a) (b) (c)

Figure 8.10 The form finding of the Bridgetown gridshell, based on a square, initially flat grid with varying point support conditions

be complex and expensive. These considerations apply to systems discussed in Chapters 12, 14 and 15.

The advantage of the strained shell lies in the fabrication and construction process. The shell is made from a series of identical splines, which are moulded to shape in their final location on the building. Candidate materials are more limited than those available for unstrained shells. Locally sourced wood is ideal for our project. Wood is capable of being bent into shape. Additionally, the stresses induced in this process will dissipate. This phenomenon is due to visco-elastic relaxation that leads to effective timber stiffness reduction under sustained load. This long-term load is due to the initial, forced curvature. The resulting induced bending stress reduces with time, making more capacity available to withstand applied loading. Additionally, timber has a very low torsional stiffness (ratio of the torsional rigidity to its length); the torsional modulus is typically one-sixteenth of the elastic-bending modulus. With low stiffness in torsion, wood is easily bent to shape in the forming process.

Finger jointing can be used to produce wood laths of any length for the strained shell. This technique involves cutting wedge-shaped 'fingers' in the ends of the pieces of wood and gluing them together. The procedure is sometimes executed with elaborate automated machinery in controlled conditions of temperature and humidity. In Barbados, where this technique is not available, we propose using skilled carpentry using a router table with a special finger-jointing cutter. To guarantee durability, wood can be treated or a naturally durable wood such as oak and larch can be chosen. Since environmental sustainability is a crucial design criterion, we must ensure that the construction wood comes from sustainable sources. In practice, this objective means that for every tree removed, a new tree is planted and managed to maturity. In this way, wood production (the growth of trees) absorbs carbon dioxide into the forest, whilst the wood removed is stored carbon in the building.

The connectors, placed wherever the laths intersect, must be compatible with the construction procedure. The connectors keep the splines in place while allowing them to rotate and distort to form the spatial surface, during construction. In a multi-layer shell, the layers can either be laid out flat and then bent into shape together (e.g. the Downland gridshell, Fig. 8.12) or

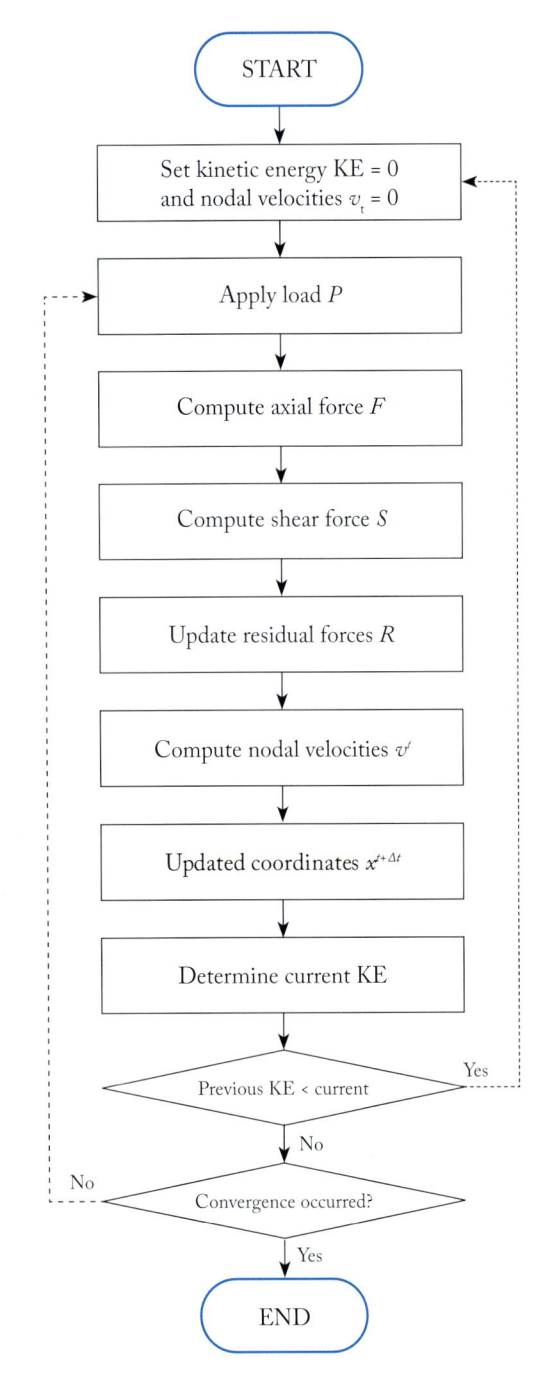

Figure 8.11 Flowchart

their length. Due to the element's twist, as it crosses the shell's surface, circular members are preferred because of their ease of fabrication. A wide range of materials offer themselves, including, steel, aluminium and laminated wood. With joints at every node, individual pieces are relatively short, but the nodes are likely to

Figure 8.12: Downland gridshell, Chichester, UK, 2002, with (top left) western red cedar cladding, (bottom left) housing a workshop and (right) detail of connectors

the first two orthogonal layers can be laid out and bent to shape, with the further layers positioned over them (e.g. the Savill Building, page 88 and Fig. 8.1).

8.7 Gridshell construction

At this point, it is clear that the construction method of the shell, whether strained or unstrained, plays a major part in producing its stable spatially curved shape. The unstrained technique enables off-site fabrication, which reduces time on site. The penalty is increased cost in manufacturing different customized nodes and node-to-spline connections. The strained technique uses a series of simple identical components but extends time on site. Often long site duration is unacceptable to the client: whilst work proceeds on the gridshell, progress on other items is on hold. On the Downland gridshell project, the main contractor left site for four months, whilst the carpenter worked on forming the shell roof. In this project, built for the Weald and Downland Open Air Museum, the long site construction programme was anticipated and accommodated. A viewing platform was made and the building construction became a temporary museum exhibit. The strained construction technique is often not viable if the duration of the site phase of construction needs to be minimized.

The forming process for the Downland gridshell project, shown in Figure 8.13 started with the scaffold construction at the level of the shell hump tops, onto which the flat mat of splines was laid out. This construction technique is in contrast with the Multihalle which used a low-level layout of the flat grid, at around 1m from floor level (see Chapters 12 and 19). This alternative bottom-up technique avoids scaffolding, but requires the rather precarious use of scaffold towers and forklift trucks. Although more scaffolding is needed in laying out the mat at

Figure 8.13 The lowering process for the Downland gridshell

high level in the top-down technique, this method is beneficial in a number of ways. First, the shell's perimeter is constructed at both the sides and the ends. Having the perimeter in place means that there is a clear destination to work towards. Second, it is clear that, as the shell is manipulated into shape, it does so under gravity. Third, in modern construction proper access platforms are needed to work safely and effectively at height. These platforms are in place from the start. After the lowering process and spatial adjustment to match the form-found shape, the gridshell is fixed at its supports and stiffened with a prestressed cable network.

8.8 Conclusion

In summary, the immediate value of using the DR process for the form finding and load analysis of reticulated shells comes from its ability to model variations in geometry and stiffness parameters as part of the design process. Very accurate geometrical information is provided for construction. However, physical modelling still has a place in the design process, either as initial sketch models or as final models to gain a real understanding of how the shell loses stability as the loads are increased.

Key concepts and terms

A **reticulated**, **grid-** or **lattice shell** is essentially a shell with its structure concentrated into individual members in a relatively fine grid compared to the overall dimensions of the structure. The wording 'lattice', 'reticulated' and 'grid'-shell are largely interchangeable.

A **strained gridshell** is a curved structural surface, made of strong yet flexible laths. These laths are combined in a grid and bent. This initial bending action strains the shell, hence the term 'strained' gridshell.

An **unstrained gridshell** is a curved structural grid surface that is, in its initial state, stress-free (apart from stresses due to self-weight).

A **spline** is, in this context, an initially straight member that is bent into a spatial, continuous curve. The word 'spline' originally denoted the flexible wooden or metal strip draughtsmen and women used to draw smooth ship lines and railway curves.

Dynamic relaxation (**DR**) was invented by Alistair Day in 1965 and is a numerical procedure that solves a set of nonlinear equations. Summarized, the technique traces the motion of the structure through time under applied loading. The technique is effectively the same as the leapfrog and Verlet methods, which are also used to integrate Newton's second law through time.

Exercises

- The Mannheim Multihalle gridshell (see Sections 4.3 and 12.4, and 19.1) and the Dutch National Maritime Museum (see Chapter 2) are both form-found gridshells. Compare and contrast these shells from a structural and constructional perspective.
- How would you approach the form finding of an unstrained gridshell such as the Dutch Maritime Museum cupola in the dynamic relaxation process?
- Form found shapes are highly dependent upon their mesh topology. Use DR to find the form of the standard grid (Fig. 6.12), only supported at its four corners. Generate another input grid with the same connectivity but rotated 45° and find the shape. What can you say about the two shapes obtained?
- The Bridgetown Office of Planning judges that the proposed pavilion developed in this chapter is too high for its urban context. How can you generate a lower shell?

Further reading

- 'Tensegrity spline beam and gridshell structures', Adriaenssens and Barnes (2001). This seminal journal paper gives the formulation of the spline element that accounts for bending.
- 'A novel torsion/bending element for dynamic relaxation modeling', Barnes et al. (2013). This journal paper expands the capabilities of the method presented in this chapter by presenting and validating a formulation that accounts for torsion and transverse moments of spline elements in dynamic relaxation.
- 'Design and construction of the Downland gridshell', Harris et al. (2003). This paper, written by the design engineers, provides a comprehensive discussion of the top-down construction of the Downland gridshell.
- 'Timber lattice roof for the Mannheim Bundesgartenschau', Happold and Liddell (1975). This must be the most comprehensive discussion on the engineering design of gridshells. The paper discusses how the structure was modelled and tested both physically and mathematically and how the models were used to determine the construction details.

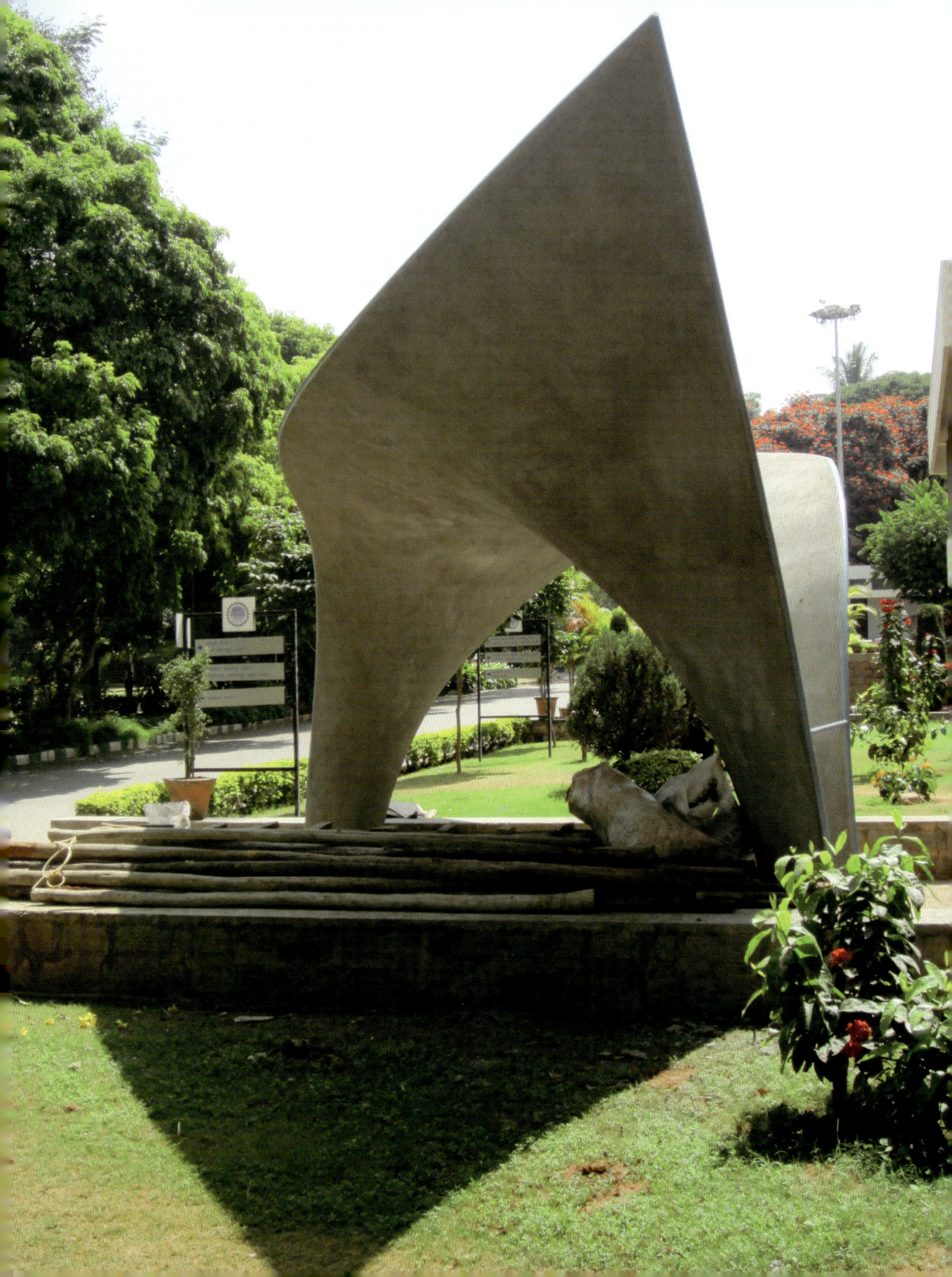

CHAPTER NINE

Particle-spring systems

Design of a cantilevering concrete shell

Shajay Bhooshan, Diederik Veenendaal and Philippe Block

LEARNING OBJECTIVES

- Summarize how particle-spring systems work.
- Compare explicit and implicit solving methods.
- Discuss the impact of using subdivision surfaces for form finding.
- Use particle-spring systems for the form finding of continuous shells, and discuss the influence of 'hanging' versus 'stretched' models.

Physical form finding using hanging chains and associated architectural design methods, as exploited by Antoni Gaudí, Heinz Isler, Frei Otto and others, are very well-established tools for the shape generation of form-active and form-passive structures, appreciated by architects and engineers alike (see Chapter 4). Their digital simulation using particle-spring simulation frameworks is also fairly established in practice following the work of early (architectural) exponents, Kilian and Ochsendorf (2005). Design methods employed in architectural practice for the formal and spatial development of geometry rely on iterative processes carried out within a relatively short timespan. Both properties are found in Particle-Spring (PS) systems, making them ideally suited to architectural design. Indeed, many algorithms and digital tools for form finding in architecture use a particle-spring framework to simulate hanging or pretensioned chains and grids.

The workflow presented here responds to the ambitions and complexities of scale, time constraints and delivery mechanisms of contemporary architectural and engineering practices. A simulation-based workflow is presented that provides intuitive control for the designer and that incorporates constraints of the production process. This approach adopts subdivision surfaces for parameterization and particle-spring systems for form finding. It resolves the dichotomy between lower resolution CAD geometry used for design and modelling and higher resolution geometries used for subsequent simulation and analysis.

First, we discuss how the definition of low-resolution initial conditions are refined using subdivision surfaces, offering the designer better control of the topology. Second, the resulting mesh is then used for the particle-spring form finding. A comprehensive mathematical description of particle-spring systems for structural design is given, including the use of explicit and implicit integration to solve for static equilibrium. Finally, we offer some details about subsequent fabrication and construction.

The brief

A canopy structure is to mark the entrance of the Computer Science department of the BMS College of Engineering in Bangalore, India. A two-week time frame for the design and execution, limited budget (c. £2,000), and the labour-intensive building economy of India are the most significant context and constraints for the project. As designers, we opt to explore the locally available skills in both tailoring of cloth and the use of Ferro-cement. We envision a shell structure built using a fabric, stressed within a frame of pre-bent steel pipe edges, acting as a guidework for a reinforcement net and wire mesh, with hand-rendered cement. The entrance area is 5.2m wide and 9.2m long. The supports need to be placed within the space.

For our brief, a wide range of low-poly meshes is subdivided to explore the design space, based on different boundary conditions. Because we wish to make use of an economic, lightweight fabric guidework and because prestressed fabrics can only accommodate anticlastic surfaces, the shell itself must be anticlastic. Furthermore, the resulting edges are used as fixed boundary conditions for the fabric, and thus also for form finding. The particle-spring method is perfectly capable of generating hanging models with loads and can be applied to other problems as well, but within the context of our brief and our chosen construction method, the form finding is limited to finding a 'stretched' surface without any loads applied. The anticlastic shell will likely feature bending to a degree dependent on the stiffness of the steel pipes, as well as the shape, weight and thickness of the shell. Considering the extreme thinness and large spans Félix Candela was able to achieve with his anticlastic hyperbolic paraboloids (see Chapter 20), the degree of bending might be a minor concern. Nonetheless, some structural analysis might be prudent to check whether the reinforced concrete has sufficient capacity.

9.1 Initial conditions

The designer usually models a structure with a predominantly quadrilateral, low-poly mesh. Designers find low-poly meshes easy to manipulate as a low-resolution 'cage' controlling a higher resolution geometry, allowing them to make global changes to form with minimal effort. Due to the malleable nature of the low-poly mesh, iterative design studies become intuitive and easy to the designer. These iterative studies are an integral part of the evolution of the design intent whilst working with simulation tools. Eventually, the low-poly mesh embeds key features of design intent including clearances, touch-down points and boundary conditions. In addition, the low-poly mesh provides control over topological features such as placement of (curvature) singularity points and holes. Control over their placement enables the designer to place these with regard to the overall tessellation of the initial surface.

9.2 Subdivision surfaces

The low-poly mesh is converted into a higher resolution mesh (high-poly mesh) using a Catmull-Clark subdivision algorithm (see Appendix D). During this process, the original set of faces and edges is subdivided, such that for each face a face point is added, and for each edge an edge point is added. Their position is an average of the neighbouring points. These increased numbers of faces and edges are easily tracked on the high-poly mesh due to the exponential and algebraic relation of their numeric identifiers to those in the original set. Further, since these edges originally described closed (face) outlines, they can be used to describe boundary NURBS patches, which are subsequently exported to downstream applications and processes. Figure 9.1 shows several design variations resulting from form finding our subdivided meshes.

Figure 9.2 shows, for the first design of Figure 9.1a, the four meshes appearing in the workflow: a low-poly mesh, a high-poly, subdivided mesh, the resulting form-found mesh and the extracted NURBS patches.

Before selecting a specific design, the next section explains in detail how particle-spring form finding works.

9.3 Particle-spring method

The principal purpose of the particle-spring method is to find structures in static equilibrium. This objective is achieved by defining the topology of a particle-spring network with loads on the particles, the masses of the particles, the stiffnesses and lengths of the

CHAPTER NINE: PARTICLE-SPRING SYSTEMS   **105**

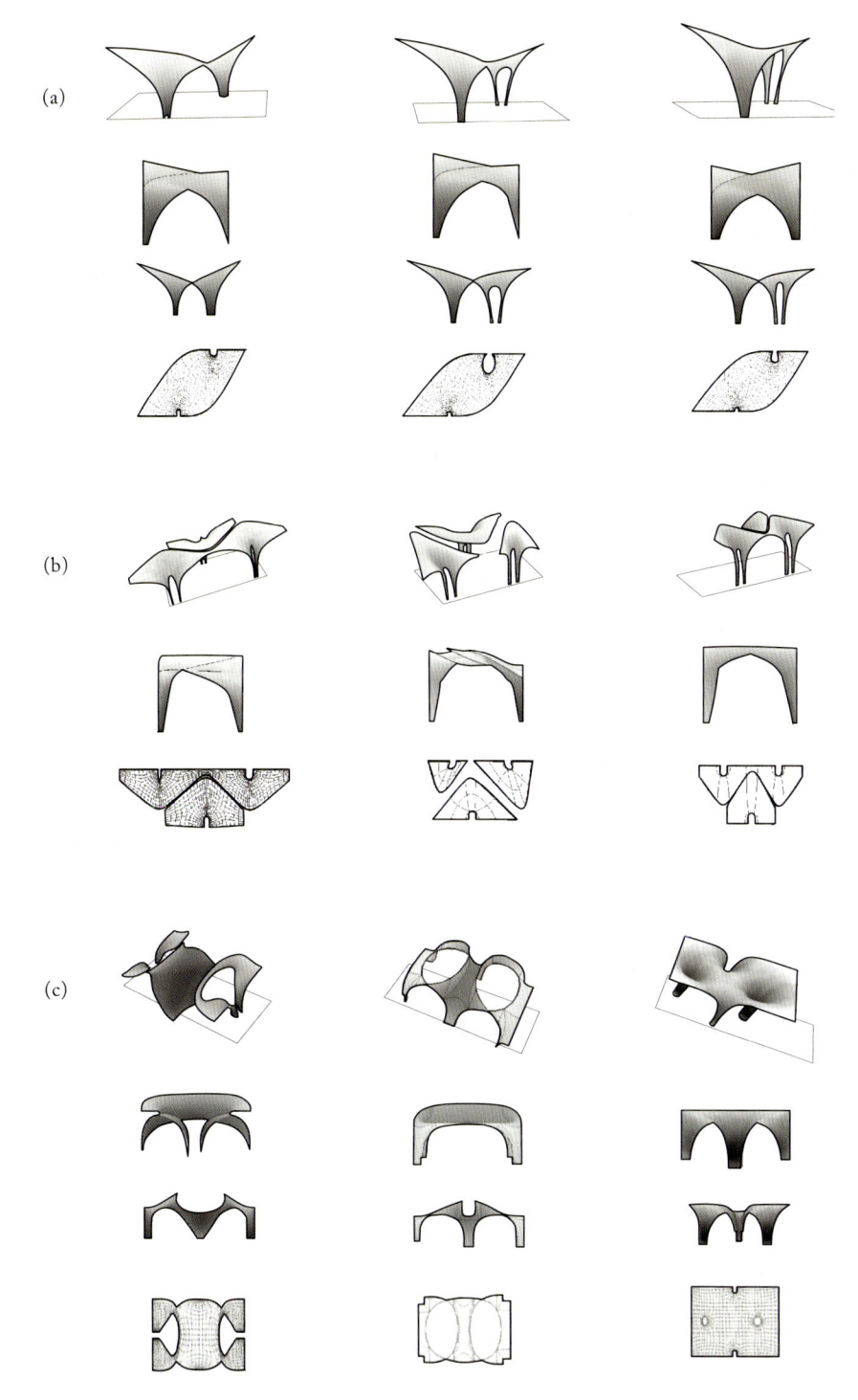

Figure 9.1 Design variations from PS model for the Bangalore shell, for (a) two, (b) three or (c) four support locations

springs, and then by attempting to equalize the sum of all forces in this system. For instance, the gravitational pull on a mass causes the displacement of the associated particle and subsequently the elongation of the attached springs. This elongation creates a counter force in the springs and stretching continues until the sum of the spring forces matches the downward force of the mass. The motion of the particle is governed by

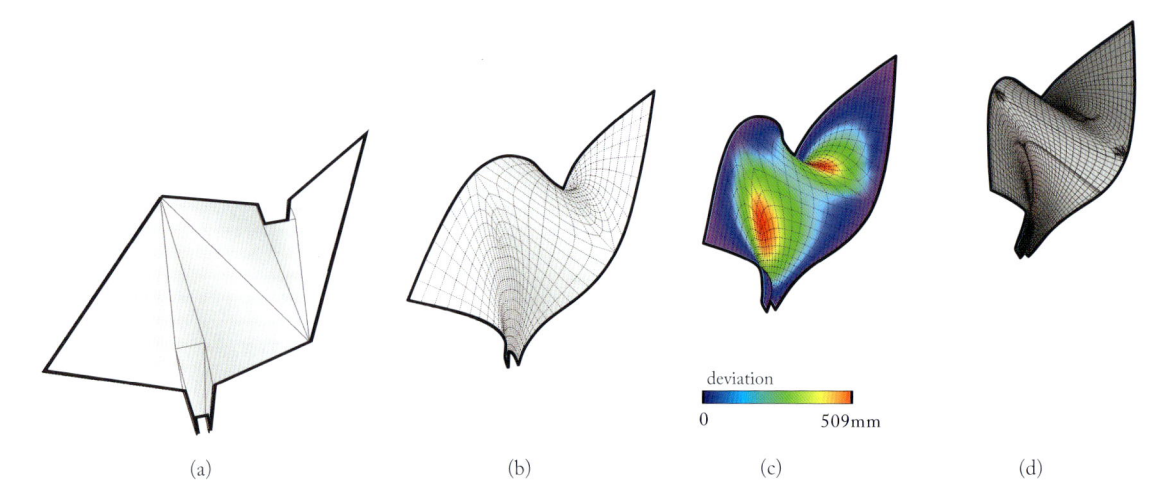

Figure 9.2 (a) Low-poly mesh showing design intentions, (b) high-resolution subdivided mesh, (c) form-found result with deviations from subdivided mesh and (d) extracted NURBS patches

Newton's second law of motion, and the force in the spring by Hooke's law of elasticity. The following are essential assumptions made within a PS form-finding framework:

- Surfaces are discretized into points and lines. The points are nodes with mass and the lines are springs connecting them.
- Upon applying forces, each node is either free to move or fixed in each direction (between zero and three degrees of freedom, corresponding to between three and zero orthogonal reaction forces).
- The internal forces (exerted by springs connecting the node to other nodes) and external forces (gravity and applied loads) act on the nodes. The result of all such interactions between nodes and springs iteratively leads to a balance of forces on each node and overall to an equilibrium shape.

9.3.1 Forces

The physics-based simulation starts from Hooke's law of elasticity,

$$\mathbf{f} = k \cdot \mathbf{e} = k \cdot (\mathbf{l} - \mathbf{l}_0), \quad (9.1)$$

where $\mathbf{f}$ are the forces, k is a stiffness constant, and $\mathbf{e}$ are the elongations, or the extent of stretching from initial, or rest lengths $\mathbf{l}_0$ to the current lengths $\mathbf{l}$. Thus, the particle-spring method is used to simulate the deformation of bodies.

For a system of m springs and n particles, the m spring forces in equation (9.1) can be expressed as internal force densities of each spring connecting two particles, such that

$$\mathbf{q}_e = k \mathbf{L}^{-1}(\mathbf{l} - \mathbf{l}_0), \quad (9.2)$$

where k is the spring stiffness, $\mathbf{L}$ and $\mathbf{l}$ the diagonal matrix and vector of the current lengths, and $\mathbf{l}_0$ the initial lengths. The subscript e refers to m force densities $\mathbf{q}$ belonging to elastic forces.

Equations (6.28) and (8.8) describe static equilibrium in a network, with n particles and m springs, and the residual forces per node in the x-direction respectively. We define the n residual forces $\mathbf{r}$ in our particle-spring network as the sum of internal spring and damping forces, and external loads. As a function of the particles' positions $\mathbf{x}$ and velocities $\mathbf{v}$, and introducing damping force densities $\mathbf{q}_d$, the residuals

$$\mathbf{r}(\mathbf{x}, \mathbf{v}) = \mathbf{C}_N^T \mathbf{U} \mathbf{q}_e + \mathbf{C}_N^T \mathbf{U} \mathbf{q}_d + \mathbf{p}_x, \quad (9.3)$$

where $\mathbf{C}_N$ is the $m \times n$ branch-node matrix (see Section 6.4.1), $\mathbf{U}$ is the diagonal matrix of m coordinates differences in the x-direction, $\mathbf{p}_x$ are the n

external loads acting on the particles, and $\mathbf{q}_d$ are the m force densities due to damping forces. Note that, in the z-direction, the loads $\mathbf{p}_z = \mathbf{m}g$, and depend on the given masses $\mathbf{m}$ and gravitational constant g. The damping force densities

$$\mathbf{q}_d = d\mathbf{L}^{-1}(\mathbf{L}^{-1}\mathbf{U})\mathbf{C}_N\mathbf{v}, \tag{9.4}$$

where d is a given damping coefficient, $\mathbf{L}^{-1}\mathbf{U}$ are the direction cosines, and $\mathbf{v}$ are the particle velocities. The relative velocity of two connected particles is $\mathbf{C}_N\mathbf{v}$.

9.3.2 Motion

To find equilibrium, the method starts from Newton's second law of motion, where for each particle i, the unknown acceleration a_i is a function of the particle's mass m_i and a net force. In a steady-state solution, the net force has to be zero, meaning that any non-zero value of force during form finding is called a residual force r_i. For the entire network with n particles, in the x-direction, we write Newton's second law, using a diagonal mass matrix $\mathbf{M}$, the residual force vector $\mathbf{r}$ and vector of accelerations $\mathbf{a}$ as

$$\mathbf{a} = \mathbf{M}^{-1}\mathbf{r}(\mathbf{x}, \mathbf{v}), \tag{9.5}$$

which is similar to equation (8.3), but formulated for the entire network rather than a single node.

9.3.3 Explicit integration

Given a known initial position $\mathbf{x}(t)$ and velocity $\mathbf{v}(t)$ of the system at time t, our goal is to determine a new position $\mathbf{x}(t + \Delta t)$ and system's velocity $\mathbf{v}(t + \Delta t)$ at time $t + \Delta t$. We then write

$$\frac{d}{dt}\begin{pmatrix}\mathbf{v}\\\mathbf{x}\end{pmatrix} = \begin{pmatrix}\mathbf{M}^{-1}\mathbf{r}(\mathbf{x}, \mathbf{v})\\\mathbf{v}\end{pmatrix}. \tag{9.6}$$

To simplify notation, we define $\mathbf{v}_t := \mathbf{v}(t)$ and $\mathbf{x}_t := \mathbf{x}(t)$. Furthermore, we use the forward difference form $\Delta\mathbf{v} = \mathbf{v}_{t+\Delta t} - \mathbf{v}_t$ and $\Delta\mathbf{x} = \mathbf{x}_{t+\Delta t} - \mathbf{x}_t$. The explicit *forward* Euler method applied to equation (9.6) approximates $\Delta\mathbf{v}$ and $\Delta\mathbf{x}$ as

$$\begin{pmatrix}\Delta\mathbf{v}\\\Delta\mathbf{x}\end{pmatrix} = \Delta t \begin{pmatrix}\mathbf{M}^{-1}\mathbf{r}(\mathbf{x}_t, \mathbf{v}_t)\\\mathbf{v}_t\end{pmatrix}, \tag{9.7}$$

or, solving for the actual velocity and position,

$$\begin{pmatrix}\mathbf{v}_{t+\Delta t}\\\mathbf{x}_{t+\Delta t}\end{pmatrix} = \begin{pmatrix}\mathbf{v}_t + \Delta t \mathbf{M}^{-1}\mathbf{r}(\mathbf{x}_t, \mathbf{v}_t)\\\mathbf{x}_t + \Delta t \mathbf{v}_t\end{pmatrix}. \tag{9.8}$$

More commonly, the slightly different, *semi-explicit* Euler method is used, where

$$\begin{pmatrix}\mathbf{v}_{t+\Delta t}\\\mathbf{x}_{t+\Delta t}\end{pmatrix} = \begin{pmatrix}\mathbf{v}_t + \Delta t \mathbf{M}^{-1}\mathbf{r}(\mathbf{x}_t, \mathbf{v}_t)\\\mathbf{x}_t + \Delta t \mathbf{v}_{t+\Delta t}\end{pmatrix}. \tag{9.9}$$

The step size Δt must be small enough to ensure stability when using this method. Note that equation (9.9) is equivalent to equations (8.6–8.7) for Dynamic Relaxation (DR). A minor distinction between the two is that DR states the velocity in central difference form $\Delta\mathbf{v} = \mathbf{v}_{t+\Delta t/2} - \mathbf{v}_{t-\Delta t/2}$, rather than forward difference form.

In this chapter, examples of explicit integration are generated using the midpoint method, which starts to calculate the acceleration, thus the residual forces $\mathbf{r}$, at the intermediate time $t + \Delta t/2$ and position $\mathbf{x}_{t+\Delta t/2} = \mathbf{x}_t + \frac{1}{2}\Delta t\mathbf{v}_t$, to then solve

$$\begin{pmatrix}\mathbf{v}_{t+\Delta t}\\\mathbf{x}_{t+\Delta t}\end{pmatrix} = \begin{pmatrix}\mathbf{v}_t + \Delta t \mathbf{M}^{-1}\mathbf{r}(\mathbf{x}_{t+\Delta t/2}, \mathbf{v}_{t+\Delta t/2})\\\mathbf{x}_t + \frac{1}{2}\Delta t(\mathbf{v}_t + \mathbf{v}_{t+\Delta t})\end{pmatrix}. \tag{9.10}$$

This approach requires the calculation of forces or positions twice per iteration though, but is more stable than Euler integration, allowing larger time steps.

9.3.4 Implicit integration

The use of implicit rather than explicit solvers in particle-spring models was presented by Baraff and Witkin (1998) and later adopted by Kilian and Ochsendorf (2005) for application to structural form finding. The implicit *backward* Euler method approximates $\Delta\mathbf{v}$ and $\Delta\mathbf{x}$ by

$$\begin{pmatrix}\Delta\mathbf{v}\\\Delta\mathbf{x}\end{pmatrix} = \Delta t \begin{pmatrix}\mathbf{M}^{-1}\mathbf{r}(\mathbf{x}_t + \Delta\mathbf{x}, \mathbf{v}_t + \Delta\mathbf{v})\\\mathbf{v}_t + \Delta\mathbf{v}\end{pmatrix}. \tag{9.11}$$

This formulation appears similar to equation (9.7) for the explicit method, but the difference between the two is that the forward Euler method's step is based solely on conditions at time t while the backward Euler method's step is written in terms of conditions at the end of the step itself, at time $t + \Delta t$. The forward method requires only an evaluation of the function $\mathbf{r}$, but the backward method requires that we solve for values of $\Delta \mathbf{x}$ and $\Delta \mathbf{v}$ that satisfy equation (9.11). Equation (9.11) is nonlinear. Rather than solving this equation exactly (which would require iteration), a Taylor series expansion is applied to $\mathbf{r}$ and the first-order approximation (i.e. Newton–Rhapson's method) is used:

$$\mathbf{r}(\mathbf{x}_t + \Delta \mathbf{x}, \mathbf{v}_t + \Delta \mathbf{v}) = \mathbf{r}_t + \frac{\partial \mathbf{r}}{\partial \mathbf{x}} \Delta \mathbf{x} + \frac{\partial \mathbf{r}}{\partial \mathbf{v}} \Delta \mathbf{v}. \quad (9.12)$$

So, for implicit integration, we now require the derivatives $\partial \mathbf{r}/\partial \mathbf{x}$ and $\partial \mathbf{r}/\partial \mathbf{v}$, which are evaluated for the state $(\mathbf{x}_t, \mathbf{v}_t)$. These derivatives are non-diagonal matrices, with every off-diagonal entry giving the dependency of two nodes; they can no longer be calculated independently. This is why particle-spring simulation is more conveniently expressed here for the entire network using matrix notation, and no longer per element or node, as was the case in DR (Chapter 8).

Substituting the Taylor approximation, as well as $\Delta \mathbf{x} = \Delta t(\mathbf{v}_t + \Delta \mathbf{v})$ into equation (9.11), then reordering, yields the linear system

$$\left(\mathbf{I}_n - \Delta t \mathbf{M}^{-1} \frac{\partial r}{\partial v} - \Delta t^2 \mathbf{M}^{-1} \frac{\partial r}{\partial x} \right) \Delta \mathbf{v} =$$
$$\Delta t \mathbf{M}^{-1} \left(\mathbf{r}_t + \Delta t \frac{\partial \mathbf{r}}{\partial \mathbf{x}} \mathbf{v}_t \right), \quad (9.13)$$

where $\mathbf{I}_n$ denotes an $n \times n$ identity matrix. This equation is then solved for $\Delta \mathbf{v}$. Provided any constraints are dealt with procedurally (i.e. enforced at each step), equation (9.13) can be transformed to a symmetric, positive definite system by multiplying the entire equation by $\mathbf{M}$, so that

$$\left(\mathbf{M} - \Delta t \frac{\partial \mathbf{r}}{\partial v} - \Delta t^2 \frac{\partial \mathbf{r}}{\partial x} \right) \Delta \mathbf{v} = \Delta t \left(\mathbf{r}_t + \Delta t \frac{\partial r}{\partial x} \mathbf{v}_t \right), \quad (9.14)$$

which is simply a system of linear equations, so of the form $\mathbf{Ax} = \mathbf{b}$.

After solving for $\Delta \mathbf{v}$, we then simply compute $\Delta \mathbf{x} = \Delta t(\mathbf{v}_t + \Delta \mathbf{v})$. At this point, we can introduce drag to influence motion and thus convergence. Whereas damping acts internally on the springs, drag is an external effect on the particles. Adding a drag coefficient b, this becomes

$$\mathbf{x}_{t+\Delta t} = \mathbf{x}_t + \Delta t(1-b)(\mathbf{v}_t + \Delta \mathbf{v}) \quad (9.15)$$

Thus, the backward Euler step consists of evaluating $\mathbf{r}_t$, $\partial \mathbf{r}/\partial \mathbf{x}$ and $\partial \mathbf{r}/\partial \mathbf{v}$; forming the system (9.14); solving the system for $\Delta \mathbf{v}$; and then updating $\mathbf{x}$ and $\mathbf{v}$.

Given a step size Δt and masses $\mathbf{M}$, we now seek the derivatives, $\partial \mathbf{r}/\partial \mathbf{x}$ and $\partial \mathbf{r}/\partial \mathbf{v}$, in our system, as explained in the next section.

9.3.5 Derivatives

For the derivative, or Jacobian matrix, $\partial \mathbf{r}/\partial \mathbf{x}$, we first note that

$$\frac{\partial \mathbf{q}_e}{\partial \mathbf{l}} = k\mathbf{L}^{-1} - k\mathbf{L}^{-2}(\mathbf{L} - \mathbf{L}_0) = k\mathbf{L}^{-1} - \mathbf{L}^{-1}\mathbf{Q}_e \quad (9.16)$$

and from equation (6.26) infer that

$$\frac{\partial \mathbf{l}}{\partial \mathbf{u}} = \mathbf{L}^{-1}\mathbf{U} \text{ and } \frac{\partial \mathbf{u}}{\partial \mathbf{x}} = \mathbf{C}_N. \quad (9.17)$$

Assuming that the derivative of damping forces w.r.t. $\mathbf{x}$ can be neglected (Choi and Ko, 2002), and that loads $\mathbf{p}$ are independent of $\mathbf{x}$, the Jacobian matrix is

$$\frac{\partial \mathbf{r}(\mathbf{x}, \mathbf{v})}{\partial \mathbf{x}} = \frac{\mathbf{C}_N^T \mathbf{U} \mathbf{q}_e}{\partial \mathbf{x}} = \mathbf{C}_N^T \frac{\partial \mathbf{U} \mathbf{q}_e}{\partial \mathbf{u}} \frac{\partial \mathbf{u}}{\partial \mathbf{x}}$$
$$= \mathbf{C}_N^T \frac{\partial \mathbf{U} \mathbf{q}_e}{\partial \mathbf{u}} \mathbf{C}_N,$$

then using the product rule of differentiation,

$$= \mathbf{C}_N^T (\mathbf{U} \frac{\partial \mathbf{q}_e}{\partial u} + \mathbf{Q}_e \frac{\partial \mathbf{u}}{\partial \mathbf{u}}) \mathbf{C}_N,$$
$$= \mathbf{C}_N^T (\mathbf{U} \frac{\partial \mathbf{q}_e}{\partial \mathbf{l}} \frac{\partial \mathbf{l}}{\partial \mathbf{u}} + \mathbf{Q}_e) \mathbf{C}_N,$$
$$= k\mathbf{C}_N^T \mathbf{U}^2 \mathbf{L}^{-2} \mathbf{C}_N$$
$$- \mathbf{C}_N^T \mathbf{U}^2 \mathbf{L}^{-2} \mathbf{Q}_e \mathbf{C}_N + \mathbf{C}_N^T \mathbf{Q}_e \mathbf{C}_N. \quad (9.18)$$

The derivative w.r.t. velocities $\mathbf{v}$, $\partial \mathbf{r}/\partial \mathbf{v}$, is often defined in structural form finding with a simple, yet stable expression (Choi and Ko, 2002),

$$\frac{\partial \mathbf{r}(\mathbf{x}, \mathbf{v})}{\partial \mathbf{v}} = -d\mathbf{I}. \quad (9.19)$$

9.3.6 Solving method

We can now, for each direction, construct the system of linear equations (9.14) of the form $\mathbf{Ax}=\mathbf{b}$; the left-hand side matrix $\mathbf{A}$ with the derivatives (9.18) and (9.19), and the right-hand side vector $\mathbf{b}$ using equation (9.3).

For small systems, a direct solving method such as Gaussian elimination or Cholesky decomposition suffices. For larger systems, typical implementations as those mentioned in Kilian and Ochsendorf (2005) use iterative methods such as the Conjugate Gradient Method (CG) or the Biconjugate Gradient Stabilized Method (BiCGSTAB). Given that the matrix $\mathbf{A}$ is symmetric, positive definite, CG is the most appropriate (Barrett et al., 1994) and employed by Baraff and Witkin (1998). BiCGSTAB, however, is mostly used for non-symmetric matrices, and would therefore constitute unnecessary overhead in our case.

Alternatively, one can solve the explicit system (9.9) using Euler's method, or increasingly higher order methods such as leapfrog integration (Chapter 8), midpoint method according to equation (9.10) or the classic 4 order Runge-Kutta (RK4) (Chapter 10). Explicit integration does not require solving a system of linear equations at each iteration, since the matrix $\mathbf{M}^{-1}$ is diagonal and no Jacobian matrix is needed. In other words, the equations per iteration can be expressed for each node separately (as in DR). Therefore, it is easier and more straightforward to implement or start with an explicit method. Generally though, implicit methods are deemed more reliable, stable (the choice of damping parameters and size of the time step are less sensitive) and thus more suitable.

The entire approach is summarized in the flowchart in Figure 9.3.

9.4 Manipulation of the results

To explore possible designs for the canopy within the workflow, we have three primary tools at hand: the control of the boundary conditions, the meshing and its topology, and the parameters of the PS system (such as spring stiffness, particle mass).

Based on the boundary conditions, the Catmull-Clark subdivision algorithm used here subdivides a low-poly mesh to generate a denser, simulation mesh.

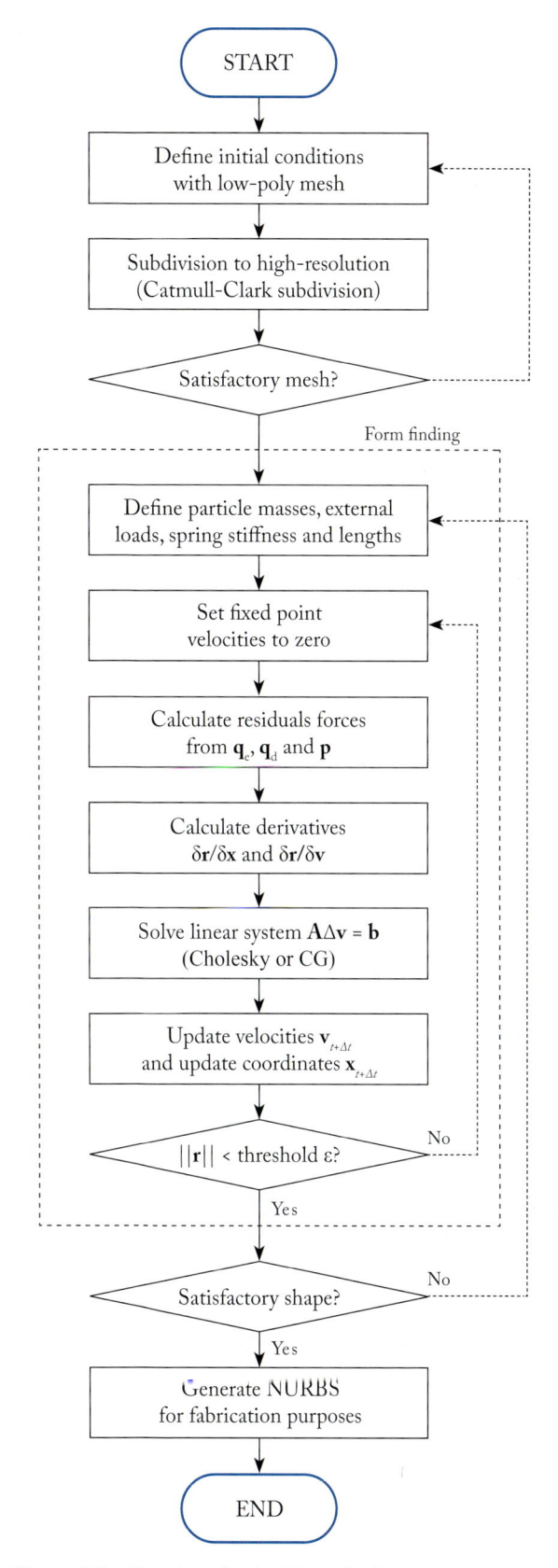

Figure 9.3 Flowchart for the PS method

Apart from topology, there are two sets of PS parameters that contribute significantly to the outcome: rest lengths l_0 and the inclusion or exclusion of gravity loads **p** within the simulation. These parameters determine whether a model is physically analogous to either hanging (high loads), or stretching (high stretch, that is, small rest lengths), a cloth.

In a stretched cloth simulation, applied to our brief, gravity is usually turned off or set to a very low value and the rest-lengths of all the springs are very low or set to zero. For such a 'zero-length spring', setting the spring stiffness becomes identical to prescribing a force density for a bar (i.e. it results in solutions identical to those from linear FDM). The simulation causes the surface to 'shrink' in area within the given boundary curve. In our examples, all vertices on the boundary are fixed and diagonal springs are not used. This type of simulation usually produces anticlastic geometries, so negative Gaussian curvature appears everywhere (Fig. 9.4a). It has to be noted that these surfaces appear similar to, but in fact are not, minimal surfaces (they are minimal squared length meshes).

In a hanging cloth simulation, the particles are allowed to 'fall' under the influence of gravity and the rest lengths of springs along the boundary edges are set to be equal to their original lengths. Further, additional diagonal springs with differential strengths might be added to ensure that faces do not distort significantly during simulation (they suggest some shear stiffness of the cloth). Usually, several points along the ground are fixed. This type of simulation often produces synclastic geometries, so (predominantly) positive Gaussian curvature (Fig. 9.4b).

9.5 Design development

As mentioned, our design must be of the 'stretched' kind since we apply a fabric guidework. Therefore, all rest lengths $l_0 = 0$ and external loads **p** = 0. Both the spring stiffnesses and particle masses were set to values of 1. Due to constraints of time and resources, the foundations for the structure need to be minimal. The first design from Figure 9.1a was selected for its relative simplicity in both form and boundary conditions. The design features two independent column footings 75cm × 75cm × 100cm (Fig. 9.5). The largest span of the canopy is 6m with 2m × 2m balanced cantilevers on either end of the symmetric structure.

Figure 9.4: (top) Stretched cloth models result in anticlastic shells (hypothetical metro station in Bangalore), while (bottom) hanging cloth models result in synclastic shells (design entry for innovation centre in China), both designs by Zaha Hadid Architects

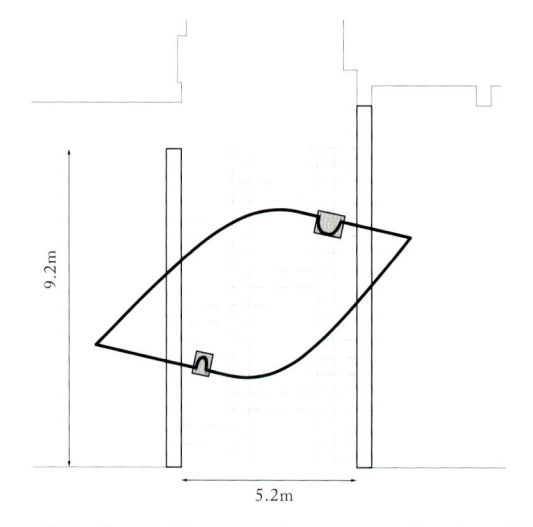

Figure 9.5 Plan and location of supports and outline of the final design

CHAPTER NINE: PARTICLE-SPRING SYSTEMS

Figure 9.6 Initial render of the design

The footprint of the prototype is 8m × 6m. The uniform shell thickness is 8cm. An initial render of the design is shown in Figure 9.6.

9.6 Fabrication

The entire construction took place within a span of two weeks, with tailoring of the fabric guidework taking only a day and a half to complete.

9.6.1 Boundary curves

The canopy's boundaries are pre-bent steel pipes (Fig. 9.7). To describe planar arcs and radii for bending, a simplified version of bi-tangent arc construction and manual reconstruction from given continuous three-dimensional NURBS edge curves was used. The boundary curves themselves are derived from the edges of the original subdivision limit surface. In this case, they provided fixed constraint points for the relaxation process.

Since the design workflow used a manipulation-friendly low-poly mesh, it proves to be convenient in ensuring that most of the boundary curves are planar and do not require a post-rationalization into arcs – if boundaries are modelled using only three vertices each, the resultant subdivided curve is always planar.

9.6.2 Cutting patterns

The geometry is decomposed into two-dimensional, planar patterns to produce the fabric guidework. Unfolding into cutting layouts is carried out in commercially available tools (Fig. 9.9). The seam lines themselves are established by dividing the form-found surface along curves of steepest descent. The resulting single piece of tailored fabric has to interface with other rigid elements only at the boundaries of the surface (Fig. 9.8). While stretching this piece of fabric, it is allowed to find a natural shape between the fixed boundaries, and any tolerance between the physical and the simulated geometry is accepted. However, compensation is necessary for the discrepancy in the unfolded surface area and doubly curved surface area as well as bias of the fabric used. This compensation is empirically established and an error-correcting directional scaling factor is used for the cutting patterns.

Figure 9.8 Cutting patterns from digital unfolding of the form-found surface

Figure 9.7 Placement of bent pipe edge curves

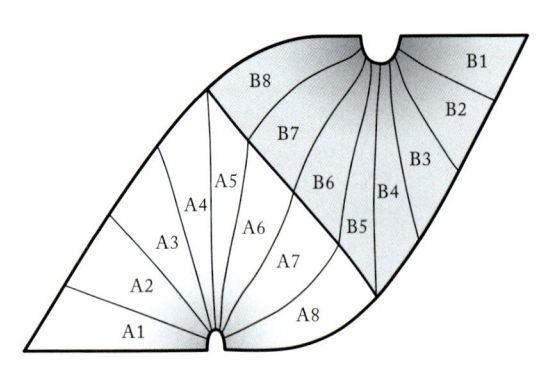

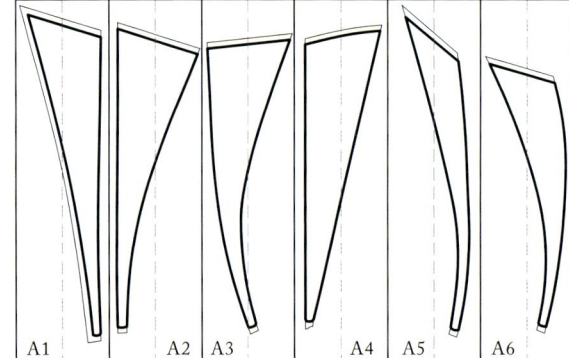

Figure 9.9 Installation of fabric guidework

9.6.3 Assembly

The pipes and the guidework have to be assembled before laying out the reinforcement steel and wire mesh. The fabric-formed and physically form-found surface with its seam lines are used as a guide to position the reinforcement steel bars (Fig. 9.10). Reinforcement consisted of 15mm bars, placed 150mm apart, and 10mm bars, at 75mm distances. The directions for placing these are established by interpolating between the tailored seam lines. Once the bars are held in place with temporary welds and ties, the mesh is stretched across the bars, and the concrete is hand-rendered onto the mesh from both sides (see page 102). This ensures that the fabric need not be prestressed to take the concrete load since the mesh and the rebar cage receives most of the load, and further concrete is manually applied in layers on either side of the surface. As such, the fabric is used very much as a physical way to describe complex geometry and not as fabric formwork for the concrete.

Figure 9.10 On-site fabric guidework with rebar in place

9.7 Conclusion

In summary, the described workflow is best suited for speedy exploration of design-parameter space with a qualitative understanding of structural behaviour, maintaining an awareness of downstream implications of design operations, and enabling a reasonable

Figure 9.11 (top) Mexico City shell with bent pipe edge curves and timber waffle formwork and (bottom) resulting concrete shell canopy

correspondence between high-resolution simulation geometry and lower resolution, manipulation-friendly, CAD geometries.

This chapter is based on the Hyperthreads project, completed in August 2011 as part of the AA Visiting School in India (see page 102). The workshop was tutored by the first author from Zaha Hadid Architects and their Computation and Design Group (ZHA|CODE), and was locally supported by Abhishek Bij, Design plus India, structural engineers CS Yadunandan and Deepak, and with construction coordination from BSB Architects India. The workflow was applied in the same year to another concrete shell canopy, built in Mexico City (Fig. 9.11).

Key concepts and terms

Particle-spring (**PS**) simulation finds steady-state equilibrium by defining a mesh topology for a network of lumped masses, called particles (nodes), and linear elastic springs (bars), and then by equalizing the sum of all forces in the system through motion. Out-of-balance forces arise due to the stiffness and geometric lengths of the springs and gravity loads acting on the particles. The solution is often obtained by higher order explicit, or by implicit integration.

Explicit integration is the numerical integration of differential equations based on known quantities (either initial values or from the last computed iteration n). The simplest method is Euler's method. Considering the Ordinary Differential Equation (ODE) $dy/dt = f(t, y)$, Euler's method is $y_{n+1} = y_n + h \cdot f(t_n, y_n)$, where h is the step size.

Implicit integration is integration based on new quantities. The simplest method is the backward Euler's method. For the same ODE, the method is $y_{n+1} = y_n + h \cdot f(t_{n+1}, y_{n+1})$. This equation cannot be solved immediately, thus requiring fixed-point iteration, Newton–Rhapson's method or similar, in each iteration.

Subdivision surfaces are a representation of a smooth surface via the specification of a coarser piecewise linear polygon mesh. The smooth surface is the limit of a recursive process of subdividing each polygonal face in the coarse mesh into smaller faces, with each recursion better approximating the smooth surface.

Exercises

- Consider the single-node problem in Chapter 6. Apply particle-spring to the same problem, and write the equations to solve for the central node P_0.
- Apply particle-spring simulation to the standard grid (Fig. 6.12) using an explicit integration method, such as semi-explicit Euler, leapfrog, or midpoint integration. What is the difference with dynamic relaxation, described in Chapter 8? Attempt the example again, but using the implicit backward Euler. Compare the speed, number of iterations and convergence of both types of integration. We can measure convergence as the norm of the residual forces at each iteration.
- Apply a four-node subdivision surface instead of our standard grid and refine it. Manipulate the four vertices of the low-poly mesh and vary the level of subdivision to change the results of our form-finding process.
- Manipulate the vertices such that they have different heights. Vary the rest lengths and particle loads and generate a few designs. Under what conditions do we obtain either synclastic or anticlastic shapes? What is the meaning of using either 'hanging' or 'stretched' shapes as the basis for a concrete shell?

Further reading

- 'Particle-spring systems for structural form finding', Kilian and Ochsendorf (2005). This seminal paper features the early application of particle-spring systems to structural form finding in architecture.
- 'Large steps in cloth animation', Baraff and Witkin (1998). This journal paper is widely cited within the field of cloth animation, and in particular for its application of implicit integration methods.
- 'Stable but responsive cloth', Choi and Ko (2002). This paper extends the work by Baraff and Witkin (1998) by including buckling phenomena, and is the source of the simple derivative w.r.t. velocities used in many PS implementations.
- 'Nucleus: Towards a unified dynamics solver for computer graphics', Stam (2009). This white paper describes the inner working of Maya Nucleus' particle-spring model, in particular how constraints are handled. In addition to our own implementations, the design in this chapter was developed using Nucleus.

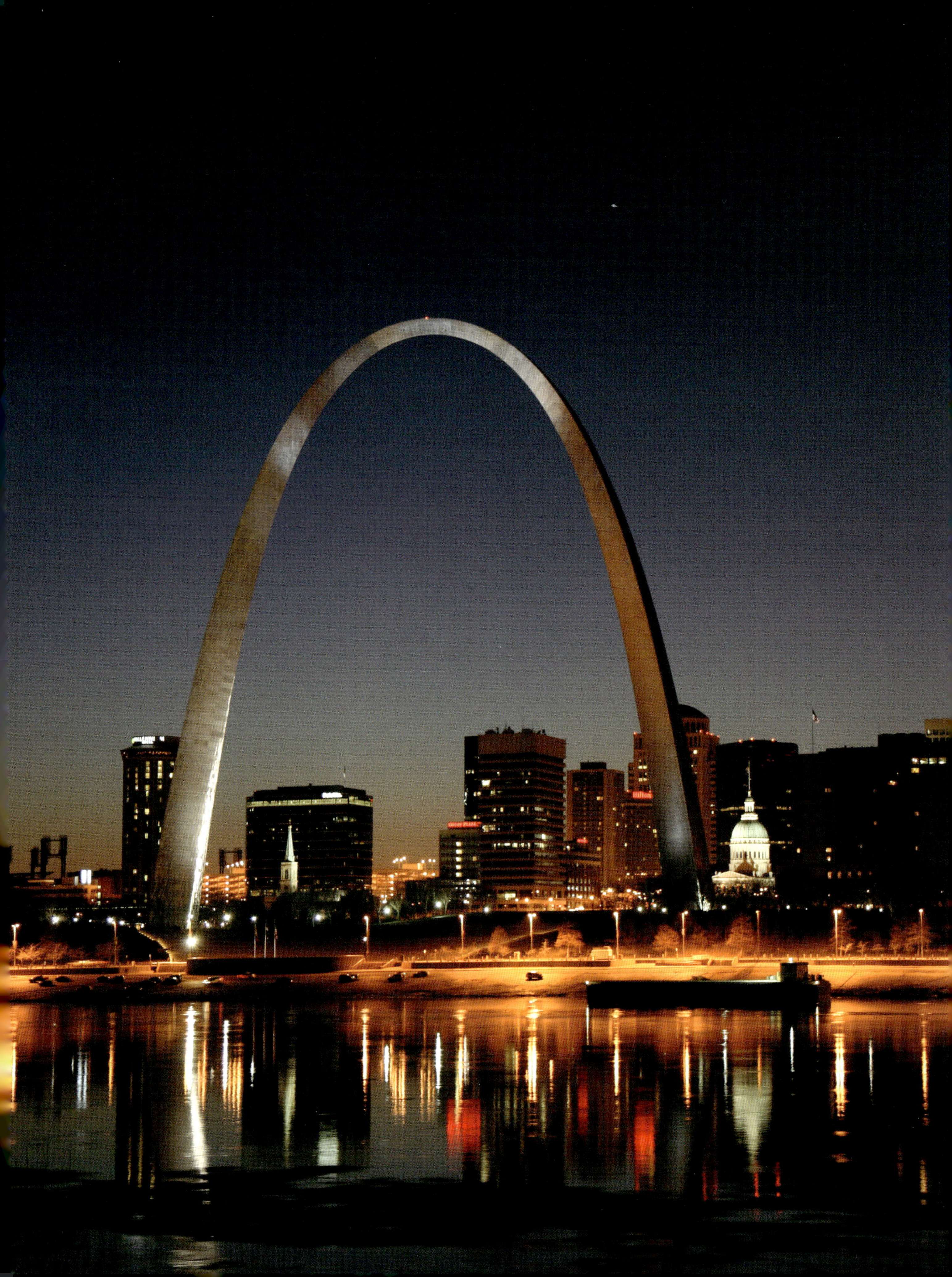

CHAPTER TEN
Comparison of form-finding methods

Diederik Veenendaal and Philippe Block

LEARNING OBJECTIVES

- Explain the different form-finding methods using a common algebraic notation and data structure.
- Apply these techniques to the same problems.
- Compare and contrast the different methods, based on the presentation of their mathematics and subsequent results.
- Discuss how to generate loads for different types of applications.

In the previous chapters, different numerical form-finding techniques have been presented to develop structurally efficient shapes for shells. The design examples might suggest that each method is suited to a particular application; the force density method to unstrained timber gridshells (Chapter 6), the thrust network analysis to unreinforced masonry shells (Chapter 7), dynamic relaxation to strained gridshells (Chapter 8), and the particle-spring systems to thin concrete shells (Chapter 9). This observation leads to the obvious question as to how these methods differ and whether they can be applied to other structural typologies. In this chapter we apply all these methods to one simple case using the same data structure and linear algebraic presentation, using branch-node matrices. This design exercise reveals similarities between the methods and provides clues as to how these techniques may be adapted to suit our purpose, why we may want to choose one over the other, or when the choice is arbitrary.

10.1 Form-finding families

Form-finding methods can be categorized in three main families:

- Stiffness matrix methods are based on using the standard elastic and geometric stiffness matrices. These methods are among the oldest form-finding methods, and are adapted from structural analysis; for example, finite element analysis.
- Geometric stiffness methods are material independent, with only a geometric stiffness. In several cases, starting with the Force Density Method (FDM), the ratio of force to length is a central unit in the mathematics. Later methods are often presented as generalizations or extensions of the FDM, independent of element type, often discussing prescription of (components of) forces rather than force densities; for example, Thrust Network Analysis (TNA).
- Dynamic equilibrium methods solve the problem of dynamic equilibrium to arrive at a steady-state

Family	Name		Year	Element type
Stiffness matrix	Natural Shape Finding	NSF	1974/1999	bar + surface (1992)
Geometric stiffness	Force Density Method	FDM	1971/1974	bar + surface (1995)
	Surface Stress Density Method	SSDM	1998/2004	surface
	Thrust Network Analysis	TNA	2007/2007	bar
	Updated Reference Strategy	URS	1999/2001	bar + surface (1999)
Dynamic equilibrium	Dynamic Relaxation	DR	1977/1977	bar + surface (1977)
	Particle-Spring system	PS	2005/2005	bar

Table 10.1 Taxonomy of form-finding methods for shells with year of first publication, and first publication in the context of shells

solution, equivalent to the static solution of static equilibrium; for example, Dynamic Relaxation (DR) and Particle-Spring (PS) systems.

Initially, numerical form-finding techniques were developed for form-active, prestressed systems such as cable-net roofs. For each form-finding method family, at least one technique has been developed for the generation of shell shapes. Table 10.1 lists these methods by name, abbreviation, first reference for the method and first reference applied to shells and whether these examples used bar or surface elements (line or triangle elements).

For references on triangle formulations for FDM and DR, and for additional reading on NSF, URS and SSDM applied to shells, see the suggested sources at the end of this chapter.

In the following sections, FDM, TNA, DR and PS will be applied to develop the shape of a simple shell. These methods use a few equivalent or analogous terms which the reader should be aware of, summarized in Table 10.2.

	FDM, TNA	DR	PS
q	force density	tension coefficient	N/A
l_0	N/A	initial length	rest length
m	branches	links	springs
n	nodes	nodes	particles

Table 10.2 Equivalent terminology in form-finding methods

10.2 A recipe for form-finding algorithms

Every form-finding procedure consists of at least the following parts:

1. A discretization to describe the (initial) geometry of the shell. The discretization can be made up of line elements, or surface elements such as triangles or quadrilaterals.
2. A data structure that stores the information on the form (geometry), connectivity of the discrete elements and forces within the shell.
3. Equilibrium equations that define the relationship between the internal and external forces. A shape resulting from form finding represents a system in static equilibrium. The internal and external forces add up to zero. Additional constraints might be placed on the equilibrium equations influencing how they can be solved numerically.
4. A solver, or integration scheme, which describes how the equilibrium equations are solved. If the system of equations is nonlinear, one typically tries to solve this system incrementally. The solver includes stopping criteria and means to measure convergence. Applicable solving methods may differ in how fast they converge or how stable they are, but assuming that they do converge, should result in the same solution if the problem and its boundary conditions are identical.

Note that for static form-finding methods, one generally uses Newton–Rhapson's method (to linearize nonlinear equations) to find static equilibrium and in each Newton–Rhapson iteration some direct (e.g.

Cholesky decomposition) or iterative (e.g. conjugate gradients) method to solve a system of linear equations. For dynamic methods, one looks towards various integration schemes, either explicit (e.g. leapfrog integration, classic fourth order Runge-Kutta (RK4)), or implicit (e.g. backward Euler) methods.

10.3 Example: a simple inverted hanging chain

The simplest example to apply form finding to, in the context of shell structures, is that of an inverted flexible line, hanging in pure tension under its own weight (see also Section 1.1).

Chapter 3 previously addressed analytical formulae for this problem. The 1965 Gateway Arch by Eero Saarinen (see page 114) was designed using an analytical formula (an inverted weighted/flattened catenary) rather than an actual hanging chain (whether self-weight is the appropriate design loading at 192m height is another matter). In this section, numerical methods are applied to the discretized form of the flexible line, a chain or catenary, to provide insight into how these methods work. In our example, the line is modelled as an initially horizontal line between simple supports, consisting of discrete line elements, and at each intermediate node, a vertical load is applied.

10.3.1 Geometry and data structure

The topology of the inverted hanging chain is described using the branch-node data structure (see also Section 6.4.1). The branch-node matrix captures the topology, or connectivity, of any bar-node network; it has rows, one for each of the m branches; and columns, one for each of the n nodes. So, it is a rectangular matrix with dimensions $m \times n$. Regardless of the method of form finding, we can use this or any type of data structure.

Here, the branch-node matrix is given directly for a simple straight line, which has been subdivided into six branches. Figure 10.1 shows this simple geometry, with nodes labelled using Arabic numerals, branches using Roman numerals. The end nodes 6 and 7 are considered fixed. The polyline is twelve units long and node 3 is at the origin $(x, y) = (0, 0)$. This model will be the initial geometry for our form-finding example. The direction of the branches can be chosen arbitrarily, but here they are chosen such that they go from nodes with a lower index to nodes with a higher index.

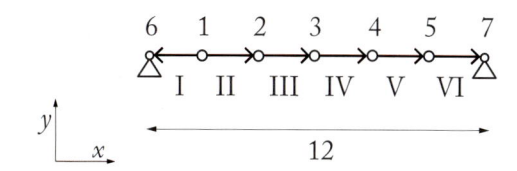

Figure 10.1 Geometry for a simple line

The branch-node matrix $\mathbf{C}$ consists of two sub-matrices, $\mathbf{C}_N$ and $\mathbf{C}_F$, for the free (or non-supported) and fixed nodes respectively:

$$\mathbf{C} = [\mathbf{C}_N \ \mathbf{C}_F] \quad (10.1)$$

For Figure 10.1, the branch-node matrix is

$$\mathbf{C} = \begin{bmatrix} 1 & . & . & . & . & -1 & . \\ 1 & -1 & . & . & . & . & . \\ . & 1 & -1 & . & . & . & . \\ . & . & 1 & -1 & . & . & . \\ . & . & . & 1 & -1 & . & . \\ . & . & . & . & 1 & . & -1 \end{bmatrix} \begin{matrix} I \\ II \\ III \\ IV \\ V \\ VI \end{matrix}$$

and the coordinate vectors

$$\mathbf{x} = [\mathbf{x}_N \ \mathbf{x}_F] = [-4 \ -2 \ 0 \ 2 \ 4 \ -6 \ 6]^T,$$

$$\mathbf{y} = [\mathbf{y}_N \ \mathbf{y}_F] = [0 \ 0 \ 0 \ 0 \ 0 \ 0 \ 0]^T.$$

The coordinate differences, vectors $\mathbf{u}$ and $\mathbf{v}$, for all m branches are a function of $\mathbf{C}$ and the coordinates $\mathbf{x}$ and $\mathbf{y}$:

$$\mathbf{u} = \mathbf{Cx}, \ \mathbf{v} = \mathbf{Cy}, \quad (10.2)$$

so that,

$$\mathbf{u} = \begin{bmatrix} 1 & . & . & . & . & -1 & . \\ 1 & -1 & . & . & . & . & . \\ . & 1 & -1 & . & . & . & . \\ . & . & 1 & -1 & . & . & . \\ . & . & . & 1 & -1 & . & . \\ . & . & . & . & 1 & . & -1 \end{bmatrix} \begin{bmatrix} -4 \\ -2 \\ 0 \\ 2 \\ 4 \\ -6 \\ 6 \end{bmatrix},$$

$$= [2 \ -2 \ -2 \ -2 \ -2 \ -2]^T$$

$$\mathbf{v} = [0 \ 0 \ 0 \ 0 \ 0 \ 0]^T.$$

With the diagonal $m \times m$ matrices $\mathbf{U}$ and $\mathbf{V}$ belonging to vectors $\mathbf{u}$ and $\mathbf{v}$, the branch lengths $\mathbf{L}$ can be calculated using the Pythagorean theorem:

$$\mathbf{L} = (\mathbf{U}^2 + \mathbf{V}^2)^{\frac{1}{2}}, \quad (10.3)$$

where the length vector $\mathbf{l}$ is the diagonal of matrix $\mathbf{L}$, so

$$\mathbf{l} = [2 \quad -2 \quad -2 \quad -2 \quad -2 \quad -2]^T.$$

At this point, all the initial geometric and topological information necessary for form finding is known: the coordinates, the lengths and the connectivity.

10.3.2 Force density method

The polyline in Figure 10.1 will now be considered as an inverted hanging chain. This means that it is subjected to gravitational forces, in the positive y-direction. When applying vertical loads $\mathbf{p}$ to the nodes of the chain, the flexible chain will not be in equilibrium in the configuration of Figure 10.1. The chain returns to a state of equilibrium if the sum of the internal and external forces on each free node equals zero. This is expressed in the following equilibrium equations,

$$\mathbf{p}_x - \mathbf{f}_x = 0,$$
$$\mathbf{p}_y - \mathbf{f}_y = 0, \quad (10.4)$$

where $\mathbf{p}_x$ and $\mathbf{p}_y$ are the horizontal and vertical component of the applied loading $\mathbf{p}$, and $\mathbf{f}_x$ and $\mathbf{f}_y$ the horizontal and vertical component of the internal forces of the bars on the nodes.

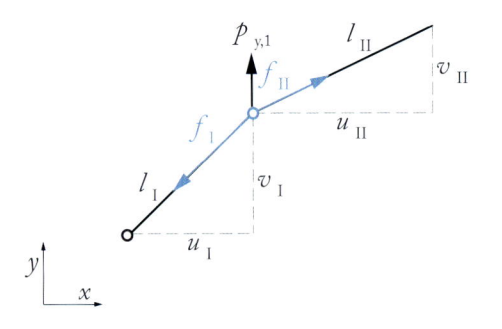

Figure 10.2 Equilibrium in node 1

The internal forces $\mathbf{f}_x$ and $\mathbf{f}_y$ on the interior nodes in equation (10.4) are calculated from the branch forces $\mathbf{f}$. From Figure 10.2, for example, it can be seen that for vertical equilibrium in node 1,

$$p_{x,1} = f_{x,1} = f_{x,I} + f_{x,II} = f_I \frac{u_I}{l_I} + f_{II} \frac{u_{II}}{l_{II}},$$
$$p_{y,1} = f_{y,1} = f_{y,I} + f_{y,II} = f_I \frac{v_I}{l_I} + f_{II} \frac{v_{II}}{l_{II}}. \quad (10.5)$$

The ratios v_I/l_I and v_{II}/l_{II} are also known as direction cosines. For the entire network, the branch forces $\mathbf{f}$ can be decomposed along x and y direction with the direction cosines, written as diagonal matrices $\mathbf{UL}^{-1}$ and $\mathbf{VL}^{-1}$, and then summed to the nodes with branch-node matrix $\mathbf{C}_N$. Equations (10.4) become

$$\mathbf{p}_x - \mathbf{f}_x = \mathbf{p}_x - \mathbf{C}_N^T \mathbf{UL}^{-1} \mathbf{f} = \mathbf{0},$$
$$\mathbf{p}_y - \mathbf{f}_y = \mathbf{p}_y - \mathbf{C}_N^T \mathbf{VL}^{-1} \mathbf{f} = \mathbf{0}. \quad (10.6)$$

The loads, prescribed only in the y-direction, representing gravitational loading,

$$\mathbf{p}_x = \mathbf{0}, \mathbf{p}_y = [1 \quad 1 \quad 1 \quad 1 \quad 1]^T. \quad (10.7)$$

The forces $\mathbf{f}$ in the branches are still unknown; FDM introduces the force-to-length ratios, or force densities, of the branches, $\mathbf{q}$, here set to 1, so

$$\mathbf{q} = \mathbf{L}^{-1} \mathbf{f} = [1 \quad 1 \quad 1 \quad 1 \quad 1 \quad 1]^T. \quad (10.8)$$

The equilibrium equations (10.4) can then be rewritten as

$$\mathbf{C}_N^T \mathbf{U} \mathbf{q} = \mathbf{C}_N^T \mathbf{Q} \mathbf{C} \mathbf{x} = \mathbf{C}_N^T \mathbf{Q} \mathbf{C}_N \mathbf{x}_N + \mathbf{C}_N^T \mathbf{Q} \mathbf{C}_F \mathbf{x}_F = \mathbf{p}_x,$$
$$\mathbf{C}_N^T \mathbf{V} \mathbf{q} = \mathbf{C}_N^T \mathbf{Q} \mathbf{C} \mathbf{y} = \mathbf{C}_N^T \mathbf{Q} \mathbf{C}_N \mathbf{y}_N + \mathbf{C}_N^T \mathbf{Q} \mathbf{C}_F \mathbf{y}_F = \mathbf{p}_y, \quad (10.9)$$

where $\mathbf{Q}$ is the diagonal matrix belonging to $\mathbf{q}$.

These form a system of linear equations (i.e. in the linear form $\mathbf{Ax} = \mathbf{b}$), because the values $\mathbf{Q}$ are given by the user. For convenience, we first compute matrices $\mathbf{D}_N$ and $\mathbf{D}_F$, from equation (10.1) and $\mathbf{Q}$,

$$\mathbf{D}_N = \mathbf{C}_N^T \mathbf{Q} \mathbf{C}_N = \begin{bmatrix} 2 & -1 & . & . & . \\ -1 & 2 & -1 & . & . \\ . & -1 & 2 & -1 & . \\ . & . & -1 & 2 & -1 \\ . & . & . & -1 & 2 \end{bmatrix} \quad (10.10)$$

$$\mathbf{D}_F = \mathbf{C}_N^T \mathbf{Q} \mathbf{C}_F = \begin{bmatrix} -1 & . \\ . & . \\ . & . \\ . & . \\ . & -1 \end{bmatrix} \quad (10.11)$$

Before solving equation (10.9) we invert the matrix $\mathbf{D}_N$, using any type of direct or iterative method; for example, with Gauss–Jordan elimination used here, or Cholesky decomposition used in Section 13.5.1. The solution is given directly,

$$[\mathbf{D}_N \ \mathbf{I}] = \begin{bmatrix} 2 & -1 & . & . & . & 1 & . & . & . & . \\ -1 & 2 & -1 & . & . & . & 1 & . & . & . \\ . & -1 & 2 & -1 & . & . & . & 1 & . & . \\ . & . & -1 & 2 & -1 & . & . & . & 1 & . \\ . & . & . & -1 & 2 & . & . & . & . & 1 \end{bmatrix} \rightarrow$$

$$[\mathbf{I} \ \mathbf{D}_N^{-1}] = \begin{bmatrix} 1 & . & . & . & . & \tfrac{5}{6} & \tfrac{2}{3} & \tfrac{1}{2} & \tfrac{1}{3} & \tfrac{1}{6} \\ . & 1 & . & . & . & \tfrac{2}{3} & \tfrac{4}{3} & 1 & \tfrac{2}{3} & \tfrac{1}{3} \\ . & . & 1 & . & . & \tfrac{1}{2} & 1 & \tfrac{3}{2} & 1 & \tfrac{1}{2} \\ . & . & . & 1 & . & \tfrac{1}{3} & \tfrac{2}{3} & 1 & \tfrac{4}{3} & \tfrac{2}{3} \\ . & . & . & . & 1 & \tfrac{1}{6} & \tfrac{1}{3} & \tfrac{1}{2} & \tfrac{2}{3} & \tfrac{5}{6} \end{bmatrix}$$

Having inverted the matrix $\mathbf{D}_N$, the coordinates $\mathbf{x}_N$ and $\mathbf{y}_N$ are calculated,

$$\mathbf{x}_N = \mathbf{D}_N^{-1}(\mathbf{p}_x - \mathbf{D}_F \mathbf{x}_F),$$

$$\mathbf{y}_N = \mathbf{D}_N^{-1}(\mathbf{p}_y - \mathbf{D}_F \mathbf{y}_F), \quad (10.12)$$

so,

$$\mathbf{x}_N = [-4 \ -2 \ 0 \ 2 \ 4]^T,$$

$$\mathbf{y}_N = [2.5 \ 4 \ 4.5 \ 4 \ 2.5]^T.$$

As previously observed in Section 6.4.4, we notice that for given values $\mathbf{q}$, the solution is entirely independent of the initial coordinates $\mathbf{x}_N$ and $\mathbf{y}_N$. One only needs to know the location of the supports $\mathbf{x}_F$ and $\mathbf{y}_F$. The solution is shown in Figure 10.3. As expected, the

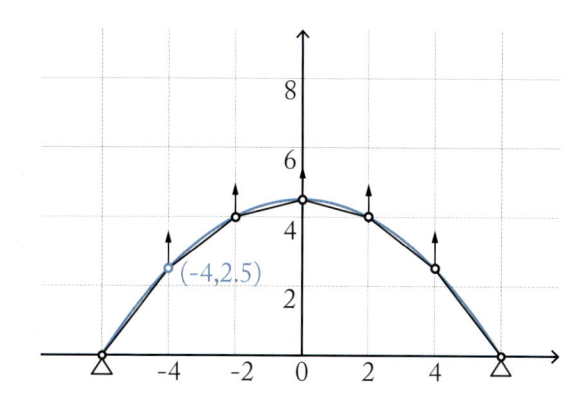

Figure 10.3 The resulting solution from the force density method for evenly spaced loads, with $p_y = 1$, force density, with $q = 1$

resulting shape is a piecewise approximation of a parabola, because equal loads are applied at equal distances in the x-direction.

Like the catenary, the parabola also presents a compression-only arch. The loads, which are evenly spaced in the x-direction, may represent an arch that is thicker towards the top, or, more plausibly, part of a bridge structure with a horizontal, suspended deck. Section 10.3.4 discusses shape-dependent loading, necessary to obtain a catenary. Figure 10.4b more clearly shows, for $q = 0.25$, that the result matches a parabola (green) and deviates from a catenary (red).

Figure 10.4b shows that by varying the force density, one obtains a wide variety of arches, scaled versions of each other; Figure 10.4a shows that the subdivision of the network influences the final result. It is therefore

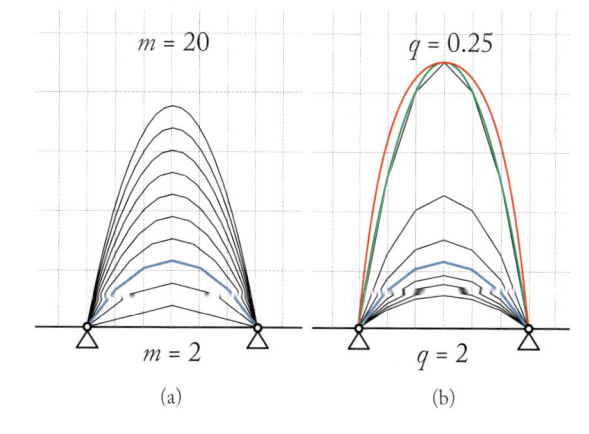

Figure 10.4 (a) Influence of number of branches m on the final result, and (b) influence of parameter q compared to a parabola (green) and a catenary (red)

difficult to fully control the size and scale of the final shape purely with these parameters, and it might thus be necessary to include a control mechanism, either through a user-interactive model in which these parameters can be varied and their influence assessed, or a constrained optimization method in which additional constraints such as a target height, or maximal or specified lengths are introduced.

Changing the value of each individual force density effectively changes the distribution of the nodes in the final shape and the relative distribution of forces in the branches. Generally, higher force densities mean that higher forces are attracted, but since the branch lengths also change, the overall impact of changing force densities is hard to anticipate. Let us randomly select force densities,

$$\mathbf{q} = [1 \quad 4 \quad 2 \quad 0.5 \quad 0.25 \quad 0.25]^T, \quad (10.13)$$

which, for the same loads $\mathbf{p}$, results in coordinates,

$$\mathbf{x}_N = [-4.98 \quad -4.72 \quad -4.21 \quad -2.17 \quad 1.91]^T,$$

$$\mathbf{y}_N = [3.68 \quad 4.35 \quad 5.19 \quad 6.55 \quad 5.28]^T. \quad (10.14)$$

Figure 10.5 compares this result with the one we first obtained in Figure 10.3.

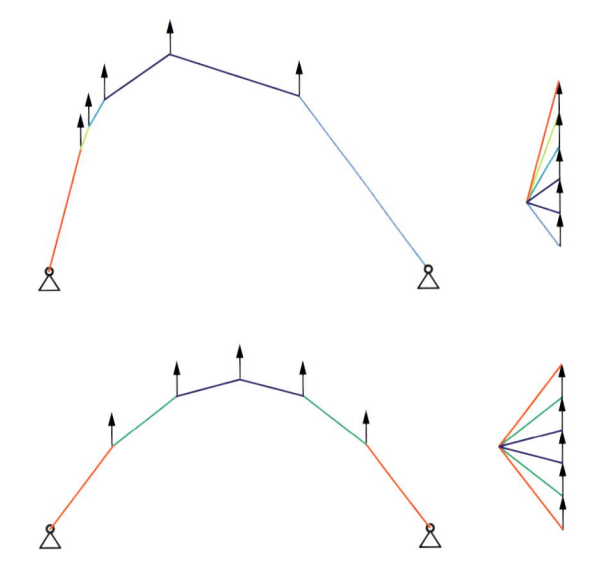

Figure 10.5 Influence of force densities on final result and corresponding force polygon

10.3.3 Thrust network analysis

The solutions in Figures 10.3 and 10.5 are the unique results for the chosen force densities $\mathbf{q}$ and the given loads $\mathbf{p}$. It is difficult, though, to anticipate the outcome of most given values for $\mathbf{q}$ and $\mathbf{p}$. TNA controls the force density values by choosing a fixed horizontal projection of the solution and by manipulating the reciprocal force diagram Γ^*, that is, the horizontal force components, or thrusts, in the branches. This approach alleviates the problem of having to decide on force density values directly. Note that in this example the reciprocal Γ^* consists of just one branch (or more correctly, all branches coincide), which corresponds to the constant horizontal thrust $\mathbf{H}$ in a funicular arch under vertical loading. Figure 10.6 shows that a single force polygon is equally valid for different results, but that the horizontal projections are different.

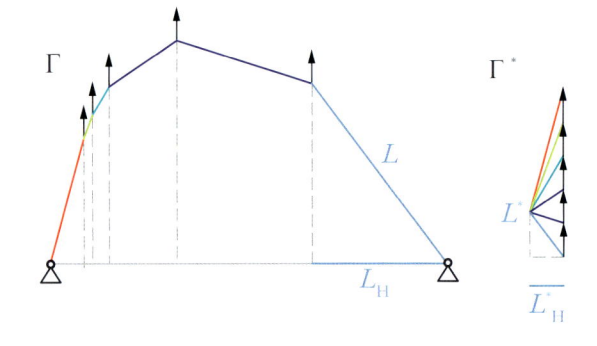

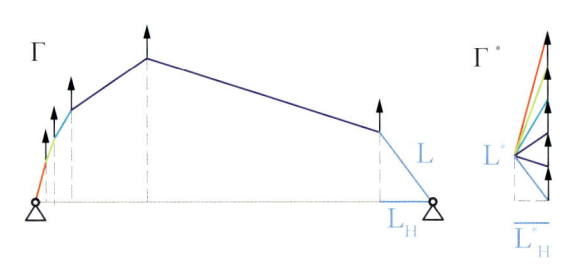

Figure 10.6 Two possible forms associated with the same funicular force polygon Γ^*

Recall from Section 7.3.3, that the force densities can be written as the ratios of branch lengths of the force and form diagrams, while introducing a scale factor $1/r$,

$$\mathbf{q} = \mathbf{L}^{-1}\mathbf{f} = \mathbf{L}^{-1}\mathbf{l}^* = \frac{1}{r}\mathbf{L}_H^{-1}\mathbf{l}_H^*. \tag{10.15}$$

Consider that because the horizontal thrust $\mathbf{f}_H = 1/r \cdot \mathbf{l}_H^*$ is the same in both examples in Figure 10.6, the force densities $\mathbf{q}$ are determined by $\mathbf{l}_H$; in other words, the horizontal spacing of the nodes. The choice of a unique spacing thus leads to an independent set of force densities $\mathbf{q} = \frac{1}{r}\mathbf{t}$.

Assuming a given horizontal spacing with $\mathbf{x}_N$ – for example, from equation (10.14) – and computing coordinate differences $\mathbf{u}$ using equation (10.2), then, from equation (10.3), but ignoring the vertical y-direction, the horizontal lengths

$$\mathbf{L}_H = |\mathbf{U}|$$

$$= \begin{bmatrix} 1.02 & . & . & . & . & . \\ . & 0.26 & . & . & . & . \\ . & . & 0.51 & . & . & . \\ . & . & . & 2.04 & . & . \\ . & . & . & . & 4.09 & . \\ . & . & . & . & . & 4.09 \end{bmatrix} \tag{10.16}$$

Then, given a horizontal thrust $\mathbf{l}_H^* = 1.02$, and a scale factor $r = 1$, the force density proportions $\mathbf{t}$ are known,

$$\mathbf{q} = \frac{1}{r}\mathbf{L}^{-1}\mathbf{l}_H^*$$

$$= \begin{bmatrix} \frac{1}{1.02} & . & . & . & . & . \\ . & \frac{1}{0.26} & . & . & . & . \\ . & . & \frac{1}{0.51} & . & . & . \\ . & . & . & \frac{1}{2.04} & . & . \\ . & . & . & . & \frac{1}{4.09} & . \\ . & . & . & . & . & \frac{1}{4.09} \end{bmatrix} \begin{bmatrix} 1.02 \\ 1.02 \\ 1.02 \\ 1.02 \\ 1.02 \\ 1.02 \end{bmatrix}$$

$$= [1 \quad 4 \quad 2 \quad 0.5 \quad 0.25 \quad 0.25]^T. \tag{10.17}$$

In this case, the result is the same as equation (10.13), but now obtained from a given horizontal projection and thrust, rather than directly prescribed with force densities.

Factor r echoes the observation from Figure 10.4a that the force density solutions for the same vertical loading are simply scaled versions of one another. However, r might be more intuitive, as higher values of r (as opposed to lower values of q), lead to higher arches (Fig. 10.7).

Following the same procedure as in Section 7.3.3, we replace the force densities in equation (10.9), with those of equation (10.15) to obtain the following equilibrium equation in the y-direction:

$$\mathbf{C}_N^T\mathbf{T}\mathbf{C}_N\mathbf{y}_N + \mathbf{C}_N^T\mathbf{T}\mathbf{C}_F\mathbf{y}_F = r\mathbf{p}_y. \tag{10.18}$$

Solving for the unknown $\mathbf{y}_N$, the resulting equation is the equivalent to equation (7.12):

$$\mathbf{y}_N = \mathbf{D}_N^{-1}\left(\mathbf{p}_y r - \mathbf{D}_F\mathbf{y}_F\right). \tag{10.19}$$

For $r = 1$, the result is the same as equation (10.14). Figure 10.7 shows the influence of varying scale factor r, noting that in each case the horizontal branch lengths indeed remain constant.

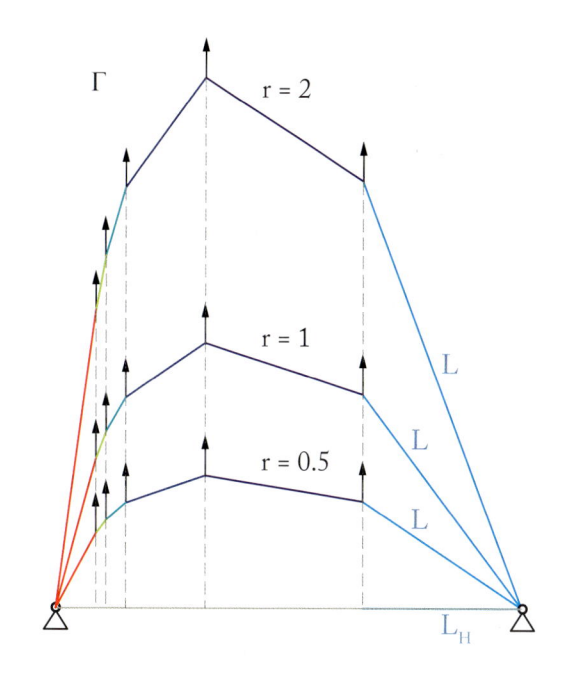

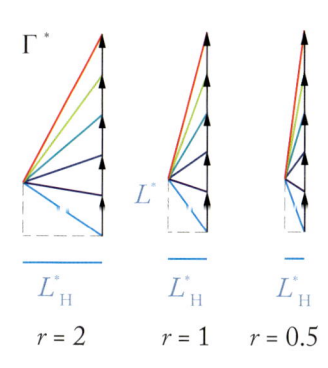

Figure 10.7 Influence of scale factor r on the final result

This example shows how TNA works in principle: it uses manipulation of projected form and force diagrams to explore funicular shapes that are in static equilibrium, under given vertical loads.

10.3.4 Shape-dependent loading

The previous examples assumed a constant load $\mathbf{p}_y = 1$. However, if the load is the self-weight of the chain, then the load is dependent on the geometry:

$$\mathbf{p}_y = \frac{1}{2}\rho g A |\mathbf{C}_N|^T \mathbf{l}, \qquad (10.20)$$

where ρ is the density in kgm^{-3}, $g = 9.81$ Nkg^{-1}, the gravitational constant, and A the sectional area of the chain in m^2 (here assumed constant, because we are looking to model a catenary).

Starting from the original line in Figure 10.1, the initial geometry has constant branch lengths and, therefore, $\mathbf{p}$ will still be constant. This means that an iterative procedure has to be introduced. At each step, the lengths $\mathbf{l}$ have to be recalculated with new coordinates from equation (10.12), using equations (10.2–10.3). The new loads $\mathbf{p}_y$ after the first iteration, using $\rho g A = 1/2$, change to

$$\mathbf{p}_y = \frac{1}{2}\frac{1}{2}\begin{bmatrix} 1 & 1 & . & . & . & . \\ . & 1 & 1 & . & . & . \\ . & . & 1 & 1 & . & . \\ . & . & . & 1 & 1 & . \\ . & . & . & . & 1 & 1 \end{bmatrix}\begin{bmatrix} 3.20 \\ 2.5 \\ 2.06 \\ 2.06 \\ 2.5 \\ 3.20 \end{bmatrix}$$

$$=[1.43 \quad 1.14 \quad 1.03 \quad 1.15 \quad 1.43]^T. \quad (10.21)$$

To determine convergence, the so-called out-of-balance, or residual, forces have to be calculated at each step. Using equations (10.9–10.19), the residual forces are

$$\mathbf{r}_x = \mathbf{p}_x - \mathbf{D}_N \mathbf{x}_N - \mathbf{D}_F \mathbf{x}_F$$
$$= [0 \quad 0 \quad 0 \quad 0 \quad 0]^T,$$
$$\mathbf{r}_y = \mathbf{p}_y - \mathbf{D}_N \mathbf{y}_N - \mathbf{D}_F \mathbf{y}_F$$
$$= [0.43 \quad 0.14 \quad 0.03 \quad 0.15 \quad 0.43]^T. \quad (10.22)$$

The procedure converges when the norm of the residuals is less than a tolerance ε. The final result is no longer in static equilibrium, but within the prescribed tolerance, at which it is considered 'close enough'. Then, with $\varepsilon = 0.01$, the following condition can be checked:

$$\|\mathbf{r}\| < \varepsilon \rightarrow \|\mathbf{r}\| = 0.63 > 0.01. \qquad (10.23)$$

The solution is not in equilibrium. A new iteration with the new loads $\mathbf{p}$, using equation (10.12), then leads to new coordinates, after which equations (10.20–10.23) can be repeated. In this case, equation (10.23) is satisfied after five iterations (Fig. 10.8a). The final shape is still dependent on the chosen force density $\mathbf{q}$ and discretization of m branches for the scale of the resulting shape, but the variation in force densities will no longer yield shapes other than a catenary, such as previously was the case (Fig. 10.5). Figure 10.8a shows the resulting catenary (in black), the first iteration (in blue) which is the parabola from Figure 10.3, and the three intermediate iterations (in grey). Figure 10.8b gives a steeper result from $\mathbf{q} = 0.55$ to more clearly show that the result is a catenary (compare to Figure 10.4).

In Figure 10.8 we observe that for vertical loads, even when updated, the nodes move only in the vertical direction. This illustrates the possible decoupling of horizontal and vertical equilibrium that TNA exploits.

10.3.5 Constant length chains

If we wish to design an arch where the segments are of constant length, the previous approaches need to be changed. In a reversal of the previous section, we now know the loads in advance as a function of constant, prescribed lengths $\mathbf{l}_P$,

$$\mathbf{p}_y = \frac{1}{2}\rho g A |\mathbf{C}_N|^T \mathbf{l}_P, \qquad (10.24)$$

whereas the projected, horizontal lengths $\mathbf{l}_H$ will vary as the nodes are expected to move in the horizontal direction as well in order to maintain constant lengths along the arch. Given the diagonal matrix of prescribed lengths $\mathbf{L}_P$, we update the current force densities at iteration i, using the diagonal matrix of current lengths $\mathbf{L}_i$, so that

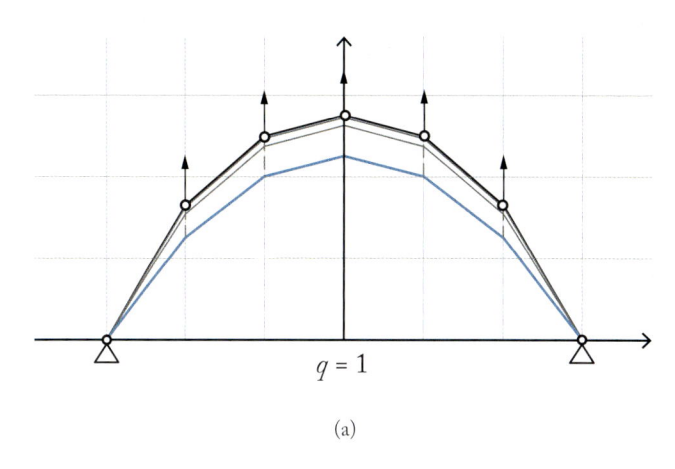

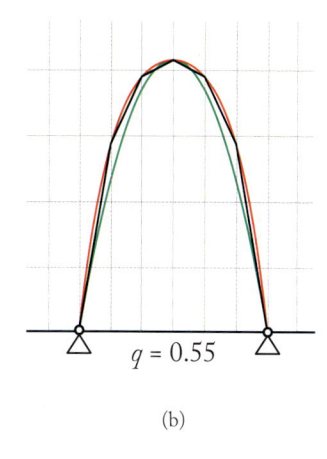

Figure 10.8 (a) Iterations of shape-dependent force density solution for $q = 1$, and (b) solution for $q = 0.55$ fitted to a parabola (green) and a catenary (red)

$$\mathbf{q}_{i+1} = \mathbf{L}_P^{-1}\mathbf{L}_i\mathbf{q}_i. \quad (10.25)$$

The constant loads in equation (10.24) then result in a catenary shape shown in Figure 10.9. If, however, the loads are based on the horizontal projection, that is, lengths $\mathbf{l}_H$, then the loads are updated at each iteration with

$$\mathbf{p}_{y,i} = \tfrac{1}{2}\rho g A |\mathbf{C}_N|^T \mathbf{l}_{H,i}, \quad (10.26)$$

resulting in a parabola instead. It is not uncommon in form finding to approximate loads using the geometry's projection. For very steep geometries the degree of approximation may be considerable, as shown in Figure 10.9b.

10.3.6 Dynamic relaxation

Dynamic relaxation solves the problem of dynamic equilibrium to arrive at a steady-state solution. This solution is equivalent to that of static equilibrium. The first difference from previous methods is the definition of the force densities $\mathbf{q}$, or tension coefficients, which now includes initial forces $\mathbf{f}_0$, an axial stiffness EA, with Young's modulus E and initial lengths $\mathbf{L}_0$. Assuming $EA = 1$ and $\mathbf{f}_0 = 1$, initially we obtain:

$$\mathbf{q} = \mathbf{L}^{-1}\mathbf{f} = \mathbf{L}^{-1}\left(\mathbf{f}_0 + EA\mathbf{e}\right)$$

$$= \begin{bmatrix} \tfrac{1}{2} & \tfrac{1}{2} & \tfrac{1}{2} & \tfrac{1}{2} & \tfrac{1}{2} & \tfrac{1}{2} \end{bmatrix}, \quad (10.27)$$

where strains $\mathbf{e} = (\mathbf{L} - \mathbf{L}_0)\mathbf{l}_0^{-1}$.

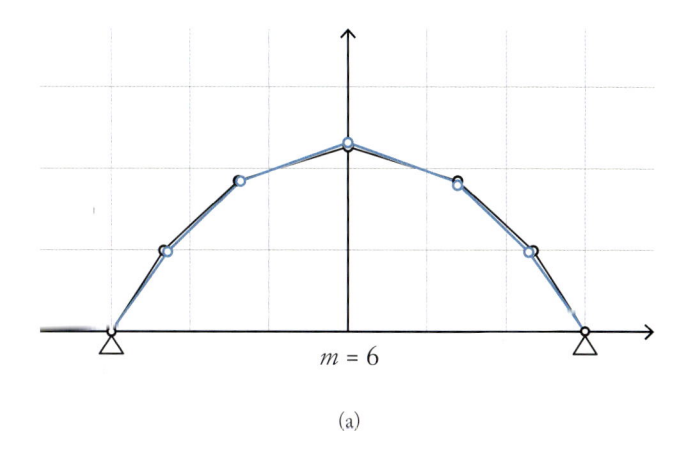

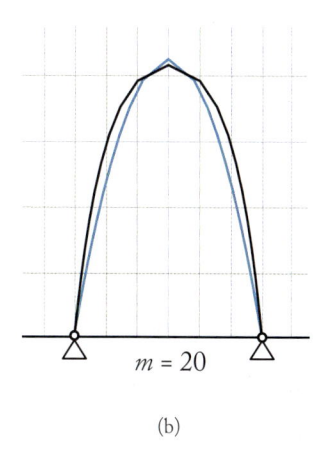

Figure 10.9 Constant lengths with prescribed loads lead to a catenary (black) or with updated, projected loads lead to a parabola (blue), respectively for (a) a shallow arch with six branches or (b) a steep arch with twenty branches

Considering that the form finding now assumes a physically meaningful initial geometry with lengths $\mathbf{L}_0$, and elastic deformation during form finding, we adapt the loads in equation (10.20) to keep the total mass constant:

$$\mathbf{p}_y = \tfrac{1}{2} A \|\mathbf{C}_N\|^T \mathbf{l}_0 = [1 \ \ 1 \ \ 1 \ \ 1 \ \ 1]^T. \quad (10.28)$$

These loads are no longer shape dependent as they are a function of the initial length $\mathbf{l}_0$, and remain the same at each step. They can be calculated only once. The residual forces $\mathbf{r}$ are still calculated using equation (10.22) at each step. Note that this is analogous to equation (8.8) but formulated for the entire network instead of one node. As in equation (8.3), using Newton's second law, and combining equations (10.22, 10.27, 10.28) (notice that the tension coefficients $\mathbf{q}$ are half the value of the example in Section 10.3.2), the residual forces on the nodes in the y-direction,

$$\mathbf{r}_y = \mathbf{p}_y - \mathbf{D}_N \mathbf{y}_N - \mathbf{D}_F \mathbf{y}_F = \quad (10.29)$$

$$= [1 \ \ 1 \ \ 1 \ \ 1 \ \ 1]^T$$

$$= \mathbf{M}\mathbf{a}, \quad (10.30)$$

where $\mathbf{M}$ is a diagonal mass matrix belonging to the mass vector $\mathbf{m}$ and $\mathbf{a}$ is the nodal acceleration vector. Applying equation (8.4) to the entire network, we obtain

$$\mathbf{m} = \tfrac{1}{2} \Delta t |\mathbf{C}_N|^T (\mathbf{L}^{-1}\mathbf{f}_0 + \mathbf{L}_0^{-1} EA) \quad (10.31)$$

$$= [1 \ \ 1 \ \ 1 \ \ 1 \ \ 1]^T.$$

Each node now accelerates as a function of the residual force and its mass. Its initial velocity for the first iteration is zero. For each subsequent iteration, the velocities are updated. As discussed in Chapter 8, the resulting movement will oscillate around a steady-state solution unless some form of damping is discussed, in our case viscous damping. Following equation (8.9), and assuming a constant damping factor $C = 0.5$ for the entire structure and time step $\Delta t = 1$, the velocities

$$\mathbf{v}_{y,t+\Delta t/2} = A \cdot \mathbf{v}_{y,t-\Delta t/2} + B \cdot \Delta t \cdot \mathbf{M}^{-1} \mathbf{r}_y \quad (10.32)$$

$$= 0.6 \cdot \begin{bmatrix} 0 \\ 0 \\ 0 \\ 0 \\ 0 \end{bmatrix} + 0.8 \cdot 1 \cdot \begin{bmatrix} 1 & . & . & . & . \\ . & 1 & . & . & . \\ . & . & 1 & . & . \\ . & . & . & 1 & . \\ . & . & . & . & 1 \end{bmatrix} \begin{bmatrix} 1 \\ 1 \\ 1 \\ 1 \\ 1 \end{bmatrix}$$

$$= [0.8 \ \ 0.8 \ \ 0.8 \ \ 0.8 \ \ 0.8]^T.$$

Thus, following equation (8.7) the coordinates

$$\mathbf{y}_{t+\Delta t} = \mathbf{y}_t + \Delta t \cdot \mathbf{v}_{y,t+\Delta t/2} \quad (10.33)$$

$$= \begin{bmatrix} 0 \\ 0 \\ 0 \\ 0 \\ 0 \end{bmatrix} + 1 \cdot \begin{bmatrix} 0.8 \\ 0.8 \\ 0.8 \\ 0.8 \\ 0.8 \end{bmatrix}$$

$$= [0.8 \ \ 0.8 \ \ 0.8 \ \ 0.8 \ \ 0.8]^T.$$

Again, these are equivalent to equations (8.7) and (8.9), but for the entire network, so expressed as vectors of size n_N. Figure 10.10 shows that the result converged after eighteen iterations. The first iteration corresponds to the coordinates calculated above in equation (10.33).

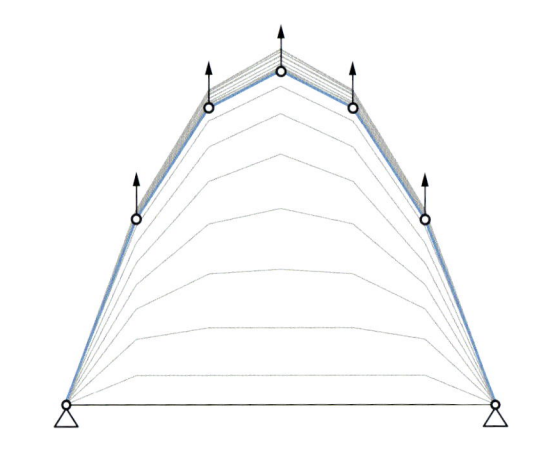

Figure 10.10 Iterations of DR

Increasing the stiffness EA, means shallower arches (Fig. 10.11a). The general solution is independent of the meshing, except for its impact on the accuracy of the solution (Fig. 10.11b).

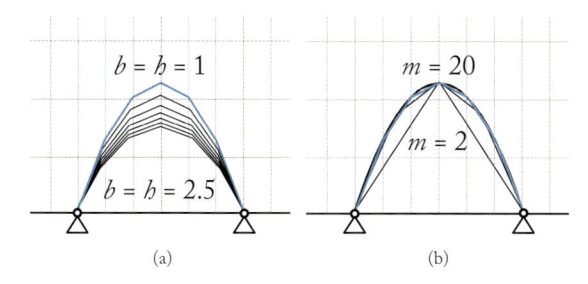

Figure 10.11 (a) Influence of dimensions $b=h$ and (b) influence of number of branches m on the final result

The numerical advantage of this solving procedure, called leapfrog integration, similar to Verlet integration, is that because the mass matrix $\mathbf{M}$ is a diagonal matrix (i.e. all off-diagonal entries are zero), the inversion $\mathbf{M}^{-1}$ does not require a costly calculation, as is the case for the non-diagonal stiffness matrix $\mathbf{D}_N^{-1}$ (compare equations (10.11–10.12) and (10.32)). More simply put, each nodal force vector is divided by the corresponding nodal mass, according to equation (8.9), and one can express the method entirely in equations per element, and per node. This is also why DR has been called a vector-based method. As a result of all this, DR generally requires more iterations to solve a problem, but the cost per iteration is lower. For very large problems, the inversion of a matrix can no longer be calculated via direct (decomposition) methods, but require more costly iterative methods (e.g. conjugate gradient methods), meaning iterations within iterations. Such methods may still outperform DR, but their implementation is obviously more involved.

10.3.7 Spline elements

Up to this point we have described DR in traditional form. We now add the spline elements discussed in Chapter 8 for the application to strained lattice shells. The nodal values for the bending moments $\mathbf{M}_b$ can be calculated from the prescribed bending stiffness EI and assuming direction and radii of curvature $\mathbf{R}$ have been calculated:

$$\mathbf{M}_b = EI \cdot \mathbf{R}^{-1}. \quad (10.34)$$

Equilibrium equations (10.9) have additional forces as a function of the bending moments $\mathbf{m}_b$:

$$\mathbf{p}_x - \mathbf{f}_x = \mathbf{p}_x - \mathbf{C}_N^T \mathbf{QCx} - \mathbf{C}_N^T \mathbf{L}^{-1} \mathbf{Cm}_{b,x} = 0,$$

$$\mathbf{p}_y - \mathbf{f}_y = \mathbf{p}_y - \mathbf{C}_N^T \mathbf{QCy} - \mathbf{C}_N^T \mathbf{L}^{-1} \mathbf{Cm}_{b,y} = 0. \quad (10.35)$$

Figure 10.12 illustrates the influence of the dimensions of a square cross section on the previous example (Fig. 10.11a) with one that includes spline elements. As the size of the cross section increases by a factor 2.5, the axial stiffness EA increases by $2.5^2 = 6.25$, but the bending stiffness EI by $2.5^4 \approx 39$. The results of Figure 10.12a model gridshells with equal initial lengths and only axial strain, suggesting hinged connections that are fixed in the final state. Figure 10.12b represents gridshells that are bent into place from an initially flat configuration.

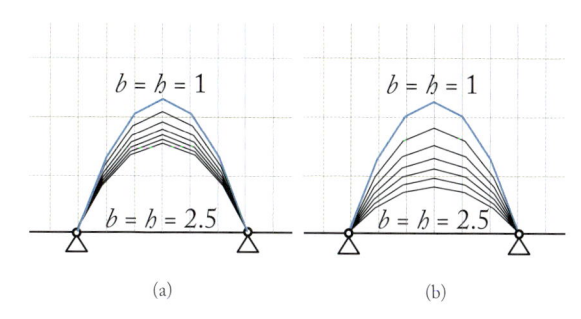

Figure 10.12 Influence of dimensions $b=h$ for DR (a) without spline elements, and (b) with spline elements

10.3.8 Particle-spring system

Another dynamic method is the particle-spring system (PS). It is in many respects similar to DR, but distinct nonetheless in how the branch forces and masses are defined, and the solvers that are employed. In this method, the branches are seen as weightless springs. The nodes are called particles. The forces in the springs depend on a spring constant, or spring stiffness, k_s, and a rest length L_0. Additional damping forces are dependent on the damping coefficient k_d, and the relative velocity of the end points along the direction of the spring. In the y-direction,

$$\mathbf{q}_y = k_s \mathbf{L}^{-1}(\mathbf{l} - \mathbf{l}_0) + k_d \mathbf{L}^{-2} \mathbf{V} \Delta \mathbf{v}_y, \quad (10.36)$$

where **V** is the diagonal matrix of coordinate differences, and the relative velocities $\Delta \mathbf{v}_y$ of the end points along the direction of the spring are:

$$\Delta \mathbf{v}_y = \mathbf{C}_N \mathbf{v}_y. \quad (10.37)$$

Comparing equations (10.36) to (10.27) we note that k_s is analogous to EA/L_0. However, it is possible in PS to define $L_0 = 0$, leading to so-called zero-length springs. Furthermore, we notice that when arriving at a steady-state solution, the second term will disappear as the velocities become zero. We also see that if rest length $L_0 = 0$, then k_s will lead to a constant force density. In that case we expect to obtain a parabola.

Equation (10.29) from DR, concerning the residual forces and Newton's second law, remains valid for PS. A drag coefficient b is also added, either by replacing equation (10.32) in the form of drag:

$$\mathbf{v}_{y,t+\Delta t/2} = \mathbf{v}_{y,t-\Delta t/2} + \Delta t \cdot \mathbf{M}^{-1}\left(\mathbf{r}_y - b \cdot \mathbf{v}_{y,t-\Delta t/2}\right), \quad (10.38)$$

$$\mathbf{y}_{t+\Delta t} = \mathbf{y}_t + \Delta t \cdot \mathbf{v}_{y,t+\Delta t/2}, \quad (10.39)$$

or equation (10.33) as a reduction of the velocities:

$$\mathbf{v}_{y,t+\Delta t/2} = \mathbf{v}_{y,t-\Delta t/2} + \Delta t \cdot \mathbf{M}^{-1}\mathbf{r}_y, \quad (10.40)$$

$$\mathbf{y}_{t+\Delta t} = \mathbf{y}_t + \Delta t \cdot (1-b) \cdot \mathbf{v}_{y,t+\Delta t/2}. \quad (10.41)$$

The loads **p** are a function of the prescribed masses **m** and gravitational constant g:

$$\mathbf{p}_y = \mathbf{m}g. \quad (10.42)$$

At this point, we note that PS, unlike DR, typically applies more advanced integration methods, such as the explicit RK4 or the implicit backward Euler methods. The former, explained here, starts with the original 'prediction' for the acceleration $\mathbf{a} = \mathbf{k}_1$, but then makes additional predictions at different time steps using new calculations of the residual forces before interpolating all four results for a final, more accurate prediction for the new geometry:

$$\mathbf{v}_{y,t+\Delta t/2} = \mathbf{v}_{y,t-\Delta t/2} + \frac{1}{6}\left(\mathbf{k}_1 + 2\mathbf{k}_2 + 2\mathbf{k}_3 + \mathbf{k}_4\right), \quad (10.43)$$

where

$$f(t,\mathbf{y}) = \mathbf{M}^{-1}\mathbf{r}(t,\mathbf{y})$$

$$\mathbf{k}_1 = \Delta t \cdot f(t,\mathbf{y}_t)$$

$$\mathbf{k}_2 = \Delta t \cdot f\left(t + \tfrac{1}{2}\Delta t, \mathbf{y}_t + \tfrac{1}{2}\mathbf{k}_1\right)$$

$$\mathbf{k}_3 = \Delta t \cdot f\left(t + \tfrac{1}{2}\Delta t, \mathbf{y}_t + \tfrac{1}{2}\mathbf{k}_2\right)$$

$$\mathbf{k}_4 = \Delta t \cdot f\left(t + \Delta t, \mathbf{y}_t + \mathbf{k}_3\right),$$

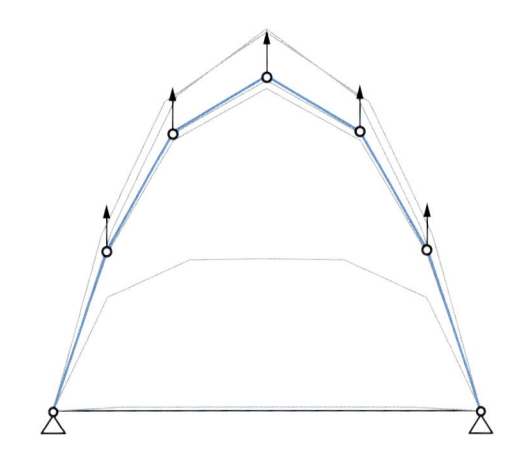

Figure 10.13 The first, every tenth and final iteration for PS with RK4

Figure 10.13 shows the result from PS with RK4 after sixty iterations (tolerance $\varepsilon = 0.01$), which requires many more iterations than DR. The parameters are $k_s = 1$, $k_d = 0.1$, $b = 0.1$, $g = 10$, $\Delta t = 0.1$. In this case, the process was underdamped, with the intermediate iterations moving beyond the equilibrium solution. Recall that PS originally came from the field of computer graphics and animation, in which iterations are a necessity to show any type of animated progression. Changing the parameters may increase convergence, but certain values may lead to divergence. This is why more stable strategies such as backward Euler allowing larger time steps and/or adapting the time step during form finding have been combined with PS.

For very large problems, such advanced Runge-Kutta methods outperform simple leapfrog integration.

10.4 Summary

Now that we have applied all the methods to the same case of a certain initial geometry, we attempt to discern their differences. As a start, we recapitulate the values that the user must or may supply. The properties, which are independent of the chosen method and need to be provided, are:

- coordinates of the supports;
- topology, connectivity of the network;
- prescribed loads **p** (or mass densities for shape-dependent loading);
- convergence tolerance (for iterative methods).

Table 10.3 shows the values that need to be prescribed by the user for each method.

The input for FDM and TNA is reduced to a bare minimum. This is an advantage, though as discussed, force densities are physically not meaningful and therefore difficult to control. Methods such as TNA attempt to indirectly determine the force densities through interactive control of (horizontal components of) forces, or through additional constraints (e.g. a best fit to a predefined target surface, see Chapter 13).

The drawback of dynamic methods, such as DR and PS, in this respect is the much larger number of parameters necessary for their control. However, in DR these parameters – for example, EA, EI, L_0 – are either fictitious values, chosen for their effect on convergence or on the resulting shape, or they are related to the material and physical properties of a structure. The latter is similar to conventional approaches in structural analysis based on the displacement, or stiffness, method. DR may therefore seem more familiar, and easier to comprehend and implement than purely geometric methods. In the case of real values, DR can also be used for static analysis directly (see Appendix A). Of course, this assumes that these properties are known or requires them to be.

Table 10.4 shows a summary of the equilibrium equations for each method. In the table, we can identify external forces, axial forces (controlled either by force density, elasticity or spring action), shear forces (for splines) as well as damping and drag forces (in PS).

From Table 10.4, we can conclude that the damping strategy of DR and PS is entirely interchangeable, as this only influences speed of convergence, not the final solution at which velocities are zero. Also, if $k_s = EA \cdot \mathbf{1}_0^{-1}$, DR and PS should yield the same solution. One can also easily implement spline elements in PS. More generally, if the force densities in each method

Method	User-prescribed quantities		SI Unit
FDM	force densities	q	Nm^{-1}
TNA	projected coordinates	$\mathbf{x}$	m
	thrust distributions (from)	$\dot{\mathbf{x}}$	N=m
	scale factor	r	-
DR	axial stiffness	EA	Nm^{-2}m^2=N
	bending stiffness (for splines)	EI	Nm^{-2}m^4=Nm^2
	initial coordinates, or lengths	$\mathbf{L}_0(x,y)$	m
	damping factor (for viscous damping)	C	-
	time step	Δt	s
PS	spring stiffness	k_s	Nm^{-1}
	initial coordinates, or rest lengths	$\mathbf{L}_0$	m
	damping coefficient	k_d	-
	drag coefficient	b	-
	time step	Δt	s

Table 10.3 Comparison of user input per method

Method	Equilibrium equations	Force densities
FDM	$\mathbf{r}_y = \mathbf{p}_y - \mathbf{C}_N^T \mathbf{QCy}$	$\mathbf{q}$ prescribed
TNA	$\mathbf{r}_y = r\mathbf{p}_y - \mathbf{C}_N^T \mathbf{TCy}$	$\mathbf{t} = \frac{1}{r}\mathbf{L}_H^{-1}\mathbf{1}_H^*$
DR	$\mathbf{r}_y = \mathbf{p}_y - \mathbf{C}_N^T \mathbf{QCy}$	$\mathbf{q} = \mathbf{L}^{-1}\mathbf{f}$
		$\mathbf{f} = \mathbf{f}_0 + EA(\mathbf{L}-\mathbf{L}_0)\mathbf{l}_0^{-1}$
DR (with splines)	$\mathbf{r}_y = \mathbf{p}_y - \mathbf{C}_N^T \mathbf{QCy} - \mathbf{C}_N^T \mathbf{L}^{-1}\mathbf{Cm}_{b,y}$	
PS	$\mathbf{r}_y = \mathbf{p}_y - \mathbf{C}_N^T \mathbf{QCy} - b \cdot \mathbf{v}_{t-\Delta t/2}$	$\mathbf{q} = \mathbf{L}^{-1}\mathbf{f}$
		$\mathbf{f}_y = k_s(\mathbf{1}-\mathbf{1}_0) + k_d \mathbf{L}^{-1}\Delta \mathbf{V}\mathbf{v}_y$

Table 10.4 Equilibrium equations in the y-direction per method

are defined in the same way, then they should also yield identical shapes. Because the forces **f** and actual lengths **l** are known after form finding, one can calculate initial lengths $\mathbf{l}_0$ for any given axial stiffness EA (or conversely, determine the axial stiffnesses for given $\mathbf{l}_0$):

$$\mathbf{l}_0 = EA \cdot \mathbf{L}(\mathbf{f} + EA\mathbf{1})^{-1}. \quad (10.44)$$

Equation (10.44) demonstrates how results from FDM are typically 'materialized' for subsequent structural analysis of different load cases. This also means that whether our form-finding result was obtained through FDM, DR or some other method, the material properties or initial lengths can be varied without disturbing shape and static equilibrium as long as **f** and **l** are kept the same. Of course, the choice of material and dimensioning will influence how the structure behaves due to different load cases and second-order effects (e.g. bending, deflection, buckling).

Table 10.5 summarizes the solving strategy per method, with TNA being a special case FDM. As discussed, FDM can be formulated as a system of linear equations if the force densities **q** do not change per iteration (as would be the case for shape-dependant loading). The dynamic equilibrium methods show the different approach in solving for equilibrium, though some similarity is noticeable with the stiffness matrix $\mathbf{D}_N$ being analogous to the diagonal mass matrix **M**.

10.5 Conclusions

At the start of this chapter, the following questions were posed:

- How do the form-finding methods presented in this book differ?
- In what cases does one choose a particular method?
- Can they be applied to other structural types?

The methods differ in two main respects: how the internal forces are defined, and how the resulting problem is numerically solved.

The definition of the internal forces depends on what information on the design is available to the

Method	Solving
FDM (nonlinear)	$\mathbf{y}_{N,i+1} = \mathbf{y}_{N,i} + \mathbf{D}_N^{-1}\mathbf{r}_y$
FDM (linear)	$= \mathbf{y}_{N,i} + \mathbf{D}_N^{-1}(-\mathbf{D}_N\mathbf{y}_{N,i} - \mathbf{D}_F\mathbf{y}_F)$
	$= \mathbf{D}_N^{-1}(-\mathbf{D}_F\mathbf{y}_F)$
DR (leapfrog)	$\mathbf{v}_{y,t+\Delta t/2} = \mathbf{v}_{y,t-\Delta t/2} + \Delta t \cdot \mathbf{M}^{-1}\mathbf{r}_y$
	$\mathbf{y}_{t+\Delta t} = \mathbf{y}_t + \Delta t \cdot \mathbf{v}_{y,t+\Delta t/2}$
PS (RK4)	$\mathbf{v}_{y,t+\Delta t/2} = \mathbf{v}_{y,t-\Delta t/2} + \frac{1}{6}(\mathbf{k}_1 + 2\mathbf{k}_2 + 2\mathbf{k}_3 + \mathbf{k}_4)$
	$\mathbf{y}_{t+\Delta t} = \mathbf{y}_t + \Delta t \cdot \mathbf{v}_{y,t+\Delta t/2}$

Table 10.5 Solving strategy per method

designer. Either one manipulates force densities to explore states of equilibrium, or one has information on the initial geometry and/or material properties, which are then deformed into shape. This seems to be the main distinction and reason for choosing a particular method.

The particular solver should be of less interest as long as it ensures convergence towards static, or steady-state, equilibrium. Only for very large problems, or for other reasons that fast and stable convergence are paramount, will they require further consideration.

We have seen that once an equilibrium state is found, material or physical properties can be changed repeatedly without disturbing shape or equilibrium. This fact, combined with the ability to manipulate the internal forces (through force density, elastic stiffness or spring stiffness, as well as loading), suggests that these methods are theoretically interchangeable.

Nevertheless, cases in which any compression-only shape of static equilibrium is acceptable are more efficiently tackled through purely geometric methods (e.g. FDM and TNA), whereas cases in which initial geometry and subsequent deformation have meaning and material properties are known, a method such as DR is straightforward and more appropriate. The explicit integration schemes in DR and PS also avoid any need for matrix algebra (solving a linear system, inverting a matrix), which may be an advantage in terms of simple implementation.

Further reading

- 'An overview and comparison of structural form finding methods for general networks', Veenendaal and Block (2012). This journal paper compares form-finding methods for the general case of discrete, self-stressed networks. It includes other methods as well, such as the stiffness matrix method, the assumed geometric stiffness method and the updated reference strategy.
- 'Computer shape finding of form structures', Meek and Xia (1999). This paper applies elastic bar elements to the form finding of thin shells.
- 'The natural force density method for the shape finding of taut structures', Pauletti and Pimenta (2008). This paper gives a triangle element for FDM. An alternative, but identical, description given by Singer (1995) in his German dissertation 'Die Berechnung von Minimalflächen, Seifenblasen, Membrane und Pneus aus geodätischer Sicht'.
- 'Concrete shells form-finding with surface stress density method', Maurin and Motro (2004). This paper offers an alternative triangle formulation, applied specifically to the form finding of shell structures.
- 'Dynamic relaxation applied to interactive form finding and analysis of air-supported structures', Barnes and Wakefield (1984). This paper is a more recent source for triangle elements in DR, which Barnes already described in his 1977 dissertation.
- 'Structural optimization and form finding of light weight structures', Bletzinger and Ramm (2001). This paper explains URS specifically in terms of tensioned membranes and minimal surfaces, but discusses the approach in a broader context of hanging models and shape optimization of shell structures.

Exercises

- FDM and TNA both use the branch-node matrix **C**. Implement DR (without spline elements) and PS for the standard grid (Fig. 6.12), using the branch-node matrix **C** obtained in previous exercises. What are the key differences in all four approaches? Attempt to obtain the same result with DR and PS. Under what conditions is this possible? Calculate the resulting force densities and use these in FDM. How could one have obtained these force densities beforehand? Instead apply TNA to create a new design, and materialize the result by finding the stiffnesses and initial lengths. Use these in DR or PS. How could one have obtained these material properties beforehand?
- Compare solutions obtained from FDM with prescribed force density $q=1$ to those of DR with zero-length springs with prescribed spring constants $k_s=1$, for the same boundary conditions and topology. Discuss the potential benefit of using zero-length springs.
- Recall from Chapter 3 the constant stress arch. Given that the stress $\sigma = F/A$ and the force density $q = F/L$, how should the shape-dependent load in equation (10.20) be changed to obtain such an arch? Implement this and compare the shape to that of the parabola and the catenary.

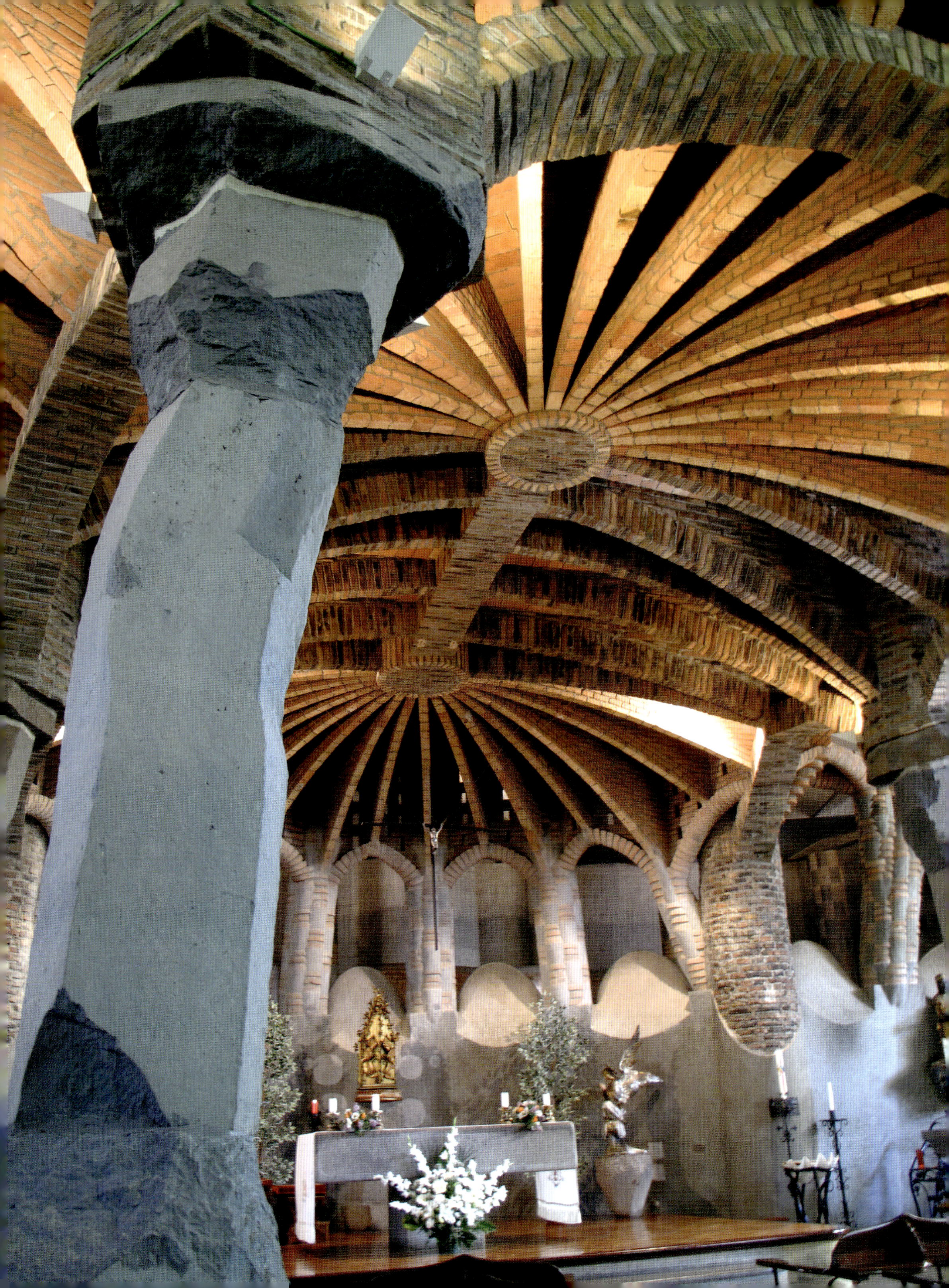

CHAPTER ELEVEN
Steering of form

Axel Kilian

The concept of steering of form is to be understood as an extension of form finding, which in turn is an extension of purely analytical approaches to structural design. Simple structural analysis is the least design related as it relies on a given design and it offers little in terms of feedback to the initial design other than identifying problem areas. Form finding directly links the form configuration with the forces in a closed loop. This approach is useful for initial form explorations and to get an interactive response to changes in settings, in both geometry and structural parameters, making it ideal for early design. This differs from traditional computational approaches in structural analysis, where the design is a given, or the result of a single set of given parameters. It is also distinct from structural optimization, where the result is a single solution based on a single set of given constraints. But in a more constrained design context or where factors other than structural performance such as constructability or architectural factors come into play, it is crucial to extend the notion of form finding towards one of steering of form towards desired design outcomes that take into account the combination of the different design goals without compromising the integrity of the result. Hierarchical design systems such as current parametric design models alone are not adequate to achieve this and neither are closed optimization systems that do not allow for the integration of other possibly non-structural constraints.

Instead, we require an alternative platform for the integration of different design factors through the shared language of form and forces. This is achieved by extending form-finding methods to steering of form and by developing additional algorithms to allow continuous manipulation of the various parameters, in order to achieve additional goals besides some structural optimum.

The examples here are using particle-spring systems, a method explained in Chapter 9, but extends beyond this particular form-finding method, and can apply to other well-known methods discussed in this book, such as the force density method (Chapter 6), thrust network analysis (Chapter 7) and dynamic relaxation (Chapter 8). In any case, steering of form offers the chance for design discoveries that would neither be apparent in the optimized solution nor in the original design intent.

11.1 Form finding

The purpose of form finding is to find force equilibrium structures. In their simplest set-up, form finding methods, both physical and digital ones, generate catenary curves, or simple hanging forms (Fig. 11.1). Physical methods have generated structures such as

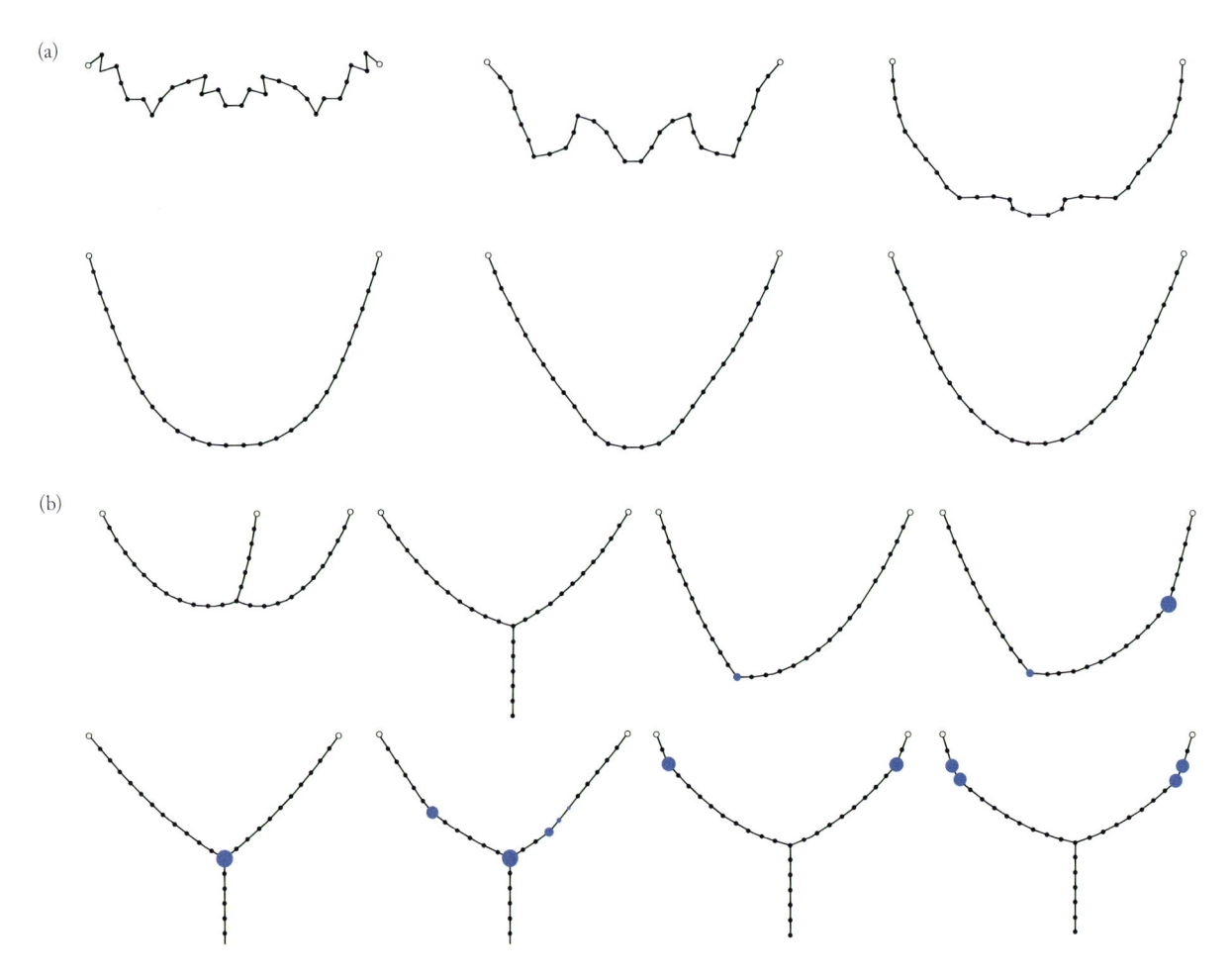

Figure 11.1 (a) A basic particle-spring chain settling into catenary form, starting from rest lengths longer that the initial straight-line state, and (b) additional point loads and attached chains

the Colònia Güell by Antoni Gaudí (see page 130). While producing structurally pure forms, the catenary form vocabulary has severe limitations in its general application to design.

In discrete systems, like a particle-spring network, form finding requires the definition of properties such as a mesh topology, spring stiffnesses and geometric lengths. Changing the parameters of the system influences the geometric shape outcome as well as the force distribution. All systems (with at least one fixed point) should eventually come to rest and provide an equilibrium solution for that specific configuration, but will respond to additional changes to the parameters. This allows the definition of form-finding problems – for example, for shell structures – to be set up for the exploration of different results. In particular, a non-hierarchical system of associations allows for edits at all levels of the definition of the design, the approach referred to here as steering of form. The term 'steering' emphasizes a more active role of the designer in the form-finding process by allowing varying degrees of 'sub-optimal' solutions. But, without any compromise, most design problems could not be addressed using form-finding techniques. Ideally, the designer is made aware of any departure from the structurally most efficient solution, but is allowed to deviate if desired for design reasons.

Figure 11.1b illustrates such exploration by different variations of simple particle-spring chains with equal weight distribution. Possible variations can be the number of springs, the length of springs and the mass of the individual particle nodes. Higher point loads are introduced and additional chains are attached. The resultant equilibrium forms expand the single catenary curve form and offer more possibilities to steer the design.

The advantage of such systems is the iterative nature of the exploration, where form is always a response to an equilibrium state of forces in the system, and every change potentially affects the entire system, hence the form will continuously adapt to unbalanced loads.

Form-finding approaches such as particle-spring systems can deal equally well with determinate and indeterminate structural systems, even if an analytical solution does not exist. This makes them particularly interesting for shell structures as all shell structures are indeterminate structures and, therefore, only one of many possible force distributions can be calculated at one time.

11.2 Steering principles

Of course, any good form-finding method will always steer the form towards a desired outcome. But, with many computational implementations, the complex inner workings of force equilibrium structures are obscured. Designers should have more possibilities of influencing and iteratively adjusting emergent forms. There are several 'steering' principles introduced here specifically for particle-spring systems, but they are valid for, or analogous to, those that govern other form-finding methods. These principles are variations in parameters such as:

- the support conditions;
- the loads on, or masses of the nodes;
- the length and strength of the springs;
- the topology of the network and the related discretization of load.

Steering here can be literally controlling form through these parametric settings, but it can be more indirect, through additional algorithms that adjust the available settings to accomplish a particular goal. For instance, one may devise a geometric form as a target and use iterative adjustments to steer the geometry towards a shape target, something that may be very time-consuming and manually almost impossible, but feasible computationally. In future developments, hybrid form-finding systems may allow for the combination of compression-only and moment-resistant loads such as shown in Figure 11.2 where the coloured sections are increasingly stiffened to depart from the pure catenary.

11.2.1 Support conditions

There are a number of possible support conditions in structures, which can be helpful in arriving at better mesh results. As the initial mesh geometry and topology is often generated flat for ease of programming and due to the unknown final shape, the geometric deformations in the form-finding process can be large. As a consequence, an initial support configuration may not be aligned with the emergent geometric equilibrium shape. In such a case, the supports should be adjusted accordingly, but within limits as the supports still need to resist the forces coming from the shell. Two simple solutions are fixed and partially fixed supports that allow for sliding in one axis or in one plane.

The simplest constraint is a pinned support which fixes some particle to an (x,y,z) position in space. Any forces trying to displace such a particle are ignored. In a system with gravity, at least one fixed support is needed for the system to come to rest in an equilibrium state.

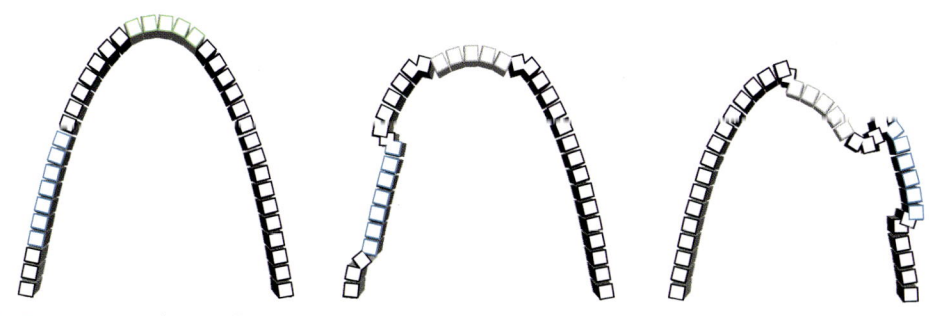

Figure 11.2 Experiment with partially moment-resistant structures in the same form-finding solver for hybrid structures

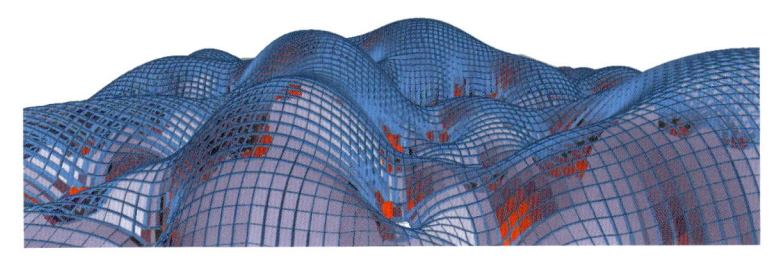

Figure 11.3 Gentle valleys and peaks of the resultant mesh, where supports were restricted to the *xy*-plane, restrained by tie-back springs

In the case of sliding supports, constrained only to the horizontal *xy*-plane, supports might simply slide together. In order to provide some resistance, tie-back springs to a specific location (such as the starting point) could be used. These may also be updated according to a specified force threshold; if the force is exceeded, their rest length may be increased. This offers more freedom for the shape to settle into equilibrium, especially if the spring mesh rest lengths are not updated, but of course at the cost of control.

Figure 11.3 shows a regularly spaced quad mesh which is attached to the support plane at random points using the described method. As the mesh is loaded, the supports adjust their position in the *x*- and *y*-directions according to the incoming forces and the resulting domes are more evenly stressed as a result.

The position of the supports may follow a higher logic. In Figure 11.4 a sine curve defines the roof support. When the sine curve is flattened, the curvature in the roof canopy is reduced as well. In this example, the spring lengths of the mesh are also adjusted in combination with the amplitude of the undulating support edge. The longer the springs, the less pronounced are the valleys and peaks.

Changing the constraints of the supports is another effective way of influencing the shell geometry. In Figure 11.5, supports are constrained in the *z*-direction and can move freely on the horizontal plane. However, once inside a given geometric boundary the particle is freed and becomes part of the shell mesh.

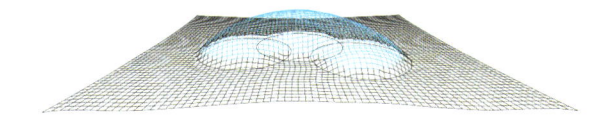

Figure 11.5 Mesh with a geometric perimeter defining when particles are either free to move or constrained to the *xy*-plane

11.2.2 Loads

The default reading of a particle-spring mesh in its simplest form is most likely that of self-weight with each particle representing the point of mass at the centre of mass of the surrounding region of material. Variations in mass between particles, unequal spacing of particles or the attachment of additional particles can be used to integrate external loads or uneven load distributions in the load-bearing structure itself.

Varying masses is one measure that can be used

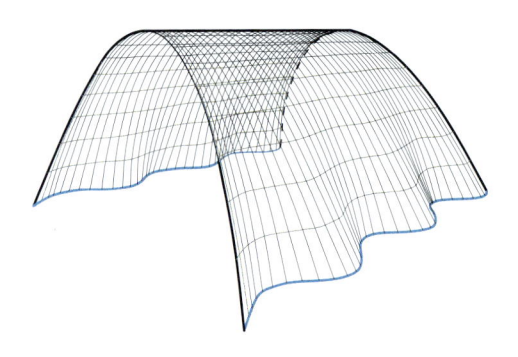

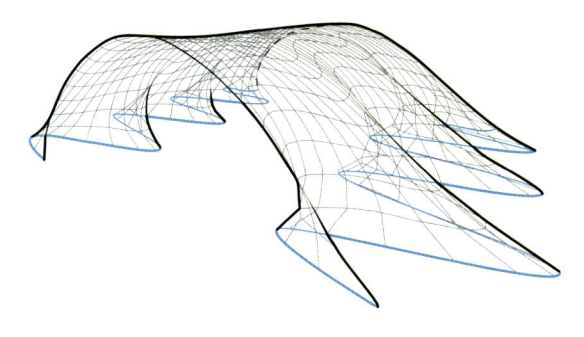

Figure 11.4 Undulating support edge of the mesh

to steer the form of the structure away from singular catenary curves (Fig. 11.1a) towards more expressive vocabulary as seen in Gothic cathedrals and Gaudí's work alike. For example, a single point load breaks the catenary curve into two separate catenary curves, and additional point loads create further catenary segments (Fig. 11.1b).

11.2.3 Spring properties and geometric targets

In particle-spring systems, the force in each spring is governed by its spring stiffness and length. Through the manipulation these parameters it is possible to direct the overall form within the limits of a given topology. One can monitor and (iteratively) adjust both quantities; for example, to get the spring to stay within a small deviation of its original starting length.

Steering of form can also be achieved by establishing fixed geometric targets in space. The mesh is adjusted by measuring the distance of the closest particles to those targets, and subsequently extending or shrinking the adjoining springs proportionally to that distance. Once the shell intersects with the geometric target the deformation is stopped.

11.2.4 Topology

The starting topology and geometry is the main interface for specifying the final form.

In the example of the simple chain, connecting additional chains mid-length of an arc will introduce additional forces and break up the singular catenary into multiple catenary curves (Fig. 11.1). This can be used to vary the formal appearance of the structure while still working with a compression-only equilibrium structure.

When using simple scripted topologies for surface structures, regular grids are often a starting point. Similarly straightforward are randomized triangulated meshes when using standard meshing approaches such as Delaunay triangulation.

Arguably, the most common default meshes are quad grids due to their simple repetitive geometry and constant node valence. But, they are not particularly well suited for form finding. Take, for instance, the case of an equally spaced rectangular grid, fixed at its four corner points, with an equal load distribution assumed using particles of equal mass. The mesh distorts extremely as it moves towards an equilibrium state with quad cells close to the supports in the corners collapsing completely into rhombic shapes. The result may seem similar to what happens when hanging a cloth in physical space, but in detail the load distribution and load density from self-weight has changed, due to overlaps and self-intersections of the mesh.

Without further force-dependent adjustments, the subsequent solutions can therefore produce unrealistic results. The force distribution could be improved if spring rest lengths or stiffnesses are adjusted iteratively to reduce excessive stretching of the springs. However, geometric distortions and self-intersection can remain a problem, and increasing spring stiffness can only be done up to a point before the solver becomes unstable.

Another improvement is a re-meshing, or updating of the masses, based on the surface area of the resultant equilibrium mesh; for example, if a shell of equal load distribution and thickness would be the goal.

Alternatively, a regular triangulated grid also prevents excessive deformation of the quad mesh due to the stiffening diagonals. Keep in mind that self-intersection can still occur. So, in any case, particular care should be taken with setting up and evaluating the results from particle-spring systems as the distortions could be overlooked.

A triangulated grid is also helpful if the mesh is supposed to visually approximate a material such as a fabric. This is because the added diagonals can mimic the shear capacity of a fabric – especially if two diagonals are added to each quad to form a cross – leading to wrinkling and folding behaviours. The diagonals also help to keep the particle distribution more even, and reduce distortion, since the cells are now less likely to collapse in the corner regions where there is a lot of shearing.

Different mesh topologies, iterative re-meshing strategies and also iterative parameter adjustments could account for the possible deformations of the geometric mesh during the form-finding process. This does require a more complex set-up. Alternatively one can set a topology and geometry close to the expected end result to reduce these deformations. Still, changing the mesh topology has a major impact on

Figure 11.6 A repetitive topology to create a cathedral-like structure, where change leads to very different design expressions and a wider range of formal responses

the resultant form, and so steering the design through topological variations is arguably the most influential of all form-steering methods. It is therefore worthwhile varying the mesh topology and observing its influence on form. Figure 11.6 shows a repetitive topology to create a cathedral-like structure, where change leads to very different design expressions and a wider range of formal responses, yet the structure is still in equilibrium. Figure 11.7 shows how one may iteratively arrive at such responses.

11.3 Structural feedback

There exist many possible ways to provide both visual and numerical details about the (structural) state of the system, embedded in the three-dimensional context of the geometry. Such feedback can help the designer assess the state of the system as it evolves and offer qualitative clues to its structural behaviour. This can be valuable in the early stages of designing shells and is crucial for steering of form.

11.3.1 Visualization of forces

Of primary importance is visualizing the forces in the system. A very effective technique for monitoring the force flow is colouring and line weight (e.g. Fig. 11.3) in correspondence to the forces present in the spring members at any time. Furthermore it is possible to generate geometric envelopes that correspond to the forces present in a member for dimensioning and fabrication purposes (Fig. 11.8). Another interesting technique is to draw the history of positions of particles over time, which gives a visual trace of the relative stability of a node in space either during form finding or during subsequent loading tests (Fig. 11.9).

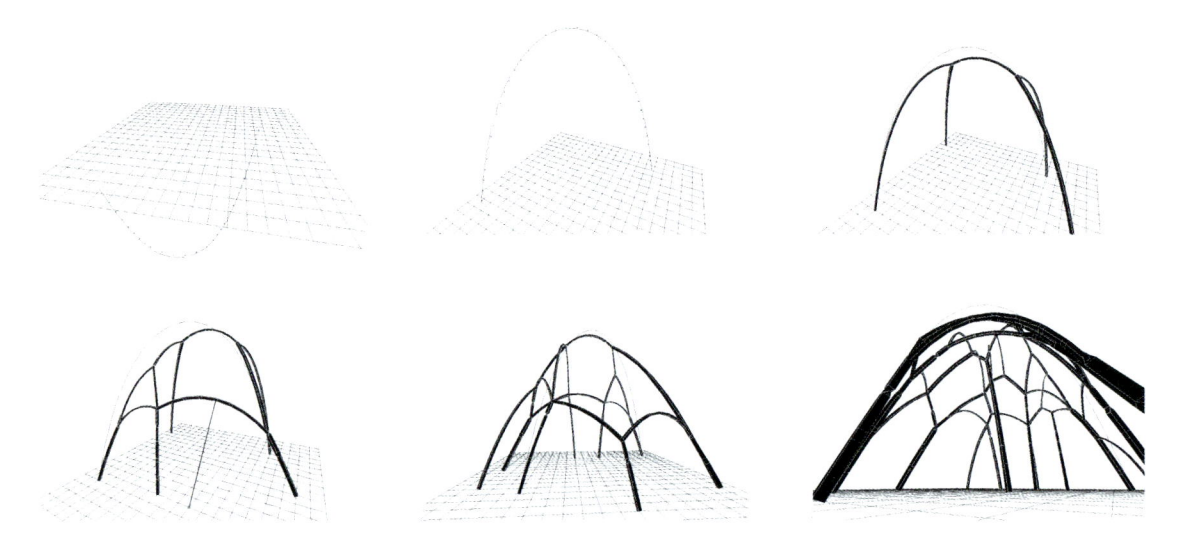

Figure 11.7 A topology created iteratively by manually adding more and more chain segments to an initial single arc

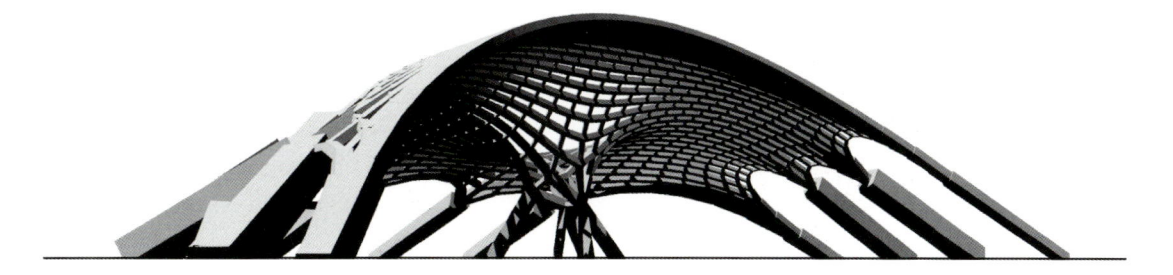

Figure 11.8 Geometric envelope based on forces present in the structure

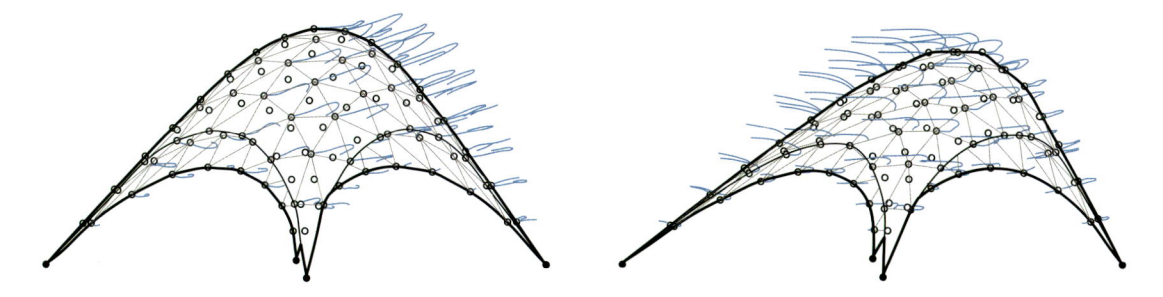

Figure 11.9 Tracing history of particles for two different load cases

11.3.2 Structural response

The addition of 'phantom forces' make the crude simulation of additional load cases possible. For instance, a lateral wind load in the form of springs pulling in the wind direction on the mesh nodes attached to massless particles in mid-space that are offset at a fixed distance to their parent mesh node allow for the testing of the mesh response to such loads. This is not a proper wind simulation but it may give an idea of the effects that external loads would have on a particular mesh geometry, how its stiffness is distributed, how its equilibrium is affected and what kind of displacements would occur, all within the same environment.

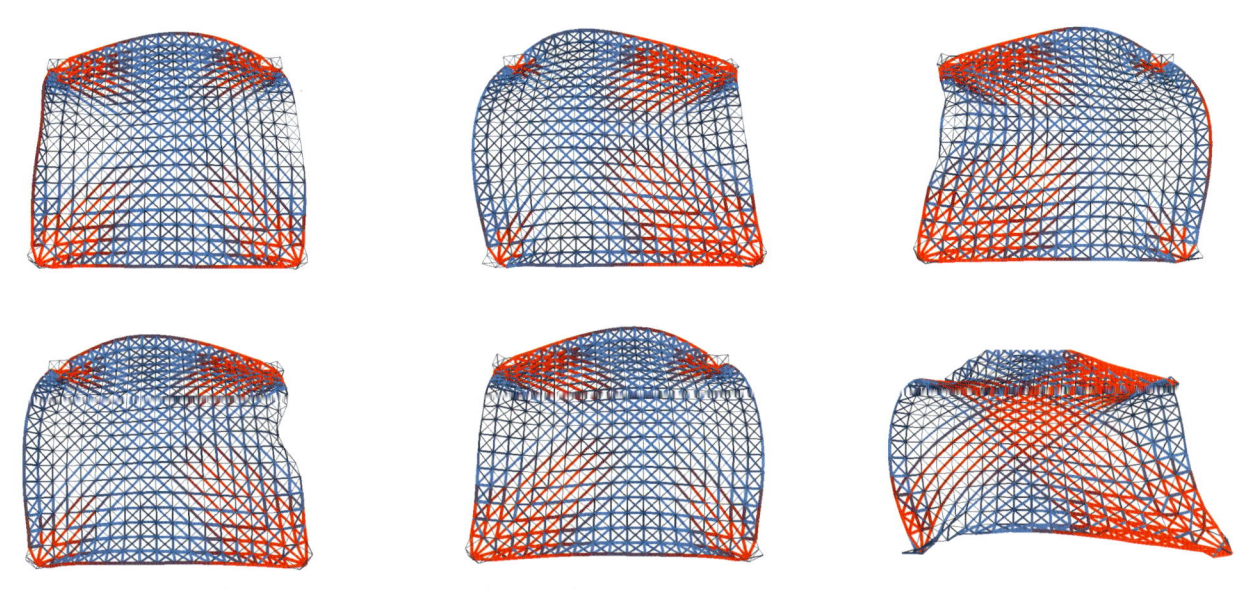

Figure 11.10 Force fluctuations triggered by shifting the fixed support positions, similar to an earthquake event

A similar, very schematic test could be done by subjecting the support particles to abrupt lateral and vertical displacements simulating movements of the ground during an earthquake and observing the resulting forces and displacements in particles and springs over time (Fig. 11.10). Such displacement history may be helpful in evaluating initial dimensioning of geometric material envelopes for the structures and help to get a sense of necessary redundancy in different areas of the shell.

11.3.3 Flow of forces

Using a dense, random-point cloud within the surface boundaries as a starting point results in a statistically uniform distribution of masses. Then one can use Delaunay triangulation to determine the mesh topology. When releasing the mesh to gravity, the loading of the springs reveals the emerging force paths through the shell as it comes to rest in equilibrium. Due to the random nature of the topology, these paths are more independent from the starting topology than the regular quad meshes and help to understand the flow of forces (Fig. 11.11).

More generative approaches may work with the metaphor of flow of forces, with forces originating from the distributed mass of the shell and 'flowing' towards the support points of the shell. A literal interpretation of this is rainflow analysis. Its hypothesis is that 'like a rainflow, loads will flow along curves with the steepest ascent on the shell surface to its supports' (Borgart, 2005). Section 7.5.1 compares the force flow derived from geometry (rainflow analysis) and from internal force distribution in a discrete network.

Topology studies may be a way to steer the design towards desired results with topologies reflecting the expected flow of forces more closely. Note that the topology itself in turns impacts the resulting geometry (Fig. 11.12).

11.4 Conclusion

For the design of form-passive structures such as shells, it is necessary to appropriate computational methods traditionally used for engineering-dominated approaches to design. This can be achieved by replacing the absolute goals of optimization with interactive, iterative approaches that make use of the same underlying algorithms. Such approaches should allow the designer to selectively deviate from optimal solutions and explore design variants, with multiple possible conflicting goals, found through discovery rather than precise analysis. Analytical methods would be computationally time intensive and, more importantly, may not have helped in discovering such a design configuration. The ultimate goal is to use the power of computational methods to achieve more integrative design solutions, and to extend the canon of forms established through conventional analytical techniques. This offers the chance for design discoveries that would neither be apparent in the optimized solution nor in the original design intention.

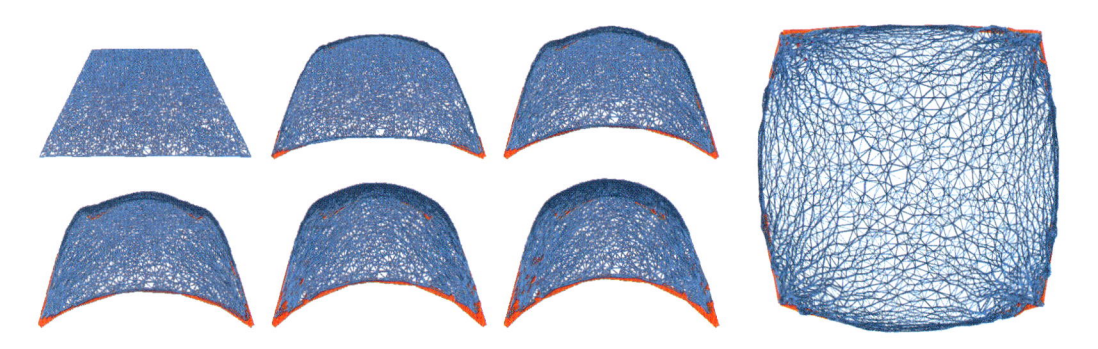

Fig 11.11 A randomized triangle mesh revealing the development of load paths towards the final equilibrium state

CHAPTER ELEVEN: STEERING OF FORM   **139**

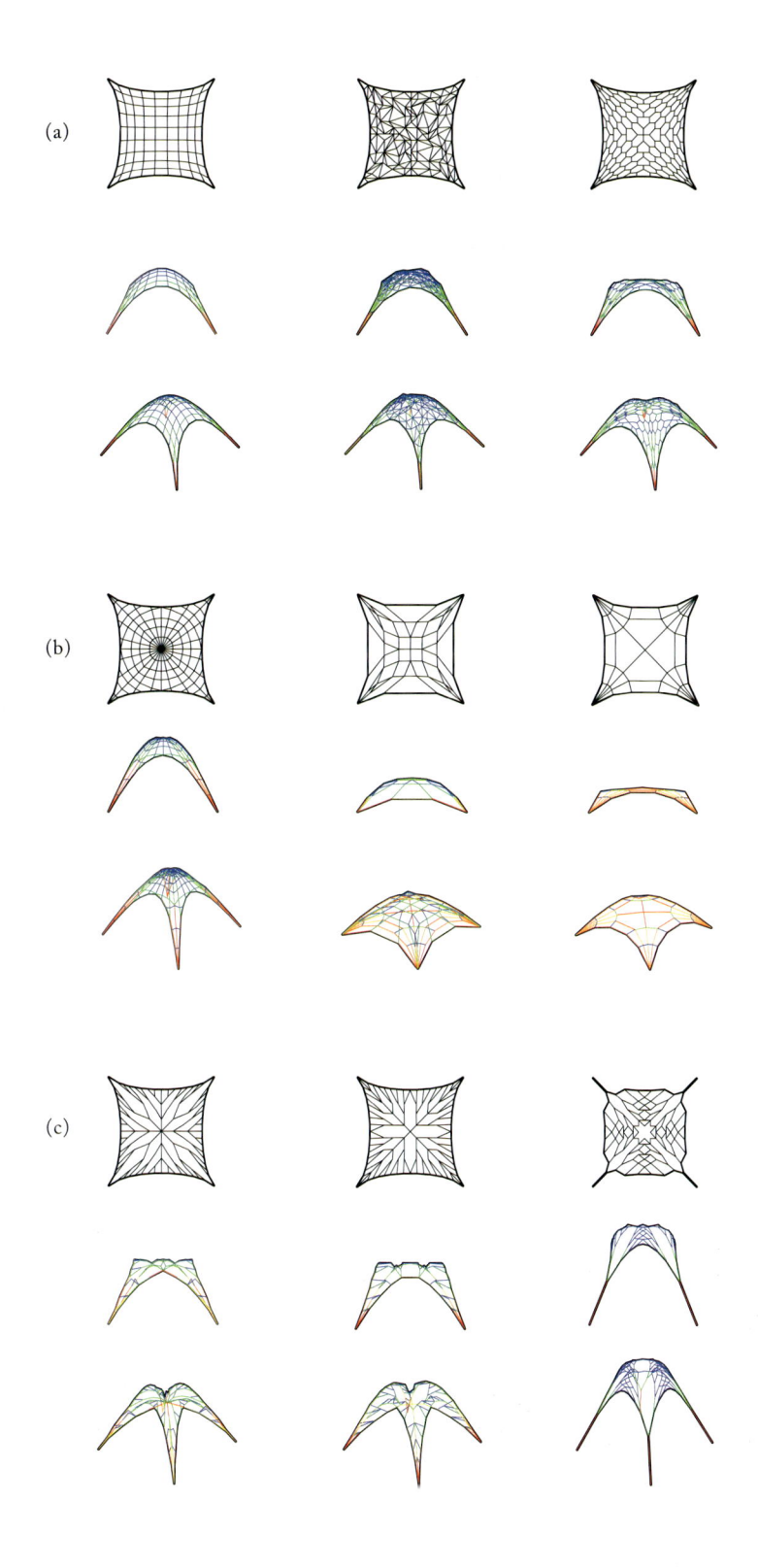

Figure 11.12 Result from particle-spring form finding for different topologies for the same plan, based on (a) conventional meshes, (b) expected flow of forces and (c) branching strategies

PART III
Structural optimization

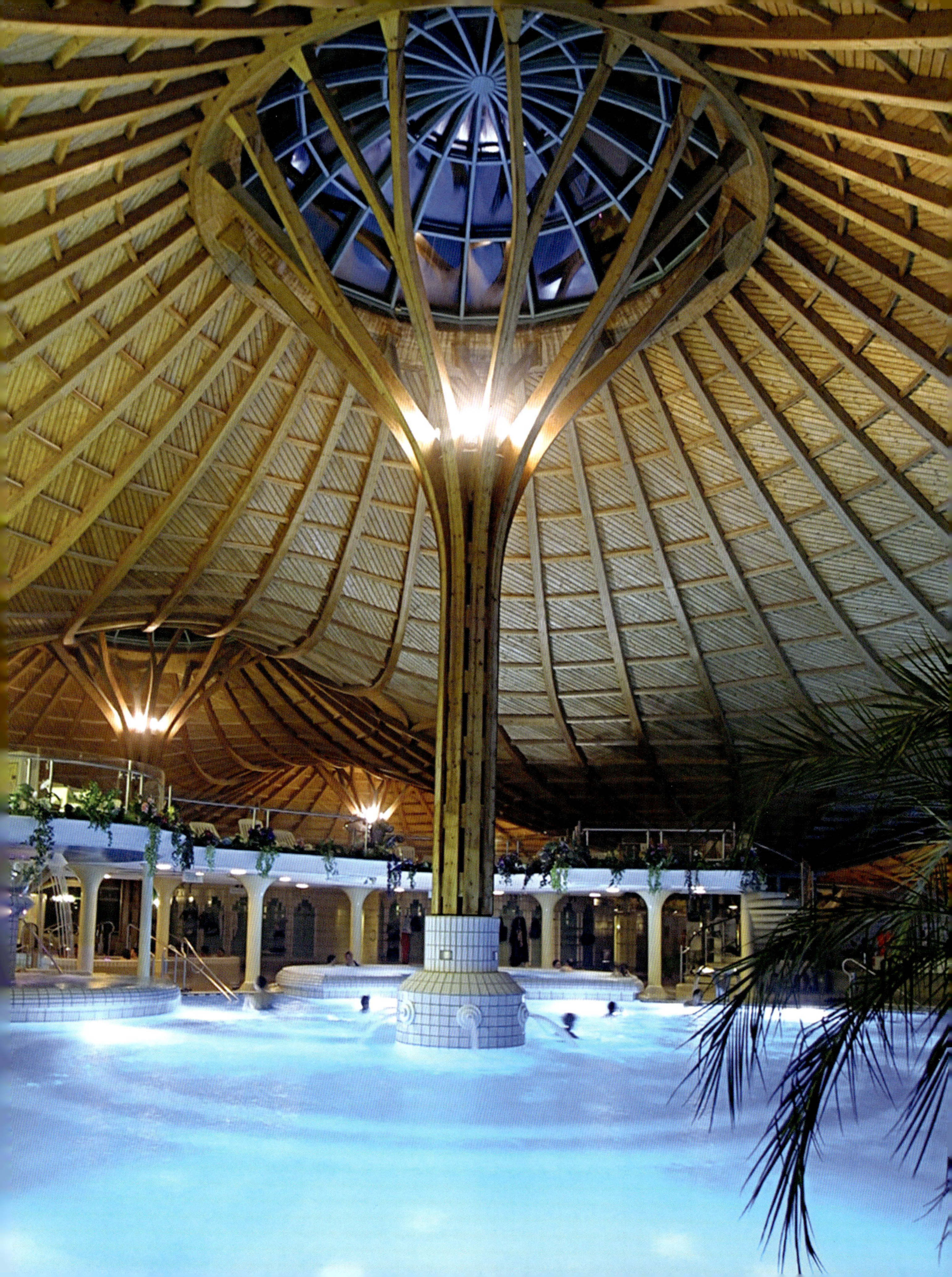

CHAPTER TWELVE

Nonlinear force density method

Constraints of force and geometry

Klaus Linkwitz and Diederik Veenendaal

LEARNING OBJECTIVES

- Describe least-squares problems (in the context of structural form finding), and the corresponding normal equations to solve them.
- Apply additional constraints to the force density method.
- Demonstrate how to incorporate material and/or fabrication constraints in the force density method.
- Find the form of a gridshell subjected to constraints, using the nonlinear force density method.

PREREQUISITES

- Chapter 6 on the force density method.

Figure 12.1 The Multihalle in Mannheim, 1976, an application of nonlinear FDM

Force densities, in their purest form, result in systems of linear equations. This is just the first step when designing a real structure. For instance, it is unlikely that our first linear form-finding result satisfies all physical and geometrical constraints. To overcome any tedious trial-and-error strategies in our selection of force densities, we require an alternative approach to effectively and systematically manipulate our design, while simultaneously creating new solutions in static equilibrium. We take advantage of the method of least squares, and present a nonlinear approach to the Force Density Method (FDM), which allows the introduction of secondary constraints.

The first author of this chapter was responsible for developing nonlinear FDM for the Munich Olympic Roofs. This very first application of the method, determining the cutting patterns of the project's cable-net roofs, showed its potential. Then, as head of Büro Linkwitz und Preuß, he adapted the method together with his team, Lothar Gründig and others, and applied it to two timber shells, which we use to demonstrate this approach: the Solemar Therme timber shell roof in Bad Dürrheim (see page 142 and Fig. 12.4); and the Multihalle timber gridshells in Mannheim (Fig. 12.1).

We apply constraints, similar to those of the featured projects, to our design brief, continuing the example of Chapter 6.

The brief

To expand its sports offerings, the municipality of Stuttgart is developing a new sports complex, which includes a swimming pool and an ice rink. A preliminary design has been made for two roof structures (Chapter 6), but several changes are necessary. First, along the central corridor between both roofs, a direct connection is made. There is an angle between both roofs at this point and the design team wants to have a smooth transition of the roof curvatures (Fig. 12.2). Second, the roof over the ice rink will be made from an initially flat grid. Therefore, the segments in the interior of this roof need to be equidistant (Fig. 12.2).

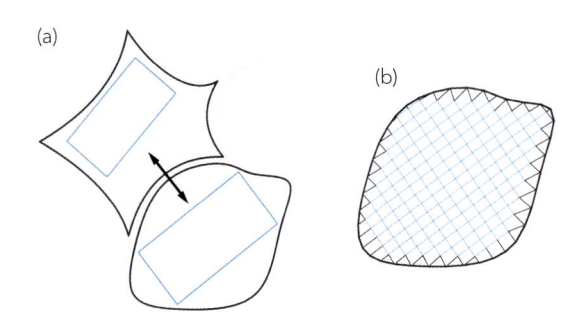

Figure 12.2 (a) Connection between both roofs and (b) the interior segments of the ice rink roof

12.1 Method of least squares

The basis of nonlinear FDM lies in the method of (nonlinear) least squares. This section introduces the basics of this method. The method of least squares finds the approximate solution of under- or overdetermined systems (number of equations <, respectively > number of unknowns). In other words, it applies to systems $\mathbf{Ax}^* = \mathbf{b}$, where the left-hand side matrix $\mathbf{A}$ is no longer square, which means there is either no or no unique solution $\mathbf{x}^*$.

12.1.1 Normal equations

Although the contemporary presentation of a least-squares problem is in an unconstrained form (see Section 12.1.2), typically nonlinear FDM has been presented in the form of problems, as simple *equality-constrained quadratic programs*. Here we show the equivalence, while deriving the solution, starting from the minimization problem as it appears historically in FDM literature:

$$\text{min.} \quad \mathbf{x}^T\mathbf{x} \tag{12.1}$$

$$\text{subject to } \mathbf{Ax} = \mathbf{b},$$

where $\mathbf{b}$ is often referred to as the observations in least squares, and in our case generally is a vector of residual forces. Applying the method of Lagrange multipliers, we introduce the Lagrangian,

$$\Lambda(\mathbf{x},\boldsymbol{\lambda}) = \mathbf{x}^T\mathbf{x} - 2\boldsymbol{\lambda}^T(\mathbf{Ax} - \mathbf{b}) \tag{12.2}$$

which we solve by finding the minimum of Λ, that is, $\nabla \Lambda(\mathbf{x},\boldsymbol{\lambda}) = 0$. We take the partial derivatives, or *optimality conditions*,

$$\frac{\partial \Lambda}{\partial \mathbf{x}} = 2\mathbf{x}^T - 2\boldsymbol{\lambda}^T\mathbf{A} = 0, \tag{12.3}$$

$$\frac{\partial \Lambda}{\partial \boldsymbol{\lambda}} = -2(\mathbf{Ax} - \mathbf{b}) = 0, \tag{12.4}$$

or in block matrix form,

$$\begin{bmatrix} \mathbf{I} & -\mathbf{A}^T \\ -\mathbf{A} & \mathbf{0} \end{bmatrix} \begin{bmatrix} \mathbf{x} \\ \boldsymbol{\lambda} \end{bmatrix} = \begin{bmatrix} \mathbf{0} \\ -\mathbf{b} \end{bmatrix}. \tag{12.5}$$

From the first condition, we obtain

$$\mathbf{x} = \mathbf{A}^T\boldsymbol{\lambda}, \tag{12.6}$$

which, substituted in the second condition, gives

$$\boldsymbol{\lambda} = (\mathbf{A}\mathbf{A}^T)^{-1}\mathbf{b}, \tag{12.7}$$

and back into the first condition results in,

$$\mathbf{x} = \mathbf{A}^T(\mathbf{A}\mathbf{A}^T)^{-1}\mathbf{b} = (\mathbf{A}^T\mathbf{A})^{-1}\mathbf{A}^T\mathbf{b}, \tag{12.8}$$

which is the solution to the so-called *normal equations*,

$$(A^TA)x = A^Tb. \quad (12.9)$$

12.1.2 Contemporary least squares

The method of least squares is usually explained by the following premise: instead of finding the solution x^* (which is not unique, or does not exist), we wish to find an approximate value x such that Ax is the best approximation of b. The errors e^* of our approximations are not known, since we do not know x^*, but can be related to the *residuals* r, which we can compute,

$$e^* = x^* - x, \quad (12.10)$$

$$Ae = Ax^* - Ax$$
$$= b - Ax = r. \quad (12.11)$$

The smaller the distances $\|b - Ax\|^2$, the better the approximation. Finding x in this context is the least-squares problem, where 'least squares' refers to the fact that the overall solution minimizes the sum of the squares of the errors (or residuals) made in the results of every single equation. A least-squares problem is an optimization problem with no constraints and has an objective which is a sum of squares,

$$\min. (b - Ax)^T(b - Ax). \quad (12.12)$$

The solution of a least-squares problem can be reduced to solving a set of linear equations, the normal equations, by multiplying both sides of the linear system $Ax = b$ with A^T, thus obtaining a square left-hand side *normal matrix* A^TA, and the system

$$(A^TA)x = A^Tb, \quad (12.13)$$

which has the analytical solution $x = (A^TA)^{-1}A^Tb$, which we also derived in equation (12.8). The matrix $(A^TA)^{-1}A^T$ is also known as the Moore–Penrose pseudoinverse A^+.

12.1.3 Weighting factors

The solution can be influenced by introducing weightings w, thus assigning relative importance to our observations b. The normal equations for a weighted least-squares problem are

$$(A^TW^{-1}A)x = A^TW^{-1}b. \quad (12.14)$$

where W, assuming the observations are independent, is a diagonal matrix of the vector of weighting factors w.

12.1.4 Nonlinear least squares

Another complication is that, in this chapter, A is a function of x, and needs to be recalculated. In other words, the problem is now a nonlinear least-squares problem. The first step is to linearize the problem according to Newton–Raphson's method, and by introducing the first derivative J, write

$$Ax = b,$$

$$Ax_0 + \frac{\partial Ax}{\partial x}\Delta x = b$$

$$J\Delta x = b - Ax_0 = r_0. \quad (12.15)$$

We are still solving a linear system $J\Delta x = r_0$ of the form $Ax = b$ to which we can apply normal equations, but instead of finding x directly, we now have iterations, in which x is updated using Δx,

$$\Delta x = (J^TJ)^{-1}J^Tr_0$$

$$x = x_0 + (J^TJ)^{-1}J^Tr_0. \quad (12.16)$$

This iterative procedure is commonly known as Gauss–Newton's method.

12.1.5 Iterative solving

Although normal equations are correct and exact, the normal matrix A^TA requires a matrix–matrix product, and has to be inverted, making them computationally inefficient and demanding. They are unsuitable for flexible implementations and large-scale problems (even more so in the force density formulation by Schek (1974), in which the constraints are all rewritten as a function of the force densities, leading to additional inverted matrix–matrix products). One solving approach, adopted for the Multihalle (Gründig and Schek, 1974), was the application of the conjugate gradient (CG) method to the normal equations. In this iterative method, it is not necessary to form the normal matrix $\mathbf{A}^T\mathbf{A}$ explicitly in memory, but only to perform (transpose) matrix–vector multiplications. Nowadays, we know this approach, for application to underdetermined systems, as applying the CG method on the Normal Equations (CGNE) (Saad, 2003). In such an approach, we first solve $(\mathbf{A}^T\mathbf{A})\mathbf{y} = \mathbf{b}$, for $\mathbf{y}$, in such a way that $\mathbf{A}^T\mathbf{A}$ is not explicitly formed, to then compute the solution, $\mathbf{x} = \mathbf{A}^T\mathbf{y}$. In the remainder of this chapter, we combine the method of least squares with FDM to deal with additional constraints; this is nonlinear FDM as it was applied to the Solemar Therme and Multihalle shell structures. Figure 12.3 shows the process of nonlinear FDM.

12.2 Notation

To present our formulations in this chapter in three dimensions, we redefine some of the matrices from Chapter 6 to include the y- and z-directions. Note that this is convenient in terms of notation, but means that the matrices become more sparse (they contain relatively more zeroes).

The branch-node matrices

$$\mathbf{C} := \begin{bmatrix} \mathbf{C} & & \\ & \mathbf{C} & \\ & & \mathbf{C} \end{bmatrix}$$

$$\mathbf{C}_N := \begin{bmatrix} \mathbf{C}_N & & \\ & \mathbf{C}_N & \\ & & \mathbf{C}_N \end{bmatrix}, \quad \mathbf{C}_F := \begin{bmatrix} \mathbf{C}_F & & \\ & \mathbf{C}_F & \\ & & \mathbf{C}_F \end{bmatrix}$$

are $3m \times 3n_N$ and $3m \times 3n$ block-diagonal matrices. The matrix of force densities

$$\mathbf{Q} := \begin{bmatrix} \mathbf{Q} & & \\ & \mathbf{Q} & \\ & & \mathbf{Q} \end{bmatrix}$$

is $3m \times 3m$, with the vector $\mathbf{q}$ still of size m. The coordinate difference matrices are stacked vertically, so

$$\mathbf{\bar{U}} = [\mathbf{U} \quad \mathbf{V} \quad \mathbf{W}]^T,$$

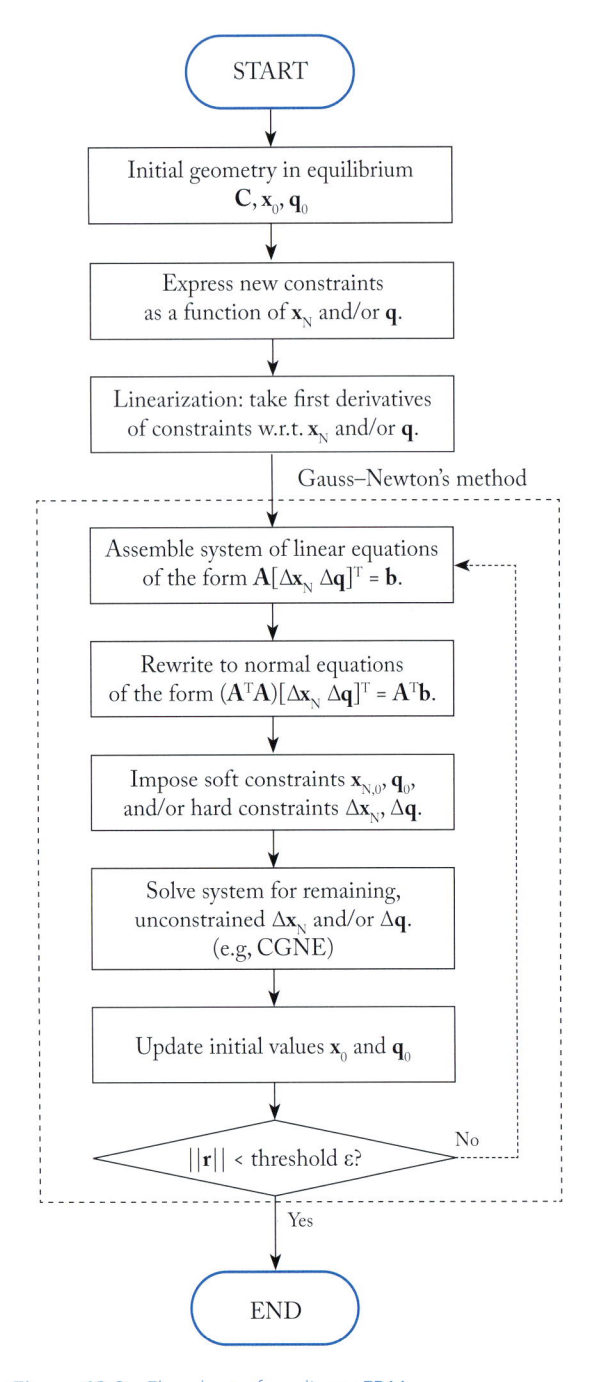

Figure 12.3 Flowchart of nonlinear FDM

of size $3m \times m$. The $3n_N$-vectors of coordinates and loads are also stacked: $\mathbf{x} := \begin{bmatrix} \mathbf{x} & \mathbf{y} & \mathbf{z} \end{bmatrix}^T$ and $\mathbf{p} := \begin{bmatrix} \mathbf{p}_x & \mathbf{p}_y & \mathbf{p}_z \end{bmatrix}^T$.

12.3 Solemar Therme

The timber shell of the Solemar Therme health spa in Bad Dürrheim, Germany (Fig. 12.4), consists of a skeleton of glued, laminated (glulam) timber beams, resting on five tree-like columns. The total surface area is 2,500m². The tree-columns are between 9.1m and 11.5m high, roughly spaced 20m apart. A doubly curved surface covers the spa, suspended as if it were a tensioned membrane, using the tree-columns as high-points and parts of the edge beams as low-points. From the 6m or 8m circular opening at the top of each tree-column, 200mm × 205mm 'meridian ribs' connect the columns and the edge boundaries in a radial pattern. They intersect with horizontally positioned 80mm × 80mm or 120mm × 140mm 'annular ribs', placed in a tangential pattern. Each rib is held together by finger jointing which can withstand both tension and compression forces.

The ideal shape of the grid-like structure would be characterized by the fact that, under self-weight, both meridian and annular ribs are subjected to axial forces, and that any bending is as small as possible. In addition, with respect to an economical prefabrication, certain symmetries in the form allow reuse of prefabrication tools and set-ups. Other constraints relate to the necessity to have uniform angles of the meridians in their connection with the top-ring of the tree-columns.

To take all these requirements into account, a strategy of form-finding, nonlinear FDM was employed which had been found to be very successful with the design of cable-net and membrane structures, such as the Munich Olympic Roofs. In a preliminary design stage, model and computer studies were executed based on prestressed equilibrium shapes, generated with FDM (in further studies the feasibility of the design was investigated).

12.3.1 Linear form finding

The process set out with a topological description, using the branch-node matrix $\mathbf{C}$, of the skeleton-like structure. Although the roof would eventually be built as a rigid structure, it was, at this stage, discretized and modelled as a network of bar elements. The resulting net was hung between the high tree-columns and bottom edge beams. The high points were fixed in the x-, y- and z-direction. The low points of the edge beams had one degree of freedom, the z-direction, with their x- and y-coordinates placed along horizontal circles (so boundaries, circular in plan). Given certain prescribed force densities $\mathbf{q}$ and external loads $\mathbf{p}$, linear FDM was carried out. Figure 12.5 shows results from linear form finding.

Figure 12.4 Solemar Therme in Bad Dürrheim, Germany

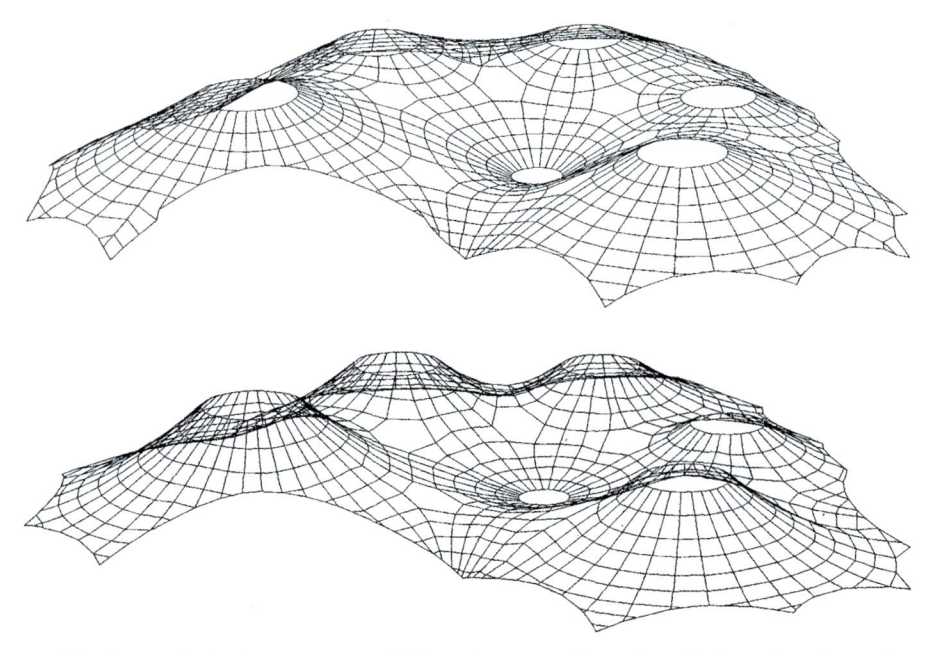

Figure 12.5 One of the first and a final computer model from the original linear form finding (Gründig, Linkwitz, Bahndorf & Ströbel, 1988)

12.3.2 Nonlinear form finding

The results of the first phase of form finding were then modified to accommodate additional constraints. The definition and application of these constraints was a trial-and-error procedure, with architect and engineer closely collaborating. The constraints, or rather design improvements, were necessary to reduce unnecessary torsion in the meridian ribs, influence the curvature of the ring ribs, observe certain heights due to spatial requirements of the spa underneath, improve the aesthetics of the connection between the ribs and tree-columns and so on. To enforce these constraints, an optimization problem was formulated as follows. Depending on the different sets of constraints imposed, target values for certain node positions or force densities were set, while the coordinates of all other points and the remaining force densities were free to change. This process rendered a number of viable designs.

The main condition in this optimization problem is static equilibrium in equation (6.28), where we introduce a function $\mathbf{g}$ with respect to $\mathbf{x}_N$ and $\mathbf{q}$,

$$\begin{aligned} \mathbf{g}(\mathbf{x}_N,\mathbf{q}) - \mathbf{p} &= \mathbf{C}_N^T \bar{\mathbf{U}} \mathbf{q} - \mathbf{p} \\ &= \mathbf{C}_N^T \mathbf{Q}\mathbf{C}\mathbf{x} - \mathbf{p} \\ &= \mathbf{D}_N \mathbf{x}_N + \mathbf{D}_F \mathbf{x}_F - \mathbf{p} = \mathbf{0}, \end{aligned} \quad (12.17)$$

where, to simplify notation, $\mathbf{D}_N = \mathbf{C}_N^T \mathbf{Q} \mathbf{C}_N$ and $\mathbf{D}_F = \mathbf{C}_N^T \mathbf{Q} \mathbf{C}_F$.

Suppose that we have additional equality constraints, such as those mentioned earlier, which we are able to express as a function of the force densities and nodal coordinates. The challenge now is to prescribe $\mathbf{x}_N$ and $\mathbf{q}$ such that these constraints are satisfied (as much as possible), while maintaining static equilibrium. These additional equations are generally nonlinear and are not fulfilled at the initial values $\mathbf{x}_{N,0}$ and $\mathbf{q}_0$, found during the initial, linear form finding. Because of the nonlinearity, we construct an iterative method, where we ask for changes $\Delta \mathbf{x}_N$ and $\Delta \mathbf{q}$, starting with $\mathbf{x}_{N,0}$ and $\mathbf{q}_0$, to satisfy linearized conditions instead. We linearize equation (12.17) according to Newton–Raphson's method (so applying a Taylor series expansion to $\mathbf{g}$, using the first-order approximation) with $\mathbf{g}(\mathbf{x}_{N,0},\mathbf{q}_0)$ written as $\mathbf{g}_0$,

$$\begin{aligned} \mathbf{g}(\mathbf{x}_N,\mathbf{q}) - \mathbf{p} &= \mathbf{g}(\mathbf{x}_{N,0} + \Delta\mathbf{x}, \mathbf{q}_0 + \Delta\mathbf{q}) - \mathbf{p} \\ &= \mathbf{g}_0 + \frac{\partial \mathbf{g}(\mathbf{x}_N,\mathbf{q})}{\partial \mathbf{x}_N}\Delta\mathbf{x} + \frac{\partial \mathbf{g}(\mathbf{x}_N,\mathbf{q})}{\partial \mathbf{q}}\Delta\mathbf{q} = 0 \end{aligned} \quad (12.18)$$

and, using $\mathbf{C}_N^T \bar{\mathbf{U}} \mathbf{q} = \mathbf{D}_N \mathbf{x}_N + \mathbf{D}_F \mathbf{x}_F$, and calling the right-hand side vector of residual forces, the residuals $\mathbf{r}$,

$$\frac{\partial g(\mathbf{x}_N, \mathbf{q})}{\partial \mathbf{x}_N}\Delta\mathbf{x}_N + \frac{\partial g(\mathbf{x}_N, \mathbf{q})}{\partial \mathbf{q}}\Delta\mathbf{q} = \mathbf{p} - \mathbf{g}_0$$

$$\frac{\partial(\mathbf{D}_N\mathbf{x}_N + \mathbf{D}_F\mathbf{x}_F)}{\partial \mathbf{x}_N}\Delta\mathbf{x}_N + \frac{\partial(\mathbf{C}_N^T\bar{\mathbf{U}}\mathbf{q})}{\partial \mathbf{q}}\Delta\mathbf{q} = \mathbf{p} - \mathbf{C}_N^T\bar{\mathbf{U}}_0\mathbf{q}_0$$

$$\mathbf{D}_N\Delta\mathbf{x}_N + \mathbf{C}_N^T\bar{\mathbf{U}}\Delta\mathbf{q} = \mathbf{r}, \quad (12.19)$$

or in block matrix form, the condition equation is,

$$\begin{bmatrix} \mathbf{D}_N & \mathbf{C}_N^T\bar{\mathbf{U}} \end{bmatrix} \begin{bmatrix} \Delta\mathbf{x}_N \\ \Delta\mathbf{q} \end{bmatrix} = \mathbf{r} \quad (12.20)$$

which is an underdetermined system of equations, since the left-hand side matrix is $3n_N \times (3n_N + m)$. In equation (12.20) the mutual interaction of form and forces becomes apparent. Despite introducing certain coordinates $\mathbf{x}_{N,0}$ and enforcing movements in $\Delta\mathbf{x}$ to be zero, we find a solution in a state of equilibrium. Similarly, by enforcing certain fixed force densities in $\mathbf{q}_0$ with corresponding changes in $\Delta\mathbf{q}$ set to zero, we are able, for whatever reason, to impose specific forces on the structure. For the Solemar Therme shell roof, for instance, the shape of the ribs was optimized at certain nodes, by moving them to the correct position and allowing for changes in the force densities, and all the other coordinates, in order to find a new geometry. Convergence will be faster if the changes are not set to zero, and the coordinates $\mathbf{x}_{N,0}$ and force densities $\mathbf{q}_0$ are considered as preferences, as soft constraints, with static equilibrium as a hard constraint. Then, the geometry is a compromise between any initial values $\mathbf{x}_{N,0}$ and $\mathbf{q}_0$, depending on the weightings.

Thus, we seek changes $\Delta\mathbf{x}_N$ and $\Delta\mathbf{q}$ such that the coordinates $\mathbf{x}_N = \mathbf{x}_{N,0} + \Delta\mathbf{x}$ and force densities $\mathbf{q} = \mathbf{q}_0 + \Delta\mathbf{q}$ satisfy the constraints. The optimization problem, subject to an equality constraint, is

$$\min. \Delta\mathbf{x}_N^T\Delta\mathbf{x}_N + w\Delta\mathbf{q}^T\Delta\mathbf{q} \quad (12.21)$$

subject to (12.19),

where $w > 0$ is a single weighting factor. By changing the weighting factor, one can reach either better reference to a prescribed, initial shape $\mathbf{x}_{N,0}$ (w is small) or greater consideration of the desired force distribution defined by $\mathbf{q}_0$ (w is large). In the case of more detailed control of the constraints, the optimization problem is

$$\min. \Delta\mathbf{x}_N^T\mathbf{W}_x\Delta\mathbf{x}_N + \Delta\mathbf{q}^T\mathbf{W}_q\Delta\mathbf{q} \quad (12.22)$$

subject to (1.19)

with $3n_N + m$ individual weighting factors, in the form of the diagonal (which assumes the factors are independent) weighting matrices $\mathbf{W}_x$ and $\mathbf{W}_q$.

The solution of the resulting normal equations, in the form of equation (12.8), is then

$$\begin{bmatrix} \Delta\mathbf{x}_N \\ \Delta\mathbf{q} \end{bmatrix} = \begin{bmatrix} \mathbf{D}_N \\ \frac{1}{w}\bar{\mathbf{U}}\mathbf{C}_N \end{bmatrix} \left(\begin{bmatrix} \mathbf{D}_N & \mathbf{C}_N^T\bar{\mathbf{U}} \end{bmatrix} \begin{bmatrix} \mathbf{D}_N \\ \frac{1}{w}\bar{\mathbf{U}}\mathbf{C}_N \end{bmatrix} \right)^{-1} [\mathbf{r}]$$

$$= \begin{bmatrix} \mathbf{D}_N \\ \frac{1}{w}\bar{\mathbf{U}}\mathbf{C}_N \end{bmatrix} \begin{bmatrix} \mathbf{D}_N\mathbf{D}_N^T & \frac{1}{w}\mathbf{C}_N^T\bar{\mathbf{U}}^2\mathbf{C}_N \end{bmatrix}^{-1}[\mathbf{r}] \quad (12.23)$$

For the next iteration one sets $\mathbf{x}_{N,0} := \mathbf{x}_{N,0} + \Delta\mathbf{x}_N$ and $\mathbf{q}_0 := \mathbf{q}_0 + \Delta\mathbf{q}$, then updates the matrices as well as the residuals and repeats the iteration until the residuals are considered small enough.

Figure 12.6 shows the final result from nonlinear FDM satisfying various constraints. The load-deflection behaviour, for typical load cases, was then checked with FE analysis.

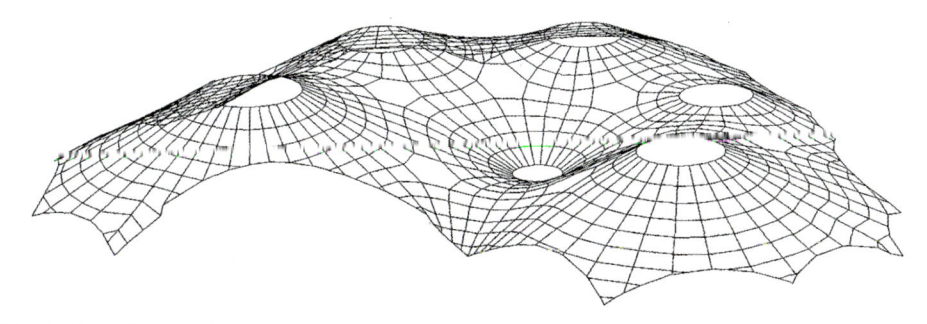

Figure 12.6 The final equilibrium shape (Gründig, Linkwitz, Bahndorf & Ströbel, 1988)

12.3.3 Prefabrication

Figure 12.7 Construction site with all ribs assembled and placement of covering boards and sheeting

Due to the discretization of the shell's surface as a spatial net, the result of the form-finding process is a network of straight centre lines, a wireframe, that idealize the actual meridian and annular ribs. This wireframe does not yet allow a unique definition of the three-dimensionally curved glulam ribs. However, their precise geometrical definition was a prerequisite for prefabrication. In addition, any torsion in the ribs had to be accounted for in their manufacture.

Figure 12.8 Prefabricated, strongly doubly curved glulam beams ready for transport to site

To define their shape, normal and tangent vectors along the ribs were derived from consecutive sets of discrete points along their paths. These vectors were determined by using neighbouring points and calculating their tangent plane. Alternatively, this information can be calculated by first tracing splines along the points of each discretely represented rib as an intermediate step. Together with the dimensions of the ribs' cross section, the edge points of the ribs were calculated. These were defined in the global coordinate system of the entire model, which was also useful for construction on site. In order to calculate the patterns of the ribs for fabrication, the coordinates were transformed to a local coordinate system for each rib. The first axis was defined by the two end points of the rib segment. The other two orthogonal axes were then defined by a best-fitting plane of the three-dimensionally curved rib. This local coordinate system allowed a convenient set-up of the manufacturing tools, minimizing the required heights above the manufacturing floor. The manufacturer was able to set up the tool paths and monitor the fabrication of around 400 individual elements of glulam timber, without any error or redundant manufacturing.

12.3.4 Connections

For the connections of the glulam ribs with cross lap joints, it is important that all their contact surfaces

Figure 12.9 (top) Meridian ribs with cross lap joints, attached to the edge beam, and (bottom) overlaid with the annular ribs

are planar, such that an adequate fit is achieved in assembly. The assembly on the construction site will be more economical if all joints are also prefabricated, avoiding any additional labour on site. The CAD modelling of the cross lap joints and the monitoring of their prefabrication were therefore of the utmost importance. The design and the realization of the joints was possible due to the close cooperation with the master carpenter and based on knowledge of the production methods available in the factory. The beams and their assembly on site are shown in Figures 12.7–12.9.

12.3.5 Application to brief

Similar to the Solemar Therme, we apply nonlinear FDM to our brief. As indicated in Figure 12.10, the roof of the swimming pool, where it connects to the adjacent ice rink, needs to follow a different curve. To this end a target surface is drawn in this area and for a part of the network, target nodes are projected. The corresponding target coordinates are given as input: the initial coordinates $\mathbf{x}_{N,0}$ are changed (blue nodes, Fig. 12.11). In addition, some changes to the topology are made and the fixed boundary nodes with coordinates $\mathbf{x}_F$ are moved to the target surface.

The subsequent nonlinear FDM then seeks to find a compromise between the (partially altered) initial coordinates $\mathbf{x}_{N,0}$ and the original force densities $\mathbf{q}_0$. For a weighting factor of $w=10$ (so relatively favouring the forces), a compromise is found after seven iterations, which is in static equilibrium within a given tolerance. Figure 12.11 shows the new network is closer to the target surface.

12.4 Multihalle, Mannheim

The previous glulam timber shell approaches a funicular shape only partially, due to a number of additional constraints imposed, and its design as an anticlastic

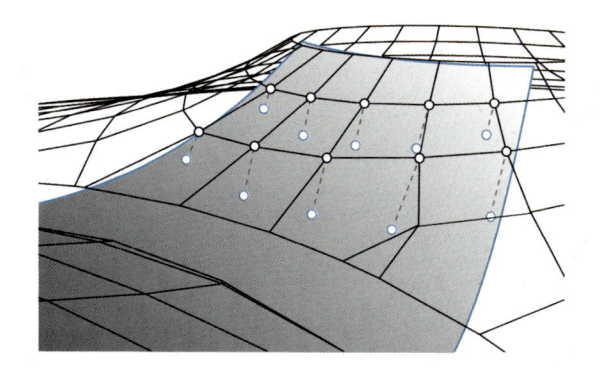

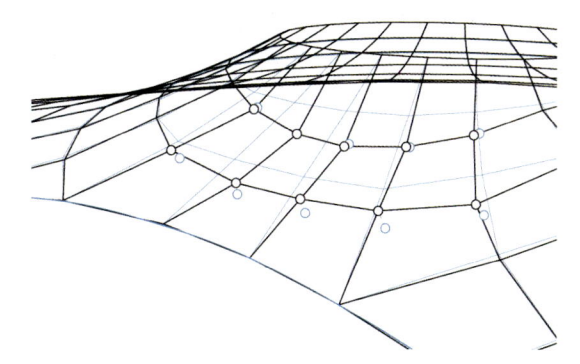

Figure 12.11 Target surface with projected initial coordinates and resulting solution from nonlinear FDM

surface. A funicular shape for a gridshell is more easily obtained from a hanging model. Here, pioneering work was achieved by Frei Otto, who, at the Institute of Lightweight Structures (IL) of the University of Stuttgart, investigated this type of shell. In 1974, he designed the Multihalle in Mannheim with a multi-layered timber grid system resting on an edge beam at the perimeter the shell. The shell spans up to 80m. The shell comprises a layered grid (two layers in each direction) of long timber beams with very small cross sections (50mm × 50mm hemlock), forming a quadrangular mesh. At their intersections the beams are pin-jointed. This feature permits arbitrary changes in the angle. This is essential when the grid was lifted up, in sequences, from its initial flat position on the ground to its final spatial positions (see also Chapter 19).

Figure 12.10 Elevation of preliminary design, showing angle between the two roofs

As the point of departure, after a couple of feasibility studies, a very accurate, scaled hanging model was built (Figs. 4.13 and 4.14), with great accuracy. This model was then measured photogrammetrically, resulting in three-dimensional coordinates of all the nodes. These model-coordinates then served as approximate values for a robust nonlinear computer calculation to guarantee compression forces throughout as well as an equidistance mesh. This calculation was executed by Büro Linkwitz und Preuß, consulting engineers, and was the basis for mapping out the geometry on site as well as a subsequent FE analysis by Ove Arup & Partners. This analysis focused on the load deflection behaviour and buckling. Büro für Baukonstruktionen Wenzel-Frese-Pörtner-Haller were assigned as checking engineers.

12.4.1 Form-finding process

Based on the physical model, an initial coarse mesh with three-dimensional coordinates **x** was chosen as the basis for form finding. The points along the edge were modelled as fixed anchors points. A first linear form finding with constant force densities **q** and constant external loads **p** was carried out, resulting in a solution with unequal bar lengths throughout (Fig. 12.12).

The coordinates, obtained from the initial linear

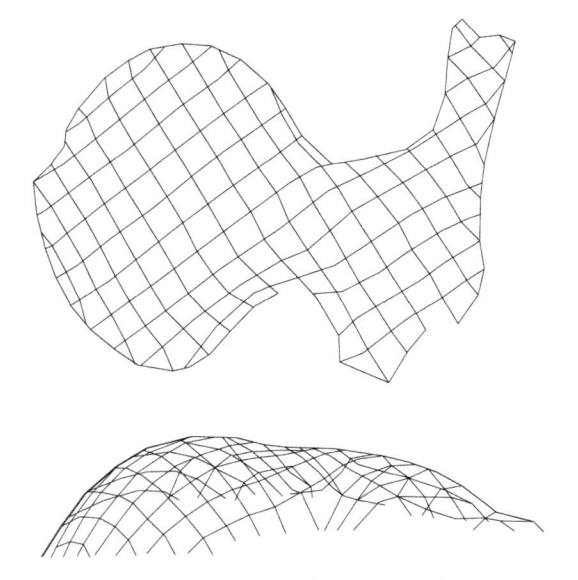

Figure 12.12 Initial linear form-finding result from coarse mesh (Gründig and Schek, 1974)

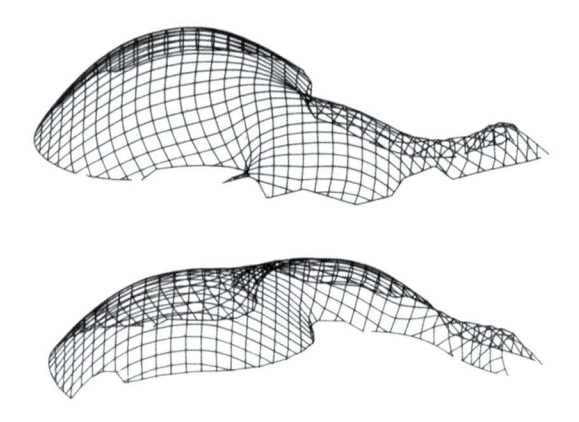

Figure 12.13 Nonlinear form-finding result from fine mesh (Gründig, 1976)

form finding, were then taken as initial coordinates $\mathbf{x}_{N,0}$ for a constrained form finding; constrained such that the interior mesh would be equidistant. Subsequently, a more refined mesh was derived from photogrammetric measurements of the physical model. Other additional constraints in the form finding included constant external forces, small deviations from the model measurements, as well as additional force conditions to avoid sudden changes in angles (caused by sudden changes of force). The result is shown in Figure 12.13.

The constraints mentioned can be expressed as conditions on the forces or force densities, or on the nodal coordinates. Compared to Section 12.3, an additional problem is the prescription of constant lengths in the inner part of network (all branches not connected to the boundaries). We wish to impose an additional equality constraint on m_I interior branch lengths $\mathbf{l}_I$, such that they approach corresponding target lengths $\mathbf{l}_{I,t}$, so the constraint is,

$$\mathbf{l}_I = \mathbf{l}_{I,t}. \qquad (12.24)$$

Linearizing the equation w.r.t. the unknowns, our additional condition equation becomes

$$\frac{\partial(\mathbf{l}_I - \mathbf{l}_{I,t})}{\partial \mathbf{x}}\Delta\mathbf{x} = \mathbf{l}_I - \mathbf{l}_{I,t}$$

$$\mathbf{L}_I^{-1}\bar{\mathbf{U}}_I\mathbf{C}_I\Delta\mathbf{x} = \mathbf{l}_I - \mathbf{l}_{I,t}$$

$$\bar{\mathbf{U}}_I\mathbf{C}_I\Delta\mathbf{x} = \mathbf{L}_I(\mathbf{l}_I - \mathbf{l}_{I,t}) = \mathbf{s}, \qquad (12.25)$$

where subscript I indicates that it applies only to the

constrained branches, and the m_I residuals **s** contains measures of discrepancy between current and target lengths. For example, the branch-node matrix $\mathbf{C}_I$ is of size $3m_I \times 3n$. Then, the solution of the following optimization problem is sought,

$$\min.\ \Delta\mathbf{x}_N^T \Delta\mathbf{x}_N + w\Delta\mathbf{q}^T \Delta\mathbf{q}$$

subject to (12.19) and (12.25)

These equality conditions, rewritten as a single system of equations in block matrix form, are

$$\begin{bmatrix} \mathbf{D}_N & \mathbf{C}_N^T \bar{\mathbf{U}} \\ \bar{\mathbf{U}}_I \mathbf{C}_I & 0 \end{bmatrix} \begin{bmatrix} \Delta\mathbf{x}_N \\ \Delta\mathbf{q} \end{bmatrix} = \begin{bmatrix} \mathbf{r} \\ \mathbf{s} \end{bmatrix} \quad (12.26)$$

where the left-hand side matrix is of size $(3n_N + m_I) \times (3n_N + m)$.

The solution of the normal equations is

$$\begin{bmatrix} \Delta\mathbf{x}_N \\ \Delta\mathbf{q} \end{bmatrix} = \begin{bmatrix} \mathbf{D}_N^T & \mathbf{C}_I^T \bar{\mathbf{U}}_I \\ \frac{1}{w}\bar{\mathbf{U}}\mathbf{C}_N & 0 \end{bmatrix}$$

$$\begin{bmatrix} \mathbf{D}_N \mathbf{D}_N^T + \frac{1}{w}\mathbf{C}_N^T \bar{\mathbf{U}}^2 \mathbf{C}_N & \mathbf{D}_N \mathbf{C}_I^T \bar{\mathbf{U}}_I \\ \bar{\mathbf{U}}_I \mathbf{C}_I \mathbf{D}_N^T & \bar{\mathbf{U}}_I \mathbf{C}_I \mathbf{C}_I^T \bar{\mathbf{U}}_I \end{bmatrix}^{-1} \begin{bmatrix} \mathbf{r} \\ \mathbf{s} \end{bmatrix} \quad (12.27)$$

As before, after updating $\mathbf{x}_{N,0} := \mathbf{x}_{N,0} + \Delta\mathbf{x}_N$ and $\mathbf{q}_0 := \mathbf{q}_0 + \Delta\mathbf{q}$, the procedure iteratively searches until the residuals are below a certain chosen threshold.

12.4.2 Construction

The need for precise form-finding results that are in a state of equilibrium become clear when discussing the method of erection. When the grid for the Multihalle was finally assembled and still laying on the ground, all the angles at the nodes between the series of the timber ribs crossing each other were more or less equal to 90°. The different shapes of the shell during the different and final stages of erection were attained only by angle-changes at the nodes. Any change in the grid angles would immediately change the form of the shell. Any deviation from the equilibrium form would also induce additional (and unwanted) bending moments. To maintain the gridshell for an extended period of time, it is absolutely imperative to prevent the grid angles from changing. This was achieved by spanning diagonal ties, made from thin steel wires, between selected pin-joints in crucial areas of the shell. Moreover, the shell has to be inspected regularly to detect local movements in time. However, experience and long-time monitoring of gridshells have shown that, due to weathering and different loads, some of the fixing cables may become slack. This causes changes in the shape, which unless corrected, have the potential of leading to critical shape changes. Nevertheless, though the Multihalle was originally intended as a temporary building, built in 1976 for the West-German federal garden festival, it remains fully functional to this day. The resulting structure and its method of construction are discussed in further detail in Chapter 19.

12.4.3 Application to brief

We apply the same procedure to the ice rink roof structure as was used for the Multihalle. The objective is to find, within the interior of the network, equidistant segments. This is necessary to consider a method of erection similar to that of the Multihalle. The segments shown in bold in Figure 12.14 have a length difference of 0.2m. Because the topology we

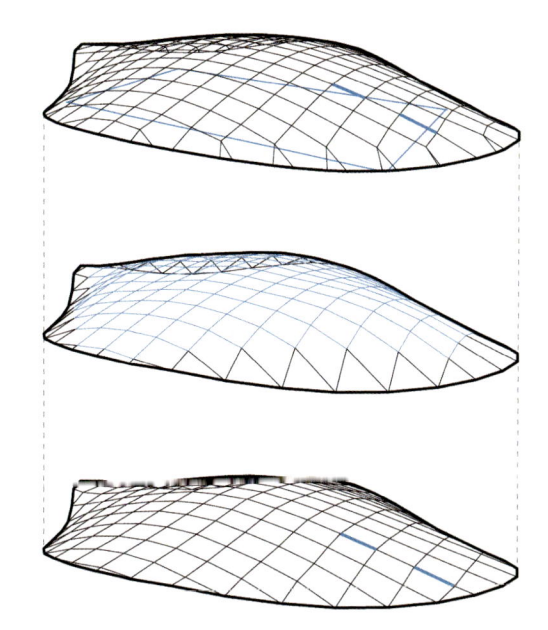

Figure 12.14 Preliminary design, interior network and topology changes and equidistant mesh from nonlinear FDM

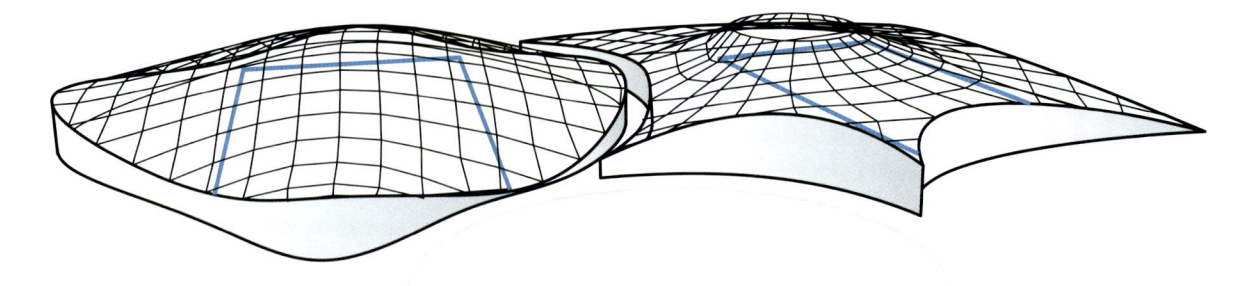

Figure 12.15 Final design of the sports complex after nonlinear FDM

have chosen requires a rectangular (not a square) mesh to fit within the surface, we specify two target lengths: 4.2m in one direction, 6.6m in the other. After our nonlinear form finding, we find an equidistant mesh with these two lengths. Both the residual forces **r** and the length differences **s** have dropped within given tolerances. The segments in bold now have a length difference of 0.0m to an accuracy of three decimal places. The resulting design for the entire sports complex is shown in Figure 12.15.

12.5 Conclusions

This chapter has presented a nonlinear, extended FDM, based on the least-squares method, as it was developed in the 1970s and later applied to the Multihalle timber gridshells and the Solemar Therme glulam shell roof.

Starting from a result from the standard FDM, one can add mechanical constraints as a function of the force densities **q** and/or geometrical constraints as a function of the nodal coordinates $\mathbf{x}_N$. After linearizing the condition equations with respect to these variables, one obtains a system of equations. Based on chosen weighting factors **w**, an optimum is then found between the initial shape and the targets set by the designer, while guaranteeing static equilibrium. This approach avoids having to manually and iteratively change the force densities, such that the constraints are satisfied as much as possible.

The approach is flexible and was successfully applied to two timber shell structures, with very different constraints. To formulate appropriate constraints, an important aspect is the careful consideration of how the structure will be fabricated and constructed. It is also important to consider that the resulting discrete, wireframe model still needs to be dimensioned and materialized, checked using FE analysis (and model testing), and finally developed into sets of precise working drawings.

Key concepts and terms

The **method of least squares** is an approach that finds the approximate solution of under- or overdetermined systems, where there is no unique solution. Instead, we find the best approximation by minimizing the sum of the squares of the residuals.

Normal equations are equations that give the standard, exact solution to least-squares problems.

A **quadratic program** is a type of optimization problem subject to equality and inequality constraints, which minimizes an objective function in quadratic form. The problems in this chapter can be reduced to unconstrained least-squares problems, because there are no inequality constraints.

The **method of Lagrange multipliers** is a standard approach that finds the optimum, the local maximum or minimum of a function, of any equality-constrained problem. It is also called the method of Lagrangian multipliers. In our chapter, it has led us to the standard normal equations.

Exercises

- Consider our standard grid (Fig. 6.12) and linearize its static equilibrium equation with respect to changes in force densities $\Delta\mathbf{q}$ and coordinates $\Delta\mathbf{x}_N$. Apply the normal equations to the linearized equations to solve for $\Delta\mathbf{q}$ and $\Delta\mathbf{x}_N$.
- For a single node in our grid, change the initial

coordinate (x_0, y_0, z_0). Find the geometry for different weightings applied to the force densities and apply five iterations of Gauss–Newton's method. What do you observe?
- Similarly we can change some initial force densities $\mathbf{q}_0$. If we wish to constrain certain member forces instead of force densities, how would you approach this?
- Subdivide the branch-node matrix $\mathbf{C}$ of our grid to an interior and exterior part. Now apply the constraint of equidistant length in the interior network and find the new geometry.
- You have been asked to design another gridshell. For construction purposes, you want to constrain the lengths in the initial, undeformed state and thus prefabricate an equidistant mesh prior to erection. How would you formulate the optimization problem with this constraint? Hint: start from Hooke's law of elasticity which gives a relation between the current and initial length.

Further reading

- 'Einige Bemerkungen zur berechnung von vorgespannten Seilnetzkonstruktionen' or 'Some remarks on the calculation of prestressed cable-net structures', Linkwitz and Schek (1971). This journal publication on FDM, written in German, did not yet mention 'force densities' explicitly, but they are intrinsic to the method. It discusses the method of least squares for application to the cutting patterns of the Munich Olympic Roofs, discussing constraints on forces and initial lengths.
- 'New methods for the determination of cutting pattern of prestressed cable nets and their application to the Olympic Roofs München', Linkwitz (1972). This conference paper at the IASS is the first English publication on FDM.
- 'Die Gleichgewichtsberechnung von Seilnetzen unter Zusatzbedingungen' or 'The computation of cable-nets in equilibrium under additional constraints', Linkwitz et al. (1974). This paper includes constraints on (final) lengths as well, as would later be used in the example of the Multihalle. The constraints are expressed in forces rather than force densities, as in the later papers.
- 'Analytical form finding and analysis of prestressed cable networks', Gründig and Schek (1974). This is the first English conference publication on the approach discussed in this paper and specifically explains the form finding applied during the original design of the Multihalle as its central example. It also discusses the nonlinear approach presented in 'The force density method for form finding and computation of general networks', Schek (1974), particularly its shortcomings with regard to computational efficiency.
- 'Formfinding and computer aided generation of the working drawings for the timber shell roof at Bad Dürrheim', Gründig et al. (1988). This publication discusses the design and construction of the Solemar Therme.
- *Iterative Methods for Sparse Linear Systems*, Saad (2003). This second edition book deals with numerical methods to solve sparse linear systems of equations, and includes a chapter on methods to iteratively solve normal equations.

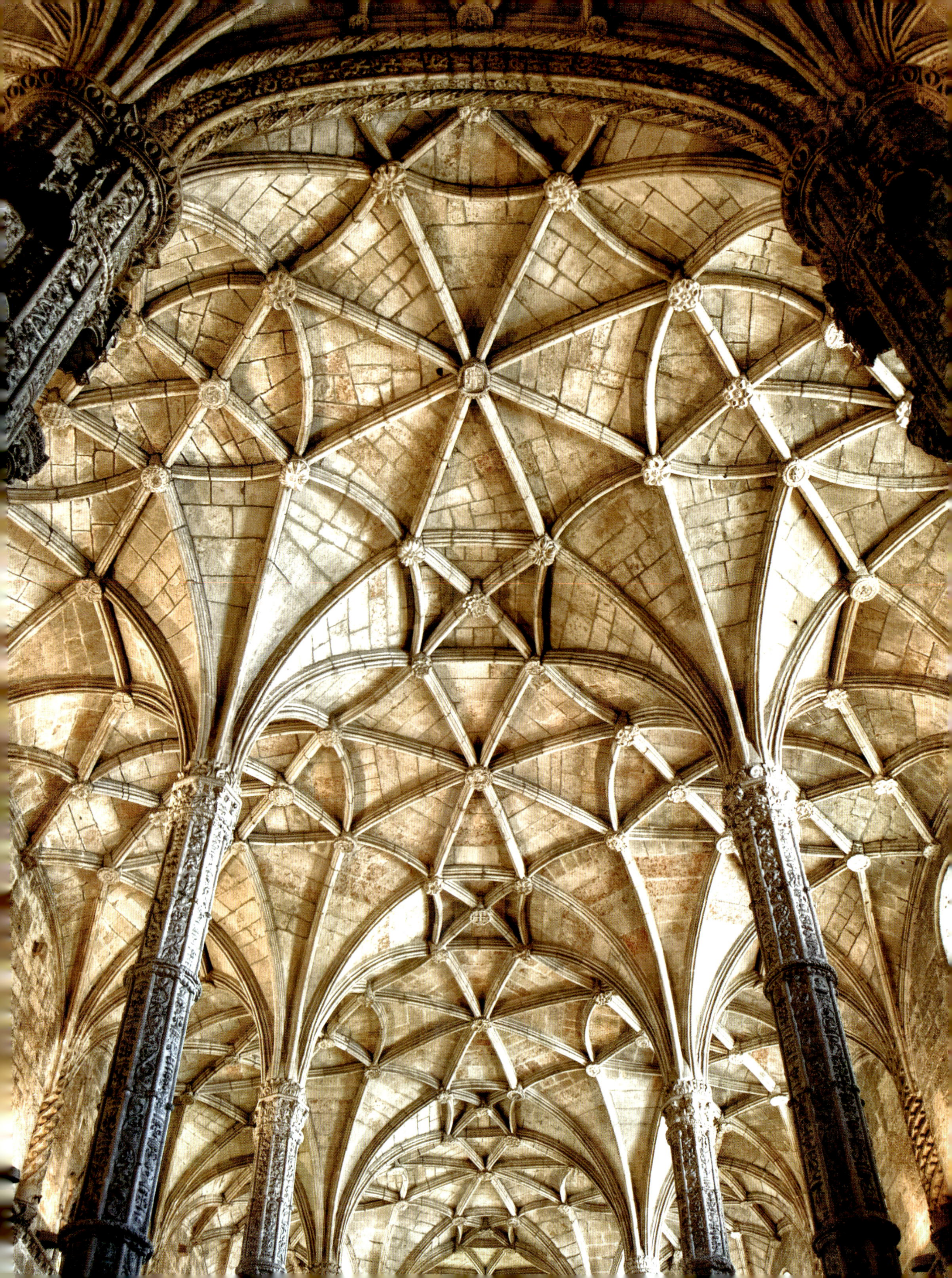

CHAPTER THIRTEEN

Best-fit thrust network analysis

Rationalization of freeform meshes

Tom Van Mele, Daniele Panozzo, Olga Sorkine-Hornung and Philippe Block

LEARNING OBJECTIVES

- Formulate the search for a funicular network in compression that is as close as possible, in a least-squares sense, to a given target surface as a series of optimization problems.
- Implement solvers for the different types of optimization problems.
- Implement this method to optimize freeform shapes; and apply this method to evaluate the stability of structures under asymmetrical loading.

PREREQUISITES

- Chapter 7 on thrust network analysis.

Chapter 7 introduced Thrust Network Analysis (TNA) as a method for designing three-dimensional, compression-only equilibrium networks (thrust networks) for vertical loads using planar, reciprocal form and force diagrams. These diagrams allow the high degree of indeterminacy of three-dimensional force networks to be controlled such that possible funicular solutions for a set of loads can be explored. By manipulating the force diagram (through simple geometric operations), the distribution of horizontal thrusts throughout the network is changed and different three-dimensional configurations are obtained.

There are infinitely many possible variations of the force diagram, each corresponding to a different three-dimensional solution for given loads and boundary conditions. This provides virtually limitless freedom in

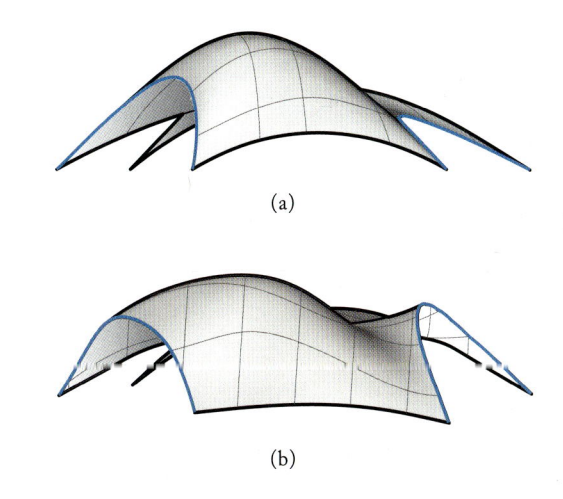

Figure 13.1 (a) The compression-only design for the pavilion as form found in Chapter 7, and (b) the new (geometrical) proposal

the design of three-dimensional equilibrium networks, but it makes it almost impossible to find the specific distribution of forces that corresponds to a specific solution, with a specific shape. For example, the required distribution of forces to achieve the upwards flaring edge of the design proposal depicted in Figure 13.1b is not obvious, and finding it is by no means straightforward.

Therefore, in this chapter, we describe how TNA can be extended to find a thrust network that for a given set of loads best fits a specific target shape. We set this up as an optimization problem and discuss the implementation of an efficient solving strategy.

The brief

Chapter 7 described the design of a vaulted, unreinforced cut-stone masonry pavilion for a park in Austin, Texas, USA, that covers the stage and spectator area of a small performance area of 20m × 15m. The client has requested modifications to the shape developed in Chapter 7 to improve the integration of the pavilion into the surrounding landscape and allow access to its top surface to provide visitors with alternative views of the site and the vault. Although the dramatic asymmetry between the two sides of the vault is a key feature of the form, the client would prefer a deeper opening on the side of the shallow main arch to let in more light and make that side of the pavilion look more open and inviting.

The original design and the new proposal are shown in Figure 13.1. Key features of the new design are thus the smoother transition between the landscape and the structure on one side, and the flaring edge on the other.

We have been asked to determine whether the new, geometrically constructed shape is feasible for an unreinforced, masonry stone structure.

13.1   TNA preliminaries

Since this chapter describes an extension of TNA, we assume the reader to be familiar with its fundamental principles as presented in Chapter 7. Here, we briefly summarize those mathematical elements, notations and conventions of TNA that are required for the optimization algorithm.

Let Γ and Γ^* be two planar graphs with an equal number of edges, m. If Γ is a *proper cell decomposition of the plane*, and Γ^* is its *convex, parallel dual*, then Γ and Γ^* are the form and force diagram of a (three-dimensional) thrust network **G** that is in equilibrium with a set of vertical loads applied to its nodes, and has Γ as its horizontal projection. Two graphs are parallel if all corresponding edges are parallel, and convex if all their faces are convex. We call two graphs or diagrams reciprocal if one is the parallel dual of the other. The force diagram of a thrust network **G** is thus the convex reciprocal of the form diagram of **G**. A proper cell decomposition of the plane divides the plane into cells formed by (unbounded) convex polygons such that:

- every point in the plane belongs to at least one cell;
- the cells have disjoint (i.e. non-overlapping) interiors;
- any two cells are separated by exactly one edge.

We can describe Γ as a pair of matrices **V** and **C**. **V** = [**x**|**y**] is an $n \times 2$ matrix, which contains the coordinates in the horizontal plane of the i-th node in its i-th row. n is the number of nodes in Γ. **C** is the branch-node matrix: an $m \times n$ matrix that contains the connectivity information of the graph of Γ (see Section 7.3.2). Note that **C** is the transpose of the incidence matrix of Γ. The edges of Γ, represented as vectors, can be extracted from **V** and **C** by computing the $m \times 2$ matrix **E** = **CV** = [**u**|**v**], which contains the coordinate differences of the i-th branch in its i-th row. Therefore, the length of the i-th edge, l_i, can be computed by taking the norm of the i-th row of **E**. **L** is the $m \times m$ diagonal matrix of the vector of edge lengths **l**. **V***, **C***, **E***, **L*** are defined equivalently for the reciprocal diagram.

The force densities **q** of the network are the ratios of the lengths of corresponding edges of Γ^* and Γ:

$$\mathbf{Q} = \mathbf{L}^{-1}\mathbf{L}^*, \qquad (13.1)$$

with **Q** the diagonal $m \times m$ matrix of **q**.

The nodes of Γ are divided into two sets, N and F, that denote the (non-fixed) free nodes and the (fixed) support nodes, respectively. The heights of the free nodes of the thrust network **G** described by Γ and Γ^* are computed as:

$$\mathbf{z}_N = \mathbf{D}_N^{-1}(\mathbf{p} - \mathbf{D}_F \mathbf{z}_F), \quad (13.2)$$

with $\mathbf{D}_N$ and $\mathbf{D}_F$ the columns of $\mathbf{D} = \mathbf{C}_N^T \mathbf{Q} \mathbf{C}$ corresponding to the n_N free and n_F fixed nodes, respectively. $\mathbf{p}$ is the vector of external loads applied at the free nodes and $\mathbf{z}_F$ are the heights of the fixed support nodes.

13.2 Formulation of the problem

Let $\mathbf{G}$ be a thrust network generated from a pair of reciprocal diagrams Γ and Γ^*, and S a target surface. Keeping the form diagram Γ fixed, our objective is to optimize the force diagram Γ^* such that the network $\mathbf{G}$ is as close as possible, in the least-squares sense, to the target S. The variables we are optimizing for are the nodes $\mathbf{V}^*$ of the force diagram. Formally:

$$\operatorname*{argmin}_{\mathbf{V}^*} \sum_i (z_i - s_i)^2 \quad (13.3)$$

subject to Γ^* is the convex reciprocal of Γ, (13.4)

where i runs over the nodes of Γ and z_i and s_i are, respectively, the height of the network and the surface at the i-th node.

Note that the heights z_i do not directly depend on the variables $\mathbf{V}^*$. However, we can compute the heights z_i from the force densities $\mathbf{q}$ using equation (13.2). The energy is thus a function of $\mathbf{q}$:

$$f(\mathbf{q}) = (\mathbf{z}_N(\mathbf{q}) - \mathbf{s}_N)^2. \quad (13.5)$$

Therefore, to find the best-fit solution, we must search for the force densities $\mathbf{q}$ that minimize the energy according to equation (13.5) and allow for a force diagram Γ^* that satisfies constraint, expressed in equation (13.4):

$$\operatorname*{argmin}_{\mathbf{q}} (\mathbf{z}_N(\mathbf{q}) - \mathbf{s}_N)^2 \quad (13.6)$$

subject to Γ^* is the convex reciprocal of Γ.

In the following sections, we describe the strategy for solving this problem.

13.3 Overview

Starting from a given target surface S, the solving procedure consists of two main steps.

1. Generate a starting point:
   a. choose a form diagram;
   b. generate an initial force diagram;
   c. optimize the scale of the initial force diagram.
2. Find a best-fit solution by repeating the following two-step procedure until convergence:
   a. find the force densities $\mathbf{q}$ that minimize energy according to equation (13.6), ignoring the equilibrium constraints;
   b. for the current force densities, find the force diagram Γ^* that is as parallel as possible to the form diagram.

We discuss each of the steps and substeps in detail in the following sections.

13.4 Generate a starting point

In this section, we discuss the generation of a starting point for the iterative part of the optimization process. First, we choose a form diagram and generate an initial force diagram, and then optimize the scale of this force diagram.

13.4.1 The form diagram

In order to be able to obtain a well-fitting thrust network for a given target, a force diagram must be chosen that is based on the target's features. Our choice of form diagram for the target surface described in the brief is depicted in Figure 13.2b. Note that in comparison with the original diagram of Chapter 7, we have added force paths that gradually divert horizontal forces to the supports before they hit the open edges. This provides finer control over the equilibrium of these edges and will, for example, allow the upward flaring edge to develop.

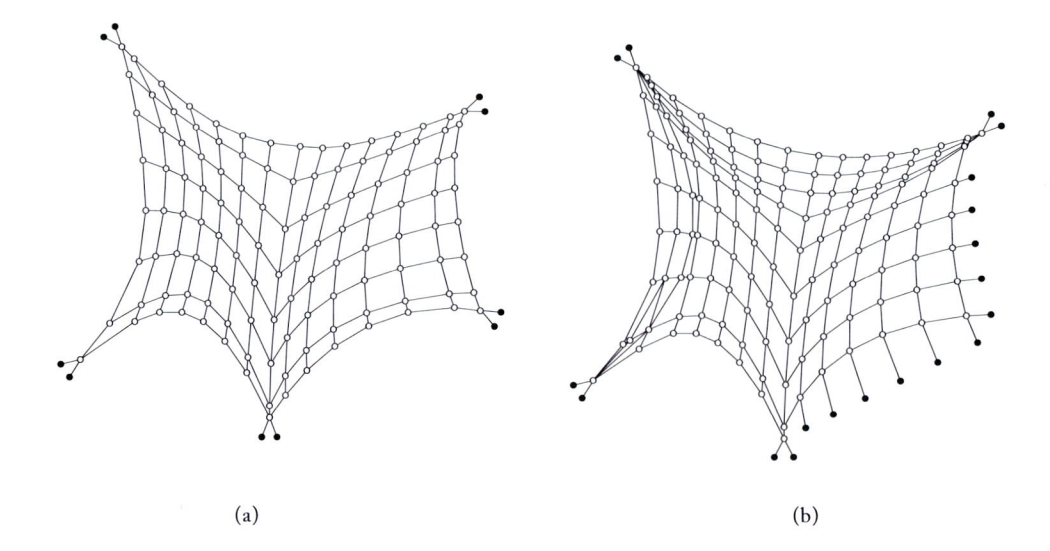

Figure 13.2 (a) The form diagram of the original design, and (b) the modified form diagram used here.

13.4.2 An initial force diagram

To generate an initial force diagram, that is, a convex reciprocal of the form diagram, we use an iterative procedure. We start with the *centroidal dual* of the form diagram, rotated 90° as depicted in Figure 13.3a. The form diagram's centroidal dual is the dual of which the vertices or nodes coincide with the centroids of the faces of the form diagram. The corresponding edges of the form diagram and this rotated dual are generally not parallel. Therefore, at each iteration of the procedure we perform the following calculations. First, we compute a set of target directions $\mathbf{t}_i$ for the edges of the new force diagram by averaging the directions of the (fixed) form diagram and the current force diagram:

$$\mathbf{t}_i^* = \left(\frac{\mathbf{e}_i}{l_i} + \frac{\mathbf{e}_i^*}{l_i^*}\right) \bigg/ \left(\frac{\mathbf{e}_i}{l_i} + \frac{\mathbf{e}_i^*}{l_i^*}\right), \quad (13.7)$$

with $\mathbf{e}_i$ the *i*-th row of $\mathbf{E}$, representing the *i*-th edge of the form diagram, and l_i its length; and, similarly, $\mathbf{e}_i^*$ the *i*-th row of $\mathbf{E}^*$, representing the *i*-th edge of the current force diagram, and l_i^* its length. Note that $\mathbf{e}_i/l_i$ is constant, since the form diagram is fixed, and thus does not need to be recalculated at each iteration. Using these target vectors, the edges of the new, 'ideal' (i.e. parallel) force diagram are thus:

$$\mathbf{e}_i^* = l_i^* \mathbf{t}_i^*. \quad (13.8)$$

Note that this new diagram cannot be properly connected, since its edges have the same lengths but different directions than before. Therefore, we search for a diagram that is similar to the ideal one, but connected, by solving the following minimization problem:

$$\underset{\mathbf{V}^*}{\operatorname{argmin}} \sum (\mathbf{e}_i^* - l_i^* \mathbf{t}_i^*)^2 \quad (13.9)$$

$$\text{subject to } \mathbf{V}_0^* = \mathbf{0} \quad (13.10)$$

Note that without the constraint in equation (13.10), there are infinite graphs that minimize energy in equation (13.9), all identical up to a translation. Fixing a single node ($\mathbf{V}_0^*$) to an arbitrary value makes the solution unique.

We repeat these steps until a convex reciprocal of the form diagram is found. The centroidal dual of the form diagram and the initial force diagram derived from it are depicted in Figure 13.3.

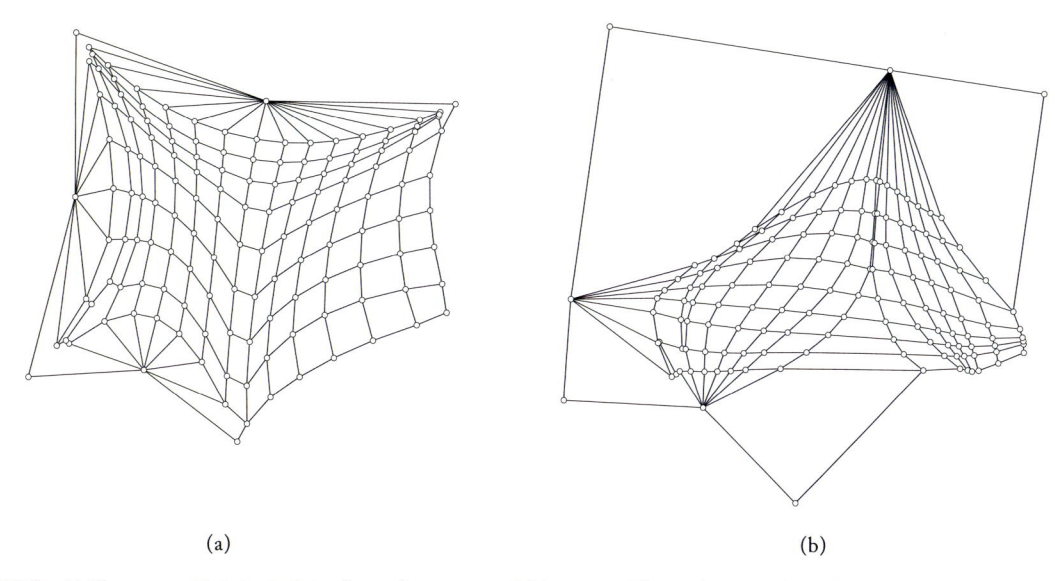

Figure 13.3 (a) The centroidal dual of the form diagram, and (b) an initial force diagram, based on the centroidal dual

13.4.3 Scale optimization

In TNA, we can change the depth of a funicular network simply by uniformly scaling all horizontal forces, which is equivalent to uniformly scaling the force diagram. Higher and lower thrusts result in shallower and deeper solutions, respectively. Therefore, before starting the optimization process, we can reduce energy according to equation (13.5), without changing the distribution of forces, simply by changing the scale of the force diagram. The optimal scaling factor r is obtained by minimizing:

$$\underset{r}{\mathrm{argmin}}\,(\mathbf{z}-\mathbf{s})^2 \qquad (13.11)$$

$$\text{subject to } \mathbf{Dz} - r\mathbf{p} = \mathbf{0}, \qquad (13.12)$$

this is a linear least-squares problem subject to linear equality constraints and can be solved using the method of Lagrange multipliers. We rewrite the problem introducing additional variables, one for every equality constraint, obtaining the following Lagrange function:

$$\begin{aligned}\Lambda(\mathbf{z}, r, \lambda) &= (\mathbf{z}-\mathbf{s})^2 + \lambda^{\mathrm{T}}(\mathbf{Dz} - r\mathbf{p}) \\ &= \mathbf{z}^{\mathrm{T}}\mathbf{z} - 2\mathbf{z}^{\mathrm{T}}\mathbf{s} - \mathbf{s}^{\mathrm{T}}\mathbf{s} + \lambda^{\mathrm{T}}(\mathbf{Dz} - r\mathbf{p}) \quad (13.13)\end{aligned}$$

with λ the Lagrange multipliers. The unique minimum of the Lagrange function is the solution we are looking for. Setting the partial derivatives of Λ equal to zero leads to the following linear system, the solution of which is the scaled thrust network and the scaling factor r:

$$\begin{bmatrix} 2 & \mathbf{0} & \mathbf{D}^{\mathrm{T}} \\ \mathbf{0} & -\lambda^{\mathrm{T}} & \mathbf{0} \\ \mathbf{D} & -\mathbf{p} & \mathbf{0} \end{bmatrix} \begin{bmatrix} \mathbf{z} \\ r \\ \lambda \end{bmatrix} = \begin{bmatrix} 2\mathbf{s} \\ \mathbf{0} \\ \mathbf{0} \end{bmatrix} \qquad (13.14)$$

Figure 13.4 shows the scaled force diagram and corresponding thrust network in comparison with the target surface.

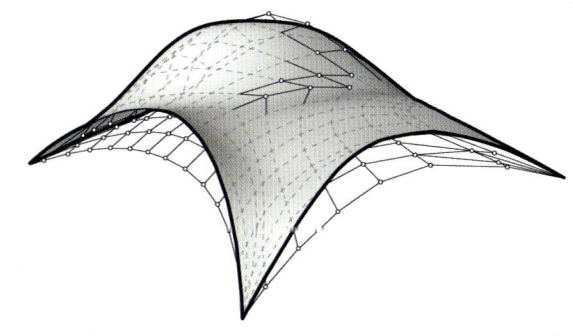

Figure 13.4 By uniformly scaling the force diagram we obtain a funicular network that is closer to the target surface

13.5 Iterative procedure

In the previous section, we have generated an initial pair of reciprocal diagrams, and rescaled the force diagram such that the corresponding thrust network is, for that distribution of thrusts, as close as possible to the target. Rescaling the force diagram has changed the depth of the funicular solution, but the overall shape has stayed the same, because the proportional distribution of thrusts remained unchanged.

During the following iterative procedure, we redistribute the thrust forces and thus change the shape of the thrust network until a better fit of the target is found. Each iteration of this procedure consists of two steps. In the first step, we optimize the force densities without taking into account the reciprocity constraint in equation (13.6) on the force diagram. In the second step of each iteration, we search for a force diagram that generates these optimized force densities and is as parallel as possible to the form diagram. We repeat these steps until a solution is found with optimal force densities and parallel diagrams.

13.5.1 Force densities optimization

To optimize the force densities, we minimize energy according to equation (13.5) using a gradient descent algorithm (Nocedal and Wright, 2000). In short, this means that we move from the current force densities to the next using

$$\mathbf{q}^{t+1} = \mathbf{q}^{t} - \lambda \, \nabla f(\mathbf{q}^{t}) \qquad (13.15)$$

with $\nabla f(\mathbf{q})$ the direction of maximum increase or decrease of f at $\mathbf{q}$ (i.e. the gradient) and λ a step length that satisfies the strong Wolfe conditions (Nocedal and Wright, 2000).

The gradient of f can be efficiently evaluated in closed form:

$$\frac{\partial f(\mathbf{q})}{\partial \mathbf{q}} = \frac{\partial}{\partial \mathbf{q}}\left((\mathbf{Z}_N - \mathbf{S}_N)^2\right) = 2(\mathbf{Z}_N - \mathbf{S}_N)\frac{\partial \mathbf{z}_N}{\partial \mathbf{q}}, \quad (13.16)$$

where $\mathbf{Z}_N$ and $\mathbf{S}_N$ are diagonal matrices corresponding to $\mathbf{z}_N$ and $\mathbf{s}_N$ respectively.

Using equation (13.2), the gradient of $\mathbf{z}_N$ can be written as

$$\begin{aligned}\frac{\partial \mathbf{z}_N}{\partial \mathbf{q}} &= \frac{\partial\left(\mathbf{D}_N^{-1}(\mathbf{p} - \mathbf{D}_F \mathbf{z}_F)\right)}{\partial \mathbf{q}} \\ &= \frac{\partial \mathbf{D}_N^{-1}}{\partial \mathbf{q}}(\mathbf{p} - \mathbf{D}_F \mathbf{z}_F) - \mathbf{D}_N^{-1}\frac{\partial(\mathbf{p} - \mathbf{D}_F \mathbf{z}_F)}{\partial \mathbf{q}} \\ &= \frac{\partial \mathbf{D}_N^{-1}}{\partial \mathbf{q}}(\mathbf{p} - \mathbf{D}_F \mathbf{z}_F) + \mathbf{D}_N^{-1}\left(\mathbf{C}_N^T \mathbf{C} \begin{bmatrix} \mathbf{0} \\ \mathbf{z}_F \end{bmatrix}\right),\end{aligned} \qquad (13.17)$$

where we used

$$\mathbf{D}_F \mathbf{z}_F = \mathbf{C}_N^T \mathbf{Q} \mathbf{C} \begin{bmatrix} \mathbf{0} \\ \mathbf{z}_F \end{bmatrix}. \qquad (13.18)$$

Finally, $\partial \mathbf{D}_N^{-1}/\partial \mathbf{q}$ can be rewritten using the identity (Petersen and Pedersen, 2008)

$$\frac{\partial \mathbf{A}^{-1}}{\partial x} = -\mathbf{A}^{-1}\frac{\partial \mathbf{A}}{\partial x}\mathbf{A}^{-1}. \qquad (13.19)$$

Applied to $\mathbf{D}_N^{-1}$ this gives

$$\begin{aligned}\frac{\partial \mathbf{D}_N^{-1}}{\partial \mathbf{q}} &= -\mathbf{D}_N^{-1}\frac{\partial\left(\mathbf{C}_N^T \mathbf{Q} \mathbf{C}\begin{bmatrix}\mathbf{I}\\\mathbf{0}\end{bmatrix}\right)}{\partial \mathbf{q}}\mathbf{D}_N^{-1} \\ &= -\mathbf{D}_N^{-1}\mathbf{C}_N^T \mathbf{C}\begin{bmatrix}\mathbf{I}\\\mathbf{0}\end{bmatrix}\mathbf{D}_N^{-1},\end{aligned} \qquad (13.20)$$

where we used

$$\mathbf{D}_N = \mathbf{C}_N^T \mathbf{Q} \mathbf{C}\begin{bmatrix}\mathbf{I}\\\mathbf{0}\end{bmatrix}. \qquad (13.21)$$

Substituting equations 13.20 and 13.17 in 13.16, and using equation 13.2, we get

$$\nabla f(\mathbf{q}) = -2(\mathbf{z}_N - \mathbf{s})^T$$
$$\left(\mathbf{D}_N^{-1}\mathbf{C}_N^T \mathbf{C}\begin{bmatrix}\mathbf{I}\\\mathbf{0}\end{bmatrix}\mathbf{z}_N + \mathbf{D}_N^{-1}\mathbf{C}_N^T \mathbf{C}\begin{bmatrix}\mathbf{0}\\\mathbf{z}_F\end{bmatrix}\right). \qquad (13.22)$$

This gives the final expression of $\nabla f(\mathbf{q})$,

$$\nabla f(\mathbf{q}) = 2(\mathbf{Z}_N - \mathbf{S}_N)\mathbf{D}_N^{-1}\mathbf{C}_N^T \mathbf{C}\mathbf{z}. \qquad (13.23)$$

In the evaluation of $\nabla f(\mathbf{q})$, we need to compute $\mathbf{D}_N^{-1}$. To avoid computing the dense inverse explicitly, we can compute $\mathbf{D}_N^{-1}\mathbf{C}_N^T \mathbf{C}\mathbf{z}$ indirectly by solving the equivalent sparse linear system:

$$\mathbf{D}_N \mathbf{x} = \mathbf{C}_N^T \mathbf{C}\mathbf{z}. \qquad (13.24)$$

Since $\mathbf{D}_N$ is symmetric and positive definite, we can

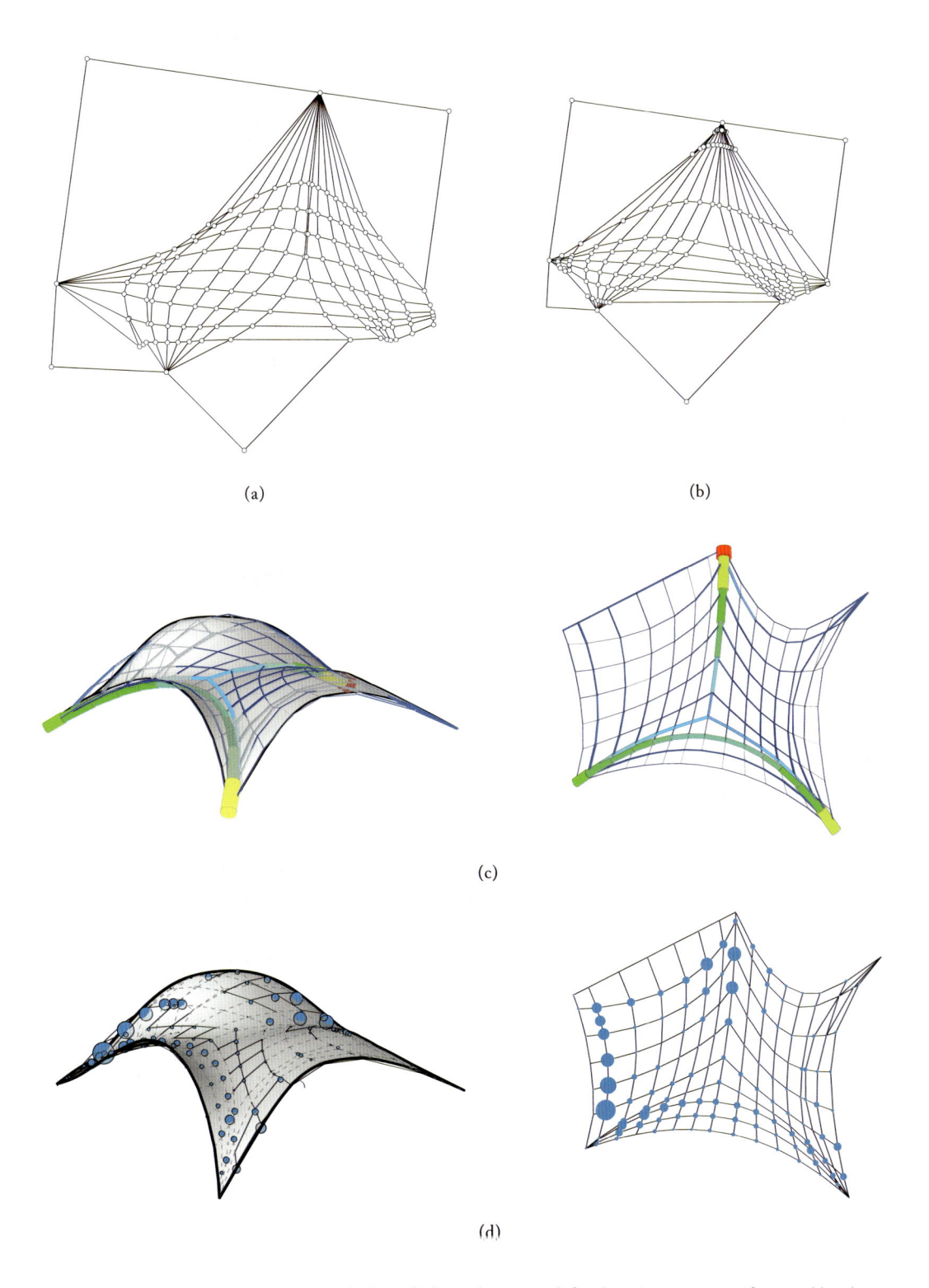

CHAPTER THIRTEEN: BEST-FIT THRUST NETWORK ANALYSIS **163**

(a)

(b)

(c)

(d)

Figure 13.5 Result of the optimization process: the best-fit funicular network for the given target surface and loads

efficiently solve the system using the sparse Cholesky decomposition (Nocedal and Wright, 2000). We first compute the Cholesky decomposition of $\mathbf{D}_N$:

$$\mathbf{D}_N = \mathbf{L}\mathbf{L}^T \quad (13.25)$$

with $\mathbf{L}$ a lower triangular matrix. Then, we solve the system of equations

$$\mathbf{L}\mathbf{y} = \mathbf{C}_N^T \mathbf{C}\mathbf{z} \quad (13.26)$$

for $\mathbf{y}$. This is done using forward substitution, since $\mathbf{L}$ is lower triangular. Finally, we find $\mathbf{x}$ by solving:

$$\mathbf{L}^T \mathbf{x} = \mathbf{y}. \quad (13.27)$$

13.5.2 Force diagram optimization

Given the current optimized force densities $\mathbf{q}$, we search for the force diagram Γ^* that is as parallel as possible to the form diagram while generating these force densities.

This procedure is similar to the one discussed in Section 13.4. We first compute a set of target directions for the edges of Γ^* using equation 13.7. Then, we generate target lengths for the edges of Γ^* using equation (13.1),

$$l_i^* = q_i l_i \quad (13.28)$$

Now we know the directions and lengths of the edges of the ideal Γ^* that generates the current force densities and is parallel to the form diagram. As before, this graph will generally not be connected. To compute a graph that is similar to the ideal one, but connected, we solve the same minimization problem as in equation (13.9).

The final result of the optimization process is shown in Figure 13.5. The figure depicts the scaled reciprocal diagram (Fig. 13.5a), which was the starting point for the optimization, and the final, optimized diagram (Fig. 13.5b). The thicknesses of the branches (Fig. 13.5c) visualize the distribution of forces in the thrust network, and the spheres (Fig. 13.5d) represent the deviation from the target surface.

13.6 Basic coding

Figure 13.6 is a flowchart that gives an overview of a complete implementation of the algorithm discussed in the previous section.

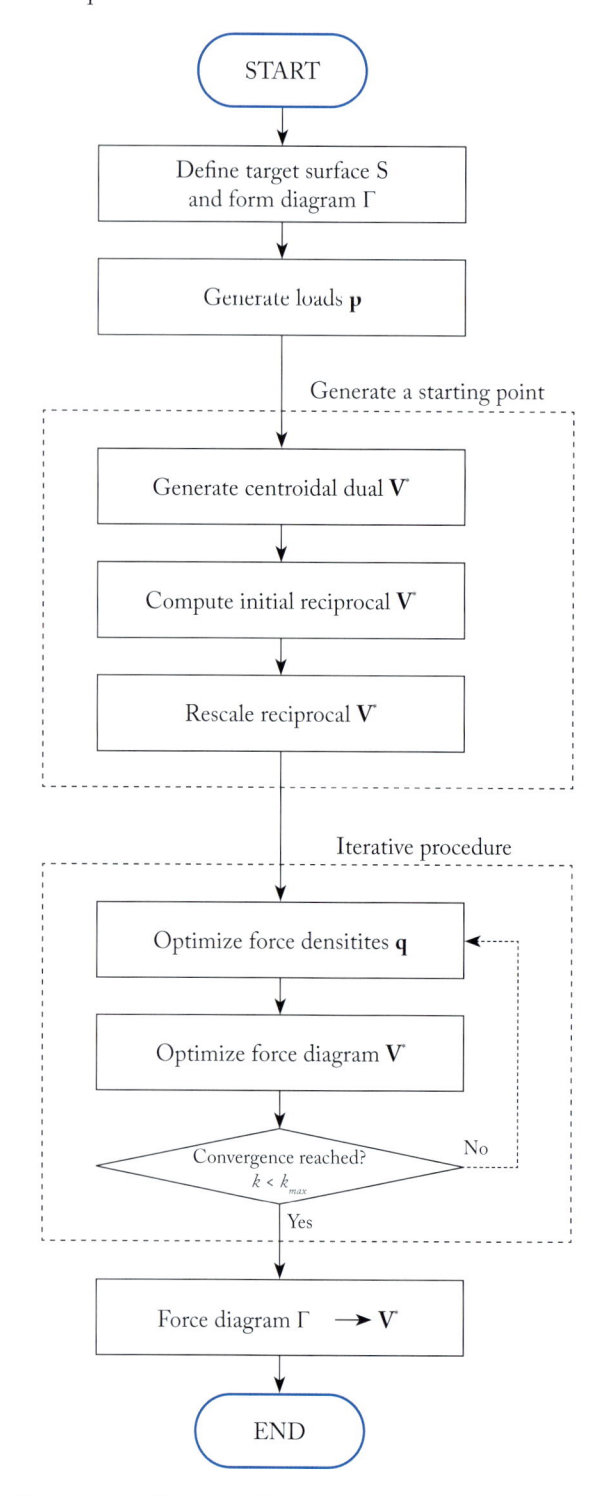

Figure 13.6 Flowchart of a complete implementation

13.7 Assessment of the proposed design

A masonry structure is considered safe if a network of compressive forces contained within (the middle third or kern of) the vault's geometry can be found for all possible loading cases (see Chapter 7).

For most masonry structures, the dominant loading case governing their design is self-weight. Therefore, to evaluate the feasibility of the proposed design, we first use the algorithm described in Section 13.5 to find the best-fit thrust network for the self-weight of the design, offset the solution with the thickness used to calculate the self-weight, and then use the algorithm to search for thrust networks contained within the kern of the new geometry for other loading cases.

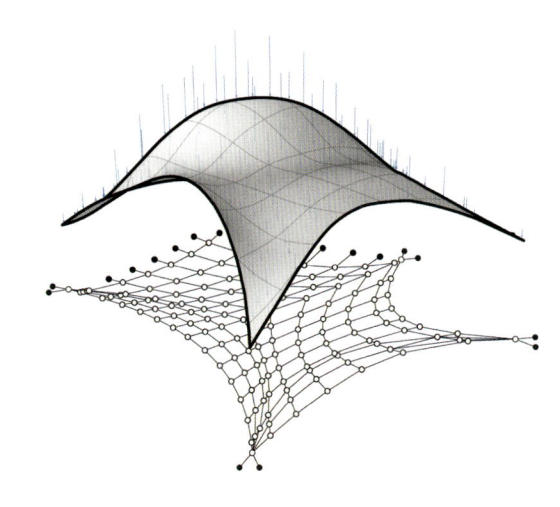

Figure 13.7 The self-weight of the structure distributed over the nodes according to their tributary areas

13.7.1 Self-weight

We can calculate the weight per square metre of the proposed design using a chosen thickness and the weight of the stone: 0.3m × 2,400kgm^{-3} = 720kgm^{-2}. The equivalent distribution of point loads on the nodes of the form diagram according to their respective tributary areas on the target is depicted in Figure 13.7. Note that the compression-only solution captures the design intent of the client very well and allows for the realization of the key features of the shape.

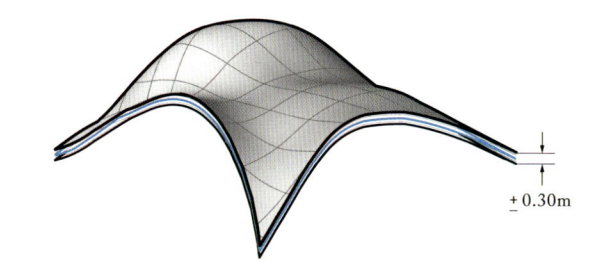

Figure 13.8 The new shell defined as an offset from the best-fit thrust network (blue) for the structure's self-weight

13.7.2 Additional live loads

For the further evaluation of all additional load cases, we define the geometry of the vault by taking the best-fit thrust network determined in the previous step and setting it off by 0.15m on both sides (Fig. 13.8). As discussed, the structure can be considered safe if we can find a thrust network within the kern of its geometry for all additional loading cases (see Section 7.1.3).

In a real project, there are many different, additional loading cases and they should all be considered. However, here, we only discuss the case resulting from the allowed public access to the pavilion's surface.

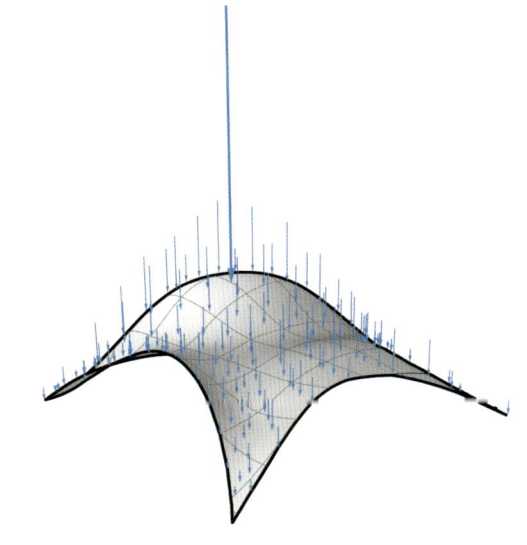

Figure 13.9 The self-weight of the vault combined with a hugely exaggerated additional point load

Typical values are 5.0kNm^{-2} for patch and 7.0kN for point loads. In this example, a point load is applied, because it has a more noticeable effect on the vault. Furthermore, a much higher point load of 100kN is used, to further emphasize the effect. Note that this is roughly equivalent to Godzilla standing on one leg on the viewing platform. The location of the additional load is depicted in Figure 13.9.

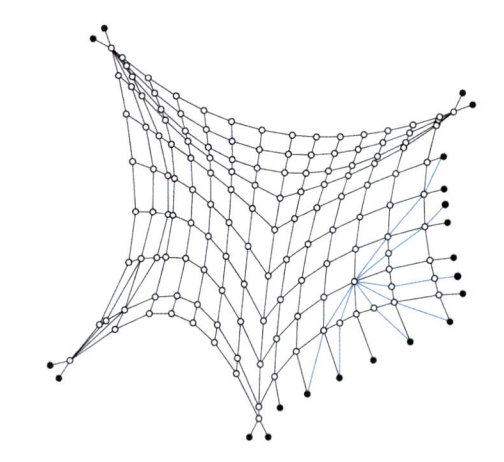

Figure 13.10 To find a best-fit funicular network for the combination of self-weight and additional live load(s) we draw a new form diagram that provides appropriate force paths

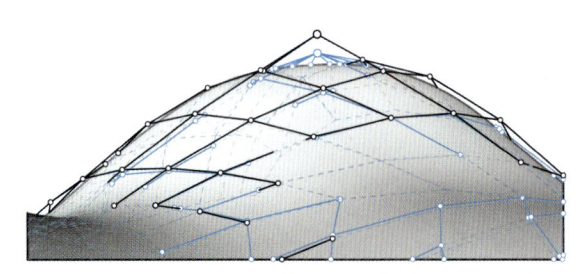

Figure 13.11 Comparison of the best-fit thrust network corresponding to the original force pattern (blue) and to the modified force pattern (black). The modified pattern clearly produces a much better fitting result

In order to find a compression-only force network that fits within the newly determined kern of the vault, we simply run the algorithm as before using point loads that represent the combination of self-weight and the additional loading.

However, as before, it is important that we start with a form diagram that provides force paths along which the loads can 'flow' to the supports. Therefore, we superimpose a force pattern on the previous form diagram that radiates from the point of application of the additional load to the supports (blue in Figure 13.10).

With this new form diagram and combined loads (self-weight and point load), we repeat the best-fit search to find the best-fit funicular network to the target surface. The result of this search is depicted in Figure 13.11. Note that, for such an extreme loading case, it is sufficient that the thrust network stays within the entire section of the vault (not just the middle third), although this would represent an equilibrium state at the onset of collapse. If such a solution cannot be found, the vault's thickness should be modified; for example, by iteratively searching for the bounding box of all loading cases.

13.8 Conclusion

This chapter has shown how to find a thrust network that best fits a given target surface for a given set of loads, formulate this search as a series of optimization problems, and use appropriate and efficient solving strategies for each of them.

The presented technique was applied to the assessment of the structural feasibility of a vaulted masonry structure with a complex, geometrically designed shape (Fig. 13.4). This entailed the search for a best-fit thrust network for the dominant loading case of self-weight, the derivation of a new geometry from this result, and the assessment of the safety of the new geometry in all other loading cases.

Another important application of the technique described in this chapter is the equilibrium analysis of historic masonry vaults with complex geometry, such as the sophisticated nave vaults of the Church of Santa Maria of Bélem at the Jerónimos monastery, completed in the early sixteenth century, shown on page 156. The approach to such an analysis is very similar to the previously discussed assessment of a design proposal. Provided that sufficient information about the geometry of the structure in its current state is available, the target surface can be taken as the surface that lies at the middle of the structure's section, and an appropriate form diagram can be derived from the structure's rib pattern and stereotomy. Otherwise, the procedure is exactly the same. The results for the nave vaults of the Jerónimos monastery are depicted in Figure 13.12.

CHAPTER THIRTEEN: BEST-FIT THRUST NETWORK ANALYSIS 167

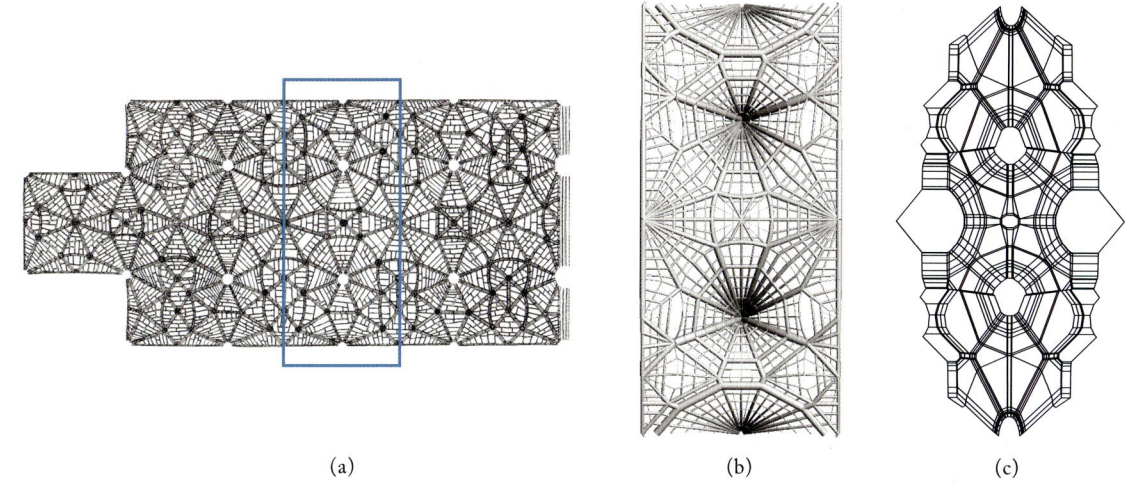

(a) (b) (c)

Figure 13.12 (a) Rib and stereotomy pattern of the nave vaults of the Jerónimos monastery. Resulting (b) form diagram with sizing proportional to the forces in (c) the force diagram

Key concepts and terms

A **graph of a network of branches and node** is a drawing that visualizes the connectivity of the network.

A **planar graph** is planar if it can be drawn on a sheet of paper without overlapping edges; in other words, if it can be embedded in the plane.

A **dual graph** is a graph with the same number of edges as the original, but in which the meaning of nodes and faces has been swapped.

The **centroidal dual** of a graph is the dual of which the vertices or nodes coincide with the centroids of the faces of the original graph.

The **convex, parallel dual** is a dual graph with convex faces and edges parallel to the corresponding edges of the original graph.

Reciprocal diagrams are two planar diagrams or graphs that are said to be reciprocal if one is the convex, parallel dual of the other. See Chapter 7 for an alternative definition.

Line search strategy is one of the two basic iterative approaches to finding a local minimum of an objective function; the other is trust region. It first finds a descent direction along which the objective function reduces and then computes a step size that decides how far it should be moved along that direction. The step size can be determined either exactly or inexactly.

A **gradient descent algorithm** is a type of line search in which steps are taken proportional to the negative of the gradient of the objective function at the current point.

Strong Wolfe conditions ensure that the step length reduces the objective function 'sufficiently', when solving an unconstrained minimization problem using an inexact line search algorithm. Strong Wolfe conditions ensure convergence of the gradient to zero.

Closed form means that a mathematical expression can be expressed analytically in terms of a finite number of certain well-known functions.

Cholesky decomposition is used in linear algebra for solving systems of linear equations. It is a decomposition of a Hermitian, positive-definite matrix into the product of a lower triangular matrix and its conjugate transpose.

Exercises

- Define a target surface and draw a form diagram according to the features (e.g. ribs, open edges) of the surface – for instance, within the plan of the standard grid (Fig. 6.12). Make sure to provide force paths that allow those features to develop.
- For a simple target surface, draw a form diagram and an initial force diagram and compute and draw the corresponding thrust network. Try to manually

- modify the force diagram such that a better fit of the simple target is obtained.
- Compare the outcome of best-fit optimizations for the same target surface, using different force diagrams (i.e. allowed force flows).
- For a simple target surface and a form diagram corresponding to the standard grid (Fig. 6.12), generate an initial force diagram and corresponding thrust network as explained in Chapter 7 considering only the structure's self-weight.
- Calculate the squared sum of the vertical distances between the nodes of the thrust network and the nodes of the target, as a function of the force densities in the edges of the network. Calculate force densities that make this squared sum smaller or, even better, as small as possible. Attempt to generate a force diagram with edges parallel to the form diagram and the length of the edges equal to the calculated force densities.
- Increase the load on one of the nodes of the network. Draw the thrust network for the current force diagram. Repeat the steps of the previous exercise until a network is found that is close to the target again.
- The architectural program for the Texas shell has changed. The architect now envisages a shell supported on the four corners and one central support. Attempt to generate such a target surface and draw a form diagram according to the features which include ribs and open edges. Hint: make sure to provide force paths that allow those features to develop.

Further reading

- *Numerical Optimization*, Nocedal and Wright (2000). This book describes efficient methods in continuous optimization, including the gradient descent algorithm in Section 13.5.

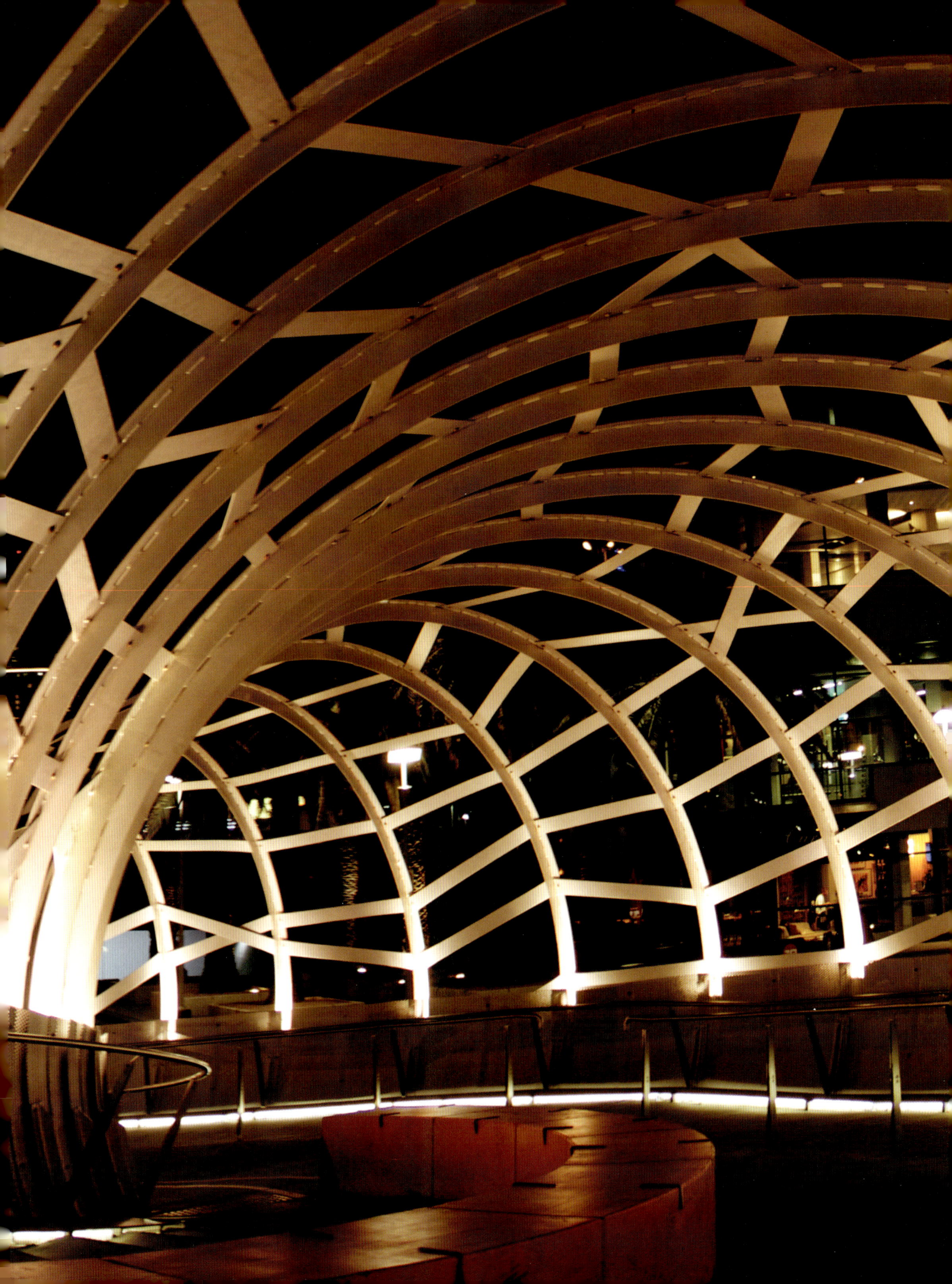

CHAPTER FOURTEEN

Discrete topology optimization

Connectivity for gridshells

*James N. Richardson, Sigrid Adriaenssens,
Rajan Filomeno Coelho and Philippe Bouillard*

LEARNING OBJECTIVES

- Convert a structural optimization problem to a mathematical formulation.
- Apply a genetic algorithm to solve an optimization problem.
- Improve the connectivity of a grid on a predetermined surface.

Figure 14.1 Science park entrance buildings, architectural interpretation.

Efficient design of gridshells has become a major challenge due to the need to account for several, sometimes conflicting, requirements including cost, but also safety and environmental impact. One particular challenge designers face is to make optimal use of, preferably renewable, resources. Generally speaking, the search for the *best* solution (according to a given criterion called 'objective function') is the central aim of optimization.

This chapter discusses the topology optimization of unstructured, or irregular, steel gridshells, where the objective is to minimize the gridshell's overall weight. An example of an unstructured gridshell is the Webb Bridge over the river Yarra in Melbourne (see page 170), though its geometry was not derived through structural optimization. For the structural optimization of the science park gridshells, a Genetic Algorithm (GA) is employed. Finite element (FE) analysis is used to evaluate structural performance and to check the constraints.

The brief

A new science park needs a series of canopies. Each canopy measures 24m × 24m in plan. We propose a series of unstrained steel gridshell typologies: supported on all four sides, one on two and one on three sides of the square base, with large openings on the free edges of the canopy. Each canopy has a total internal height of 5m and the openings provide at least 12m × 3m of unobstructed access. The client has agreed to a concept of the architectural design of the canopies, shown in Figure 14.1. The grid systems are clad in either translucent or opaque plate material, creating interplay between diffuse natural light and shading.

14.1 Evolutionary algorithms

Genetic Algorithms (GAs), a subclass of evolutionary algorithms, are search techniques inspired by evolutionary genetics and follow the Darwinian law of natural selection or *survival of the fittest*. In other words, in a random population of potential solutions, the best individuals are favoured and combined in order to create better individuals at the next generation. For structural optimization, the use of GAs is very attractive, as explained in Appendix C.

When considering structural problems, we can represent the structural variables in data structures often called 'chromosomes'. In the case of structural optimization, these variables depend on the type of optimization considered. Normally, the chromosome will consist either of member sizing, nodal position (shape) or topology variables, or of some combination of these three types. The capability to handle variables of different types at once is one of the great strengths of GAs as an optimization procedure. GAs allow, for example, for combination of continuous shape variables with discrete topology variables in a single chromosome representation.

A large number of chromosomes, each representing one structure, are considered by the GA at any given time. These chromosomes are called 'individuals', and make up the current 'population', of the GA. In single-objective optimization, the chromosome is associated with a value (such as the mass), of which we would like to find the optimal value. The function mapping the chromosome to this value is called the objective, or fitness function. Furthermore, there may be structural properties associated with the chromosome we wish to know in order to ensure that the structure adheres to certain structural requirements. For example, the maximal stresses in the structure should not exceed a certain limit, ensuring satisfaction of the stress constraint. The mapping of the chromosome to these constraint values often occurs by way of structural analysis such as the FE method, called at least once per iteration of the GA.

GAs process populations of individuals. The size of the population considered can be chosen by the user, often based on experience. Successive populations are generated from the previous population by way of three genetic operations: selection, crossover and mutation (see Appendix C).

It is worth noting that the parameters controlling these genetic operations will be of great importance to the efficacy of the algorithm. So far, no universal method exists to optimally choose these parameters, and the experience of the user will play a dominant role in the choice of values for the parameters.

14.2 Optimization method

This section outlines the general framework of the algorithm used in the optimization of the gridshell

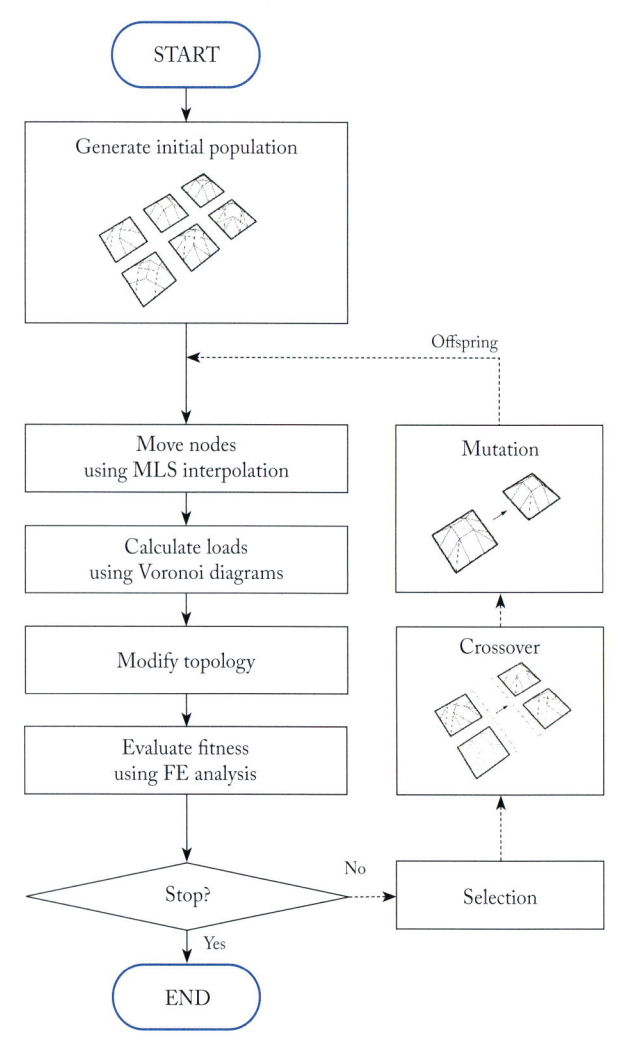

Figure 14.2 Genetic algorithm applied to gridshells

topologies (Fig. 14.2). Circular steel sections, with an outer diameter of 88.9mm and a thickness of 8mm, are chosen as the gridshell members.

The objective is the minimization of the gridshell's overall weight:

$$\min_{\mathbf{x}} f(\mathbf{x}) = \sum_i \rho A_i l_i x_i \qquad (14.1)$$

subject to constraints discussed in Section 14.2.4, where ρ is the density, and A_i, l_i and x_i are respectively the cross section area, length and design variable associated with element i. The value of the objective function is found by simple geometry, taking the density of steel to be $\rho = 7,850 \text{kgm}^{-3}$.

14.2.1 Topological design space

Each structure comprises forty-nine connection points or 'nodes' at which the steel bars are welded together. The form-finding approach, based on dynamic relaxation for gridshells, illustrated in Chapter 8, was used to determine the global form of the canopies. By pinning boundary nodes on certain sides of the gridshells and leaving others free, we achieve different shapes that can accommodate different architectural programs. Three typologies are presented in Figure 14.3.

The first form is supported on all four sides. The second form is a barrel vault, supported on two opposite sides. The third form is supported on three of the four sides, giving the gridshell a wide opening on one side.

In our approach, the 'ground structure' defines the allowable connectivities: the topological design space. The ground structure specifies the upper limit of the design space and expresses all allowable nodal connectivities.

When implementing discrete topology optimization problems, one constraint is particularly problematic (Richardson et al., 2012). The kinematic stability of a discrete structure, such as a gridshell, is intimately linked to the topology variables. While virtually all other constraints are present in sizing and shape optimization, the kinematic stability is exclusively of interest in topology optimization. The characteristics (including the dimensions) of the stiffness matrix **K** in the FE formulation are affected by changes in topology, that is, in the node connectivities, since the number of elements and nodes can vary. Significantly, too, unlike other constraints on the structure, the relative size of the solution set defined by this constraint in relation to the total search space decreases dramatically with an increasing number of degrees of freedom in the system.

Gridshells tend to have many nodes and relatively few connections between the nodes. The allowed connectivities should be such that the shell does not become too thick structurally (Fig. 14.4). This means that randomly generated structures are almost always kinematically unstable. This constraint is a binary one, unlike others which can be represented by a real number. It is therefore difficult to meaningfully penalize the fitness of the individual proportionally to the value of this constraint. Randomly generating the initial population leads to convergence problems: the algorithm struggles to find solutions which can be suitably evaluated.

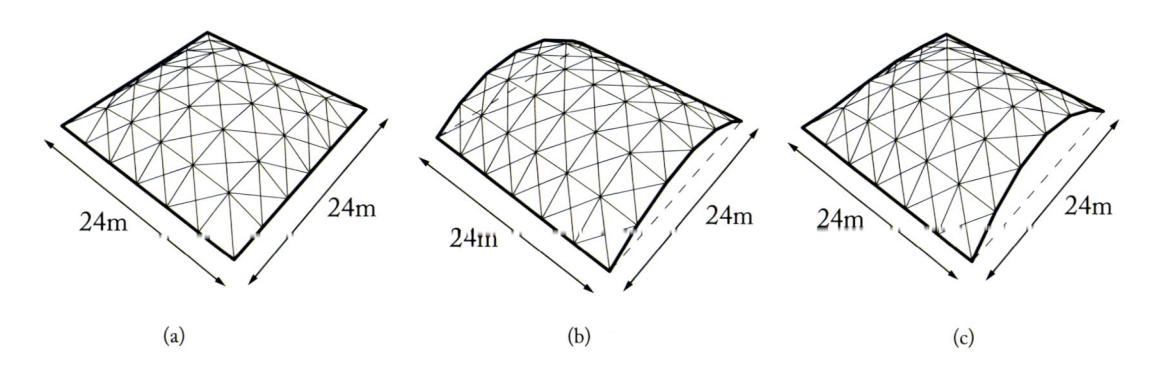

Figure 14.3 Three canopy typologies: ground structures 1, 2 and 3. The topologies of these structures are used as the allowable connectivities for the computation of the three optimized structures

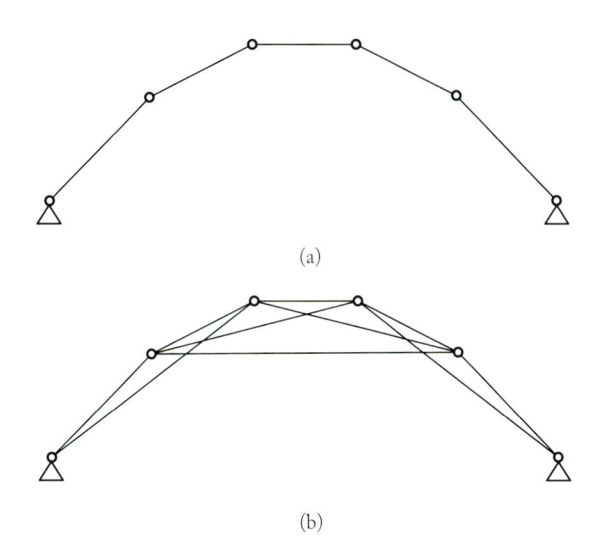

Figure 14.4 Cross sections of (a) a single layer gridshell and (b) a gridshell with overlapping connectivities

One way of dealing with this problem is to provide the GA with a certain percentage of kinematically stable individuals in the initial population. To generate these individuals, various algorithms can be developed. It is, however, important to introduce sufficient diversity into the initial population to prevent the algorithm from focusing on only a limited part of the search space. The algorithm implemented here is illustrated in Figure 14.5. In this case, a stable core structure is first produced. Next groups of elements are added onto the stable core to incorporate other nodes. Once all nodes are connected the procedure is stopped.

14.2.2 Geometric design space

The shape variables relate to x- and y-coordinates of the nodes. These shape variables can be either continuous or discrete. The vertical z-coordinates are constrained to a surface interpolated through the original gridshell nodal positions. In this way the overall equilibrium form is maintained, even though strictly speaking the shape changes at the nodal level. A Moving Least Squares (MLS) (Lancaster and Salkauskas, 1981) interpolation scheme is used to find an appropriate z-coordinate for the given x- and y-coordinates as stipulated by the shape variables. Limits are placed on the shape variables to avoid overlapping of nodes or nodes switching position, by limiting them to less than half of the original distance between the nodes.

14.2.3 Load calculation

Building codes tend to prescribe a range of loading cases that need to be considered in the analysis of a structure. These include asymmetric loading, distributed loading and point loads. From the point of view of the designer, the choice of loading to consider in the optimization process is important. In this case it is chosen, for the sake of consistency, to consider an evenly distributed loading consisting of an external static load and an approximation of the self-weight of the entire canopy. The building's cladding transfers the loads to the nodes of the gridshell. The forms considered are relatively flat, without excessively steep sides and the structure's self-weight is assumed to be relatively low. It is reasonable to approximate this by an evenly distributed load. We wish to approximate the loading on each node. This loading will be proportional to the horizontal projection of the surface area carried by the node, and is therefore a function of the geometrical positions of the nodes, which are variable. An automated scheme, such as a Voronoi decomposition of the horizontal plane projection of the structure, is essential to be able to

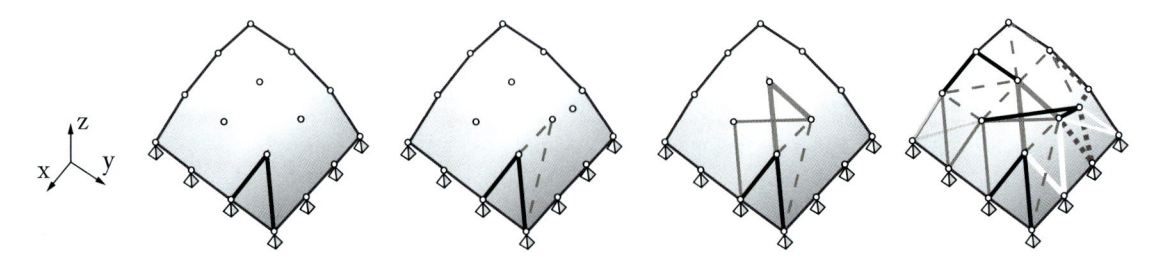

Figure 14.5 Generation of kinematically stable members of the initial population

calculate the loading at each point in the design space. A Voronoi diagram is a spatial decomposition based on distances between a set of points (using Delaunay triangulation), such as the nodes in the structures we are considering. A horizontal projection of the nodal positions is used to approximate the loading on the structure. Once the Voronoi polygons corresponding to each node have been assembled, their areas are calculated and the loading transferred to an equivalent point load applied to the node in the FE analysis of the structure.

14.2.4 Constraints

Three constraints are considered in the following designs, namely the maximal stresses in the elements, the local buckling of the elements, and the total deflection of the structure. All three of these values can be obtained by an FE analysis. The global buckling of the structure is not addressed since this requires a costly nonlinear calculation. However, once the optimal structure subject to these three constraints has been found, it is essential that the structure also be checked against nonlinear global buckling.

During the optimization, the maximal stress in the structure was constrained using the relation $\sigma^{max} - |\sigma_i| \geq 0$, where σ^{max} is the maximum allowable (yield) stress of the material, and σ_i is the stress in element i.

The local Eulerian buckling constraint is

$$-(\sigma_i^{cr} + \sigma_i) \geq 0, \qquad (14.2)$$

where

$$\sigma_i^{cr} = \frac{-\pi^2 E_i I_i}{A_i l_i^2 K^2}, \qquad (14.3)$$

where, for element i, E_i is the Young's modulus of the material, I_i the second moment of area of the cross section, A_i the area of the cross section, l_i the length and K the buckling constant associated with the connections between elements. In the calculations, the Young's modulus of steel was taken to be $E = 200 \times 10^9 \text{Nm}^{-2}$ and the strength $\sigma^{max} = 355 \times 10^6 \text{Nm}^{-2}$. Since we are dealing with a preliminary design method, where the exact rigidity of the joints is unknown, we choose an effective length factor, or buckling constant, of $K = 0.9$,

between the pinned condition ($K = 1.0$) and one side pinned, one side fixed ($K = 0.8$).

The total deflection of the structure is constrained to $\delta_z^{max} = 1/200$ of the minimum span (24m). This constraint is $\delta_z^{max} - \delta_z \geq 0$, where δ_z is the vertical deflection of the structure. Finally, the nodes are constrained to a range of 2m around their position in the initial design, for both x- and y-coordinates of node j: $(1.0m - |\Delta x_j|) \geq 0$ and $(1.0m - |\Delta y_j|) \geq 0$, where node j is not on the fixed boundary of the structure.

14.2.5 Analysis procedure

In order to achieve a smooth interface between the optimization and analysis procedures, the parameters outputted within the optimization loop, representing a single structure, are interpreted by an intermediate script. This script prepares the necessary FE analysis inputs, calls the FE program, and interprets the FE outputs which are in turn taken as responses by the optimization controller. Numerous analysis procedures can be employed to evaluate individual structures as long as this automation is made possible. In this way, the procedure can easily be adapted for more advanced simulation types, other than the linear elastic analysis employed here. For this analysis, we made use of truss elements with three degrees of freedom per node.

14.2.6 Evolutionary optimization

Once the algorithm has assessed the fitness of the individuals in the population, it will consider whether the stopping criterion has been met (a satisfactory convergence of the algorithm). If an unsatisfactory condition exists, the algorithm will perform another iteration. Changes are made to the population by selecting individuals to be retained, discarding others, introducing new individuals, mutation of chromosomes and mating of pairs to produce offspring. The exact mechanism is dependent on the method selected and the parameter values chosen.

14.3 Optimized connectivity

In order to illustrate some of the important aspects of discrete topology optimization, we allow the loading (combined self-weight and external loading) to vary

Solutions	Loading (kNm^{-2})	Mass of structure (kg)	Active constraint
	3	5,098	limit on node positions
	5	5,161	local buckling
	7	5,630	buckling
	8	6,393	buckling

Table 14.1   Overview of results of topology and shape optimization for gridshell supported on all sides, for various loadings

Solutions	Structure	Mass of structure (kg)	Active constraint
	barrel vault	7,330	deflection and buckling
	canopy with opening	6,048	local buckling

Table 14.2   Overview of results for a topology and shape optimization of a barrel vault and canopy with one opening for a 3kNm^{-2} loading

Parameter	Four sides pinned	Two sides pinned	Three sides pinned
Number of topology variables	19	39	70
Number of shape variables	6	14	27
Population size	500	800	1,200
Stable initial population size	400	640	960
Crossover rate	0.8	0.8	0.8
Mutation rate	0.4	0.2	0.2

Table 14.3 GA parameters for the gridshell problems with 3kNm^{-2} loading

between 3kNm^{-2} and 9kNm^{-2} and observe the effect the loading has on the final shape and topology of the structure. The optimal solution in each case is found at the extremity of the feasible solution set. The constraint bounding the feasible solution set at this point is often called the 'active constraint'. In Table 14.1 an overview of the optimal solutions for a selection of the loadings is given. For 9kNm^{-2} the buckling constraint was violated for all possible values of the design variables, therefore no feasible solution could be found.

For practical purposes, one loading value (3kNm^{-2}) is chosen to optimize the topology and shape of the remaining gridshells. Results of the optimization of the two remaining canopy typologies are found in Table 14.2. To achieve these results, the GA parameters in Table 14.3 were found to produce good results.

14.3.1 Discussion

In Figure 14.6, the selected topologies are shown in a possible configuration for the science park implementation.

Significant reductions in the mass of the structures can be achieved through topology (and shape) optimization of the gridshells. In Table 14.4, the un-optimized 'ground structures', shown in Figure 14.3, are compared to the optimized structures shown above. These values also correspond to a loading of 3kNm^{-2}.

Figure 14.6 Architectural representation of canopies selected after optimization

The ground structure designs meet the criteria for this loading case. Note that the constraint values are near the limit permitted in two of the examples. From an analysis point of view these initial structures may be seen as very acceptable for implementation. However, these intuitive designs can be vastly improved, as can be seen from the mass reduction achieved through optimization, in this case between 35% and 50%. The active constraint plays an important role in the solutions. In the case of the shell supported on four sides, this can easily be seen. For small loadings (around 3kNm^{-2}), the limits on the nodal positions are the active constraints, since the stresses in the structure are too low to cause buckling or large displacements, or are low compared to the strength. In terms of the shape variables, this represents the least-mass configuration. Increasing the loading to 5kNm^{-2} activates the buckling constraint. Several of the nodes shift position relative to the previous case to account

Ground structure	Mass (kg)	Stress constraint	Deflection constraint	Local buckling constraint	Mass reduction after optimization (%)
1	9,956	0.1112	0.0538	0.759	48.8
2	11,260	0.2403	0.789	0.8325	34.9
3	10,484	0.258	0.847	0.96	42.3

Table 14.4 Comparison of optimized and ground structure gridshells

for this. However, the topology is unchanged. For a loading of 7kNm^{-2}, the topology changes dramatically, forming two bands of material with less material at the corners. This may be due to the fact that the load path is shortened, by allowing for more load to be transferred to the supports in the middle of the edges of the footprint. By doing so, a stiffer structure, with shorter elements (less susceptible to local buckling) emerges. When a loading of 8kNm^{-2} is applied, the topology is modified so that this form is made more prominent with the addition of four new elements. The loads are redistributed so that the critical elements in the previous case do not buckle under this higher loading.

14.4 Conclusions

The approach adopted here can be seen as a filtering process on the shape variables, or an approximation of the global shape optimization. No global shape optimization is carried out, but rather the form of the structure is chosen a priori through form finding since this eliminates many poor solutions from the search space. By considering the shape and topology variables, large savings can be made in the material required for the structures. The configuration of the gridshell elements is quite dissimilar to the traditional approach of using a repeated pattern of regularly spaced elements.

It should be emphasized that the method presented here is intended for preliminary design. Further investigation can be carried out on the structures to assess their feasibility as designs in practice. Often objectives such as cost or environmental impact are more appropriate for structural design. These objective functions may be more complex to define, but the GA framework lends itself well to this adaptation. One major simplification is the use of pinned connections between the elements in the calculations. This is justifiable since we assume only one load case, which the form finding ensures will induce only axial forces in the members. In practice, welding of the steel sections will provide additional moment stiffness in the joints when other load cases, such as asymmetric loading, are considered. This will give the shells greater resistance to global buckling and other issues relating to the shell's stability. More detailed analysis can be carried out using commercial FE software with several other load cases and combinations, modelling it as a more realistic shell with moment connections.

From an architectural standpoint, the structural optimization dictates forms which invite speculation about the design process itself. Given the use of the buildings, this architecture is a good fit for the site. In the context of a science park, the structures could serve as an introduction to the field of optimization for visitors to the park.

Key concepts and terms

Discrete topology optimization is a form of structural topology optimization in which the variables considered are taken from a discrete set. In structural optimization discrete topology optimization often deals with truss-like structures, where the existence or non-existence of individual elements is controlled by binary variables.

A **genetic algorithm** (**GA**) is a class of algorithm inspired by the principles of natural selection. The algorithm mimics the processes of genetic mutation, crossover and selection to iteratively improve the performance of solutions to the optimization problem. Along with evolution strategies and genetic programming, GAs are instances of evolutionary algorithms.

An **objective function** in optimization is a function which represents the quantity to be either minimized or maximized. In structural optimization this is often the mass of the structure or the maximal displacement of the structure.

Constraints define the feasible region of an optimization search space. A constraint is a condition which must necessarily be satisfied in order for a solution to be considered valid. Typically in structural optimization constraints such as a limit on the maximum stresses in a structure are encountered.

Kinematic stability is satisfied when enough connectivity between the nodes exists that no mechanisms are present in the structure and can therefore be seen as a constraint on the topology of a structure.

A **Delaunay triangulation** of a set of points is

a triangulation of the convex hull of the points considered, such that none of the points lie within the circumcircles of the triangles.

A **Voronoi diagram**, also known as a Dirichlet tessellation, is a spatial decomposition. This diagram divides space into a number of polygons, or Voronoi cells, based on distances between a set of points, such that each polygon contains exactly one of those points and every vertex in a given polygon is closer to its generating point than to any other. A Voronoi diagram is dual to the Delaunay triangulation.

Moving least squares (**MLS**) is a regression method used to approximate or interpolate a continuous function (such as a smooth surface) known only at a limited set of sample points. The method uses a weighted least-squares measure, assigning a higher influence to the samples belonging to the vicinity of the point to be constructed.

Exercises

- The current approach deals with topology optimization, and we wish to include the sizing of the cross-sectional area of the elements in the problem. How would you introduce these variables into the genetic algorithm chromosome parameterization? How is the chromosome for each structure changed?
- Most gridshells are lightweight structures. Their structural behaviour is largely influenced by wind load. How could the optimization problem be reformulated to consider the effect of wind?
- Optimize the grid connectivity for the standard grid (Fig. 6.12), based on a previously form-found surface shape. Apply a gravity load of $1 kNm^{-2}$ in your optimization.

Further reading

- *Topology Optimization: Theory, Methods and Applications*, Bendsøe and Sigmund (2003). This book discusses continuum as well as discrete topology optimization.
- *Genetic Algorithms in Search, Optimization, and Machine Learning*, Goldberg (1989). This book is the main reference on genetic algorithms.
- 'Genetic algorithms in truss topological optimization', Hajela and Lee (1995). This journal paper represents an important early investigation on discrete structural topology optimization and the use of genetic algorithms to solve these problems.
- 'Truss topology optimization by a modified genetic algorithm', Kawamura et al. (2002). This journal paper discusses kinematic stability in topology optimization of discrete structures.
- *Genetic Algorithms + Data Structures = Evolution Programs*, Michalewicz (1996). This book is one of the main references on evolutionary algorithms.
- 'Multiobjective topology optimization of truss structures with kinematic stability repair', Richardson et al. (2012). This journal paper discusses kinematic stability of truss structures, and the effect of the kinematic stability on multi-objective genetic algorithms for topology optimization.

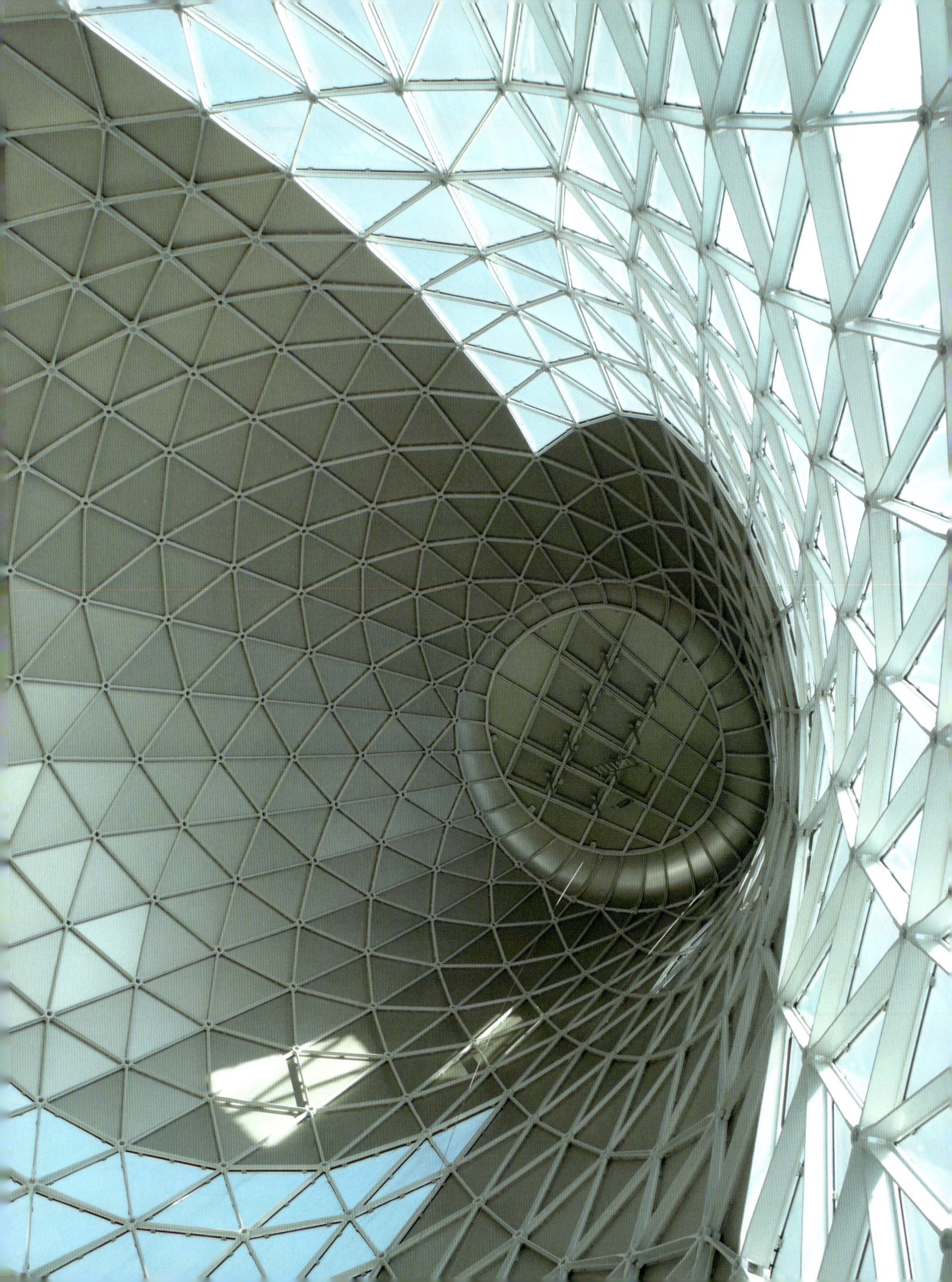

CHAPTER FIFTEEN

Multi-criteria gridshell optimization

Structural lattices on freeform surfaces

Peter Winslow

LEARNING OBJECTIVES

- Discuss how the structural efficiency of gridshells can be approached as a multi-objective design problem.
- Review how computer graphics algorithms can be applied to generate and manipulate grid layouts on freeform surfaces.
- Implement an optimization solver that finds a Pareto optimal set of designs.
- Create optimal gridshell patterns on predetermined surfaces.

Creation of efficient gridshell structures to support complex surface forms, such as the roof of the New Milan Trade Fair, Italy (see Fig. 15.1 and page 180), is a major challenge. The widespread availability of NURBS modelling software currently allows designers, architects and sculptors to precisely create a huge range of different geometries. However, it is not always clear how to create an efficient gridshell structure to support a given freeform shape. The challenge is compounded by the many competing requirements and performance objectives that are often associated with architectural engineering projects – for example, buckling, deflection, multiple load cases, cladding,

Figure 15.1 Roof of the New Trade Fair, Rho-Pero, Milan, Italy, 2002–2005, by Massimiliano and Doriana Fuksas architects, with a grid mapping onto a complex surface

constructability and aesthetics – and at the end of the process design freedom needs to remain with the architect and the engineer.

One approach for tackling such freeform engineering design problems is to break them down into three stages:

1. Surface form – conceptual shape by the designer.
2. Grid layout – defining member layout (also referred to as 'rods') on a given surface.
3. Member size – section sizes are chosen once geometry is defined.

This approach differs from, say, a 'hanging chain' form-finding approach in which surface form and grid layout are interconnected. This logical, three-stage approach is relatively widespread among engineering practice working on bespoke architecture, and tools exist for tackling each stage. The focus of this chapter is specifically on the 'grid layout' stage, that is, design and optimization of a lattice or grid of structural members. It is assumed that the surface geometry is given – for example, by a sculptor or concept form-finding process – and that detailed member sizing will be carried out later. By way of example, Figure 15.2 shows a structural grid that has been mapped onto a given, fixed, surface form.

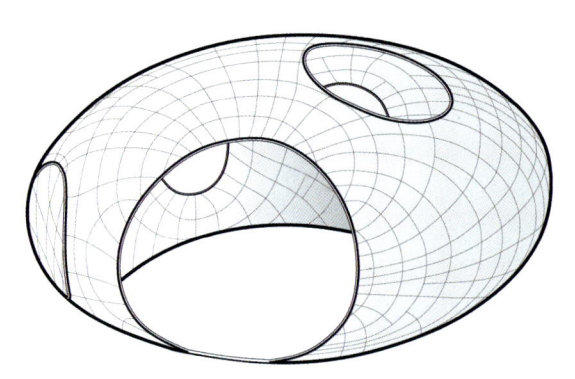

Figure 15.2 Example grid mapping on a freeform surface

Many gridshells are built from steel or timber sections. However, the generic approach presented in the following pages can be considered applicable to a wider range of materials; for example, ribbed reinforced concrete, ribbed steel shells, or fibre reinforced polymers. The key challenge is to lay out material strength and stiffness in response to multiple performance criteria, whilst not constricting the designer to a single design solution which is optimal in theory but unworkable in practice.

The brief

A local sports ground needs a canopy with a span of 54m and a height of 10m to provide weather protection to spectators. The doubly curved surface is given a priori and shown in Figure 15.3. The primary structure will consist of a single layer grid of steel tubes. The ends of the arching surface are fixed against translation. Our aim is to develop an efficient, regular layout for the lattice structure on this given surface.

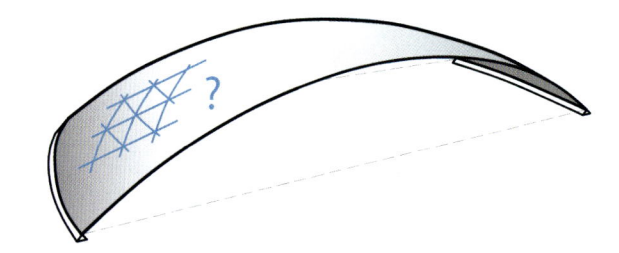

Figure 15.3 Mapping and optimization of grid on a given surface

15.1 Pareto optimal grid layouts

A wide range of techniques do exist for the creation and optimization of gridshell structures, some of which can be found elsewhere in this book. Since the example in Figure 15.3 is based on a given surface geometry, well-known methods such as hanging chains and membrane inflation are not applicable. Instead, a dedicated grid layout mapping and optimization tool is required.

Traditionally, optimization and form-finding approaches focus on a single important criterion. For instance, an equal mesh net mapping technique will create a grid in which every member has equal length but node angles vary, see Figure 15.4. This would not be cost effective for any project where the nodes are more expensive than the members, and there is no guarantee that such a grid is structurally efficient under multiple load cases.

In practice, many real structural engineering design problems do have a number of competing objectives. Even pure optimization algorithms are

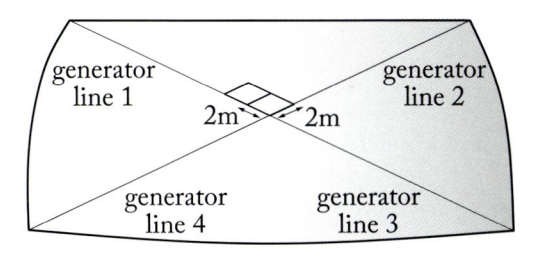

Figure 15.4 Equal mesh net grid layout, in plan

traditionally used to find a single optimal answer to single objective problems; for example, member sizing optimization. Novel alternative methods must be sought for problems such as grid layouts which have a wider influence on aesthetics, constructability and structural performance; a single solution from a 'black box' is not sufficient. Therefore, the approach for this case study will be to find a Pareto optimal set of different grid designs, thus allowing the engineer to see and understand the trade-offs between competing objectives.

Figure 15.5 plots eight different (hypothetical) designs in objective space. A design is said to be Pareto optimal if it is better than every other design for at least one of the performance objectives. So, for instance, design C is better than every other design in the population for *either* objective 1 *or* objective 2. In some cases, it is better for *both* objectives (see design D). Multi-objective solvers find the Pareto optimal set of designs (A, B and C) through a series of iterative or evolutionary steps. However, is design A better than design C? That depends on whether the designer perceives objective 1 to be more important than objective 2; the choice lies with the designer, which is a key strength of multi-objective design optimization.

Prior to delving deeply into the details of grid layout geometry – member lengths, nodal coordinates, connectivity and so on – it is useful to first explore the advantages that varying the grid layout can bring.

15.2 Objectives and constraints

The eventual aim is to optimize the layout of steel tubes on the surface. Long-span shell design tends to be governed by stiffness behaviour, therefore two design objectives are considered for the first stage of this example: first, to minimize peak deflection due to dead load; and second, to maximize the eigenbuckling load factor under wind and dead loads acting together.

This optimization is subject to four types of constraints and requirements: first, the grid is triangulated, consisting of steel tubes, where main members are 200mm diameter and 5mm wall thickness, and triangulation members are 100mm diameter with 5mm wall thickness; second, the target spacing between nodes in the grid should be 1 to 3m (due to cladding requirements); third, the structural members should follow smoothly curving paths to give visual and structural continuity, as per Figures 15.1, 15.2 and 15.3; and fourth, the grid connectivity should be regular, but lengths of members and positions of nodes can be varied.

The dead load consists of the self-weight of the grid of steel tubes (which is a function of the grid geometry, so will vary) and a cladding and services load of 1.5kNm^{-2}. The ultimate wind load is defined in Figure 15.6.

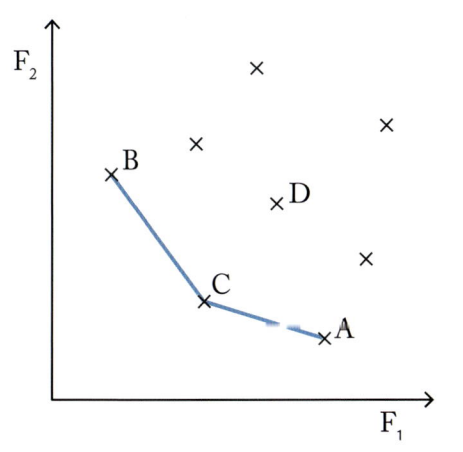

Figure 15.5 Bi-objective design space, showing the Pareto frontier, and where F_1 is the first, and F_2 is the second design objective

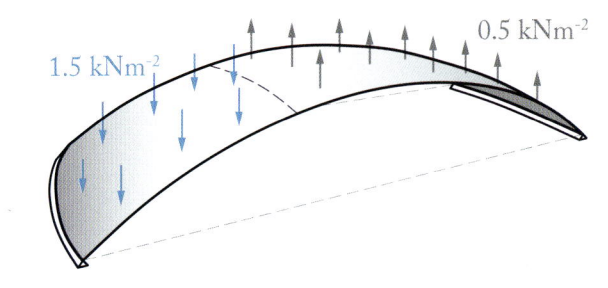

Figure 15.6 Wind pressure loading (normal to surface)

In order to tackle this design optimization problem, a number of key steps are made:

1. Obtain the surface geometry (already given).
2. Define suitable grid geometry design parameters.
3. Create a complete grid layout geometry from these design parameters.
4. Evaluate the grid against the performance objectives.
5. Improve the grid by a multi-objective optimization algorithm.
6. Repeat steps three to five.

15.3 Unit cell

This section is about understanding and exploring the building of our gridshell, the 'unit cells'. For our example, a grid is used which consists of such regular repeating unit cells, as is the case for many gridshell structures. Although the spacing and the angle between members may vary, the connectivity is as shown in Figure 15.7. If the angle between the rods, α, is reduced then the grid structure will get stiffer in the x-direction because the primary rods are more closely aligned with the x-direction. Conversely, if α is increased, then the grid structure will get stiffer in the y-direction. At this stage it is not necessary to think about the (x, y) coordinates of every node; the important geometrical parameters for this regular grid are only the angle α and the perpendicular distance between the rods, spacing L.

Consider this grid on a macroscopic scale, and replace many members and nodes by a piece of continuum shell, of equivalent stiffness, made from a hypothetical anisotropic material. This is analogous to Fibre Reinforced Polymer (FRP) laminate plate theory (Kueh and Pellegrino, 2007), where an orthotropic material stiffness matrix is calculated, instead of modelling individual fibres. When designing a larger piece of structure, this material can be aligned with the directions in which it is most needed.

One might imagine that a triangulated grid is relatively isotropic, so it is worth investigating whether there is a significant benefit to varying rod angles. In order to quantify the magnitude of potential anisotropy, we need to further consider just the unit cell. Taking the unit cell and applying a deformation

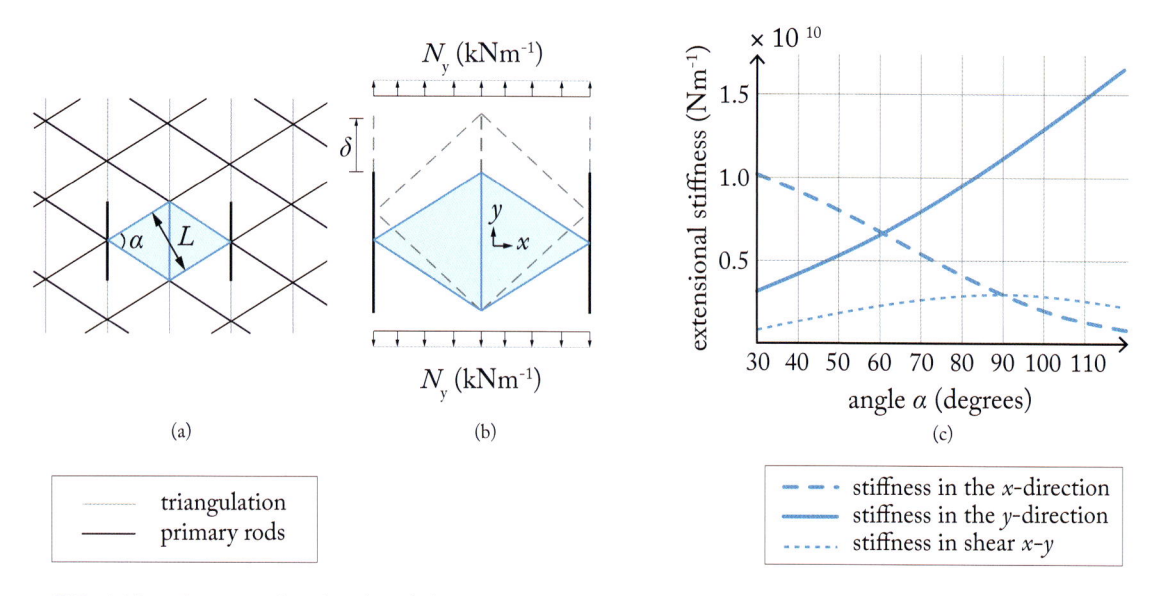

Figure 15.7 (a) Regular triangulated grid, with (b) unit cell extracted and stretched in the y-direction in order to determine (c) macroscopic membrane stiffness, as a function of rod angle α

(Fig. 15.7), enables the macroscopic stiffness to be found in different directions.

The method of periodic boundary conditions (Kueh and Pellegrino, 2007) can be used more formally to determine the equivalent continuum stiffnesses for a unit cell. This is typically implemented using a beam finite element model of the unit cell, which has a large number of constraint equations in order to precisely apply the different boundary conditions and deformation modes. It is a rigorous numerical method to achieve what has been shown diagrammatically in Figure 15.7. Some illustrative membrane stiffness results for a range of different angles α are shown in Figure 15.7. Note that exactly the same considerations apply to the bending stiffness of the unit cell.

From this figure it can be seen that changes in the angle between the primary sets of rods lead to significant changes in the equivalent anisotropic stiffness of the unit cell. For instance, increasing α from 60° to 90° gives approximately 60% lower stiffness in the x-direction but a 60% higher stiffness in the y-direction. Thus there is substantial scope to tailor the rod directions and 'tune' the stiffness properties of the triangulated gridshell.

As an aside, it is interesting to note that the equivalent (anisotropic) continuum material could now be applied to a shell finite element mesh on a given surface, that is, mimic the behaviour of a gridshell by using a continuum shell with user-defined material (Winslow et al., 2010). This has parallels with Free Material Optimization (FMO) (Bendsøe and Sigmund, 2003), and means that one can begin to explore the structural behaviour prior to looking at geometry in any great detail. However, there are clearly practical limitations precisely because there is no explicitly defined node or rod geometry. Note that FMO is related to the homogenization method (Chapter 17). FMO works directly with the design of the material tensor, after this process one may use inverse homogenization to look for the composite that realizes the previously optimized material tensor.

15.4 Geometry generation

This section presents a method which takes rod angles as key input parameters and generates full gridshell geometry, of sufficient fidelity for further optimization and beam finite element analysis.

15.4.1 Prerequisites

The first step is to discretize the given surface (initially represented with NURBS, see Figure 15.3) as a triangulated surface with triangular facets. Practically, one way to do this is to use a proprietary FE surface meshing tool (although it is never directly used for any FE analysis). This discretization facilitates the use of discrete differential geometry to synthesize and manipulate grid geometry on the surface.

It is vital to have a suitable surface coordinate system (surface parameterization) which allows us to map a two-dimensional grid into three-dimensional space. If the starting point for our shell structure was a NURBS surface, then it will inherently have a (u,v) surface coordinate system – and that is practically all that is often required to calculate the (u,v) coordinate of each node in the triangulated surface mesh. However, depending on how the surface was created by the designer, this coordinate system may be too distorted, non-continuous (e.g. multiple NURBS surfaces stitched together), or may not exist (e.g. if the surface is a triangulated mesh created by digital scanning of a physical model).

Surface parameterization has been the subject of considerable study in computer graphics, for reasons of texture mapping and surface re-meshing. Ideally we would like a length-preserving (and thus also angle-preserving) parameterization but in practice this is not possible for freeform surfaces (only if the surface has zero Gaussian curvature). A conformal parameterization (angle-preserving) mapping is adopted instead.

We start by using a relatively simple method to define directions for the two sets of rods on the surface. At any point on the surface we need two

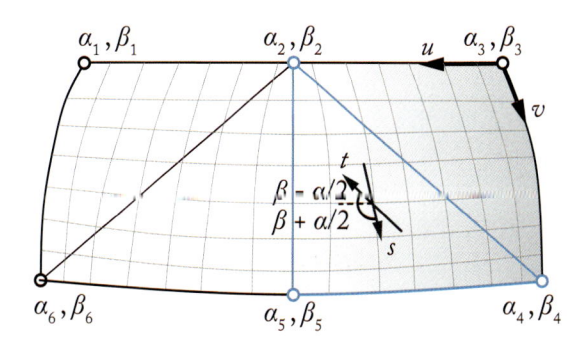

Figure 15.8 Rod angles as design parameters, interpolated over each of the four regions

parameters: the angle between the rods, α; and the rotation of the structural grid relative to the surface coordinate system, β. Therefore, parameters α and β are defined, at a small number of key points on the surface, by shape functions used to give a piecewise linear interpolation. For six key points, this gives twelve (geometrical) design parameters (Figure 15.8).

At any point on the surface the directions for the primary rods, **s** and **t**, are defined relative to the original surface coordinate system (u,v) (Figure 15.8).

15.4.2 Rod path plotting

After defining two direction fields on the entire surface for the primary sets of rods, the original surface coordinate system is disregarded. It is now necessary to plot continuous rod paths for these two directions in order to create a complete structural geometry, using discrete differential geometry techniques.

Figure 15.9 shows rod paths sketched onto one of the rod direction vector fields. One way of achieving this is to pick a start point for a single rod path and plot it out in a step-wise manner such that it is tangential to the vector field. However, if the vector field has non-zero divergence (or just a small amount of noise), it is very difficult to control the spacing between rod paths. Mathematically this is done by finding two suitable scalar potential functions on the surface Ω

$$u = U(x_\Omega, y_\Omega, z_\Omega),$$
$$v = V(x_\Omega, y_\Omega, z_\Omega), \quad (15.1)$$

such that the contours of these functions are tangential to the two rod direction fields.

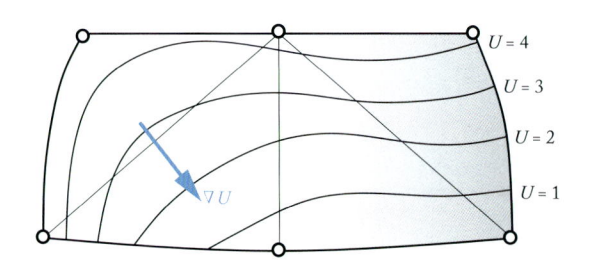

Figure 15.9 Potential function U defined on surface, fitted to first set of rod direction vectors

The functions U and V, according to equation (15.1), define a mapping from the surface Ω in R^3 to the planar (u,v) domain (R^2). The scalar value of U is the local u-coordinate, and the scalar value of V is the local v-coordinate.

If, at any point on the surface, the contour lines of U are tangential to the rod direction vectors, **s** and **t**, then the gradient of U is perpendicular to the rod direction vectors (dot product is zero). Therefore if the following equations are satisfied

$$\int_\Omega (\nabla U \cdot \mathbf{s})^2 dA = 0$$
$$\int_\Omega (\nabla V \cdot \mathbf{t})^2 dA = 0 \quad (15.2)$$

then contours of U and V would represent 'perfect' rod paths.

This type of approach is used in computer graphics for subdividing surfaces based on principal curvature lines (Ray et al., 2006) and for principal stress trajectories (see Chapter 16). However, before attempting to solve these equations, an additional scaling term is added to promote an even distribution of rod paths over the surface.

If the magnitude of the gradients of U and V were both unity, then there would be no distortion or variation of rod spacing, $L = 1$. However, for any vector fields, **s** and **t**, on a given surface it will almost never be possible to find rod paths which precisely follow the vector directions and also have constant spacing (only if the vector fields happen to be divergence free and the surface is developable). Therefore the aim is to minimize a global 'energy' functional, which is a measure of spacing distortion and direction distortion. So,

$$\min. E_1 = \int_\Omega ((\nabla U \cdot \mathbf{s})^2 + \omega(\|\nabla U\|^2 - 1)^2) dA, \quad (15.3)$$

$$\min. E_2 = \int_\Omega ((\nabla V \cdot \mathbf{t})^2 + \omega(\|\nabla V\|^2 - 1)^2) dA, \quad (15.4)$$

where ω is a scalar weighting value (typically between 0 and 10) which controls the trade-off between direction and spacing L of the rod paths. The given surface, which was previously converted to a triangulated mesh, has p nodes and q triangular facets. Therefore, equations (15.3) and (15.4) can be evaluated as a sum over all triangles T in the surface mesh,

$$E_1 \equiv \sum_{T=1}^{q} \int_T ((\nabla U \cdot s)^2 + \omega(\|\nabla U\|^2 - 1)^2) dA. \quad (15.5)$$

The function U is piecewise linear over any given surface, that is, a value of u will be defined at each node p in the triangulated surface mesh, with a linear interpolation over each triangular facet. As a result, ∇U is constant over each triangle, and if $\mathbf{s}$ is calculated at the centroid of each triangle (Fig. 15.10), the 'energy' functional becomes

$$E_1 \equiv \sum_{T=1}^{q} \left(((\nabla U)_T \cdot s_T)^2 + \omega(\|(\nabla U)_T\|^2 - 1)^2 \right) A_T, \quad (15.6)$$

where A_T is the triangular facet area.

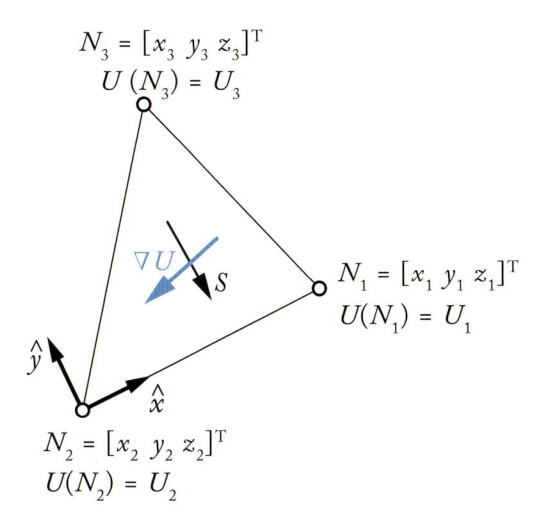

Figure 15.10 Definition of triangular facet, T, for rod path plotting. Each corner node has a position in global coordinate system N and an associated value of the piecewise linear scalar potential U

In practice, the convergence of this set of nonlinear equations is highly dependent upon the starting point. Therefore, for the purposes of this exercise, the starting point is found using a slightly different linear formulation,

$$\min. E_1 \equiv \sum_{T=1}^{q} \|(\nabla U)_T - s_T\|^2 A_T, \quad (15.7)$$

which can be solved reliably using the Conjugate Gradient method with Jacobi Preconditioning (PCGM) (Press et al., 2007). The solution from this set of equations (the u value at every node of the triangulated mesh) is used as the starting point for either equation (15.6) or a more stable nonlinear formulation,

$$\min. E_1 \equiv \sum_{T=1}^{q} \left(\|(\nabla U)_T - s_T\|^2 + \omega(\|(\nabla U)_T\|^2 - 1)^2 \right) A_T, \quad (15.8)$$

either of which can be solved using the quasi-Newton Broyden–Fletcher–Goldfarb–Shanno (BFGS) method (Press et al., 2007).

It is then repeated with vector field $\mathbf{t}$ and scalar potential V, to give values (u,v) at every node in the triangulated mesh. Due to the piecewise linear nature of the scalar potentials the contour lines can be extracted numerically and plotted, as shown in Figure 15.11.

From this primary grid, it is necessary to calculate the location of nodes and populate the connectivity matrix for the primary beam finite elements. The well-structured surface coordinate system (u,v) means that it is then straightforward to add secondary structures: triangulation members, edge beams, infill panels and so on. The end result is a high-fidelity model ready for finite element analysis.

The parameter ω from equation (15.6) gives control over the spacing L of the rod paths, see Figure 15.12. If ω is considered as a design variable, then the design space can be significantly increased. This may be useful if we might need near-parallel rod paths for fabrication reasons, or if perhaps some areas of the gridshell are working much harder than others, scope to significantly vary the spacing L is beneficial.

15.5 Designing and optimizing a grid structure

The previous section has shown how to synthesize full grid geometry, starting from sparsely defined rod directions. But what should those directions be? One approach would be to pick these directions manually, as shown for the freeform surface in Figure 15.13. This in itself could be considered as an interesting sketching and design tool.

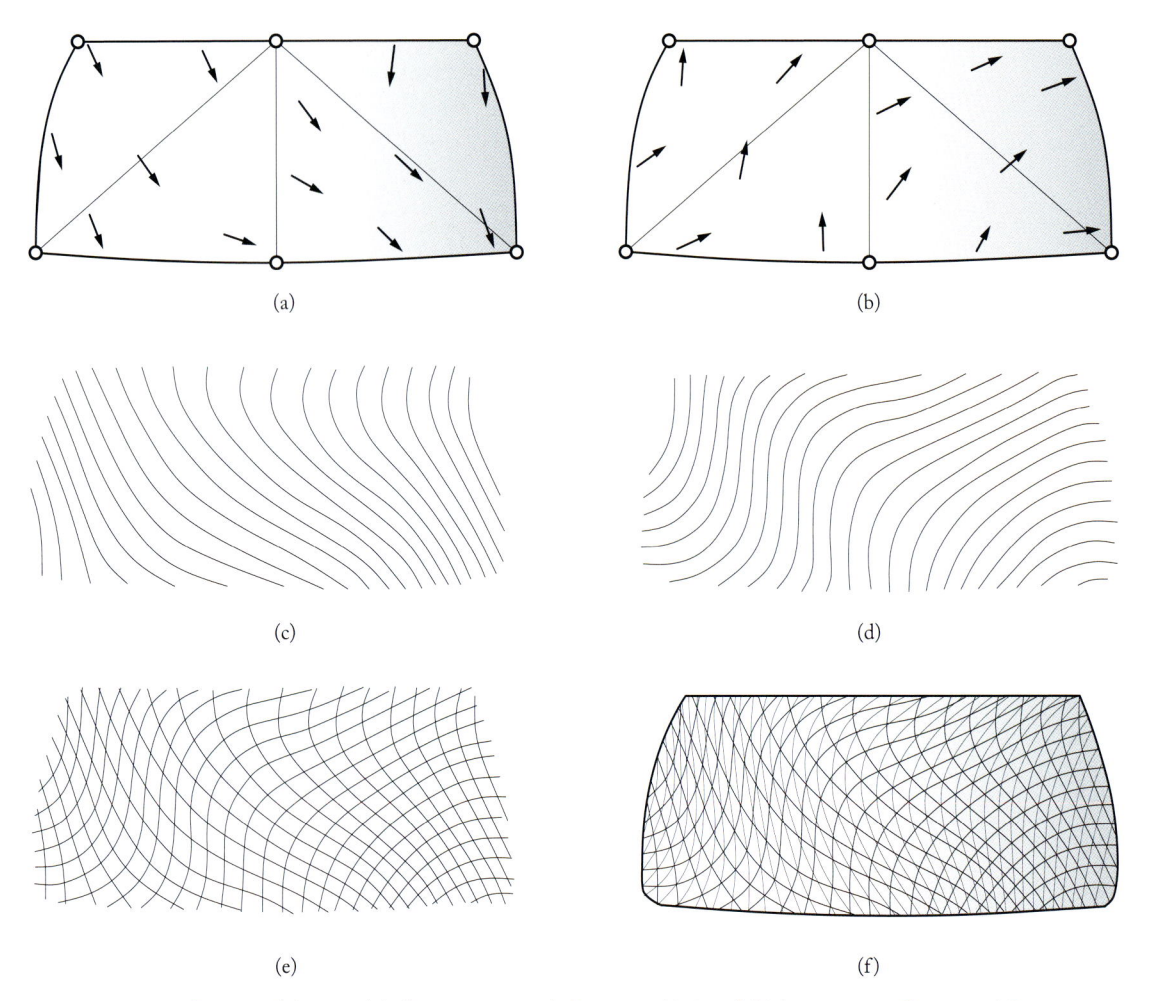

Figure 15.11 Process for assembling gridshell geometry, rod directions (a) 1 and (b) 2, contours of potential functions (c) U and (d) V, (e) primary grid from contour overlay and (f) complete structure, including triangulation rods and infill panels

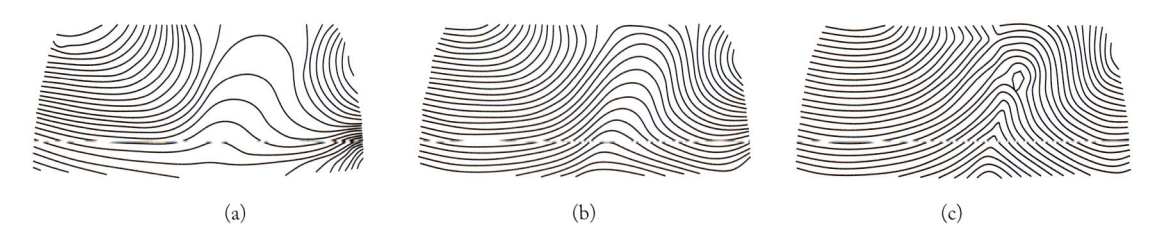

Figure 15.12 Contour function U fitted to rod directions (a) $\omega = 0.01$, very good direction control, poor spacing control; (b) $\omega = 1$, good direction control, good spacing control; (c) $\omega = 10$, poor direction control, very good spacing control

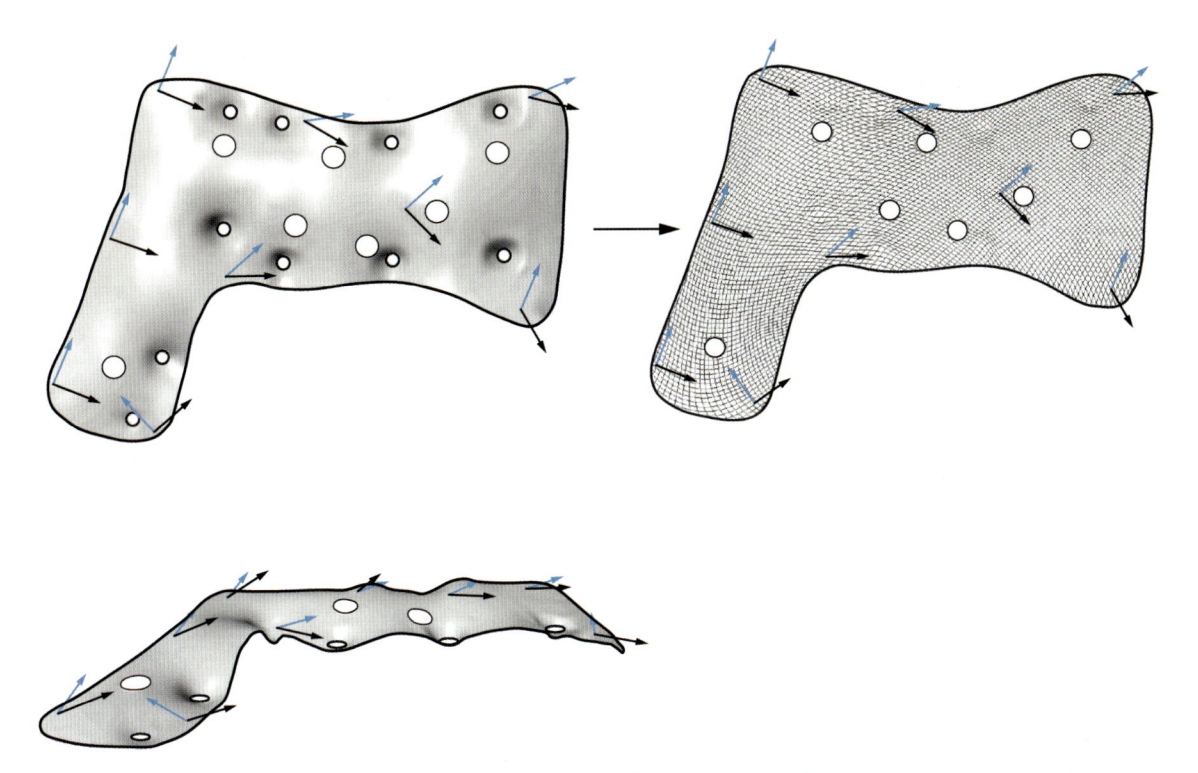

Figure 15.13 Automatic generation of complex grid geometry on freeform surface, from sparse user-defined rod directions

15.5.1 Multi-objective genetic algorithms

An alternative approach is to use an optimization algorithm, which will iteratively refine the rod directions in response to the multiple design objectives (defined in the brief). A Multi-Objective Genetic Algorithm (MOGA) is used for this purpose. A wide range of MOGAs exist (Deb, 2001). Most work on broadly similar principles:

1. Generate an initial population of designs.
2. Evaluate performance of the designs using finite element analysis.
3. Select the best designs in the population as parents for the next generation, using two metrics:
   a. Pareto optimality: is the design Pareto optimal? If not, the algorithm will rank designs by how close they are to Pareto optimality.
   b. Clustering: how similar is this design to other designs in the population? Promoting diversity in the population leads to better exploration of the design space.
4. Create 'child' designs by taking design variables from two parents and carrying out crossover and mutation.
5. Evaluate performance of the child designs and then add to population of designs.
6. Repeat steps three to five to evolve the population over a number of generations. For example in Figure 15.5, designs D and E could be parents to create the improved child designs A and C.

The non-dominated sorting genetic algorithm II (NSGA2) is used here; it is widely considered to be the high-quality benchmark by which other multi-objective algorithms are measured. Our implementation (Bleuler et al., 2003) includes functions for selection of parents and for mutation and crossover in order to create child designs using a real-valued representation (as opposed to bit-strings).

15.5.2 Performance evaluation and practical implementation

Structural performance objectives are evaluated for large numbers of designs; for example, by establishing an automated link to an off-the-shelf finite element package, via the Application Programming Interface (API) or batch mode. Any analysis method that is available in a typical engineering design office will then be available to the optimization routine. For other projects, the performance objectives of concern to the engineer may be very wide ranging – for example, geometrical, constructability or material mass criteria – this is considered in Section 15.5.4. These objectives can be calculated directly by user-written functions.

The flowchart representing the optimization software program is shown in Figure 15.14, in this case for a population of 100 designs.

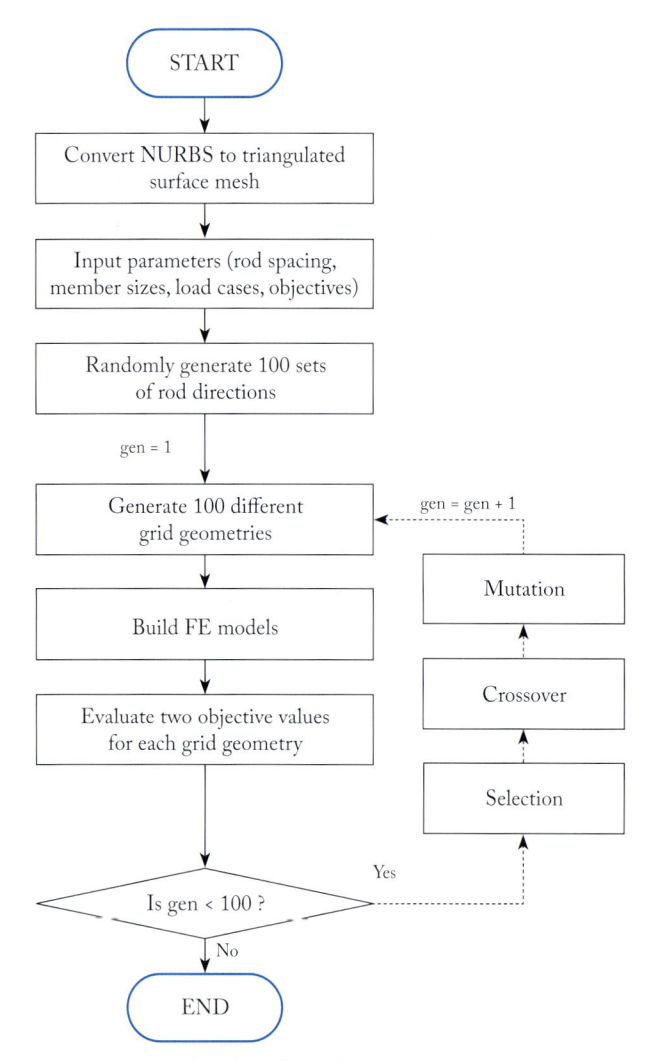

Figure 15.14 Flowchart for software program

15.5.3 Case study preliminary results

Using a population of 100 designs, variable crossover probability of 75% and variable mutation probability of 5%, the MOGA was run for 100 generations. At each generation every design in the population is evaluated individually, as described in Section 15.5.2. The results for the performance objectives, deflection and buckling can be seen in Figure 15.15.

At each generation in Figure 15.15 there is clearly a trade-off between the two competing objectives: some designs are better at resisting buckling due to the asymmetric wind load, whilst others are better at resisting deflection under dead load. Even after a large number of iterations, the designer still retains control over the design, that is, they can choose between the 100 efficient structural designs, based on perceived importance of the multiple objectives and aesthetic appeal of individual designs. Two designs (Figs. 15.15a and 15.15b) from the 100 generation are shown, in plan. For comparison purposes, a non-optimized conventional design is also shown, consisting of straight rods which are ±30° from the original (u,v) NURBS surface coordinate system (Fig. 15.15c).

After ten generations of evolution, the conventional design has been surpassed and after 100 generations the performance gains are 66% lower deflection and up to 35% higher buckling load. The mass is (nominally) held constant at 41 tonnes for all designs in this optimization process. Note that even the conventional gridshell geometry has every rod a slightly different length and every node at a slightly different angle – due to the surface double curvature and the (u,v) parametric distortion – so may not be easier to construct than the structurally optimized designs.

15.5.4 Case study advanced results

The design can be taken a step further by adding new design variables and performance objectives. This enlarges the design space and incorporates more practical design considerations. The brief becomes: minimize four objectives:

- total mass of steel;
- the inverse of the buckling load due to wind;
- deflection under dead load;
- warping of worst case quadrilateral panels;

with the following design variables:

- rod direction angles α and β at each of the six key design points;

CHAPTER FIFTEEN: MULTI-CRITERIA GRIDSHELL OPTIMIZATION 191

- target rod spacing L (0.5m to 3m);
- rod smoothing parameter ω, see Figure 15.12;
- diameter of primary rods (0.05m to 0.2m);
- diameter of secondary triangulation rods (0.05m to 0.1m).

The optimization algorithm is run for 500 generations with a population of 100; results from the final generation are shown in Figure 15.16.

These graphs show a final population with a wide spread of designs which are either Pareto optimal

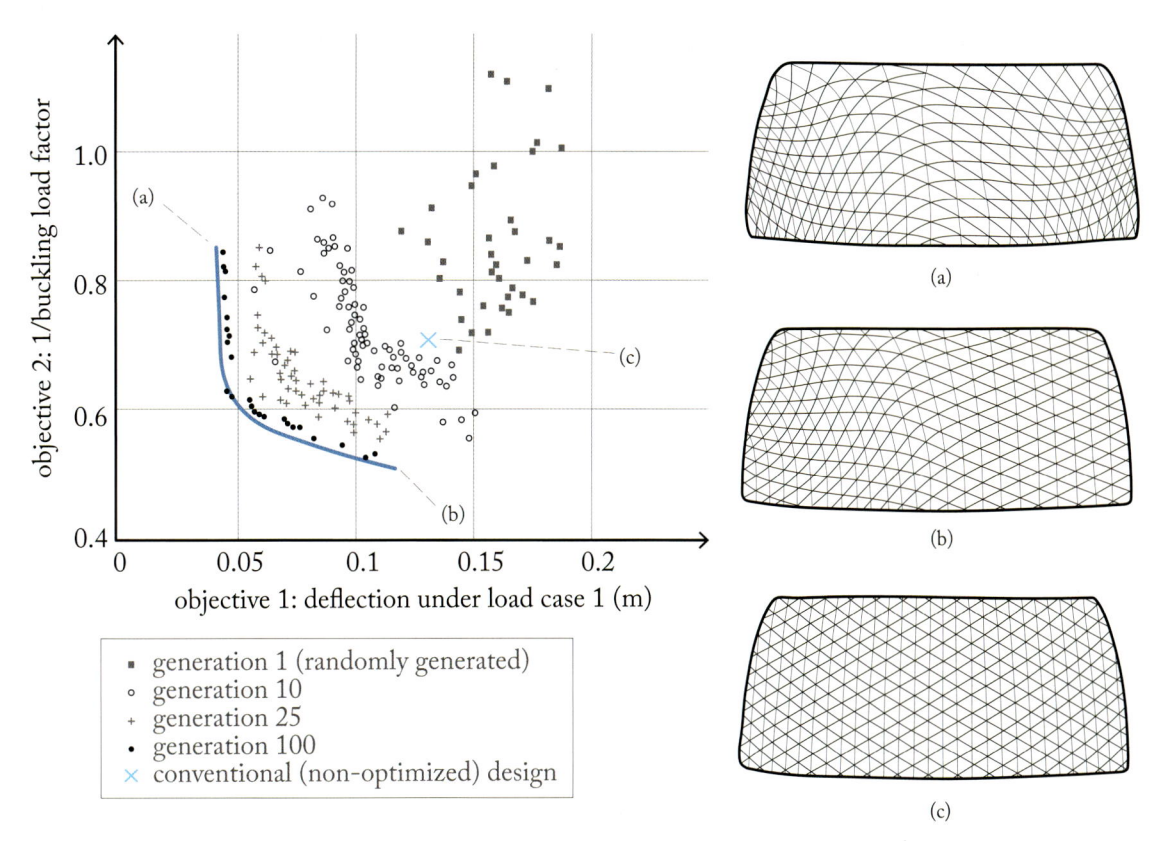

Figure 15.15 Results from the multi-objective optimization process

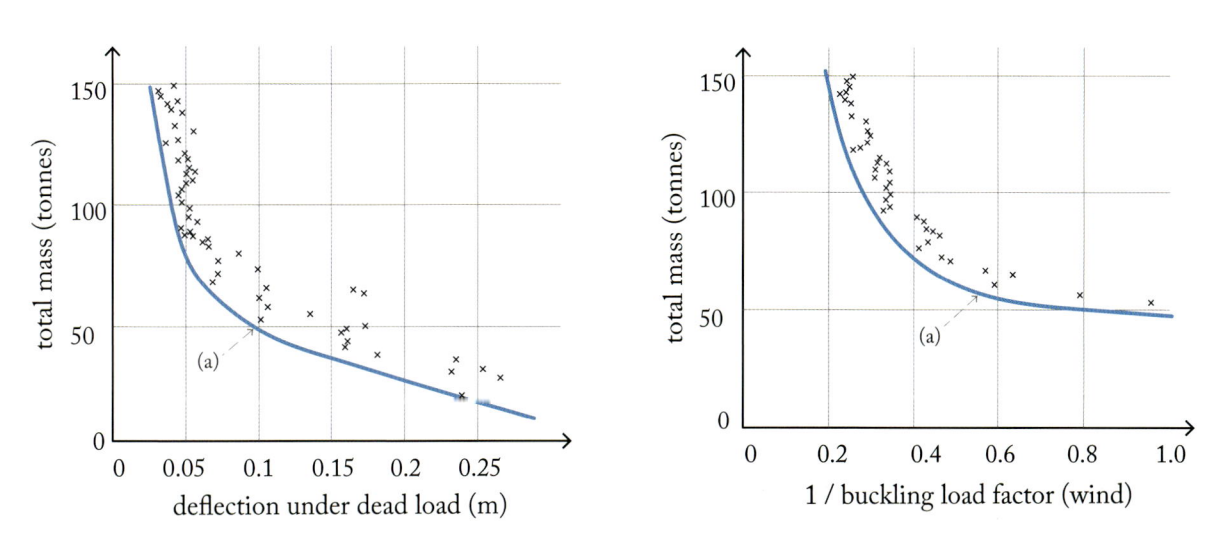

Figure 15.16 Results from four-objectives optimization, where (a) is the optimal trade-off surface. Marker size is directly proportional to panel warping objective

or near Pareto optimal (although the four-dimensional objective space makes it hard to visualize on a two-dimensional page). A diverse pair of designs is shown in Figure 15.17. The engineer and architect can explore trade-offs between performance objectives, and have the freedom to pick a structure from this population before moving on to more detailed design. The final design would need further stress checks and a nonlinear buckling analysis.

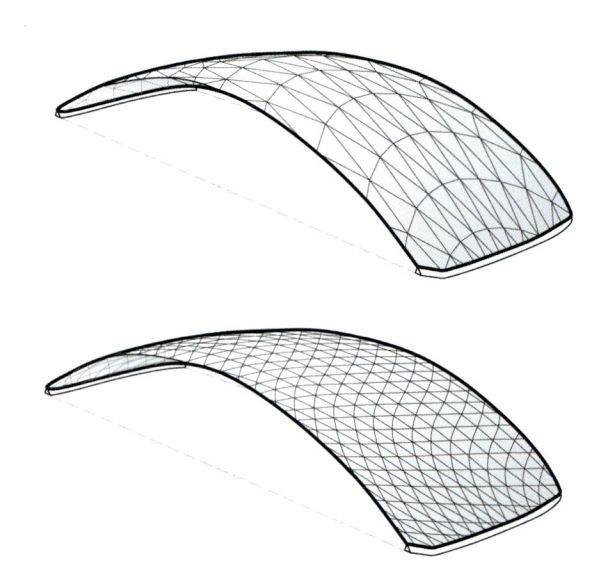

Figure 15.17 Two designs from the Pareto optimal trade-off surface

15.6 Conclusions

This chapter has shown how surface parameterization techniques can be used to create and control the layout of a grid structure. This acts as a tool for designers to create efficient, optimized structures to support a given surface form. The guiding principle is that if a structural material or building block is not isotropic, then it should be arranged so as to make the best use of its properties.

Multi-objective optimization techniques find a Pareto optimal set of designs, thus the designer can better understand trade-offs between different performance objectives, even for complex problems. In contrast to many other optimization techniques, the designer is not forced to pick a single 'optimal' design, but instead can choose from the final diverse, optimized population.

Key concepts and terms

Multi-objective optimization is the process of simultaneously optimizing or improving more than one performance objective.

A **Pareto optimal set** is a group of structural design solutions, each of which is better than all other design solutions with respect to at least one of the (multiple) performance objectives.

Clustering, in the context of this chapter, is the phenomenon where a large number of designs within the genetic algorithm population become very similar (such that the architect or engineer would not view them as distinct designs). Instead, specific parts of the algorithm are needed to promote diversity.

Surface parameterization is the two-dimensional (u,v)-coordinate system which is defined on our three-dimensional shell surface, that is, it is a one-to-one mapping between two- and three-dimensional domains. Imagining contour lines of u and v drawn on a shell surface then, a **length-preserving parameterization** would show u-lines and v-lines forming a square grid (only possible on flat or singly curved surfaces), and an **angle-preserving parameterization**, or **conformal parameterization**, parameterization would show u-lines and v-lines intersecting at right angles but the lines themselves could be curved.

Non-uniform rational basis spline (**NURBS**) is a mathematical means of defining freeform surfaces defined by a set of control points, which allow a user-friendly creation and modification of shapes. The degree and knot matrix are other parameters describing a NURBS. It is widely used in modelling software and could be considered as the standard way of describing and modelling freeform shapes in computer-aided design (CAD).

A **unit cell** is a basic structural unit which, when repeated many times over, forms the grid structure.

Exercises

- The advanced case study in this chapter had four objectives. What could additional structural objectives relating to the grid layout be? How could this

grid layout optimization be integrated with optimization of surface form and member sizes? Discuss how implementing these objectives influence the optimization and its results.
- Imagine a tool which allows the user to sketch rod directions at a few key locations on the surface, and then automatically generates a complete gridshell structure. How might this be implemented? More advanced methods for defining rod directions could be considered; for example, Fisher et al. (2007).
- If the rod directions would be defined at a larger number of points on the case study surface, what effect would this have on the optimization process in terms of speed of performance and performance of the generated designs?
- Define and implement an optimization solver that finds a Pareto optimal set of designs for two objectives.
- Using a predetermined NURBS surface with a 10m × 10m footprint, find the optimal grid layout.

Further reading

- 'Mesh parameterization: theory and practice', Hormann et al. (2007). This paper describes surface parameterization techniques and aspects of discrete differential geometry.
- *Multi-Objective Optimization using Evolutionary Algorithms*, Deb (2001). This book is widely regarded as the definitive book on the topic.
- 'ABD matrix of single-ply triaxial weave fabric composites', Kueh and Pellegrino (2007). This paper describes the process of using periodic boundary conditions to represent the behaviour of a woven composite using an homogenized stiffness matrix (Kirchhoff plate).
- 'Multi-objective optimization of freeform grid structures', Winslow et al. (2010). This paper presents details for optimization of gridshell layouts using a type of surrogate FE analysis model, with anisotropic shell elements rather than a lattice of beam elements.
- *Topology Optimization: Theory, Methods and Applications*, Bendsøe and Sigmund (2003). This book covers all aspects of topology optimization, by two of the leaders in this field.
- 'Periodic global parameterization', Ray et al. (2006). This publication describes the detailed mathematics of this advanced parameterization technique.
- *Numerical Recipes 3rd Edition: The Art of Scientific Computing*, Press et al. (2007). This book is a great reference for anyone writing algorithms or code in C/C++, it has mathematical derivations and working code for a wide range of problems.

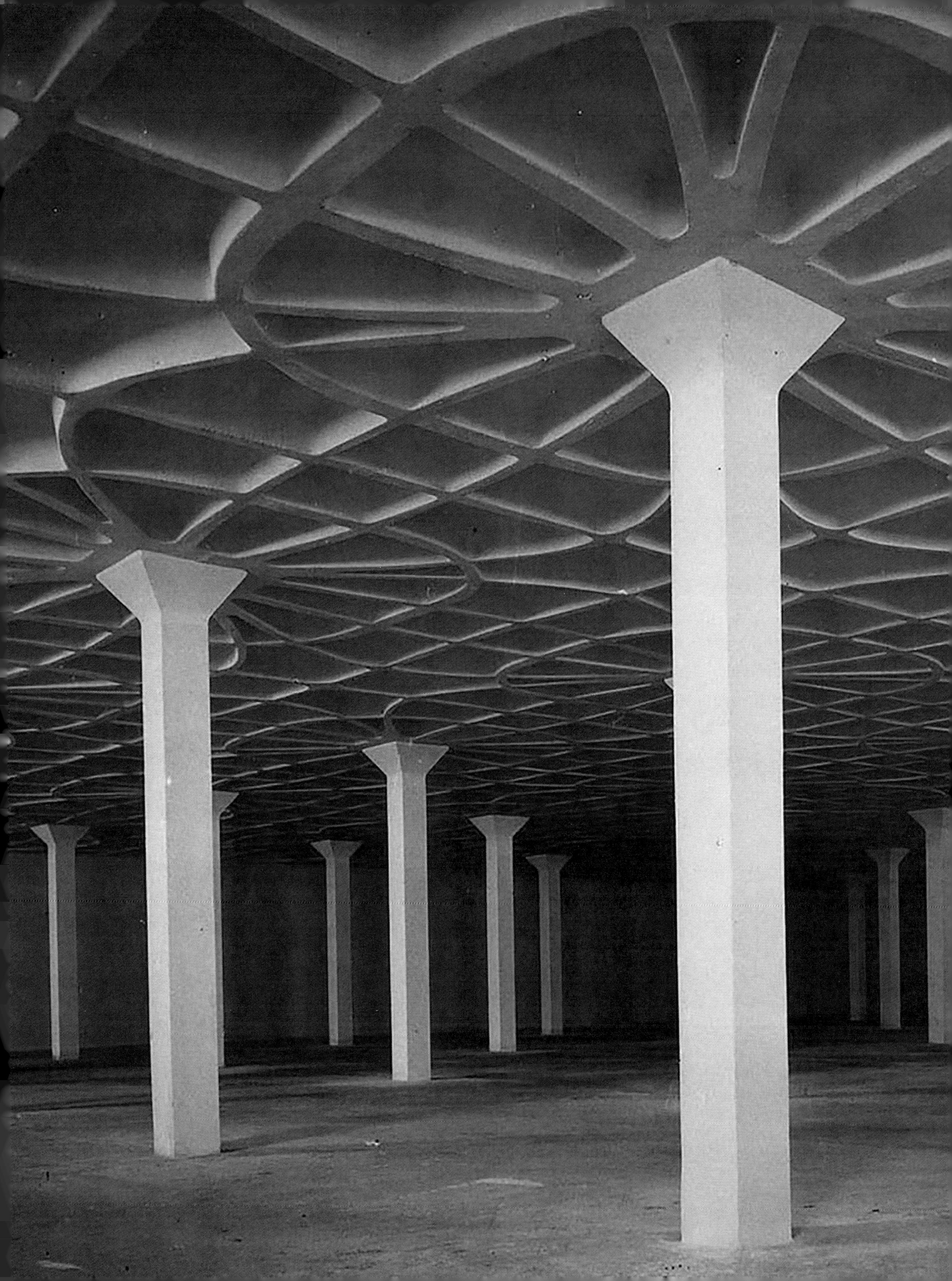

CHAPTER SIXTEEN
Eigenshells

Structural patterns on modal forms

Panagiotis Michalatos and Sawako Kaijima

LEARNING OBJECTIVES

- Discuss how to apply shape modifiers to, and structurally optimize, the shape of a shell.
- Apply the eigenvectors, calculated from the Laplacian matrix, as shape modifiers.
- Use the principal stress vector field to trace curves as the basis for a structural pattern.

The problem of shape optimization through shape modification, and that of pattern generation that determines the placement of ribs, reinforcements, or openings, have attracted considerable attention.

This chapter explains sequential use of shape and pattern optimization, explaining how to (structurally) optimize a shell shape, and subsequently find an orthogonal surface pattern, in our examples derived from the principal stresses. Both processes consider the intrinsic structural and geometric properties of the shell, with the dual objective of improving structural performance, while at the same time achieving a degree of aesthetic consistency between structure and its formal expression. In addition, we offer ways in which designers can exercise control over the optimization definition, and hence, the outcome of the optimized shape, based on their design objectives. This is to avoid optimization based on a singular relationship between the form and the structure, and to obtain solutions that integrate the architectural objectives and structure effectively.

Shape optimization utilizes standard optimization algorithms such as simulated annealing. Given an initial shell, the problem is to find a deformation of that shell described by a normal displacement at each point. The space of possible deformations is infinite and in order to reduce the input parameters of the optimization algorithm, we first create a set of o basis displacement functions f_i with desired properties. The final outcome will be a linear combination of these displacement functions, $a_1 \cdot f_1 + a_2 \cdot f_2 + ... + a_o \cdot f_o$, and the optimization algorithm needs only to search for their o coefficients a_i. The basis displacement functions cover either the entire shell, or part of the shell, in order to provide control over the areas of the shell we wish to optimize.

Structural patterns are networks of curves whose geometry is intrinsically linked to the structural properties of the surface, and they indicate paths of desired material continuity or reinforcement. The patterns are determined by superimposed vector fields such as principal stress directions. An early precursor to this philosophy is Pier Luigi Nervi's 1951 Gatti Wool Mill, shown on page 194. Aldo Arcangeli, one of the engineers at Nervi's office, proposed following the isostatic lines of the principal bending moments of a floor slab for its rib structure (Nervi, 1956). This chapter presents all the algorithms involved, defined over orthogonal grids. This simplifies the presentation considerably, improves the optimization speed, and provides the additional advantage that the input can be in the form of images, literally painted by the designer. However, all the processes would work equally well, albeit slower, for arbitrarily discretized domains.

Throughout this chapter we generate imagery, called 'eigenshells', which constitute an integrated environment for the interactive investigation of such systems. The term 'eigenshells' is a play of words. 'Eigen' is a German word, roughly translating to 'characteristic', 'own', or 'self' and appears in the mathematical term 'eigenfunction', which is what we used in order to generate the basis displacement functions. In addition, this term describes the formal outcome, displacement and pattern, which is driven by intrinsic geometric and structural properties of the shell itself.

The brief

The local zoo has organized a design competition for the entrance to their new rhinoceros habitat. The entrance has a building on either side and it is possible to support our canopy on both the ground and the two buildings. The decision is made to design a quadrilateral gridshell with glazing to provide as much daylight as possible. At this stage, different footprints and boundary conditions are explored for the gridshell: a square, triangular, circular and more arbitrary footprint, touching the ground and/or connected to the adjacent buildings.

16.1 Outline of the process

Shape optimization and structural pattern optimization can be performed individually or collectively. Here, we assume the collective case, where we address the improvement of a global shell geometry using shell optimization, followed by structural patterning over the improved shell. The outline of the proposed procedure is shown in Figure 16.1.

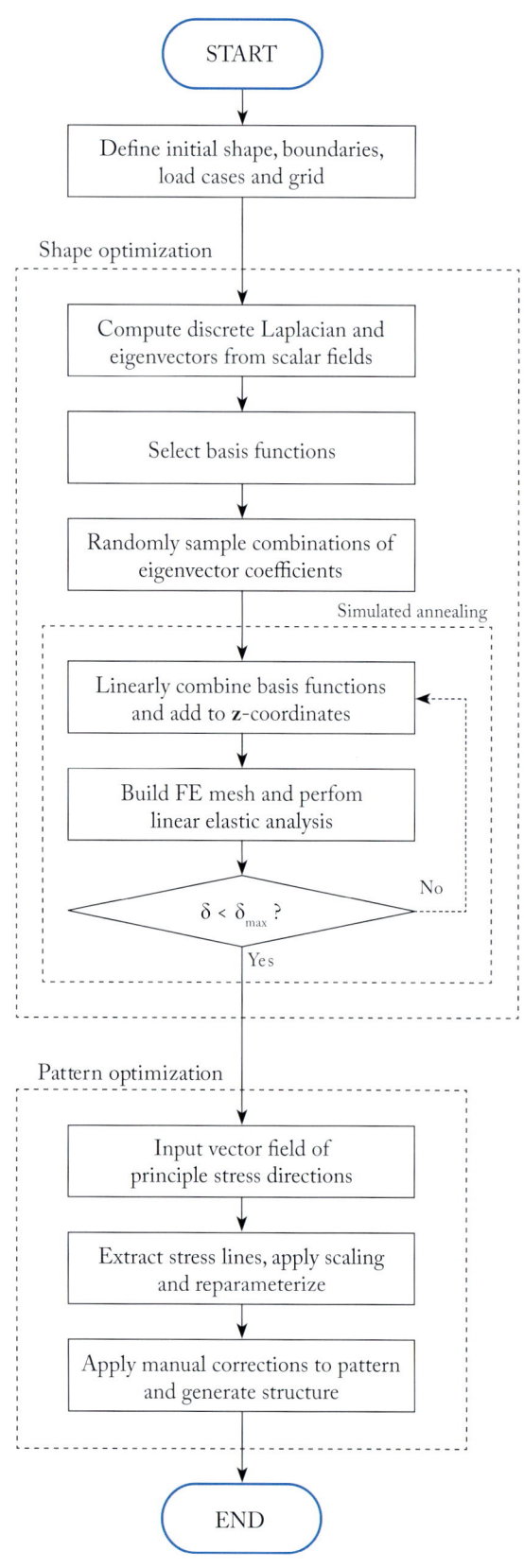

Figure 16.1 Flowchart for eigenshells and structural patterning

16.1.1 Shape modification

Before discussing the specifics of the structural problem, we consider a generic way of describing our problem, which is not related to structural properties, that is, to find a 'nice' set of shape modifiers for a given shape, which will allow us to search a solution space using some optimization algorithm and generate a pattern on the result, so that we can control the directionality and scaling according to some criteria.

In general, it is not practical to optimize a shape by assigning a displacement variable to every node of the geometry (described as a mesh) because it leads to systems with numerous variables for optimization, and, more importantly, most algorithms will yield solutions characterized by considerable geometric noise (Fig. 16.2).

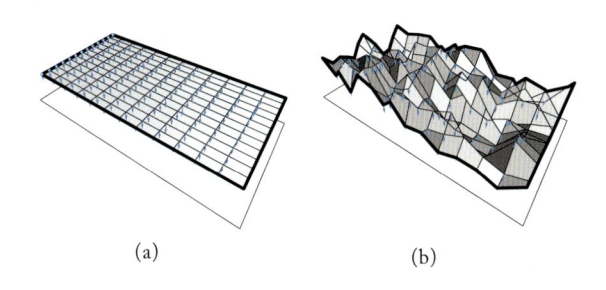

Figure 16.2 Result of applying simulated annealing to optimize (a) a cantilevering plate, using each node's displacement as a separate optimization variable, yielding (b) geometric noise

Therefore, it is desirable to define appropriate ways to deform a shape, which can cover a wide range of solutions with the least number of variables when searching for an improved shape. With information regarding the definition of the shape to be optimized – for example, if it is defined by some parameterized function or NURBS – one could use the parameters of the function or the control points of the NURBS surface as optimization variables. In our case, we opt to find a set of shapes that are intrinsically linked to the geometry of the shell, which allow the designer to select the types of deformations that are visually satisfactory. For this purpose, we used the eigenfunctions of a discrete Laplacian defined over the shell (Figs. 16.3 and 16.4). These functions define scalar fields over the shell and form a series of standing waves of increasing spatial frequency. The shape of the shell is modified by changing the linear combination of these eigenfunctions. The eigenfunction computation of the Laplacian matrix is carried out using the Arnoldi algorithm, which is a common method to compute eigenvalues. The domain of the eigenfunctions does not have to coincide with the shell; this allows for partial optimization (Fig. 16.5a). Furthermore, these eigenfunctions can be made to vanish at the boundaries, which is an ideal property in cases where we want to preserve the geometry of the perimeter of a shell, that is, impose boundary conditions (Fig. 16.5b).

Given the number of variables involved in shape optimization problems, stochastic methods such as genetic algorithms and simulated annealing have been widely adopted in the past. For the shape optimization, we selected simulated annealing owing to its simplicity of implementation, assuming a single objective function.

16.1.2 Structural analysis

The problem becomes specifically structural because we use a Finite Element (FE) analysis step in the objective function of the optimization process, and we employ the principal stress directions to drive the pattern. The technique would remain valid if we were to use some solar analysis results for the optimization phase and some fabrication/geometric constraints for the patterning phase.

16.1.3 Principal stress directions

Structural pattern optimization deals with the generation of structural grid-like patterns that specifically follow the principal stress directions. In this case, we adopted a method used in computer graphics – the most active field of research in computational geometry. The selected algorithm (periodic global parameterization algorithm with curl reduction (Ray et al., 2006)) not only allows us to generate consistent networks of orthogonal curves following the stress directions, in effect, quadrangulating the surface, but also allows designers to control the scaling of the pattern by manipulating a simple scalar field (Fig. 16.6). The decoupling of pattern directionality and pattern scaling offers interesting design opportunities and makes it possible to map

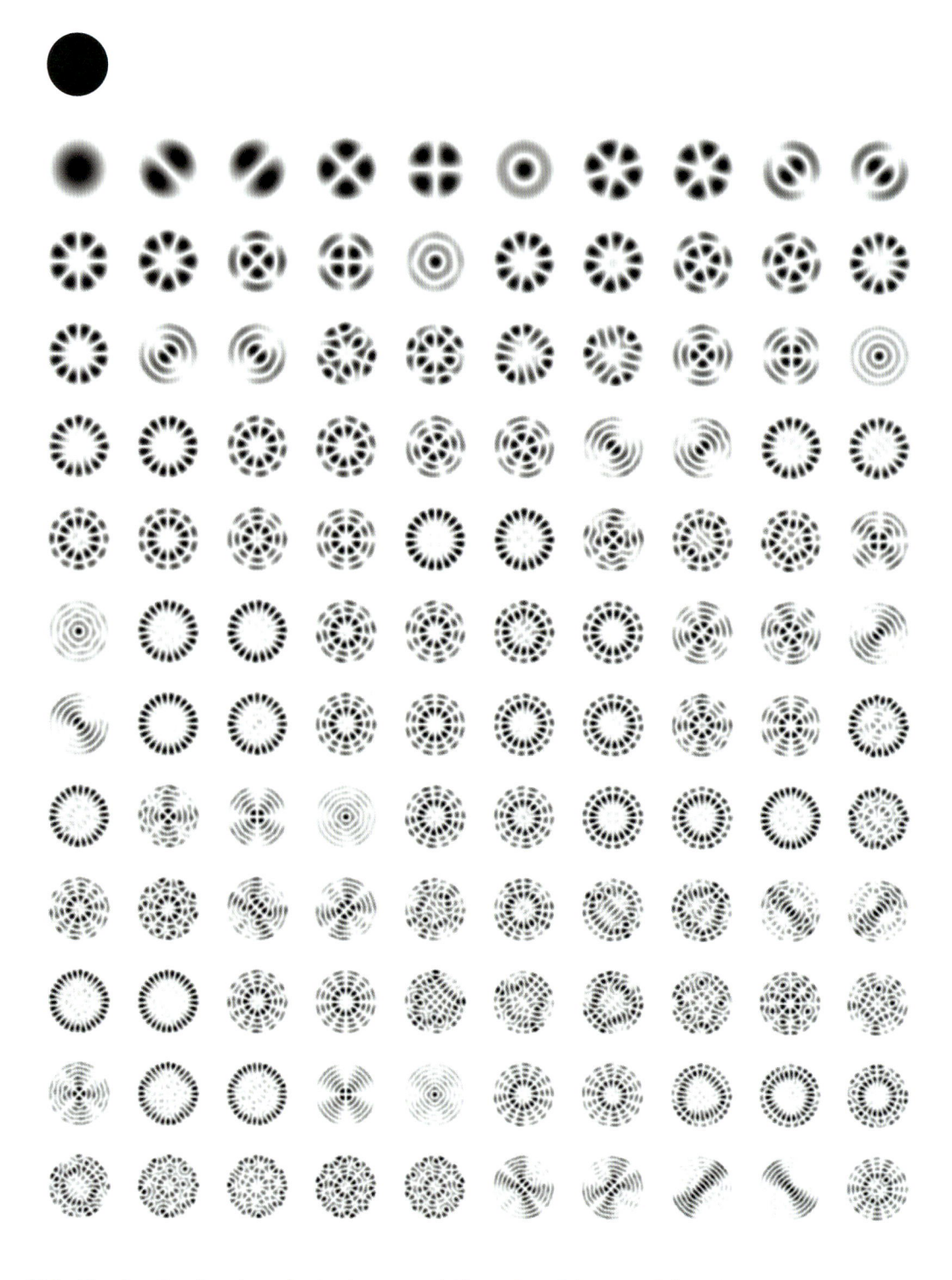

Figure 16.3 First few eigenfunctions of a circular potential. The scaling of the waves follows the gradient with higher frequencies to the right of the disc

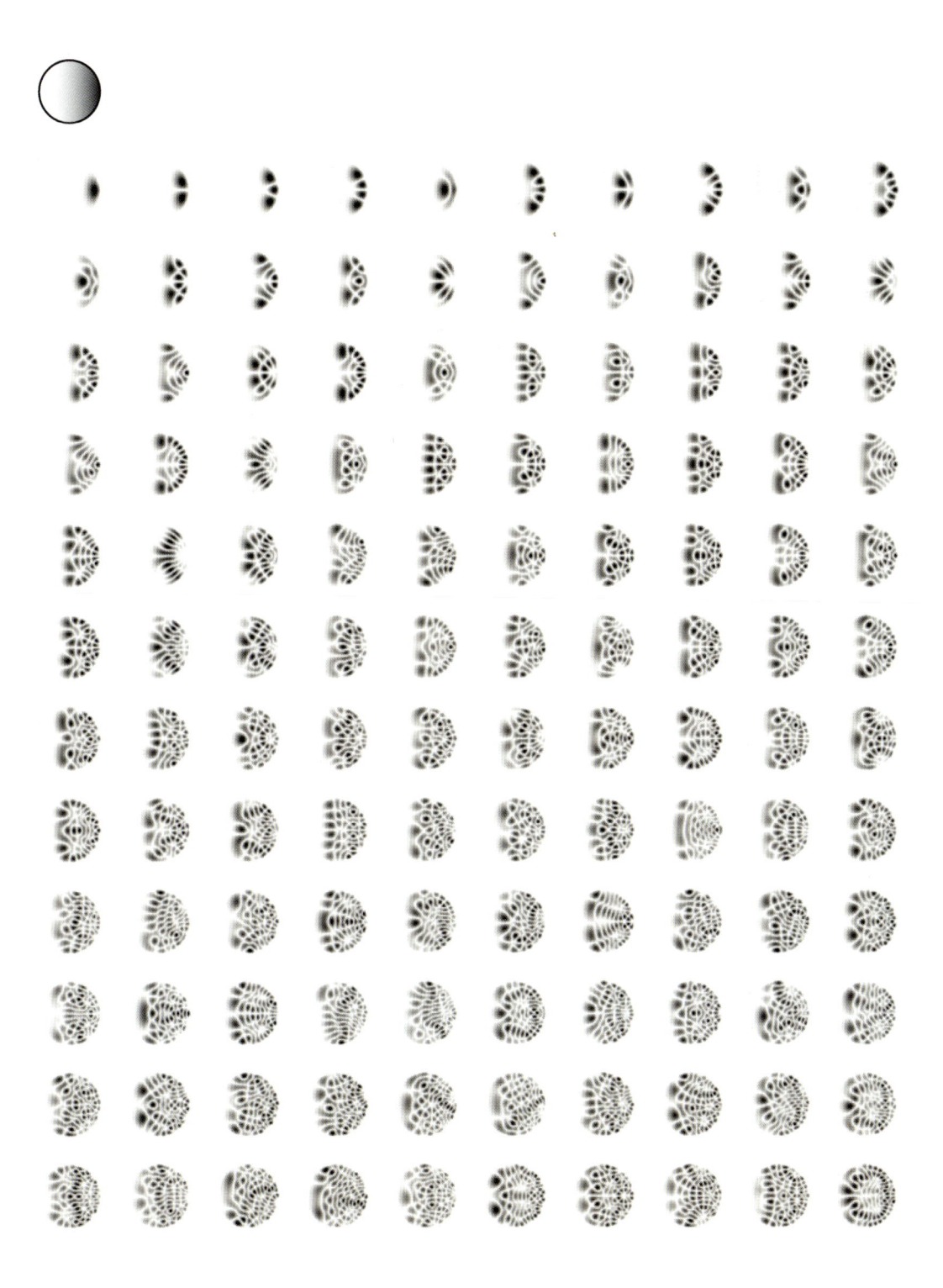

Figure 16.4 First few eigenfunctions of a circular potential with a linear gradient. The scaling of the waves follows the gradient with higher frequencies to the right of the disc

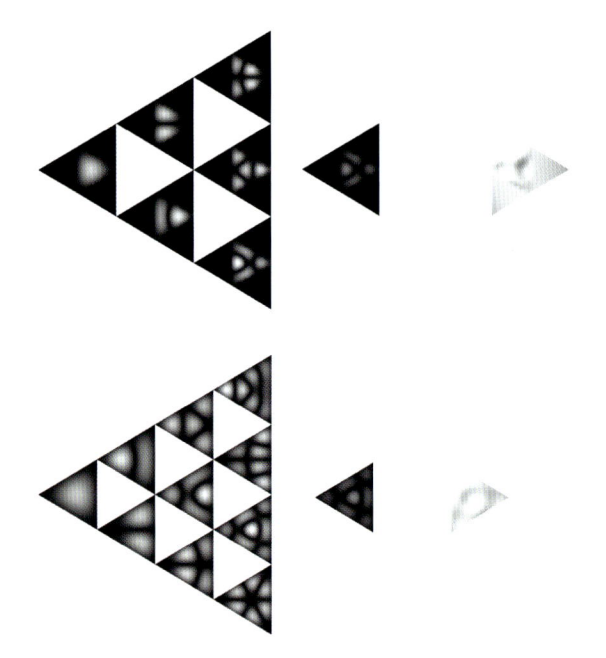

Figure 16.5 Shape optimization (top) within a subdomain (smaller triangle within larger triangle), and (bottom) full optimization with vanishing boundaries

different structural or design parameters to the two aspects of the pattern. For example, it is conceivable that the directionality of the grid will be controlled by the principal stress directions, and the scaling by the magnitude of stress, lighting requirements, or fabrication constraints.

16.2 Input

The most important primary inputs in our method are the initial geometry of the shell and the support conditions. For simplicity, we present examples in which the only load case is the self-weight. However, the method could handle any other load case or combination of load cases. In addition, the designer can define two scaling scalar fields over the given shell (either as a projected image or as colour per vertex information). The first field determines the region of the shell domain that will be a candidate for shape optimization as well as the relative scaling of shape oscillations. The second field determines the relative scaling of the pattern. Secondary inputs will be generated from a static analysis of the shell structure. These are either global results, such as maximum deflection used in the shape optimization phase objective function, or local element results, such as principal stress directions used in the structural pattern generation stage.

16.3 Shape optimization

Here, we use the terms 'displacement' (as in displacement map) for modifications to the initial shell shape and 'deflection' for the displacement of nodes as a static analysis result.

In general, the shape optimization phase can be summarized as follows. We wish to find a height

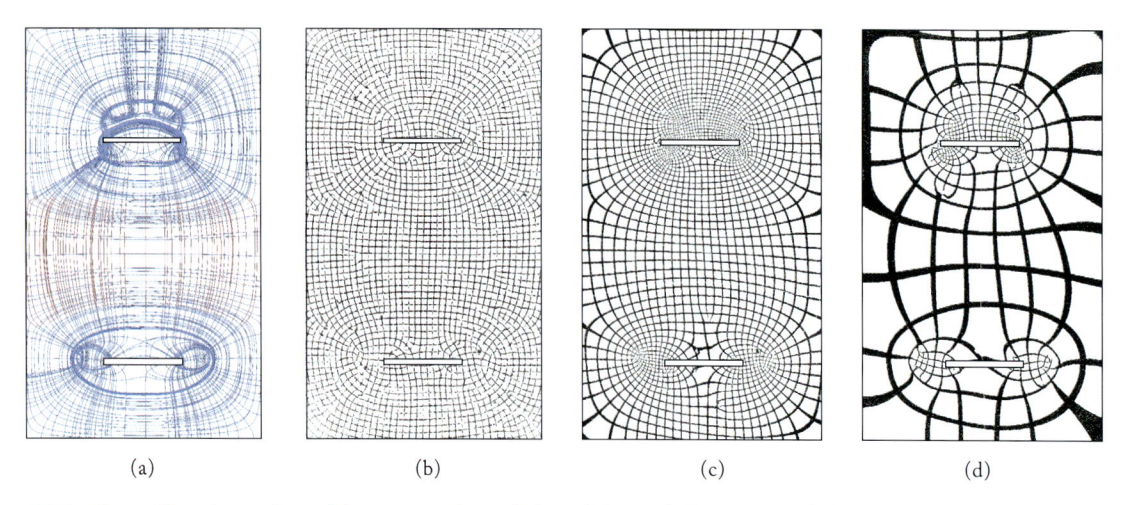

(a)   (b)   (c)   (d)

Figure 16.6 Controlling the scaling of the pattern through that of the underlying vector field. (a) The stress lines of slab with two linear supports, (b) pattern with no scaling, (c) scaling roughly proportional to stress and (d) exaggerated scaling

displacement field for the given shell or a region of the shell that may or may not preserve z-coordinates on the boundaries, and minimizes a certain objective function related to the structural performance of the shell. The region for shape optimization is defined as a scalar field over the shell. The nodes of the $m \times n$ grid will have values ranging from 0.0 to 1.0. Areas with a potential value of 0.0 will form the potential well, that is, the region that the optimization basis function will fill, and 1.0 will remain fixed.

The problem formulated consists of three parts: optimization variables, objective function and optimization.

16.3.1 Optimization variables

'Displacement field' refers to some discrete scalar function that moves each node on our shell upward or downward along the z-axis. The most generic formulation would allow each point on our grid to vary independently, resulting in an optimization search space of 2,500 dimensions for a 50×50 grid. This would usually yield some random solution with considerable noise (wide variations in height from one node to the next, see Fig. 16.2) that is often undesirable as an architectural solution. Therefore, we introduce several ways to control the optimization results. We consider noise reduction, introduction of simpler displacement functions, optimization area specifications and variable scaling.

Noisy solutions contain high frequencies of height variation. From signal analysis, we know that in order to avoid such frequencies, it is better to operate in the frequency domain of a signal, cut off the high frequencies, and convert the signal back to the time or space domain. Another approach is to consider the problem as a search for a shape modifier that is a linear combination (weighted sum) of simpler displacement functions of acceptable smoothness (frequency ≈ scale of height variation). We enforce an additional constraint on these functions, that is, they should vanish near boundaries, which may or may not coincide with the shell boundary. Thus, we can search for local modifications of the shell or modifications that preserve the shell boundaries. We may also want to include solutions that exhibit high frequencies on certain parts of the shell and lower frequencies on others.

All the requirements stated above can be fulfilled if the shell displacement field is composed of the eigenfunctions of a certain discrete Laplacian operator. The discrete Laplacian is a very large square matrix with as many elements in a row as there are nodes on our shell (see Section 16.6.1). This matrix corresponds to the connectivity matrix of the underlying FE mesh. The eigenfunctions of the Laplacian matrix if plotted back on the original geometry, form standing waves of increasing frequency, similar to the modes of vibration of a membrane with a given boundary. The eigenfunctions are ordered by increasing frequency so that we can select the first few eigenfunctions (low frequency = smoother functions → less curvature variations); they vanish near the boundary of the potential well, and from the viewpoint of design, they reflect the intrinsic properties of the shell shape.

In order to modify the shape of the shell, we simply need to multiply each eigenfunction by some weight factor and add them together in order to obtain a displacement field for the coordinates of the nodes of the shell. In addition, we can add a variable scalar field to the Laplacian matrix. This field will modulate the frequencies of the standing waves so that these become denser in regions of low potential values.

16.3.2 Objective function

Now that we have a generic description of the shape of our shell, we need to define an objective function, that is, an evaluation criterion for each design outcome. At present, for simplicity, we shall employ the maximum deflection calculated by a linear elastic analysis as a rough measure of stability, and as a single objective function. Since the selected optimization algorithm does not depend on the specifics of the objective function, any other more refined criteria of structural performance would also suffice.

Here, the structural analysis is performed using an FE model with four-node shell elements that are a combination of a Mindlin plate element and a standard membrane element.

16.3.3 Optimization

As mentioned, we want to find the coefficients of the selected height modifier eigenfunctions that minimize the deflection of the shell. By using the eigenfunctions, we need to find only a few coefficients instead of searching a large space with every node's height as a parameter. Figure 16.7 shows the FE results based on some optimized, linear combination of the first few eigenmodes shown. From all variations generated by simulated annealing (varying only ten coefficients), these exhibited the lowest maximum deflection.

16.4 Structural patterns

After the shell's shape has been fixed, the next problem can be summarized as follows. What sort of network of curves could we use to add ribs or reinforcements, or to distribute many small openings over a shell?

Thus far, many surface discretization schemes used by designers are dependent on the effective parameterization of NURBS surfaces and other explicitly parameterized surfaces. In general, NURBS have skewed local coordinate systems; hence, isolines of such surfaces form skewed networks of curves that may introduce problems for fabrication (very narrow angles at connections with difficult joints).

16.5 Principal stress vector fields

It is better to start, in generating a network of curves, from the principal stress directions, which define orthogonal pairs of vectors at each point on the shell along which the shear stresses vanish. This means that all the intersections occur at 90°. This may facilitate the fabrication of joints between elements. However, as is the case with most vector fields, it is quite difficult to extract a consistent network of curves from them. This is further complicated because the principal stress directions (like the principal curvature directions on a surface) are not proper vector fields, but the eigenvectors of tensor fields; thus, these eigenvectors come in pairs of pure direction (no sense), and if they were to be treated as ordinary vector fields, one would obtain unexpected results such as abrupt flipping of sense and 90° rotations from one point to the next.

Another obvious candidate field is the principal curvature directions, which would be ideal for the discretization of a shell into panels with minimal twist. Moreover, it would be possible to control local field

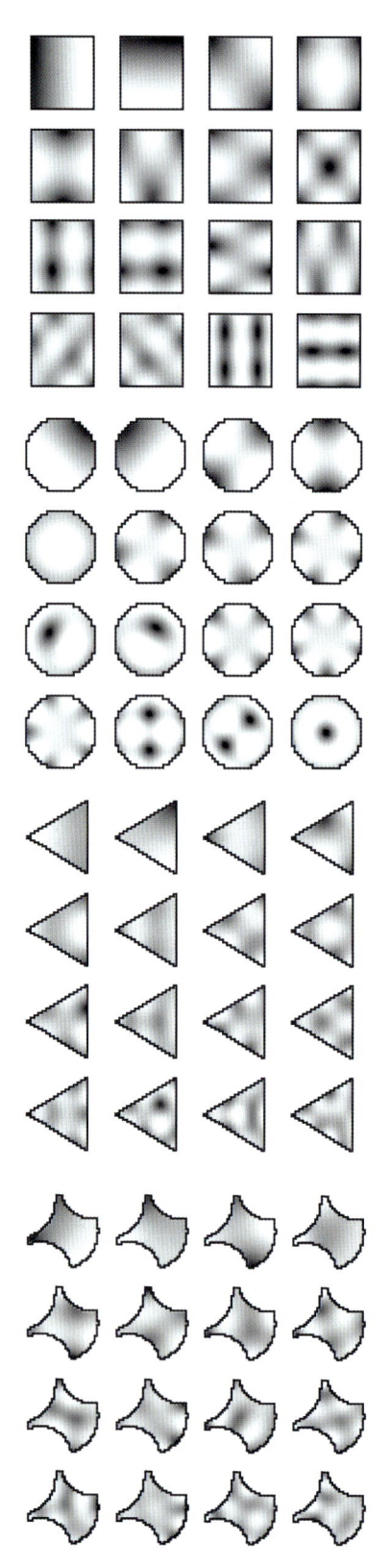

Figure 16.7 FE results from linear combination of eigenmodes and given support heights for a square, triangular, circular and arbitrary plan

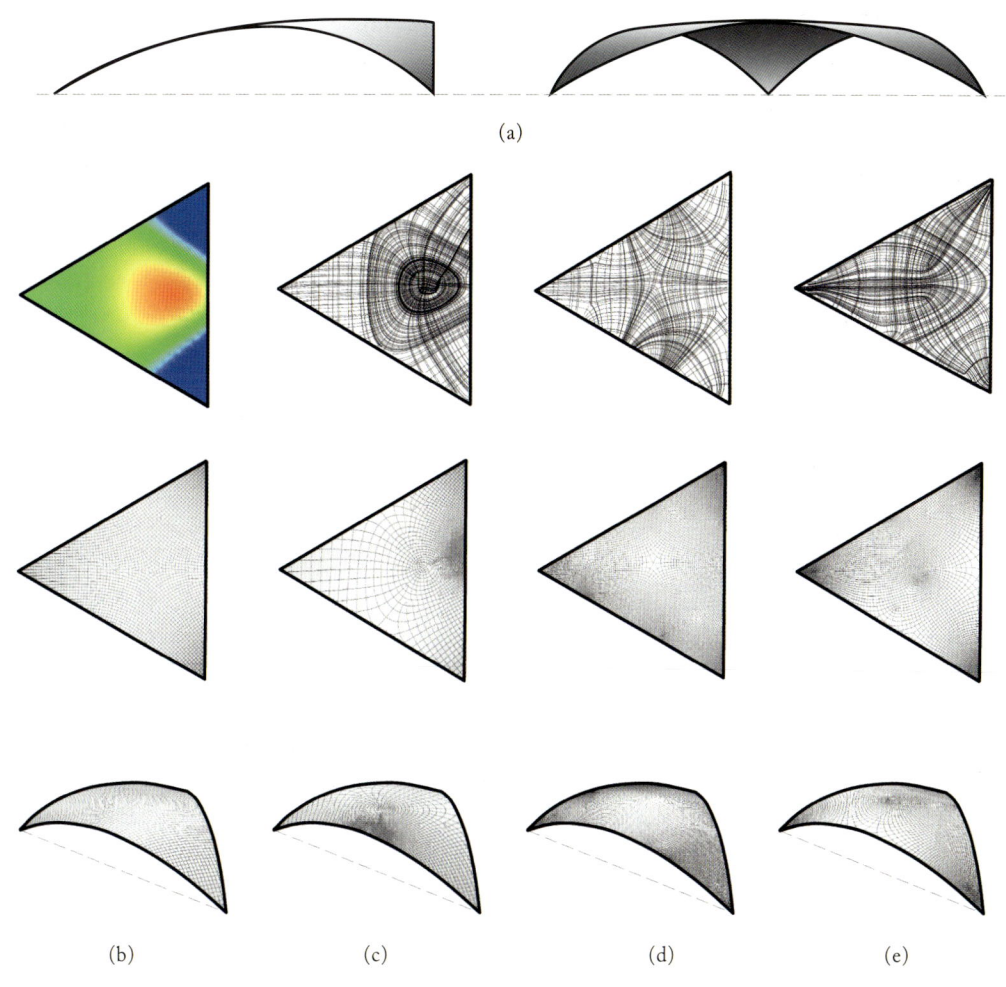

Figure 16.8 (a) Asymmetric shell with patterns derived from (b) principal curvatures, (c) gradients, (d) principle moments and (e) principal stresses

directionality through handles that can be manipulated by the user in order to define the local scaling and direction. The algorithm could mark out regions that are not structurally critical and in which such design considerations may override the structural pattern.

Figure 16.8 shows different patterns derived from the same shell shape. The first two, principal curvatures and gradients, are geometric, while the second two, principle moments and principal stresses, are mechanical, and taken from the FE analysis.

The first step in constructing some geometry from the stress vector field is the interactive selection of principal stress lines on the given shell, followed by their post-rationalization to obtain, for example, a proper network of ribs. This is an intuitive approach that gives the designer maximum control. From the viewpoint of implementation, it will look like a standard integral curve problem (using either the Euler or Runge-Kutta methods), but one where special care is required to handle the possible 90° or 180° flipping in the field, always following the best candidate direction (this does not work very well near singularities, but one can detect these regions and terminate the integration).

16.5.1 Reparameterization

A rather more robust approach involves determining how to reparameterize a surface so that its new coordinate curves would run along these vectors (at least locally), and then, simply mapping a grid on the surface using the new map. This is exactly what the periodic global reparameterization algorithm does.

Using this algorithm, one obtains a set of texture coordinates per vertex per facet (and not just per vertex), which enables the mapping of any pattern on the surface. There is only one limitation for the pattern, that is, it has to be invariant under 90° rotations; otherwise, inconsistencies may appear along edges with flipped first and second principal directions on either side.

This method requires the principal stresses defined on the nodes rather than on the shell elements; thus, one can try to either extract stresses at nodes from one's analysis module or run the reparameterization on the (centroidal) dual graph of the grid, where the centres of the quadrilaterals (quads) become nodes.

An advantage of this method is the decoupling of pattern scaling (a scalar field) and pattern alignment (orthogonal pair of vector fields). This flexibility makes it possible to determine the relative scale of the output pattern by inversely scaling the input vector fields. The scalar field that determines the scaling of the pattern can be determined by the designer, according to the project requirements or design objectives. In addition, the description of a pattern as a vector and a scaling field allows for the integration of different concerns in the same pattern through vector field blending.

After the execution of the algorithm is complete, we can extract the isolines of the new parameterization by simply mapping a quad pattern on the surface using the new (u,v) coordinates.

16.5.2 Singularities

Another interesting property of such field-induced networks of curves from the viewpoint of design and fabrication is the existence of singularities in these fields. Three types of singularities appear in such fields, and they would require special care by the designer. Singularities are a mathematical and computational nuisance. However, they provide fundamental information about the field that generates them (they give it a structure). Hence, the interpretation of singularities in the detail design and fabrication stage is a design opportunity rather than a nuisance. They appear as focal points and lines that hold the entire structure together, visually and often literally, like a pole in a woven basket or an upholstery button.

16.6 Implementation

For this particular problem, the data structure is a grid of square cells. The information stored at the nodes is location (initial x, y, z, and possibly, displaced z), all the usual structural analysis inputs (material property, boundary conditions, loads), a pair of vector fields that will be used in the latter pattern generation phase, and whether the node is a part of the shell (within the boundaries). The grid cells are linked to shell FE definitions, from which we can extract FE linear elastic analysis results (in particular, the principal stress directions).

Regardless of the geometry of the shell itself, we are going to work in an orthogonal domain big enough to hold the shell in plan. The use of a grid that extends to the boundary box of the shell may seem computationally wasteful at present; however, its usefulness will become apparent later. The $m \times n$ discrete square grid is a raster representation of our system, and the number of cells in the system is $k = m \cdot n$. Grid **G** has cells G_{ij}, where $i = 1,...,m$ and $j = 1,...,n$.

Figure 16.9 shows the triangular shell, the grid extending beyond the geometry of the shell, the

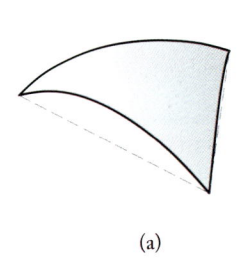

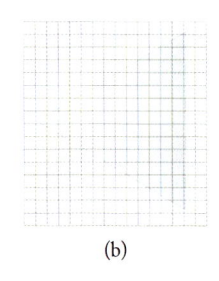

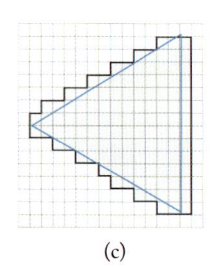

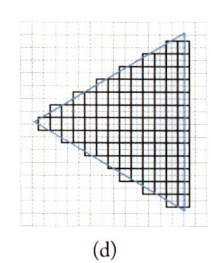

Figure 16.9 (a) Triangular shell resulting from the (b) initial plan shape and extended grid mapping, (c) the grid with values for the Laplacian and (d) the FE mesh

subgrid used to compute the Laplacian and the FE mesh.

The following $k \times k$ matrices need to be calculated: the discrete Laplacian matrix **L**, the potential matrix (boundaries and scaling) **P**, and the combined matrix **H** = **L** + **P**. Matrices **L** and **H** are symmetric and **P** is also a diagonal matrix. In order to conserve memory, **P** is stored as a vector of length k.

16.6.1 Discrete Laplacian

For simplicity, we construct the discrete Laplacian Δ of a regular rectangular grid containing the shell. The boundaries of the shell will be enforced (if needed) by adding a scalar field **P** representing the shell as a potential well which will be added to the Laplacian (low values inside the shell and high values outside). Here, we have a grid of constant cell size so the discrete Laplacian ∇ looks like the graph Laplacian **L** of the grid. We could construct the same matrix **L** for irregular grids (meshes) of arbitrary graph topology by using the discrete Laplace–Beltrami operator. The computational overhead is not significant and the slowest step (calculation of eigenfunctions) depends only on the number of nodes, not on the complexity of the graph.

The discrete Laplacian **L** is easy to compute as we are working on a grid graph (the shell boundary will be a potential function so it will not affect the form of the **L** matrix). Basically, we need to set the diagonal elements L_{ii} equal to the degree of that cell, that is, the number of connected neighbours, and for each row, set the indices corresponding to the neighbours of the cell $L_{ij} = L_{ji}$ to -1. Each cell L_{ij} may have four, three or two neighbours depending on whether it is an interior, edge or corner cell. In other words, the entries of **L**,

$$L_{ij} = \begin{cases} \deg(n_i) & \text{if } i = j \\ -1 & \text{if } i \neq j \text{ and } n_i \text{ is adjacent to } v_j \\ 0 & \text{otherwise} \end{cases}$$

where $\deg(n_i)$ is the degree of node i.

The graph Laplacian can also easily be constructed from the branch-node **C**, discussed in Section 6.4.1,

$$\mathbf{L} = \mathbf{C}^\mathrm{T}\mathbf{C}, \quad (16.1)$$

where, for a 3×3 grid,

$$\mathbf{L} = \begin{bmatrix} 2 & -1 & . & -1 & . & . & . & . & . \\ -1 & 3 & -1 & . & -1 & . & . & . & . \\ . & -1 & 2 & . & . & -1 & . & . & . \\ -1 & . & . & 3 & -1 & . & -1 & . & . \\ . & -1 & . & -1 & 4 & -1 & . & -1 & . \\ . & . & -1 & . & -1 & 3 & . & . & -1 \\ . & . & . & -1 & . & . & 2 & -1 & . \\ . & . & . & . & -1 & . & -1 & 3 & -1 \\ . & . & . & . & . & -1 & . & -1 & 2 \end{bmatrix}. \quad (16.2)$$

In order to calculate **P**, note that each cell in the $m \times n$ grid **G** corresponds to one diagonal element in **P** (since the diagonal of **P** has $k = m \times n$ entries). For each element P_{ii} along the diagonal, where $i = 1,\ldots,k$, set it to 0.0 if the corresponding cell's G_{ij} centre falls within the boundary of the shell or 1.0 if it is not. Here, any value p between 0.0 and 1.0 can actually be used if the eigenfunctions are to resemble waves with variable frequencies (denser waves near cells with higher P_{ii} values). Stored as a vector, in this case,

$$\mathbf{p} = [0 \ 0 \ 0 \ 0 \ p \ 0 \ 0 \ 0 \ 0]^\mathrm{T}. \quad (16.3)$$

The combined matrix **H** is simply the sum of **L** and **P**. We now want to calculate the eigenvectors of **H**. There are k eigenvectors for **H** but for our purpose we only need the first few. The reason for this is that the eigenvectors represent displacement functions that look like standing waves within the boundary of the shell. If we want to limit the shapes to smooth functions with limited number of oscillations, then we can assume that these displacements will be some linear combination of the first few eigenvectors. There are several algorithms, such as the Arnoldi algorithm used here, for calculating the eigenvectors **v** of a matrix. We store the o number of eigenvectors **v** with k values in a $k \times o$ matrix **V**. The k values represent displacement values for the k cells in the original grid. The o eigenvectors **v** are now used for the basis displacement functions f_i, with $i = 1,\ldots,o$. The optimization algorithm searches for the o coefficients a_i that are used to linearly combine them to describe the shape of the shell.

Figure 16.3 shows the eigenfunctions for a circular domain. In Figure 16.4 a linear gradient has been applied by changing the values of **P** from 1.0 to

0.0 along the grid. Figure 16.5 shows how, for the triangular domain, the first ten eigenvectors can be mapped back on the shell. The boundary can be limited to a subregion of the shell (Fig. 16.5, left) below by modifying **P** so as only the elements of **P** corresponding to cells within the subregion are set to 0.0.

16.6.2 Finite element model

For the FE model we take the centres of the cells that fall within the input shell boundary and form a grid of shell elements connecting them (Fig. 16.9d). The x- and y-plane coordinates of each node in the FE model are known. The z-coordinate will be the result of adding a weighted sum of the eigenvectors to the original elevation of its point projected on the input shell.

For the optimization phase we can use either a simulated annealing or genetic algorithm or any other stochastic optimization method. The optimization variables are the coefficients of the eigenvectors in the weighted sum that determines the shape deformation. As for the objective function, here for the sake of simplicity we just use the maximum deflection of the shell.

The FE model, however, has fewer nodes than there were cells in the original grid and we need to know how to map the new FE nodes to the original cells. The objective function for a single step in the optimization loop would be: given a test solution of o coefficients **a**, eigenvectors **V** and the k z-coordinates of the grid points z_G corresponding to the initial grid projected on the input shell. First, build the FE mesh, and then calculate the coordinates **z**, where each node in the FE mesh corresponds to the cell from the original grid, so $z = z_G + Va$.

After applying boundary conditions and loads, structural analysis is performed, which returns the maximum deflection as the output of the objective function.

16.7 Tracing integral curves of eigenvector fields

Given an FE mesh with shell elements Q and assuming that we know the principal stress vectors s_1 at the centre of each shell element $\mathbf{p}_j$, we want to trace a single curve along the first principal stress direction $\hat{\sigma}_{1,j}$ starting from a point P_0.

1. Create an empty list of points **P** to hold the stress line points.
2. Find the element Q_j closest to $\mathbf{p}_0$ and set $\mathbf{p} = \mathbf{p}_j$.
3. Append **p** to **P**.
4. Set $\sigma_{dir} = 1$ and $j_0 = j$.
5. While the number of points in **P** < maximum number of steps:
   a. If $\sigma_{1,j} \cdot \sigma_{1,j_0} > 0$ then $\sigma_{dir} = -\sigma_{dir}$.
   b. Find the edge Q_E of Q_j intersected by the ray $r(t) = \mathbf{p} + \sigma_{dir}\hat{\sigma}_{1,j}t$ and set **p** to the intersection point.
   c. Append **p** to **P**.

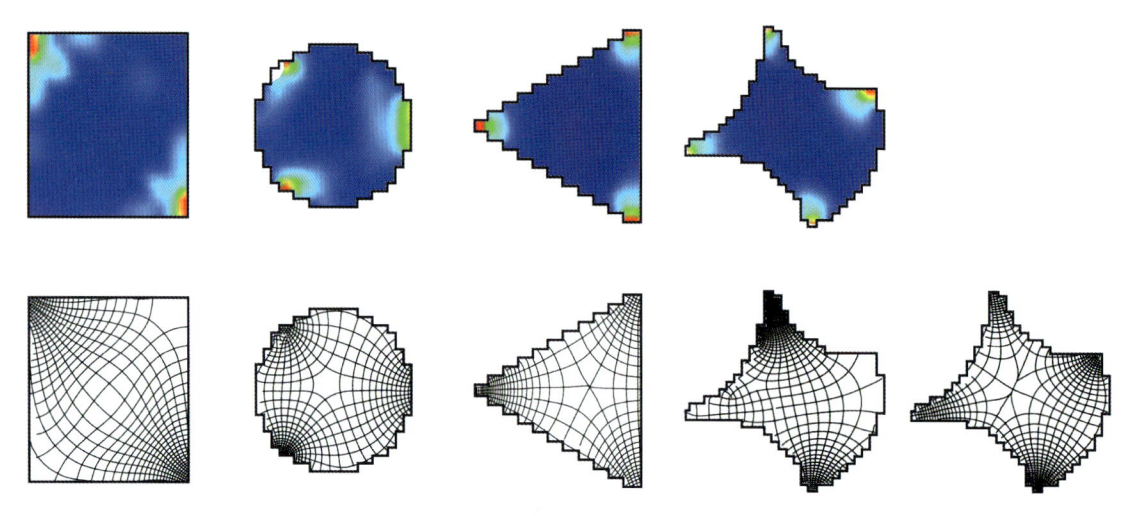

Figure 16.10 Structural patterns derived from principal stress vectors in FE analysis

d. If Q_j has a neighbour Q_h along the intersection edge Q_E:
   i. then $j_0 = j$, $j = h$ and return to step 5a
   ii. else break.

Step 5a is necessary because eigenvector fields tend to flip from point to point so we need to keep track of the last direction we moved along. Figure 16.10 shows the patterns derived for our four design plans, based on the optimized shapes and their FE results.

This simple algorithm gives a rough piecewise approximation of the stress lines. The stress lines are smoother if instead of moving directly to the edge of each shell element, little steps are made along interpolated stress directions from the corner nodes of each quads (similar to a simple Euler integration scheme, taking care to account for the occasional flipping of direction). For general meshes, interpolating the stress tensors and extracting the eigenvectors of the interpolated tensor is difficult as the stress tensor for each shell element is expressed in the local coordinate system of the element. However, because of the fact we are using an orthogonal grid and provided that our shell does not have very steep slopes, we can use simple bilinear interpolation to get an approximation of the interpolated stress tensor at each point.

Extracting stress lines requires some kind of seeding strategy for the initial points which can lead to inhomogeneous distributions of curves. The extracted pattern can be improved in several ways. Careful seeding of start points for stress lines can result in a more even distribution of curves. Alternatively one can use an algorithm such as periodic global reparameterization in order to extract isoparametric curves that follow the stress lines.

16.8 Application to brief

The preliminary designs for the entrance to the rhinoceros habitat has yielded a wide range of possibilities, due to both our freedom in choosing different footprints and the option to have high support points. Figure 16.11 shows the resulting gridshell for the square, triangular and circular shapes, all demonstrating open edges that flare up. Figure 16.12 illustrates that the same footprint can yield very different shapes

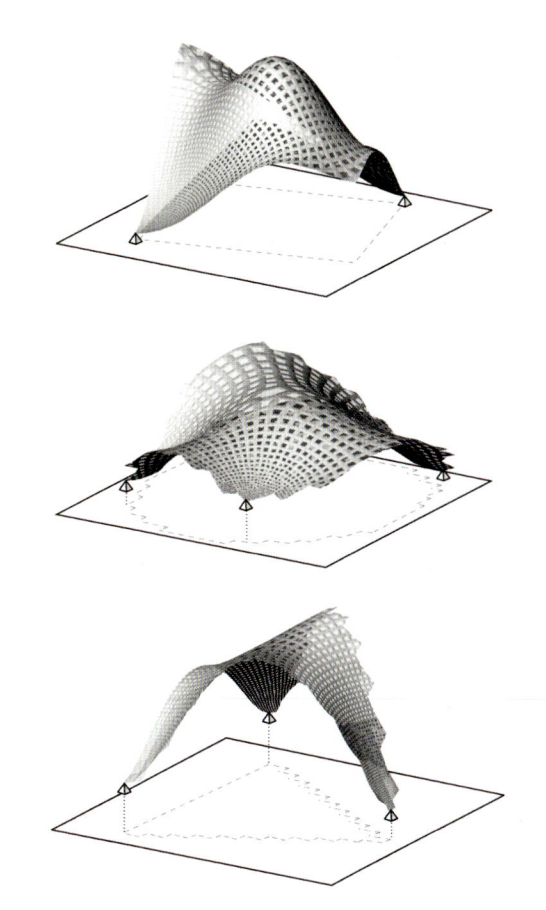

Figure 16.11 Resulting structures from shape and pattern optimization for the square, circular and triangular plans

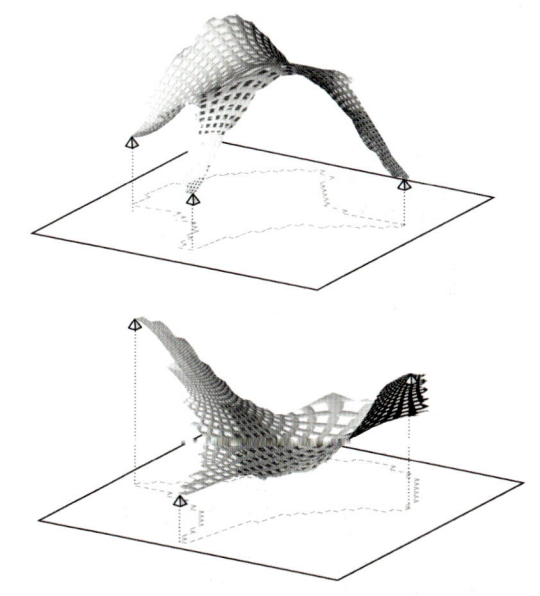

Figure 16.12 Two resulting structures for different heights of the boundary conditions, for the same arbitrary plan shape

depending on the boundary conditions, as well as resulting in a different structural pattern.

16.9 Conclusion

Computer simulation methods have opened up new possibilities for design and research by introducing environments in which we can manipulate and observe. For instance, architects utilize three-dimensional modelling tools to simulate architectural geometries, and engineers use FE software to simulate structural behaviour. Simulation tools make certain aspects of architecture efficient, but they have brought new types of challenges into the field. One such challenge is the structuring of so-called complex geometries often described as surfaces or shells. These forms are often conceived in an environment where gravity, scale and material are absent but analysed in a model where geometries are frozen and static. As a result, there exists little understanding between the two disciplines in solving the design problem and arriving at a well-negotiated form.

In fact, computer simulation involves the modelling of a reality of something, as an abstraction, in order to facilitate understanding towards a specific aspect of interest. Simulation is a manipulation of a model that enables us to perceive interactively, the relationship between parts, as well as its overall implications, all of which would be difficult to observe otherwise. Generally, architects and engineers look into different aspects of reality, thus the models they develop as well as manipulate hold different structures and controls.

In this chapter, we proposed a way of integrating aspects of both architecture and engineering models.

Key concepts and terms

The **Laplacian** is a differential operator Δ that features in many mathematical descriptions of physical processes, such as the heat and wave equation. It is defined as the divergence of the gradient of a function. Its eigenfunctions are associated, for example, with vibration modes on a membrane and other related phenomena.

The **Laplace–Beltrami operator** is a generalization of the Laplace operator for curved surfaces.

Discrete Laplacian or discrete Laplace operation is the discrete version of the Laplacian operator in the form of a matrix. In its simplest form it can be given by the graph Laplacian **L** which is related to the connectivity matrix of a graph.

Eigenvectors are, given an $n \times n$ matrix **A**, n-vectors **v** which satisfy the equation $\mathbf{Av} = \lambda\mathbf{v}$ where the eigenvalue λ is a scalar. For example, if the matrix **A** is 3×3 and describes some non-uniform local scaling of geometry, these vectors will point to the directions of the maximum and minimum stretch. What the above equation tells us is that these vectors correspond to special directions in space that the matrix **A** is applied to, their direction does not change. The eigenvectors of the stress tensor are the principal stress directions. The eigenvectors of the curvature tensor are the principal curvature directions and always form orthogonal pairs of vector. For example, around a cylinder the maximum eigenvector at a point will be tangent to the circle that passes through this point and the minimum eigenvector will be parallel to the line generator.

Eigenfunctions are basically the infinite dimensional equivalents of eigenvectors. If instead of a matrix one has a differential or integral operator **K**, then its eigenfunctions will be the functions **f** that satisfy the equation $\mathbf{K}(\mathbf{f}) = \lambda\mathbf{f}$ for some scalar λ. Grossly simplifying for the sake of explaining, one can think of operators such as Δ as a matrix of infinite rows and columns. In the discrete case the rows and columns become finite.

Spectral methods are special methods used in numerical analysis that take advantage of the fact that certain problems are easier to solve in the frequency domain. There is interest in the application of these methods in computer graphics as they provide novel ways of thinking and manipulating geometry.

Exercises

- Given the standard grid (Fig. 6.12), instead of using eigenfunctions, we first apply simpler displacement functions, from two superimposed

waves $a_1 \cdot \cos(f_1 \cdot x) + a_2 \cdot \cos(f_2 \cdot y)$, where x and y are the coordinates of each mesh vertex, a_1 and a_2 the amplitudes of the two waves and f_1 and f_2 the frequencies. Select boundary conditions and loads. Apply an optimization algorithm (such as simulated annealing or GA) with a_1, a_2, f_1 and f_2 as optimization parameters and the maximum deflection from FE as the objective function.
- From previous exercises we know the branch-node matrix **C**. Compute the discrete Laplacian of the generated grid, the graph Laplacian **L** and extract the eigenvectors. Selecting one of them (they have the same number of components as the number of points in the original grid) and use it in order to displace each grid point along the z-axis.
- Use a linear combination of the first few eigenfunctions to create a composite displacement function. Run the same optimization using the coefficients of the linear combination as parameters.
- Given an FE model of our optimized result, trace stress lines from the centres of all the quadrilaterals and colour them according to the average stress along each path. If instead you have a very fine grid, selectively (either manually or computationally) remove stress lines that have very low mean stress value or are very close to other stress lines. After you finish with the strategic thinning of the stressline bundles, simplify the remaining ones and add ribs along their paths.

Further reading

- 'Periodic global parameterization', Ray et al. (2006). This paper presents an algorithm for the generation of quad meshes aligned to an input vector field. The same algorithm will allow us to map any pattern on a curved surface aligned to any input vector field.
- *Advanced Topics in Finite Element Analysis of Structures*, Asghar Bhatti (2006). This book contains a typical technical treatment of finite element analysis with many examples of step-by-step computation and example code.
- 'Discrete Laplace–Beltrami operators for shape analysis and segmentation', Reutera et al. (2009). This paper is a comprehensive overview of the current state of spectral methods on discrete surfaces. There is a good discussion of the pros and cons of the different discrete approximations to the Laplace–Beltrami operator for discrete manifolds. The Laplace–Beltrami operator lies at the heart of many interesting algorithms for the analysis, manipulation and representation of meshes.

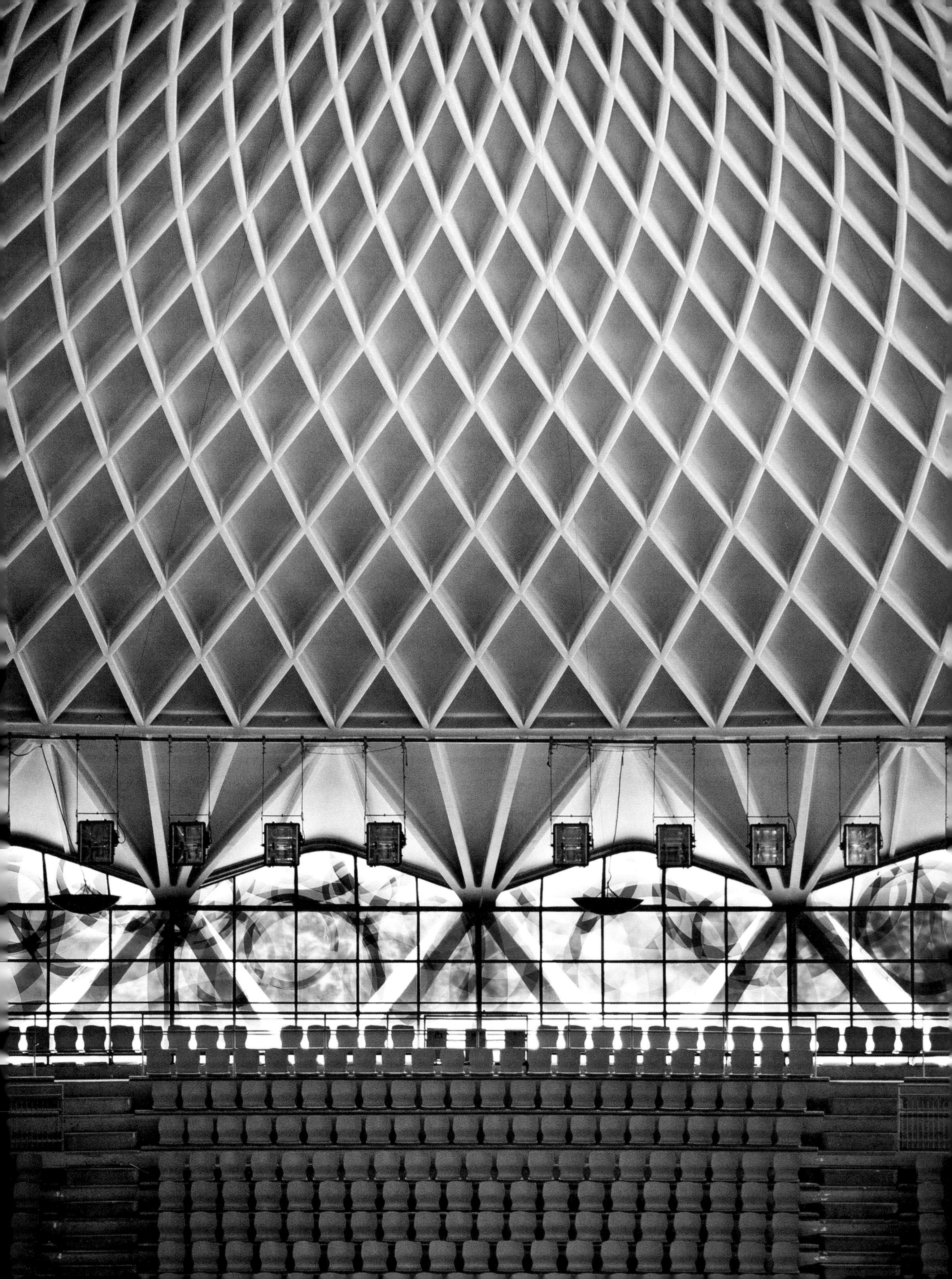

CHAPTER SEVENTEEN

Homogenization method

Distribution of material densities

Irmgard Lochner-Aldinger and Axel Schumacher

LEARNING OBJECTIVES

- Explain the interaction between the form generation of a shell and the support conditions.
- Discuss the homogenization method as an optimization technique for the generation of both surfaces and surface topologies.
- Apply the method to develop surface shapes for thin shells.
- Use this method to develop a structurally efficient topology on a predetermined surface.

Figure 17.1 Point-supported Palazzetto dello Sport, Rome, 1957–8, by Pier Luigi Nervi

Shells are form-passive structures; they exhibit a direct interaction between form, loading, support conditions and structural behaviour, and resist forces through their form. While they can be very efficient in their structural performance, at the same time they can react very sensitively to loadings or support conditions that do not match their shape. Loads and support conditions are determining factors in the process of shape generation of shells.

Support conditions (point or linear) have a clear impact of the optimal form for a shell in terms of their global geometry and grid layout. We explore the influence of support conditions on the generation of shell geometries, using the homogenization technique. Our design exploration has two phases:

1. Surface shape: for a given footprint and within a given design space (three dimensions) we carry out three-dimensional topology optimization for a series of support conditions to find efficient shell surface shapes.
2. Surface topology: within a given three-dimensional surface (established in the previous phase), we carry out topology optimization to establish an optimal layout.

The brief

A private client wants to build a small, iconic equestrian arena. The proposed multi-purpose arena has a floor

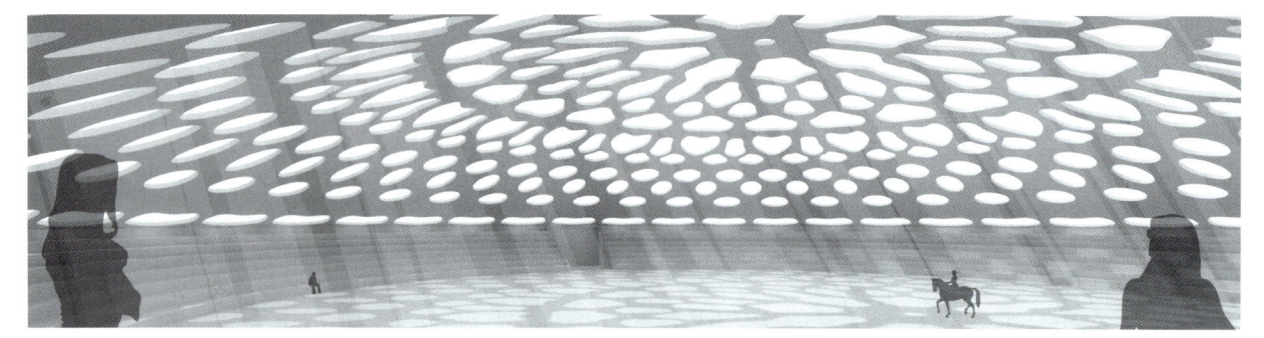

Figure 17.2 Interior view of a point-supported roof based on the homogenization method

plan of 900m². We start our design exploration with basic structural studies of shell structures. We develop the footprint in response to the structural geometry studies, so there is no a priori compulsory floor layout.

The final design resulting from the study presented here is shown in Figure 17.2. The support conditions and footprint are reminiscent of the Palazzetto dello Sport in Rome, by Pier Luigi Nervi (1891–1979), shown in Figure 17.1 and on page 210.

17.1 Objectives

In generating shell shapes, we aim to minimize mean compliance, or 'flexibility', which describes the energy of internal forces stored during the deformation of a structure. The studies carried out can be characterized as 'compliance design'. Our design study will show that when developing structural geometries using methods of structural optimization, this approach produces shell-like structures. The minimization of compliance produces, in case their shape is developed correctly, very efficient shells with high stiffness. The model set-up and optimization formulation are as follows:

- definition of the design space;
- definition of the optimization objective or more specifically minimal compliance;
- definition of the optimization constraints – the volumetric fraction of the design space allowed to be filled with material (since otherwise the optimization algorithm would simply fill up the whole design space).

The optimization method and geometry studies described in this chapter aim to:

- analyse and evaluate the interaction of shell shape and its performance for a series of different boundary constraints;
- quantitatively describe how shell geometries compare in terms of structural efficiency.

We carry out comparative studies by using the homogenization method as one method for topology optimization, while varying the support conditions. The shape development is carried out in two steps: the development of an overall surface shape of the structure within a given design space, and the development of a layout on that, or any other given surface.

17.2 Methodology

This section describes the background of the homogenization method for the structural optimization. We start from a well-defined optimization problem. A straightforward approach is the use of a parameterized geometry description in a CAD program that we use for shape and topology. Generally, one needs a large amount of computational time to solve real problems. As a result, fast pixel-based topology optimization methods, and more specifically, the homogenization method were developed to reduce this computational time.

17.2.1 Design space

The basic idea is the subdivision of the structure in many small domains (pixels). The optimization problem is to find an optimal structure, described by pixels, or voxels, each with, or without material (a discrete, binary, 0-1, problem). In standard approaches, one pixel is one finite element. After this optimization, we get a rough design proposal which we have to interpret for the generation of a real structure. The result gives information about the topology and about the shape. One of these pixel-based methods is the so-called homogenization method, which is a very common approach used in industry. The homogenization method (Bendsøe and Sigmund, 2003) reformulates the binary problem into a continuum problem that optimizes the density of a porous material for every pixel. After the optimization, there are pixels with the density of '0' (no material), pixels with the density of 1 (fully solid material) and pixels with the density between 0 and 1 (porous material). The pixels subject to optimization form the design space. The remainder of the problem space consists of either space with fully solid material, such as predetermined enclosures, or space free of material, resulting from design restrictions, such as openings.

In many cases, the reduction of the calculation problem by setting up symmetry conditions is helpful, in order to reduce computational time. This can also be done by coupling of the design variables. This is necessary, if the goal is to obtain a symmetrical shell.

17.2.2 Problem definition

In the large range of possible objectives and constraints, the group of functions, which can be described by integrals over the whole structural domain, are the easiest possibility for handling in automatic optimization loops. Therefore, we formulate the sensitivities directly based on the results in the individual finite elements (design pixel). The problem with using local functions (e.g. the maximum value of local stresses) instead of domain integral-based functions is the possible jumping of the position of the maximum value in the structure from iteration to iteration, which makes the use of the calculated sensitivity information more complicated or unreliable. It is thus a good idea to reformulate the objectives and constraint functions as an integral over the whole structural domain Ω (Dems, 1991), in the form of the functional

$$G = \left[\frac{1}{\Omega} \int \left(\frac{\sigma_v}{\sigma_0} \right)^n d\Omega \right]^{\frac{1}{n}}, \quad (17.1)$$

where σ_v is the local Von Mises stress value and σ_0 the allowed stress value. With $n>1$ the large local stress peaks get larger influences, so that satisfying allowable stress conditions can be ensured more easily.

The easiest formulation of objective and constraint functions is the use of the deformation energy; for example, using mean compliance. The mean compliance is described by the integral over the product of boundary stresses (respectively volume forces) and the corresponding displacements. We start from the principal of minimum total potential energy, defined as $\Pi = U - W$, where the internal energy

$$U = \int_\Omega \varepsilon^T \mathbf{C} \varepsilon d\Omega \quad (17.2)$$

and the external energy, in terms of the boundary stresses and displacements, for a structure with given boundary stresses $\mathbf{t}_{(\Gamma_N)}$ and boundary displacements $\mathbf{v}_{(\Gamma_D)}$,

$$W = \int_{\Gamma_N} \mathbf{v}^T \mathbf{t}_{(\Gamma_N)} d\Gamma_N \\ - \int_{\Gamma_D} \mathbf{v}^T_{(\Gamma_D)} [\sigma] \mathbf{n} d\Gamma_D + \int_\Omega \mathbf{v}^T \mathbf{f} d\Omega, \quad (17.3)$$

where $\mathbf{C}$ is the stiffness matrix, ε the strains, $\mathbf{v}$ the displacements, $[\sigma]$ the Cauchy stress tensor (normal and shear stresses), $\mathbf{n}$ the normals to the boundary and $\mathbf{f}$ the volume forces. These three terms are integrals over the restraint stress (Neumann) boundary Γ_N, the restraint deformation (Dirichlet) boundary Γ_D and the whole structural domain Ω. In the case of linear elasticity, the mean compliance is the same as the deformation energy U.

17.2.3 Homogenization scheme

The initial model for the shell structure is set up as a volume of small micro-cells that take into account the quantification of the stress–strain behaviour. The porous material of these micro-cells is described by parts with, and parts without material (Fig. 17.3).

The first step is the homogenization of the material for a faster computation. The task is to calculate the coefficients of the stiffness matrix for a porous material. The coefficients of the stiffness matrix depend on the hole size a_1, a_2 and the hole orientation angle θ. The stiffness matrix, in the case of our orthotropic material,

$$\mathbf{C} = \begin{bmatrix} C^{11} & C^{12} & C^{13} \\ C^{21} & C^{22} & C^{23} \\ C^{31} & C^{32} & C^{33} \end{bmatrix} = \begin{bmatrix} C^{11} & C^{12} & 0 \\ & C^{22} & 0 \\ sym & & C^{33} \end{bmatrix}, \quad (17.4)$$

which for a solid, isotropic material,

$$\mathbf{C}^{\text{solid}} = \frac{E}{1-v^2} \begin{bmatrix} 1 & v & 0 \\ v & 1 & 0 \\ 0 & 0 & \frac{1-v}{2} \end{bmatrix}. \quad (17.5)$$

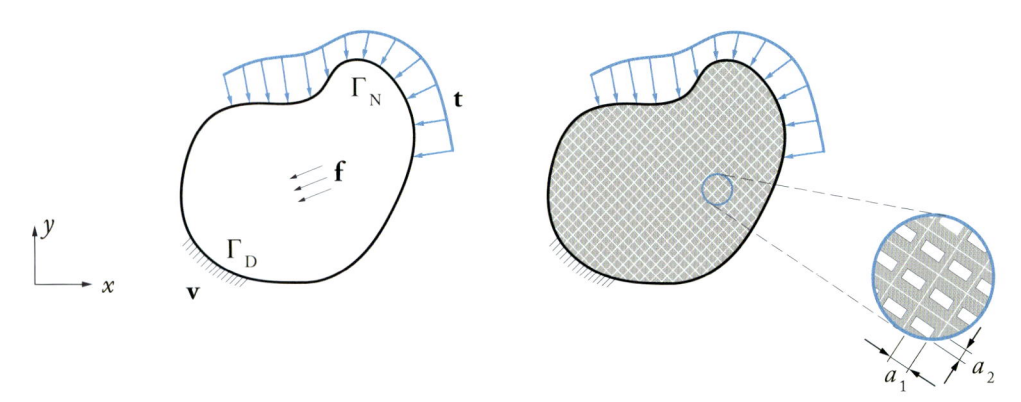

Figure 17.3 Defined loads and boundary conditions of the shell, and rectangular holes in a quadratic micro-cell

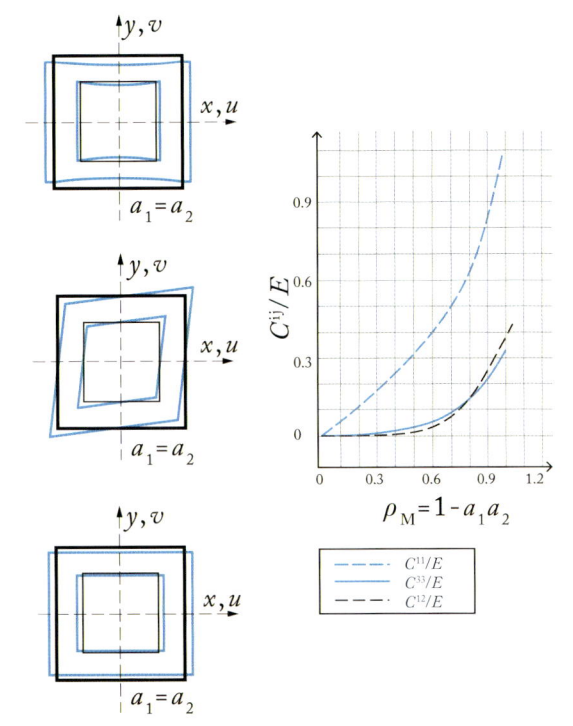

Figure 17.4 Stiffness calculation of a micro-cell depending on the porosity for homogenization of the porous material

The construction of this matrix can be done numerically by using a simple finite element model of a micro-cell (Fig. 17.4). In this study, the finite element model has a size of 10mm. The finite element side length is 0.2mm. The Poisson's ratio v of the solid part is 0.3. Figure 17.4 shows the mechanical models for the calculation of the coefficients of the stiffness matrix depending on the porosity $\rho_M = 1 - a_1 a_2$ (Fig. 17.3). The coefficients of the stiffness matrix C^{11}, C^{12} and C^{33} are calculated by several porosities (range of porosity from 0 to 1). This is the basic for fitting polynomials for a close analytical description of the porosity influence.

The mass is a linear function of the porosity $\rho_M = 1 - a_1 a_2$. The combination of the nonlinear stiffness functions and the linear mass function helps to get values 0 for no material and 1 for full material. In the optimization iteration step, design variables which are nearby 0 are motivated to run to 0, because the mass reduces with only a moderate reduction of stiffness. Design variables which are nearby 1 can increase the stiffness fundamentally with a moderate increase in the mass. Considering the porous material, the originally discrete optimization problem is then solvable using an optimization algorithm that operates in the continuum.

It is also possible to work with more simple approaches, where

$$\mathbf{C} = \mathbf{C}^{\text{solid}} \rho_M^3. \quad (17.6)$$

This third-order polynomial is similar to the polynomial generated by the homogenization method.

An advantage of the homogenization method is the possibility to physically interpret the nonlinear stiffness-density behaviour.

17.2.4 Sensitivity analysis

A fast calculation of the sensitivities for each element concerning the size (a_1, a_2) and the orientation Θ requires analytical equations. The classical approach is the minimization of the mean compliance (deformation energy U) considering a mass constraint. Therefore, the sensitivities concerning the hole sizes are given in analytical terms from the volume, or area, of the finite element Ω and the partial derivatives $\partial/\partial a_K$ of $\mathbf{C}$ with $k = 1,2$. For each finite element E, with area Ω^E,

$$\begin{aligned}\frac{\partial U}{\partial a_K} &= -\frac{1}{2}\Omega \varepsilon^T \frac{\partial \mathbf{C}}{\partial a_K}\varepsilon \\ &= -\frac{1}{2}\Omega\Big[\frac{\partial C^{11}}{\partial a_K}\varepsilon_{11}^2 + \frac{\partial C^{22}}{\partial a_K}\varepsilon_{22}^2 \\ &\quad + \frac{\partial C^{33}}{\partial a_K}\gamma_{12}^2 + 2\frac{\partial C^{12}}{\partial a_K}\varepsilon_{11}\varepsilon_{22}\Big]. \end{aligned} \quad (17.7)$$

The sensitivities concerning the orientation of the rectangular pores in one finite element can be determined as follows (Pedersen, 1989):

$$\begin{aligned}\frac{\partial U}{\partial \Theta} = -\frac{1}{2}\Omega\Big[&(\varepsilon_I - \varepsilon_{II})\sin 2\phi((\varepsilon_I + \varepsilon_{II}) \\ &(C^{11} - C^{22}) + (\varepsilon_I - \varepsilon_{II}) \\ &\cos 2\phi(C^{11} + C^{22} - 2C^{12} - 4C^{33}))\Big],\end{aligned} \quad (17.8)$$

with the principal strains ε_I and ε_{II}. The angle ϕ denotes the difference between the rectangular orientation $^E\Theta$ and the orientation of the first principal strain. The shown scheme for the analytical sensitivities is valid for the membrane behaviour of three-dimensional shell structures. If analytical sensitivities of the bending behaviour are important, we can use equations (17.7) and (17.8) layer-wise. Alternatively, the calculation of the coefficients of the stiffness matrix depend on the porosity of a micro-cell such as in Figure 17.4.

17.2.5 Practical scheme

The overall procedure of the homogenization method is shown in Figure 17.5.

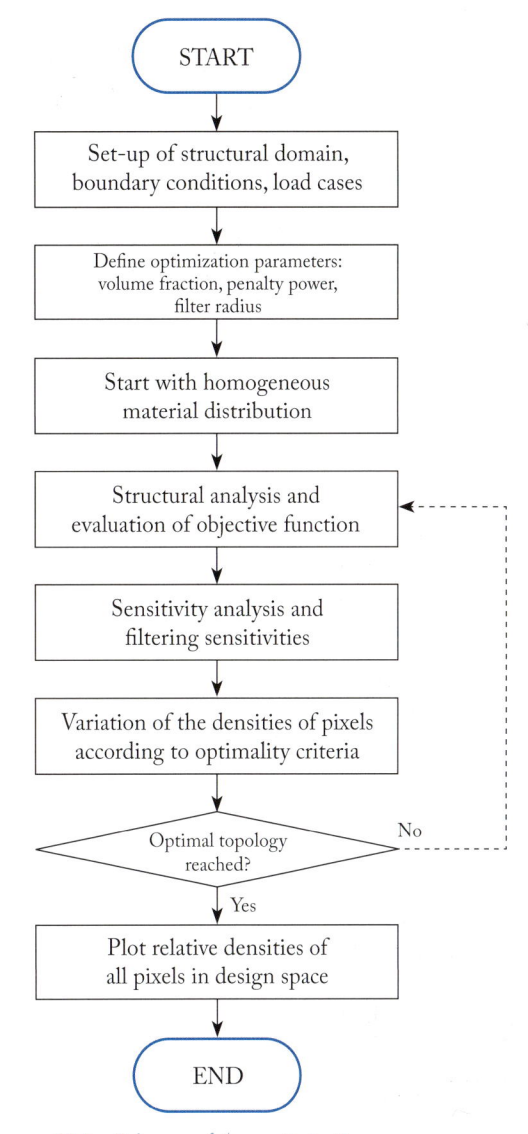

Figure 17.5 Scheme of the optimization process

After identifying the available design space, and setting up a finite element model, the design variables are defined. Typically, for each finite element, each pixel, a property that describes the local density, or porosity, of the material is the design variable. The variation of the density of all pixels is done by some optimization or optimality criteria algorithm.

The optimization is influenced by three process parameters: the volume fraction, the penalty power and a filter radius. The volume fraction defines the amount of structural volume permitted to be implemented into the design space as a fractional value. The value must be between 1 (full material) and 0 (no material). Values between 0.3 and 0.6 generally lead to feasible optimization results. If the value is too high, the result is very solid; if the value is too low, the optimization does not produce continuous structures. The penalty power defines the penalty value for the convergence of relative density. The value has to be larger than 1 in equation (17.6).

The filter radius defines a mesh-independent filter which prevents checquerboard, that is, disconnected, patterns (0 → no filtering, ≥ 0 → declaration in 'number of elements') (see Section 17.2).

The structure resulting from the optimization algorithm is a design proposal, which needs to be interpreted to generate a real structure. For example, we have to take into account that the homogenization method cannot consider stability problems such as buckling. The model is also heavily discretized, and needs to be turned into a real structure in a CAD system.

One could follow the development of the design proposal in each iteration, while at the same time following the value of the mean compliance of the actual design. This value decreases as the optimization proceeds, which indicates that the optimization algorithm redistributes the material within the design space in a more efficient way in each step, therefore reducing the compliance of the structure with the amount of material remaining the same (within the predefined constraints).

17.3 Two-dimensional example

As an example, a simple structural model was generated for a comparative optimization run. The objective of the optimization algorithm is the minimization of the mean compliance. The square design space is subject to a uniformly distributed gravity load. In both cases, the constraint is a volume fraction of 0.4 (i.e. 40% of the design space allowed to be filled with material). The structure is supported by one fixed and one sliding support (Fig. 17.6a) or two fixed supports (Fig. 17.6b). In this simple example, the definition of the material properties, the Young's modulus E, the Poisson's ratio v and the element stiffness matrix are fixed. In this case, once the decrement between two compliance values is lower than 0.01, the algorithm stops.

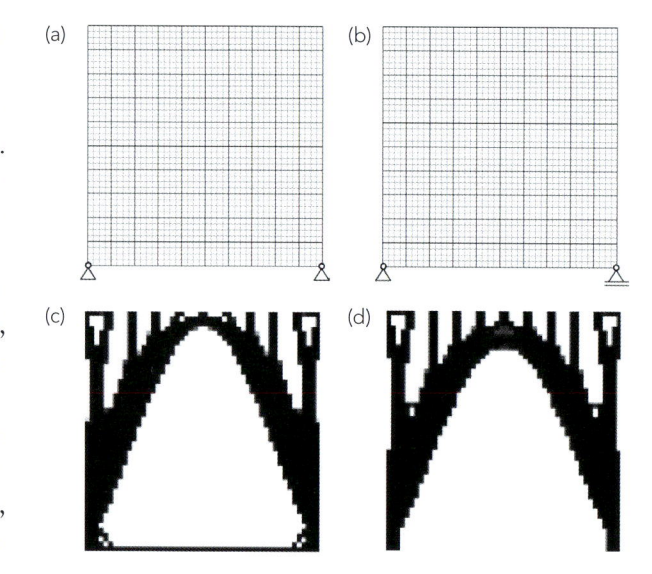

Figure 17.6 Square design domain with distributed loading at the top for (a) fixed supports, (b) simply supported, and optimized result for an arch bridge (c) with and (d) without tie member

Figure 17.6 shows the result of the optimization process, which converges after seventeen iterations reaching a compliance value of 67kNm in case (a), and after nineteen iterations reaches a compliance value of 82kNm in case (b).

The design proposals are very similar, except for a tie member being established in case (b), which is obviously necessary due to the sliding support and the horizontal force component of the arch. The compliance values are directly related to this: the structure with a sliding support will undergo higher deformations, with a larger force ratio remaining inside the structure (instead of being transferred to the supports), resulting in a higher compliance value.

The next two sections describe the homogenization method applied to three-dimensional volumes and curved two-dimensional surfaces as the design space. These can be regarded as simple extensions of the two-dimensional example given here: either the (x, y) grid is extruded to the third dimension z, or it is mapped to the surface (u,v)-coordinates (Fig. 17.7).

17.4 Generation of continuous shells

The design development for the brief now leads us to a three-dimensional challenge: for a given footprint and

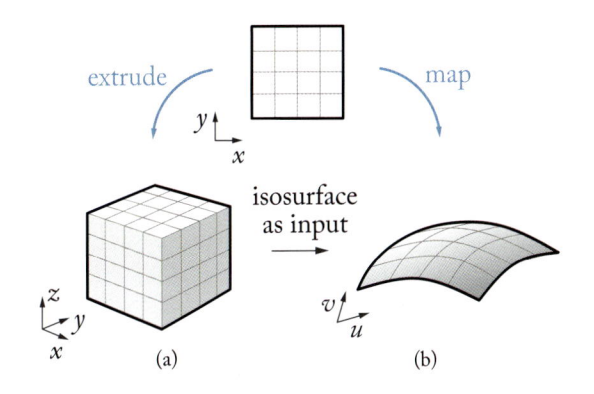

Figure 17.7 The two-dimensional (x,y) grid (a) extruded to a three-dimensional domain (x,y,z), or (b) mapped to a surface (u,v)

a height restriction, what are the boundary conditions, shape and topology for a curved surface structure enclosing our equestrian arena?

The material distribution within a given volume is our first design step. First, we focus our attention on a square footprint, typical of equestrian arenas. Within a given cuboid, with uniformly distributed gravity loading on the top face, we want to generate a structural shape with low compliance (or high rigidity) with supports along the perimeter of the system. In other words, the optimization objective is the minimization of compliance. We define the design constraint, the volume fraction, as 40%. In other words, 40% of the volume is allowed to be filled with material. The choice of the volume fraction value is up to the designer and should be subject to precedent comparative studies. The resulting material distribution is then used to extract a surface structure for further design development. This is done by selecting an 'isosurface plot', a surface of equal material density, from the material distribution of the volume. Before carrying out a series of studies where we change the location of the supports, we first decide on the density value for the extracted isosurfaces. This is done on the basis of a single result for given boundary conditions.

17.4.1 Extracting surfaces

Figure 17.8 shows the structural model for the optimization and the design proposal produced by

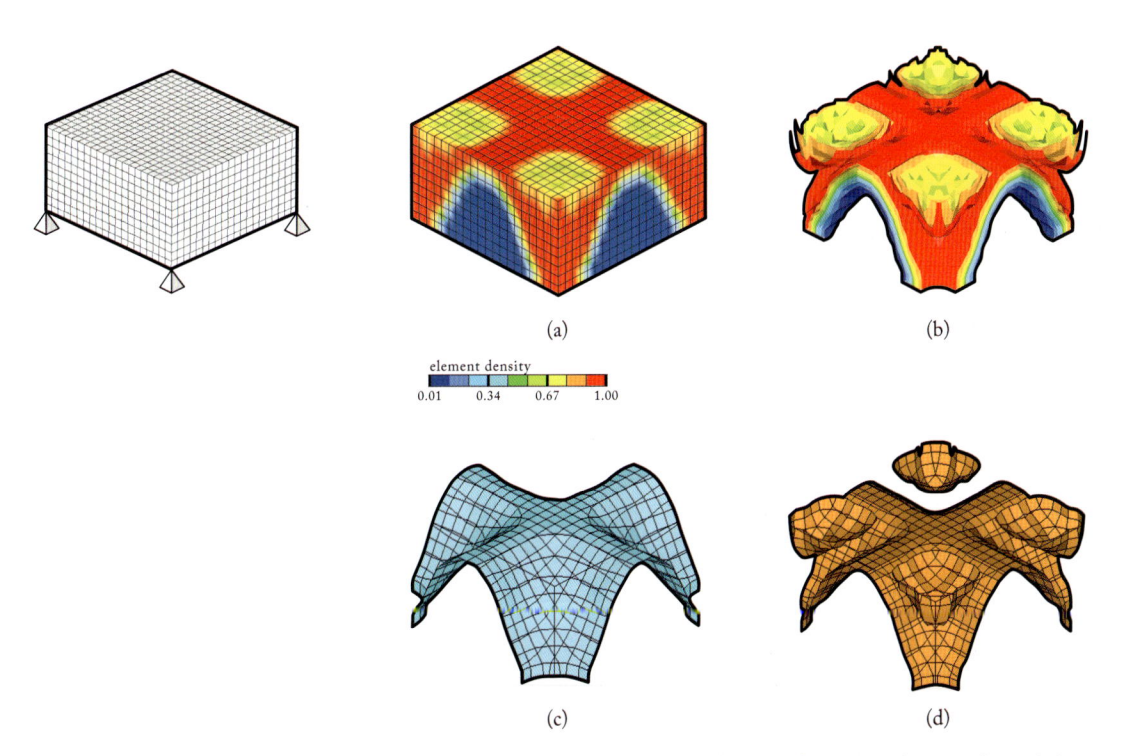

Figure 17.8 (a) Distribution of density values between 0 and 1 as contour plot; (b) isosurface plot of areas of equal density; (c,d) plot of one chosen density value of 0.4 and 0.7 respectively

the optimization algorithm. The optimization result ('design proposal') shows the distribution of material (40% of the design space) within the cuboid, with areas of low density (0, dark blue) to high density (1, red). The areas with high density are allocated at the top surface of the volume (where the loading is applied) and the four bottom corners (where the supports are).

At this stage, we evaluate the material distribution for different density values, by plotting isosurfaces, and choose one value for further design development, which is a decision up to the designer and needs to be carefully considered: areas with densities around 0.8 are the ones where material is most efficiently used, however, this may not lead to a continuous structure; areas with low densities of around 0.2 may lead to very solid structures. Figure 17.8c shows an isosurface plot for an isolated value of approximately 0.45, and Figure 17.8d for 0.85. Both plots show the distinctive feature of a cross-vault-like, doubly curved structure.

The choice of a density value for the further development of the design proposal is an essential step after performing the optimization algorithm. In our case, it is feasible to eliminate intermediate values that are necessary to transmit the loading from the top surface of the cuboid onto the optimized structure. For the design study with varying support conditions, the density value of 0.45 is chosen since it delivers a descriptive image of the optimization result for the following comparative study.

The optimization algorithm converges after fourteen iterations until the stopping criterion (the decrement being lower than 0.01) is fulfilled, with the graph of the objective function shown in Figure 17.9. The optimum is reached quickly after four iterations, while the following iterations hardly feature any changes in the compliance value. This phenomenon occurs as the algorithm redistributes the intermediate values (the 'porous' material) within the structure, once it has approached its optimum topology. This redistribution does not influence the compliance value but results in a clearer readability of the design proposal. When comparing the shape of the design proposal after four iterations to the final result, they are very similar with small deviations of the areas with density close to 1.

17.4.2 Comparing support conditions

The following study uses the design domain and loading as shown in Figure 17.9. We use the same optimization objective function (minimum compliance) and design constraint (volume fraction of 40% of the design space).

The support conditions of the structural model is systematically varied. Through this parametric study we would like to understand the interaction between the support conditions and the structural shape produced by the optimization algorithm and its associated compliance value.

Replacing the point supports in the corners (model 1) by line supports along two of the edges (model 2), the optimization algorithm produces the single curved vault shown in Figure 17.10. This shape with symmetrical line supports produces a symmetrical design proposal; except for slightly doubly curved areas near the free edges.

Comparing the graphs setting out compliance (objective function) versus number of iterations, we can conclude that:

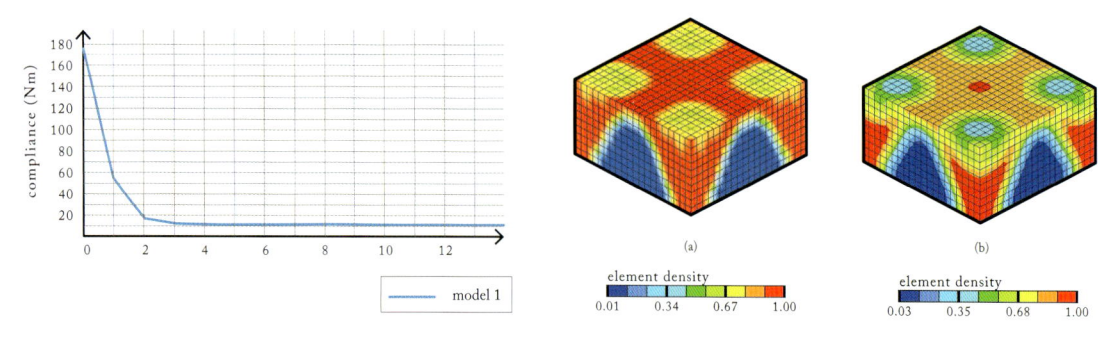

Figure 17.9 Compliance during the iterations and design proposals produced by iteration 4 and iteration 14

CHAPTER SEVENTEEN: HOMOGENIZATION METHOD   219

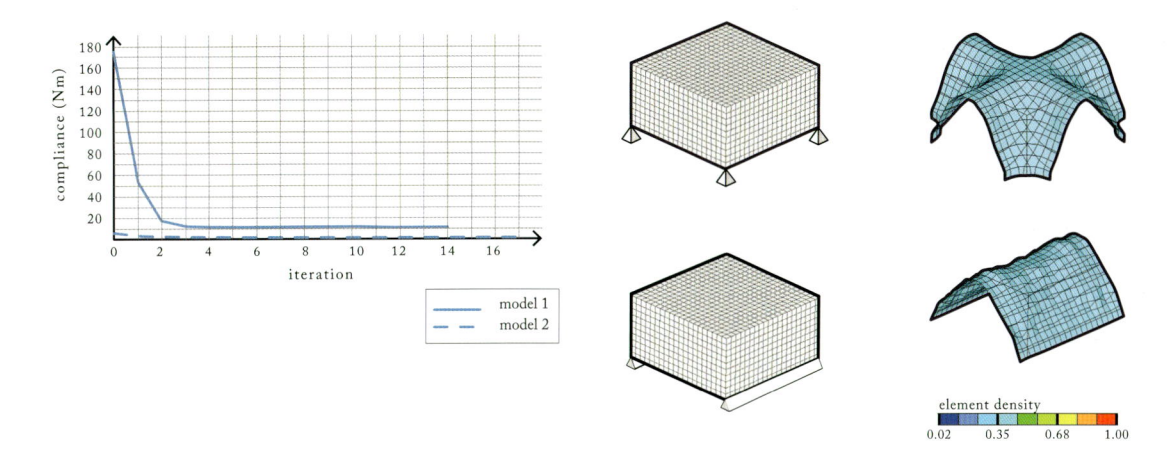

Figure 17.10 Compliance during the iterations and resulting design proposals for point supports in the corners (model 1) and with linear supports (model 2)

- As the algorithm starts, the structure with line supports along two sides of the volume (model 2) has a very low compliance value compared to the structure with point supports (model 1). This observation makes sense: in a continuously supported structure, the loads are transferred directly to the supports with low compliance, low deformations, high rigidity.
- The compliance values of both structural models converge quickly to their optimum, followed by many iterations with no significant change of compliance value resulting from the redistribution of material within the previously found optimum, which is an effect described in Section 17.4.
- The final compliance values of the two structural models, once the optimization has converged, still show a difference in value. However, the ratio between the two values has been reduced to about 1/20 compared to the difference in the beginning of the optimization. The structural model with point supports in the edges 'catches up' in compliance by inducing increased double curvature and thus higher geometric stiffness. Nevertheless, its compliance is higher, that is, its rigidity is lower compared to the structure with continuous supports. Their structural performance can be directly compared through their objective graphs.

Continuing the scheme of parametric structural models with constant optimization objectives, we establish a systematic variation of the support conditions in order to compare the resulting design proposals and their associated compliance values. Figure 17.11 shows the diagrams for the variation of the support conditions, ranging from continuous line supports along all bottom edges of the design space to local line supports at two sides. We exclude single point supported structures from this study since the point supports induced singularities in the FE calculation and produced results difficult to compare. The support conditions and the optimization result, including both the geometry of the design proposal and their associated final compliance value, are shown in Figure 17.11.

Our study shows how, using the homogenization method, the support conditions of a structure influences the generated shape and the structural performance of a structure in general. With more supports, the loads are transferred more directly, related to lower compliance, higher rigidity, or higher structural performance. Structures with a larger number of supports result in lower compliance values reached, shown most clearly in the comparison of model 3 (circumferential supports) and model 6 (small local supports) in Figure 17.11. Furthermore, single or double symmetry of the support conditions leads to single or double symmetry in the geometry of

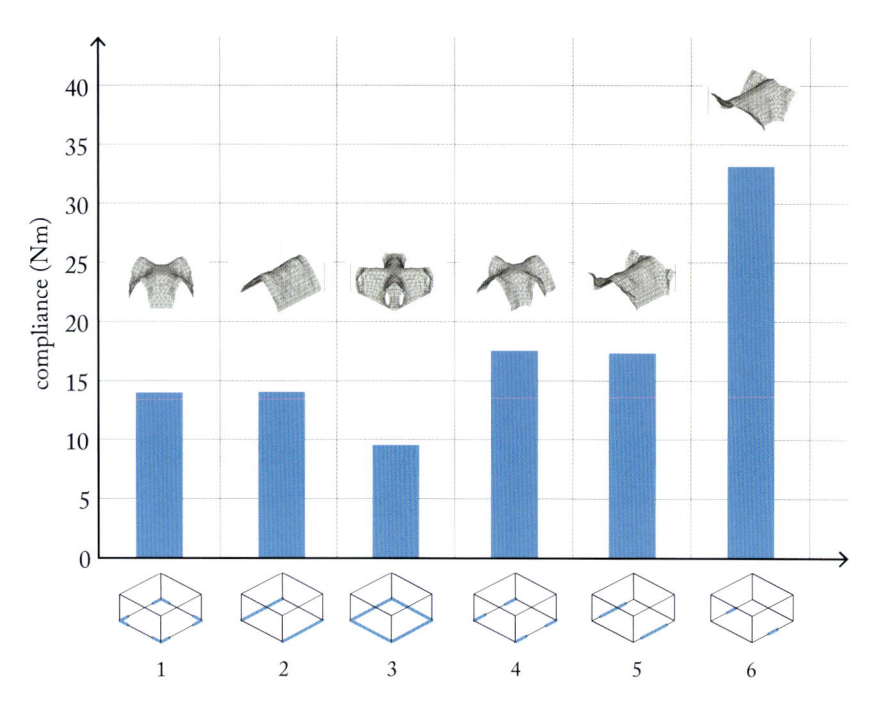

Figure 17.11 Diagrams illustrating the cuboidal design space and structural systems of models 1–6: bold lines indicate the supports of the structure

the design proposal. The benefits of optimization studies lie in the numerical and visual comparability of geometries, evaluating both the structural performance and aspects of shape and aesthetical qualities.

17.5 Generation of surface topologies

As a second design step, we carry out studies on the material distribution within a given curved surface. The resulting forms are non-continuous, gridshell-like surface structures. The previous optimization study produced a surface within a given volume, which can serve as input for this second step. The topology optimization can also be carried out for a given (curved) surface. This tool leads us to study for geometries of gridshells. In this study, the influence of the overall shape of the given design space (e.g. single or double curvature) and of the support conditions are of interest.

17.5.1 Conical vault

We keep the same problem formulation: minimization of compliance, with a given volume fraction (0.4) of the design space. We start with a curved design space on trapezoidal plan, a single curved vault, mathematically described as a cone section.

Linear supports along the edges produce the design proposal in Figure 17.12a, which shows the flow of forces to its base. Replacing the linear supports by point supports at each edge produces an arch-like structure which transfers its loads to the corners (Fig. 17.12b).

The graphs of the objective function show, similarly to the studies carried out with the volume model, that the model with point supports undergoes higher deformations. The optimization takes off with a higher value, and the value reached after convergence is also higher than the compliance value of the model with linear supports (Fig. 17.12).

The optimization graphs show the so-called 'minimum member size control'. Since the optimization algorithm needs intermediate density values (between 0 and 1) to ensure convergence, it happens that the 'design proposal' produced is not 'discrete'. This reveals the existence of many intermediate values of densities between 0.3 and 0.7, making it difficult to interpret the design proposal. The 'minimum member size control' induces a 'penalty function', redistributing

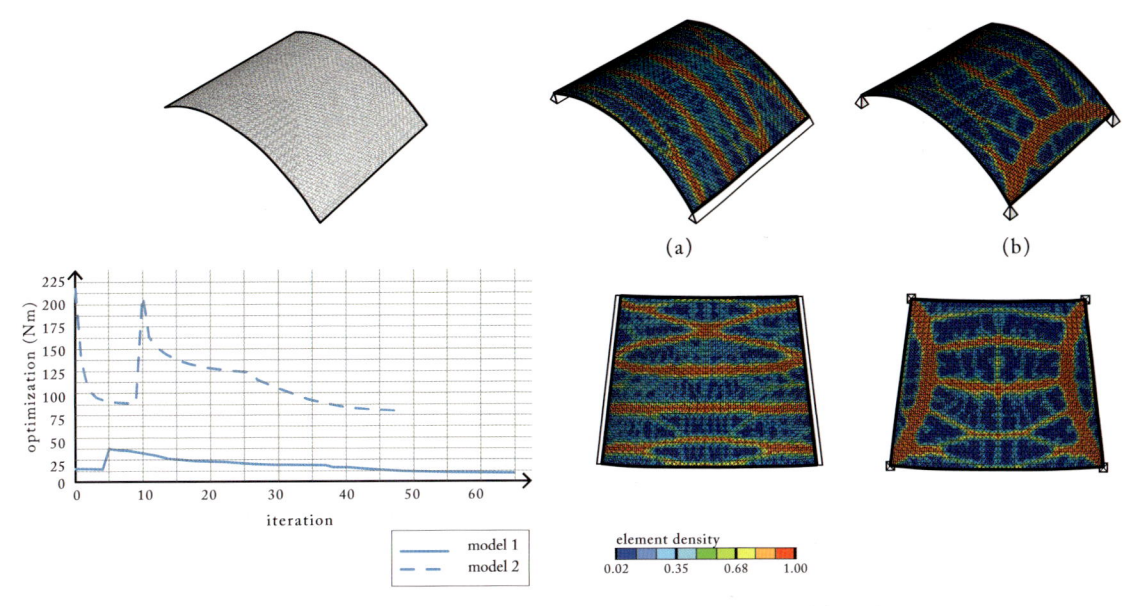

Figure 17.12 The conical design space and optimization result (a) with line supports and (b) with point supports

intermediate densities towards lower and higher values respectively. This step results in a jump in compliance, since the concentration of material on the one hand and the creation of voids on the other hand produces a structure with less rigidity. However, after the jump in compliance value, the optimization algorithm then approaches a minimum below the value reached before.

17.5.2 Spherical dome

We now expand the complexity of the initial design space surface to a doubly curved surface. One-quarter of a partial sphere is modelled using symmetry boundary conditions along the inversion lines. The support conditions are varied as before, modelling numerous point supports along the perimeter or single point supports respectively, as shown in Figure 17.13.

These topology optimization studies show how the layout is adjusted to the shell support conditions, related to the example shown in Figure 17.1.

17.5.3 Practical development of the design

The design proposal generated by the optimization algorithm does not yet produce a structure, but leaves us with a geometry that needs to be interpreted and translated into a constructable shell. This interpretation can take construction constraints into account, such as repetitive geometries, simplification of curvatures and others. The design proposal suggests where the material is most efficiently used within a structure. Using the example of a doubly curved gridshell with frequent supports, we develop the geometry of a topologically optimized gridshell in the following steps.

In the first step, we choose a density value of the design proposal, between 0 and 1. This choice is up to the designer and is usually determined by the need for a feasible continuous structure (Fig. 17.14).

A useful procedure is to export this geometry and to draught the final layout of the structure, while the design proposal serves as a reference point for the three-dimensional geometry. In this case, we consider repetitive patterns based on radial symmetry.

Our final form appears very ornamental, in some aspects organic (Fig. 17.15). Some aspects of the shape, such as void outlines with bends around mid-height or sharp kinks near the top of the shell, may lead to discussions and many details would certainly have to undergo further steps in optimization. However, as an idea of an optimized structure, it does show a certain logic, and above

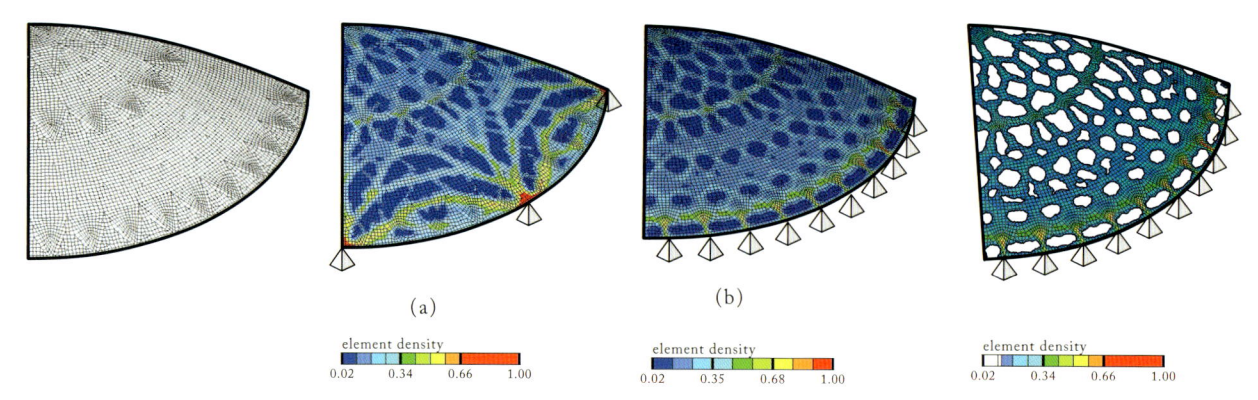

Figure 17.13 Topology optimization of a circular shell structure with (a) highly discrete to (b) very dense point supports

Figure 17.14 Choosing a density value for the further development of the design

all, optimization studies like this support structural understanding.

The geometry of the gridshell resembles the shape of diatoms (see Fig. 17.16), a major group of microscopic algae, documented by Ernst Haeckel (1834–1919) in the late nineteenth century (Haeckel, 1904). Nature shows us optimized geometries, determined by the necessity to develop lightweight structures. The diatom skeleton is the result of an optimization procedure running over millions of years, and is still being optimized due to the continuing struggle between crabs and diatoms, with the crabs improving their pincers and the diatoms reinforcing their skeleton subsequently. Both of them require principles of lightweight structures since they need to carry their pincers and protective shield respectively. Natural structures are therefore good 'role models' in the design of efficient structures.

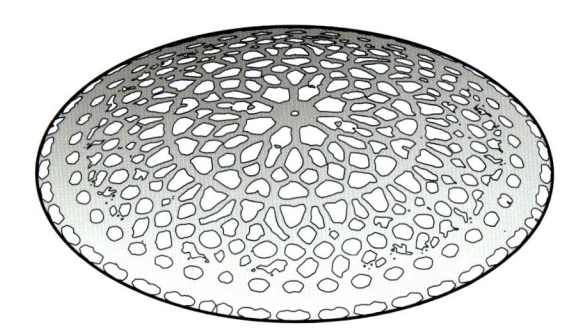

Figure 17.15 Geometry designed by interpretation of the design proposal

17.6 Conclusion

Using mathematical tools for structural topology and shape optimization can be a useful tool for finding first design concepts. But we need to be careful: a

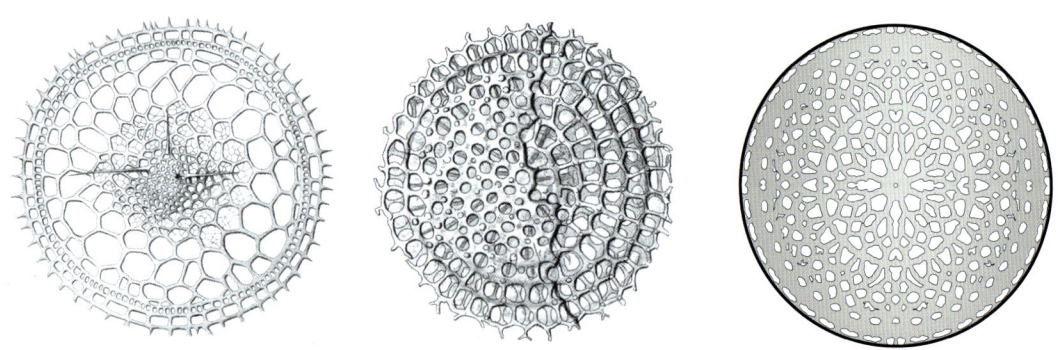

Figure 17.16 A resemblance between diatoms (Haeckel, 1904) and our design proposal

wrong definition of the optimization problem leads to wrong optimization results. It is not in every case so obvious as it is in the examples shown here. It is a necessity to discuss and interpret the results to find good ideas for their technical realization.

Structural optimization tools can be a useful support in the design of structural geometries. It does not, however, replace structural understanding. It is the responsibility of the designer to set up the optimization properly and to interpret results carefully.

Key concepts and terms

The **homogenization method** is a form of topology optimization. This approach distributes material within a given design space by defining a porous material and optimizing the density of that material within this space.

A **pixel** or **voxel** is a unit of subdivision of the two- or three-dimensional design space respectively, which is correlated to the material distribution during the optimization process. After the optimization, the design proposal produces pixels without material (density 0), full with material (density 1) and intermediate values between 0 and 1 ('porous material' needed by the optimization algorithm.

The **volume fraction** is the ratio of the design space allowed to be filled with material by the optimization algorithm, ranging between 0 and 1 (corresponding 0% and 100%).

Exercises

- The Palazzetto dello Sport (1958), designed and built by Pier Luigi Nervi for the 1960s Rome Olympic Games, has been lauded for its intricate ribbing pattern. Can you relate this pattern to the supports of the shell?
- Compare the pattern of the Palazzetto dello Sport with the one found in this chapter for similar support conditions. What do you observe? In what ways is it optimized? What are the benefits of Nervi's chosen pattern? How could the optimized pattern have informed Nervi's design?
- Construct a design space of 10m × 10m × 10m. Use the homogenization method to develop the shape for a shell with four corner supports and one central support.
- Design a NURBS surface with a 10m × 10m footprint. Apply the homogenization method to develop an efficient gridshell topology on this surface. Solving the problem is similar to the simpler two-dimensional problem, considering that the surface's (u,v) coordinates can be mapped to a plane (x, y) grid.

Further reading

- *Topology Optimization: Theory, Methods and Applications*, Bendsøe and Sigmund (2003). This book is an overview of several methods of topology optimization. Some of the explained methods are still in use today and are extended in interesting ways.
- *Optimierung mechanischer Strukturen*, Schumacher (2013). This publication gives an introduction in algorithms and strategies for solving structural optimization problems.
- 'First- and second-order shape sensitivity analysis of structures', Dems (1991). This publication describes the possibilities for simplifying optimization functions in order to find analytical sensitivities and robust optimization loops.
- *Kunstformen der Natur – Kunstformen aus dem Meer*, Haeckel (1904). This book is a documentation of Haeckel's systematic catalogues of shapes in nature, with a differentiation between natural shapes in general, and the specification on creatures in the sea. Haeckel's explicit illustrations serve as a continued inspiration for different professions. The 2012 edition combines two of Haeckel's major works.
- 'On optimal orientation of orthotropic materials', Pederson (1989). This journal paper deals with the efficient orientation of anisotropic material with respect to the structural strain and describes a method depending on one non-dimensional material parameter and the two principal strains.

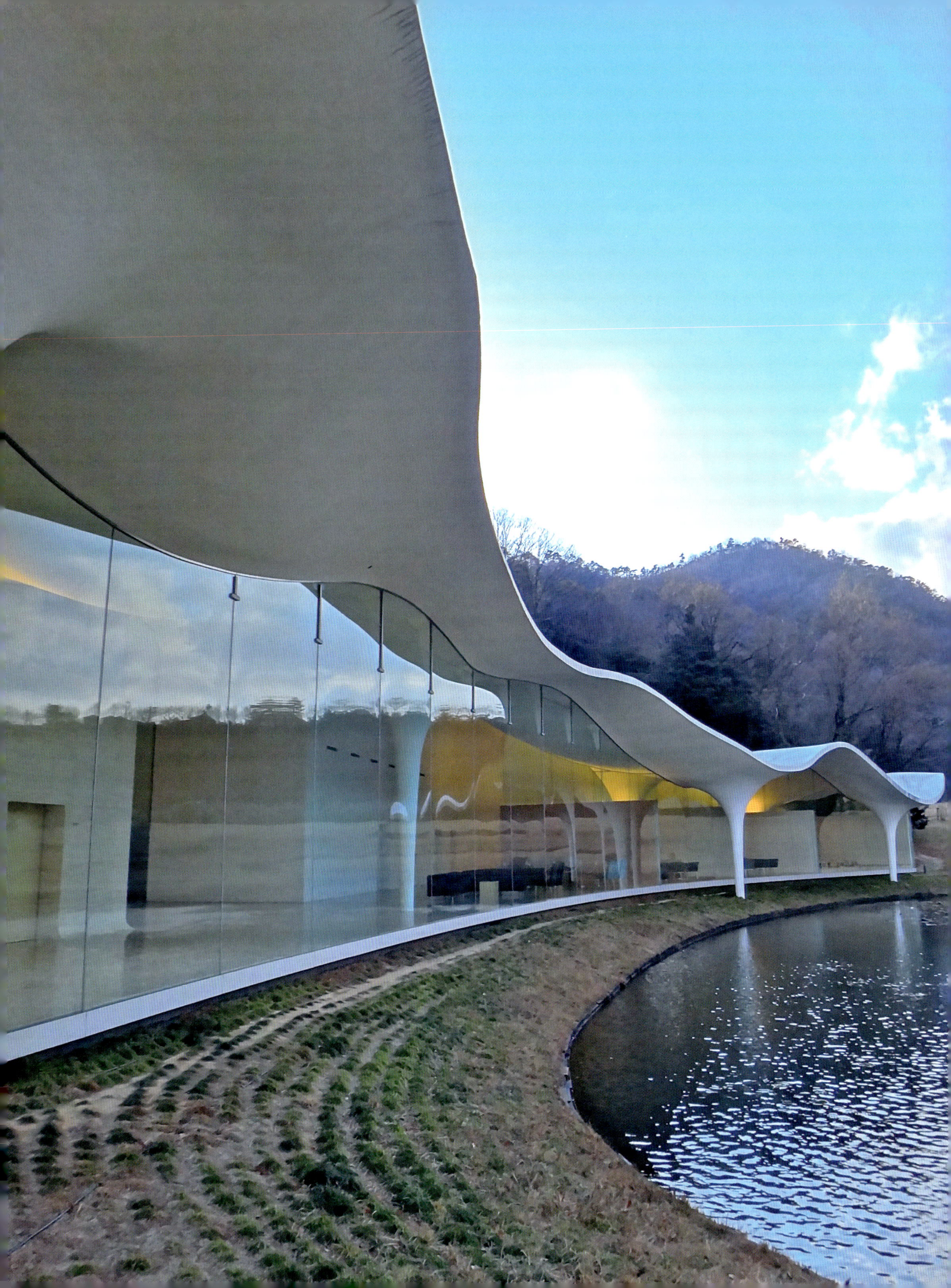

CHAPTER EIGHTEEN

Computational morphogenesis

Design of freeform surfaces

Alberto Pugnale, Tomás Méndez Echenagucia and Mario Sassone

LEARNING OBJECTIVES

- Discuss the conditions in which structural optimization can become a method of form exploration.
- Construct a parametric model, search space and structural fitness function for shape optimization.
- Generate shell shapes within a search space by means of a genetic algorithm.

Since the introduction of digital technologies, computational design and optimization has been gradually replacing practice based on physical models. Even if born as a set of mere solution strategies, architectural and structural design optimization seems also to be effective in supporting conceptual design. This aspect is highlighted through two design exercises of computational morphogenesis using Genetic Algorithm (GA). Two key aspects emerge from such processes: the possibility of freely formulating objective functions apart from structural performances, and the opportunity for designers to actively interact throughout the computational process.

Generalizing the concept of optimization, it is not just a matter of finding structural geometry. It is suitable to deal with every architectural problem in which a specific performance can be formulated numerically, that is, with an objective or fitness function to minimize. Technical issues such as light shading, acoustics of concert halls and construction problems of gridshells are just a few of the several performance criteria to which optimization can be applied. Optimization focuses on iterative improvements of candidate forms. In this chapter we show how the designer can follow and interact throughout the optimization process. In architectural design, such a characteristic can be more important than reaching an actual optimum. The issues involved are not just structural in nature, but in fact are more complex and require an architect's experience and creative intent. Optimization algorithms can be used in a more exploratory manner whether an element of interaction with designers is introduced or not. The computational power and speed of the process are used to inform and interact with them. From being simple solution tools, optimization techniques become effective support for conceptual design. The process of form improvement turns into form exploration. Taking advantage of optimization as a design instrument of form exploration is called computational morphogenesis.

The brief

A local park requires a new visitor's centre, to be built next to a small river. It requires a small footbridge to provide access from the other side. We design both structures as continuous shells, with a doubly curved surface. The bridge has a span of 24m and a width of 4m. It is simply supported at the short edges and is free along the long edges (Fig. 18.1). The visitor's centre is heavily inspired by, and a freely redesigned homage to, the crematorium of Kakamigahara (Fig. 21.5 and page 224). This building, designed in 2004–2005 by Toyo Ito and Mutsuro Sasaki, has a concrete freeform roof structure, which has been optimized by means of Sensitivity Analysis (SA), an iterative gradient-based optimization technique discussed in Section 21.2.

18.1 Basic implementation

The computational morphogenesis of a shell bridge and the Kakamigahara crematorium, discussed in the brief, is set up and described in a step-by-step procedure involving the following:

1. The formulation of the design problem in parametric terms with boundary conditions, other constraints, design variables, and search space.
2. The description of the structural performance with a fitness (objective) function.
3. The study of the problem by mapping the search space into the objective space as a 'fitness landscape'.
4. The use of a genetic algorithm to explore the solution space and identify optimal forms.
5. The interpretation of results of the computational process.

In order to perform a basic optimization process, we need the integration of three different digital tools: a geometry modeller (usually a CAD application), which can provide parametric control on shapes; an FE (finite element) solver; and an optimization algorithm.

18.1.1 Parametric model and search space

The shell bridge, 24m × 4m in plan, is constrained along the short sides of the shell and is free along its

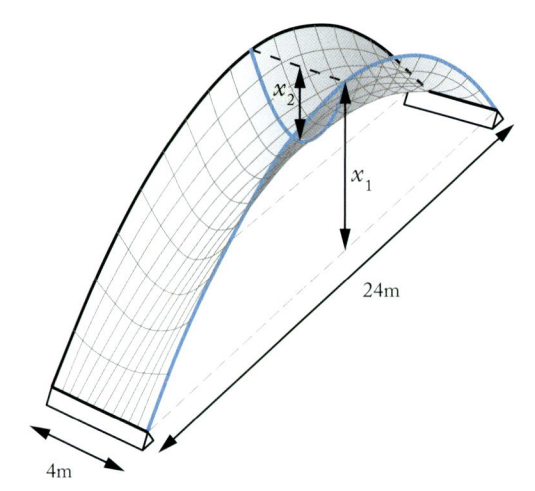

Figure 18.1 Parametric definition of the design variables for the shell bridge

long edges. The surface is generated by two orthogonal section curves. They are parabolas defined by one parameter each: the heights x_1 and x_2 (Fig. 18.1).

The side vertex of the transverse curve lies on the longitudinal curve's middle point and the constrained edges of the surface are always straight segments. The other transverse sections are parabolas with decreasing height, from the vertex to constrained sides.

This parametric definition of the surface guarantees that, by varying the values of x_1 and x_2, four configurations can be obtained: a completely flat surface (Fig. 18.2a); positively, doubly curved surfaces (Fig. 18.2b); negatively, doubly curved surfaces (Fig. 18.2c); and cylindrical surfaces (Fig. 18.2d).

Since these two parameters effectively control the overall shape of the bridge, x_1 and x_2 are chosen as the parameters. For both we have established a domain spanning from -40m to +40m. This means that the search space of this problem is two-dimensional and can be represented as a grid of values from -40 to 40 in its x and y axis.

The parametric definition proposed above is then implemented in the geometry modeller. When the GA calls for a shape in terms of a set of x_1 and x_2, the CAD modeller generates the corresponding NURBS (non-uniform rational basis splines) surface, providing the object to be evaluated. For the FE analysis, a discrete model has to be generated. The geometry is discretized into a mesh, composed of shell or beam elements, depending on the type of structure. For this

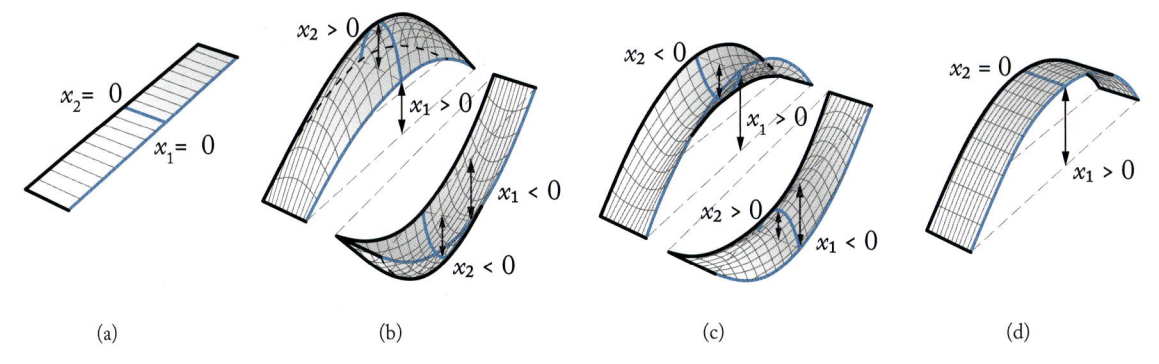

Figure 18.2 The four possible shape configurations with parameters values

exercise, the shell is simplified as a gridshell of comparable mechanical properties.

The construction of an FE model for structural optimization presents some differences with the ones used in normal analyses. Complex models require time-consuming calculations, which represent bottlenecks in the flow of operations. The first requirement of FE models for optimization is to be simple, with a number of elements strictly necessary and with a mesh correctly defined to evaluate the pertinent aspect of structural behaviour. Even with a powerful hardware set-up, the repetition of hundreds or even thousands of analyses might transform the optimization process into an extremely long task.

18.1.2 Structural fitness (objective) function

Displacements, strains, stresses and strain energy are basically the effect of a load condition on an elastic structure. A stiff structure will show small displacements and strains, while a strong structure will result in relatively small stresses, and both will have small strain energy. Displacements are a vector field, stresses are a tensor field, locally defined, and the strain energy is a scalar value, computed as an integral over the whole structure. Such quantities can be adopted as a measure of the structural performance. However, their differences will drive the optimization process to search for different optimal solutions.

For this exercise, the maximum displacement of the whole structure is chosen as the fitness function to be minimized by the GA. As opposed to the strain energy, nodal displacements can reveal local, as well as global, weaknesses.

18.1.3 Fitness landscape

Considering that the optimization problem proposed is defined by only two parameters (design variables), we can further explain the work done by the GA using a graphical representation. We first map out the solution domain of x_1 and x_2 by taking a two-dimensional parameter grid with grid points $P(x_1,x_2)$. By assigning a z value to each point of the grid, we convert it into a three-dimensional landscape in which z represents the fitness calculated for the shape (individual) corresponding to that grid point P. For example:

- for $P(0,0)$ the maximum displacement δ_z is 285mm;
- for $P(-26,1)$ the maximum displacement δ_z is 21mm;
- for $P(40,40)$ the maximum displacement δ_z is 43mm.

By repeating this operation for the entire solution domain, we end up with a fitness landscape. Figure 18.3 shows the fitness landscape, which has many local minima, two global minima and an area of global maxima.

18.1.4 Genetic algorithm parameters

Appendix C describes the importance of a correct formulation of parameters and operators of GAs, in order to properly perform in any given problem. For this exercise, the GA uses a binary coding of design parameters with mutation and elitism operators. It terminates after twenty generations with a population size of 100 individuals.

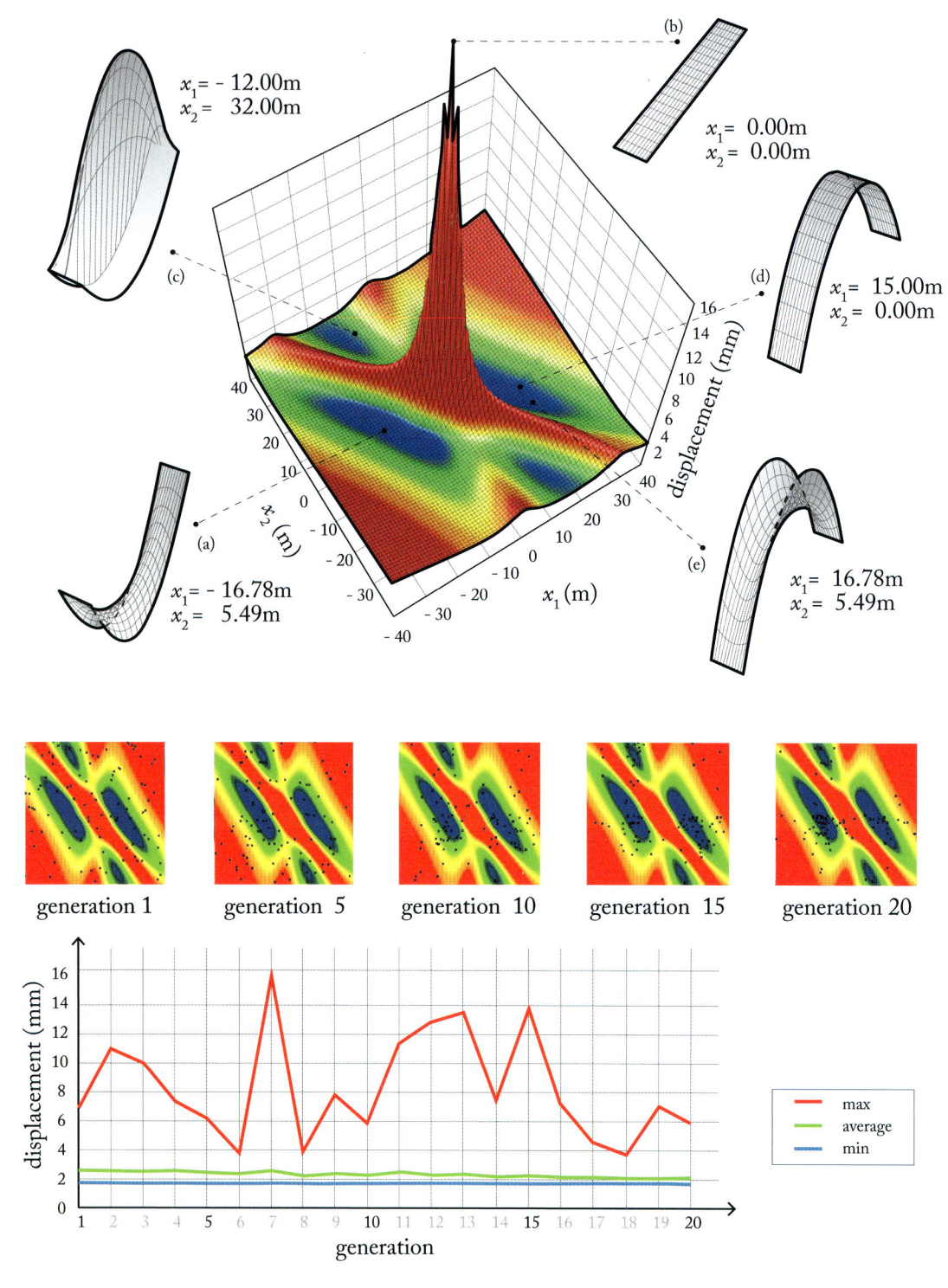

Figure 18.3 Fitness landscape, GA progress and relevant shape configurations

The binary representation of design parameters, forming the chromosome of every individual, has been proved to extract the largest amount of information. However, different kinds of representations have been studied, such as simply using the real values. When the number of design parameters is small, as in the case of this exercise, the representation scheme or coding strategy is relatively influential on the algorithm performance.

The two variables (x_1 and x_2) are here coded into eight digit binary numbers or genes. Such genes are combined into one chromosome with the x_1 value positioned first and the x_2 as second. The single point crossover operator is used. (For further details about the crossover operator, see Appendix C.)

18.1.5 Search, optimization and result interpretation

In Figure 18.3, we see a three-dimensional representation of the fitness landscape for this exercise. Such a graphical tool allows us to study the exploration performed by the GA within the solution domain and to evaluate its efficiency by mapping – generation by generation – the optimization progress (see the plan views at the bottom of Figure 18.3). The figure also shows different configurations of the bridge structure in relevant points of the fitness landscape. It is of particular interest to see how the shapes of the local minima differ from one another, even if they possess similar fitness values.

Optimization aims to find global minima but we have seen with this exercise that, by using a morphogenetic approach, even other sub-optimal candidate solutions might be worth considering.

Figure 18.3a is the global minimum as found by the GA after twenty generations. It is a hypar with a maximum displacement of 16.9mm. Because of the symmetrical nature of the problem, we can say that Figure 18.3e is the symmetric opposite to Figure 18.3a and becomes a global minimum as well. Figure 18.3b represents the global maximum, a flat surface with a maximum displacement of 285mm. Other configurations of interest are Figure 18.3c, a local minimum with very tall parabolas forming an irregular hypar, and Figure 18.3d, a near optimal cylindrical configuration.

18.2 Advanced application

In order to show the potential of a computational process of morphogenesis, such a technique has been applied to an existing building, starting from the significant issues of the architectural program and generating alternative project configurations. The building is the crematorium of Kakamigahara, a thin-shell, reinforced-concrete roof designed by architect Toyo Ito in collaboration with structural engineer Mutsuro Sasaki. The characteristics of this project make it a good testing ground for investigating how computational morphogenesis can enhance the work of architectural designers.

The spatial concept of the building is based on a freeform smooth roof that suggests 'rolling hills' and 'slowly fluctuates above the site, like clouds'. Further, the roof is pulled in some locations to the ground to form a set of vertical supports. The dips and bumps of the shell, as well as the position of the columns, do not follow any criterion of geometrical regularity (Fig. 18.4) as the lack of direct relation between columns and internal walls suggests. Clouds and hills, as many natural shapes, are always characterized by some kind of irregularity and randomness, as well as the internal stochastic operators of the GA. The search algorithm can then reproduce directly this aspect of the concept.

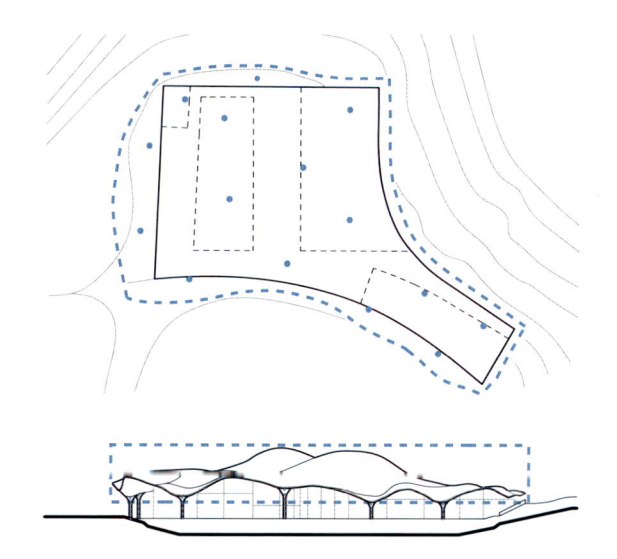

Figure 18.4 The crematorium of Kakamigahara by Toyo Ito and Mutsuro Sasaki. Plan view and section of the original project with highlighted boundary conditions, fixed supports and solution domain of the morphogenetic process

In a traditional approach, the design process starts from an architect's sketch – just a few lines on paper – to outline initial suggestions and ideas. Those sketches are translated afterwards into more complex shapes, also taking in account the third dimension and the actual development in space. Such a stage could involve dramatic simplifications, perhaps to reach a more manageable shape. However, it could also preserve the original richness of the idea by means of suitable representation of the surface. A CAD model of the building roof is then produced and submitted to the structural consultants in order to check and possibly improve its structural behaviour. The stiffness, and consequently the efficiency of the concrete shell, can then be investigated through a sensitivity analysis (Section 21.2) by locally modifying the surface geometry and checking the influence of changes on the displacements under gravity load. The structural optimization takes as an input a specific initial spatial concept iteratively searching for better configurations, following what we previously defined as a process of form improvement.

The availability of optimization techniques such as GAs opens the way to another design approach: instead of defining a starting shape to modify, the architect can provide a parametric model, defined in terms of boundary conditions, the design variables and the solution domain. This implies a clear consciousness of the major architectural aspects of the project, as well as of the ones which are less relevant. With reference to the crematorium roof, the exact position, size and height of bumps seems to be less important than the general 'bubbling' or 'waving' effect. This could be better controlled by global parameters, such as the number of bubbles, their average size and width-to-height ratio.

Once the relevant architectural aspects of the roof are defined in the parametric model, the GA becomes free to explore possible variations of the geometric system, providing the designer with different shapes, increasingly improving some structural performance criterion.

18.2.1 Parametric definition

In the proposed application, the crematorium roof shape is represented by a third-degree NURBS surface. This allows easy modifications by directly acting on the control points, whose position defines the polynomial coefficients. The generation of a two-parameter NURBS surface requires three to four boundary curves and is defined by a grid of control points which, except for the ones laying on the boundary, do not belong to the surface itself.

The plan view of the building of the crematorium has an irregular shape, roughly defined by five sides. They must be converted into a four-curve boundary to generate a NURBS surface. This process affects the overall shape, as well as the control point positions on the surface shape. The number of control points defining the grid can be decided by the designer and is the most important issue. Their coordinates represent the degrees of freedom of the problem. A large number of points increases the dimensionality of the solution domain (search space), but can also generate newer solutions, far removed from the original concept. Furthermore, control points can be used to simulate specific situations, such as the presence of constraints. In Figure 18.5, two different control point networks, generated from different starting curves, are shown in plan view. The blue circles represent the fixed points, which correspond to the tops of the columns.

Once those boundary conditions are defined, the vertical coordinates of control points are assumed as design variables. In other words, they are constrained to a unique plan projection while their height, which generates the 'waves', is free.

The solution domain of the problem, or search space, contains all the surfaces sharing the prescribed plan projection and fixed points, while the vertical position of each control point is defined by a range of values. It should be underlined that such a parametric definition and the range of variability of the design parameters substitute what has been previously called the 'initial shape'.

18.2.2 Search space

The success of the morphogenetic process largely depends on the definition, or design, of the solution domain (search space). In optimization and search problems, scalar variables are bounded in a range with a lower, upper and mean value. The domain is an n-dimensional hypercube, but when the variables

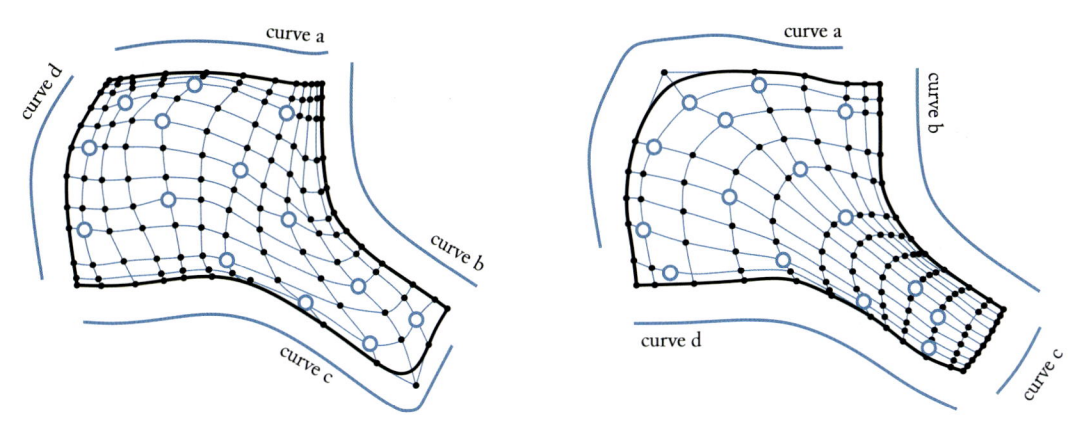

Figure 18.5 Two different control point networks based on different boundary generative curves

represent point locations in the three-dimensional space, it becomes the bounding box of all the possible shapes. This plays a role from an architectural point of view.

When the domain is just a narrow layer, the points defined by the mean values form a mean surface that can be considered similar to an initial shape of optimization processes – the final result will be relatively close to that configuration (Fig. 18.6).

On the contrary, with a wide domain, the GA process is able to generate forms without any relation to the domain shape, except for being included in it. In computational morphogenesis, the concept of initial shape does not make any sense. The process can produce, or make emerge, unexpected results. Furthermore, wider domains allow finding more sub-optimal solutions, which can be of interest to architectural designers. Investigating sub-optimal solutions is a powerful way of understanding relations between form and structure.

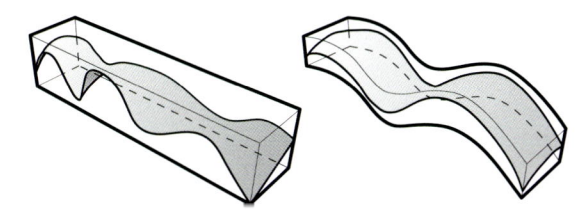

Figure 18.6 Wide and narrow solution domains (search spaces)

18.2.3 Evaluation of the structural behaviour

After the abstract definition of the set of potential solutions, a measure of the structural performance must be chosen in order to drive the optimization process. The definition of the fitness function involves two main aspects. First, the designer has to choose or define what performance is more significant for the building, and how to evaluate it quantitatively. Second, a computational model for the simulation of the performance must be developed. The whole behaviour of a concrete shell under dead load, serviceability loads, snow, wind and earthquakes, as well as the assessment of a suitable safety level are tasks too complex to be included in the search process. The state of deformation, expressed by internal strains and displacements, is usually considered a valid measure, because in curved shells the stiffness is strictly related to the shape. In search algorithms, the fitness function is a scalar, to be minimized or maximized, depending on the problem. Hence, it should be a functional of the displacements or strains (a functional returns a scalar from a vector input, or a scalar field from a vector space). One possibility is provided by the strain energy, which is an integral over the shell of a function of the strains. In a general approach, the functional can be given by the norm of the strains or of the displacements fields, mainly when they are discretized in finite vectors of values, as nodal displacements. A p-norm assumes different values depending on the p exponent. Two cases are of interest: $p=2$, the Euclidean norm, and $p=\infty$, the so-called maximum, or infinity norm. Applied to nodal displacements, the maximum norm can reveal local situations of stiffness lack, due to the shell shape. We choose this measure in the application.

The second step in the fitness evaluation involves the choice of the computational tool for the performance analysis. In case of a performance related to the mechanical behaviour, an FE solver can be used to analyse the model and to evaluate the performance. The use of an FE solver requires some preprocessing steps, such as the creation of the mesh, the definition of the material properties, and application of the boundary conditions (loads and constraints). Here, the NURBS surfaces are meshed as 50×50 four-node, flat-shell elements, with a thickness of 0.30m, a Young's modulus E of 30,000MPa and a load of 1kNm^{-2}. The constraints are fixed supports above the columns.

18.2.4 Morphogenetic process and results

Figure 18.7 represents the flowchart of the optimization process. Candidate solutions are first described as NURBS surfaces, then converted as structural meshes for the FE analysis, and finally coded as binary chromosomes for the GA.

In Figures 18.8 and 18.9, the iterative procedure of structural morphogenesis has been summarized, showing, every tenth generation, the plan view with vertical displacements of the roof structure on the left side of Figure 18.8, and the related perspective view of its spatial configuration on the right. In a first step, the edges of the structure, as well as its central part, present large vertical displacements, with a maximum displacement δ_z = 52mm. The shell curvature is generally moderated and the algorithm explores for another thirty generations of new complex curved surfaces that become increasingly wavy and gradually reduce these points of structural weakness. Next, in the last forty iterations, the algorithm converges towards a single good solution, only refining some local parts of the structure. At the end of the evolutionary process, the maximum vertical displacement of the final shape is about ten times lower, δ_z = 5mm, than the maximum displacement of the best individual of the first generation.

18.2.5 Discussion

The morphogenetic process shown in this chapter demonstrates how the GA operates through *exploration* and *exploitation* of design solutions, and how they can be put to good use in architectural and structural design problems.

The *exploration* capabilities of the GA are used to generate and consider as many different design options as possible. Depending on the problem and its formulation, a single population in the GA can contain solutions of similar fitness values but very different shapes. When the architectural concept is translated into a domain of potential shapes, all such different shapes only implicitly exist, while, once generated and proposed by GA's exploration, they become alternatives that the architect can consider. For example, Figure 18.10 shows three equally valid options following from the morphogenetic process.

The *exploitation* capabilities of the GA are used in order to guarantee the solutions proposed by the GA are optimal, or near optimal solutions. In the fitness landscape shown in second part of this chapter we saw that different solutions of similar fitness values gather in families: around local minima. While not all of them represent the global optimal solution, the exploitation process reaches the best fitting of each family, providing the architect with near optimal suggestions. We saw that the double curvature configuration was the global optimum, but that the cylindrical configuration was not far off in its fitness value.

The exploration in the GA guarantees wider search and also diversity; the exploitation provides optimal or near optimal solutions. Some GAs employ operators specifically to guarantee such diversity (niching operators). Sometimes, in order to obtain valid alternatives with high fitness values, we need to launch the GA several times with different genetic parameters.

As we can see in Chapter 21, the approach used by Sasaki focuses on exploitation and optimization of an initially well-defined solution. The exploration of alternative design solutions is performed out of the computational process, without any consideration of optimal structural behaviour. The optimization purpose is to search for optimal shapes that differ the least from Toyo Ito's design proposal. We can say that the exploration in this case is being done exclusively in an architectural context by Ito, and the exploitation is done later by Sasaki in a more reduced search space. Such an approach leaves more freedom to the architect, but it limits the benefits of structural search to the

CHAPTER EIGHTEEN: COMPUTATIONAL MORPHOGENESIS 233

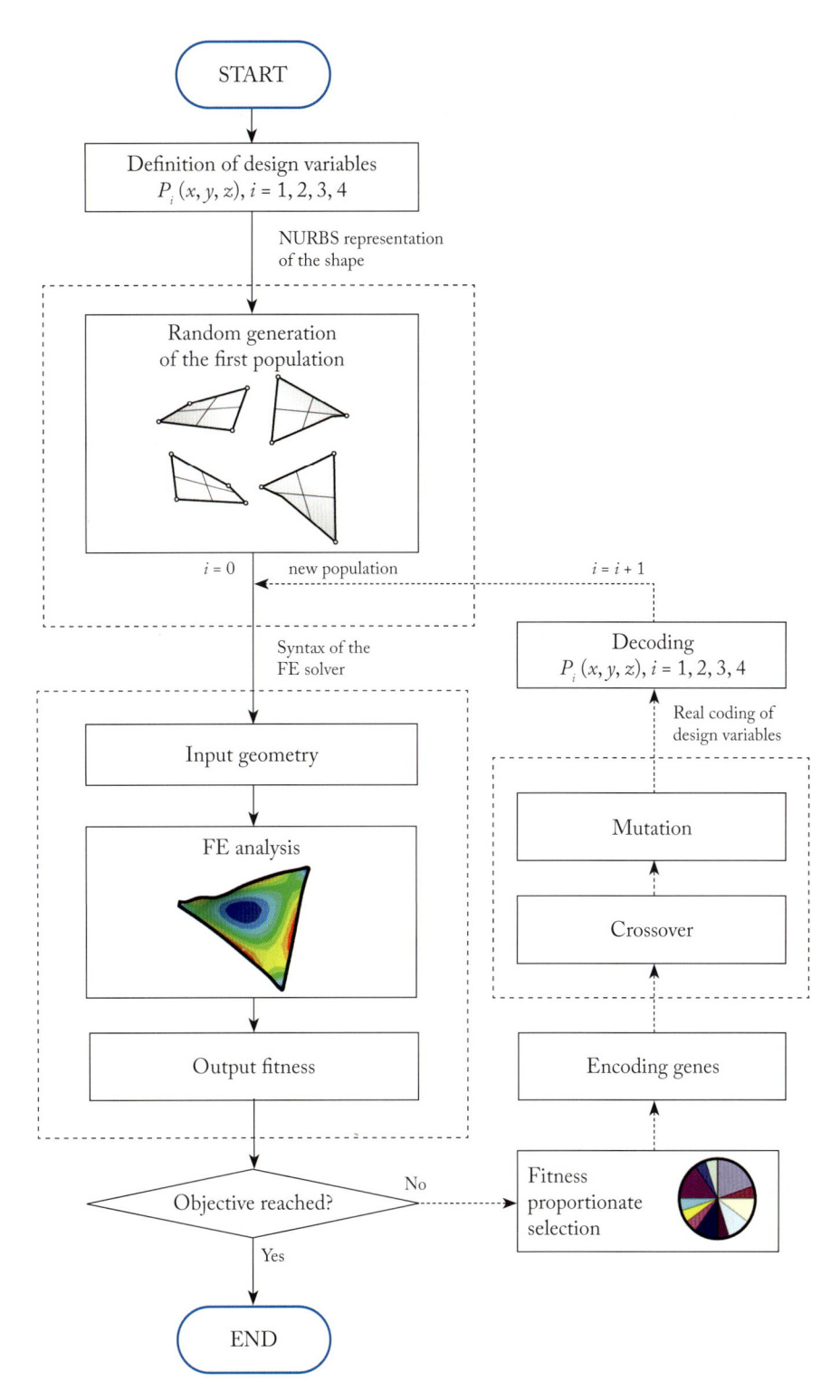

Figure 18.7 Flowchart of the structural procedure of morphogenesis by means of a GA

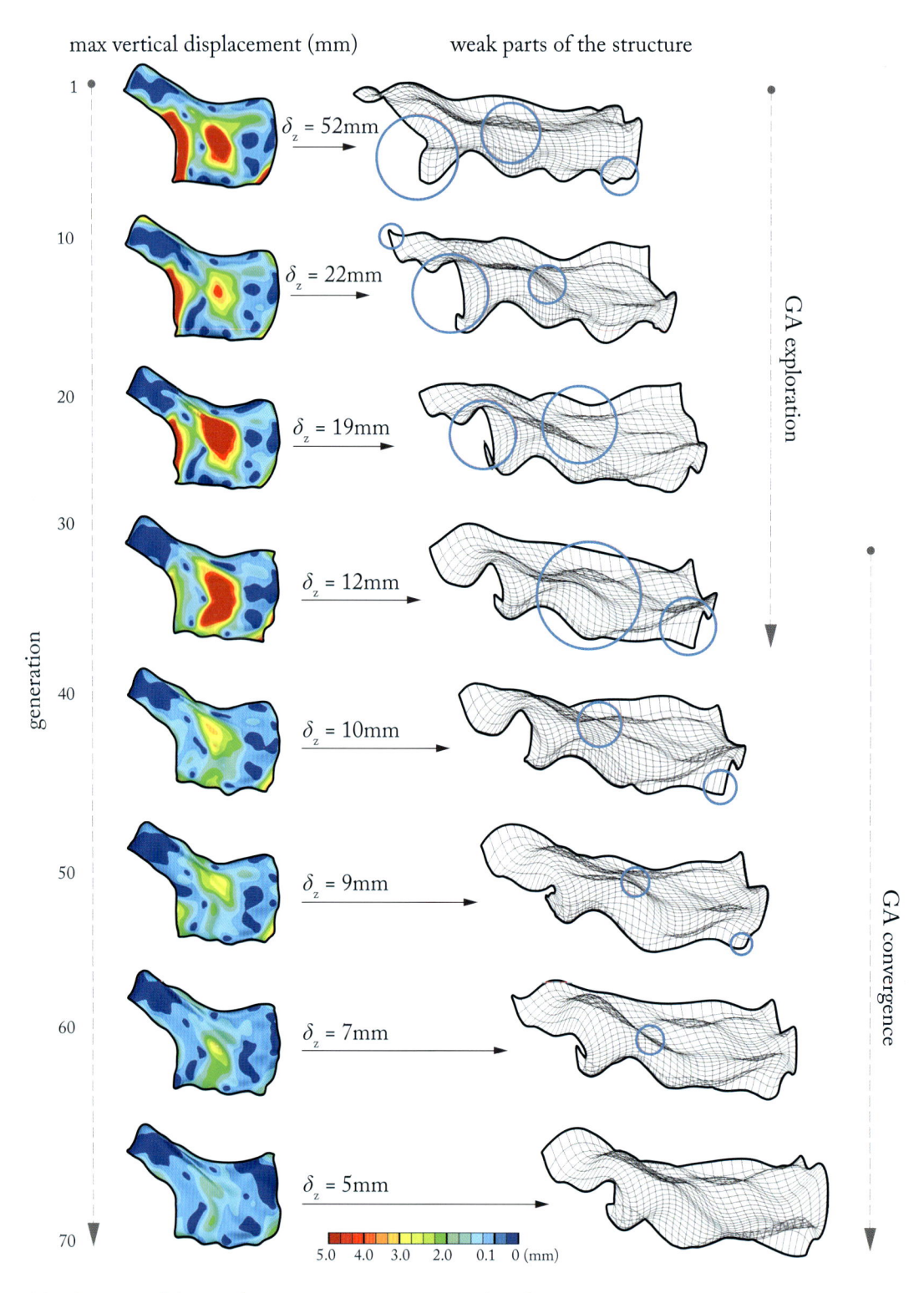

Figure 18.8 Overview of the morphogenetic process: geometrical configurations and structural behaviour at different evolutions phases of the procedure

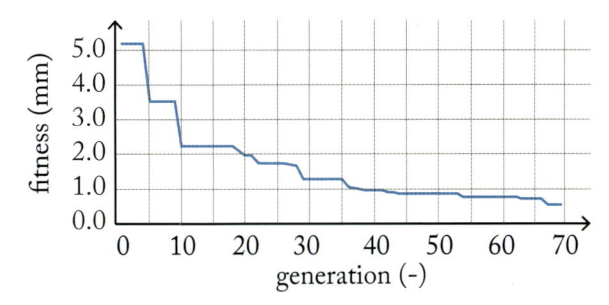

Figure 18.9 Fitness evolution during the morphogenetic process

refinement of the project, as in classical optimization. The morphogenetic design approach, however, assumes the search techniques as design aids, capable to help the designer even at the early stage, enhancing their capacity to develop performing and efficient shapes.

Key concepts and terms

Computational morphogenesis is a design process that takes advantage of the two main features of evolutionary algorithms: exploration, of a wide set of possibilities, and exploitation, of the best solutions generated, in analogy with natural evolutionary processes. The designer strongly interacts with the algorithm, providing a continuous input to the evolutionary search.

Sensitivity analysis is, generally speaking, the evaluation of how a model prediction is influenced by inputs, obtained by varying the input and checking the corresponding output variation. In deterministic optimization, processes can be used to drive the algorithm to follow the most 'efficient' path in the solutions domain.

Fitness landscape is a graphical representation of the fitness values as a function of the design variables. It can show global, as well as local minima. It can also play a central role in the interaction between designer and evolutionary algorithm, allowing the development of an autonomous knowledge and judgement on the fitness function.

A **parametric definition** uses elements, or objects, and relationships among them, that is, the invariants of every design process. For instance, a masonry wall could be parametrically described by setting up length, width and height of every brick, plus their spatial interrelations. Designers generate topological spaces rather than metric ones and they are free to concentrate on numerical variations of the system parameters, within discrete or continuous domains.

Scripting is the development of small snippets of codes, called 'scripts', that only work under the presence of more sophisticated supporting commercial software. Scripting was born in the 1960s with the aim of automating long and repetitive operations. It has become a way of extending and customizing the standard possibilities offered by programs to better fit specific user requirements. The most common CAD applications implement simple scripting environments.

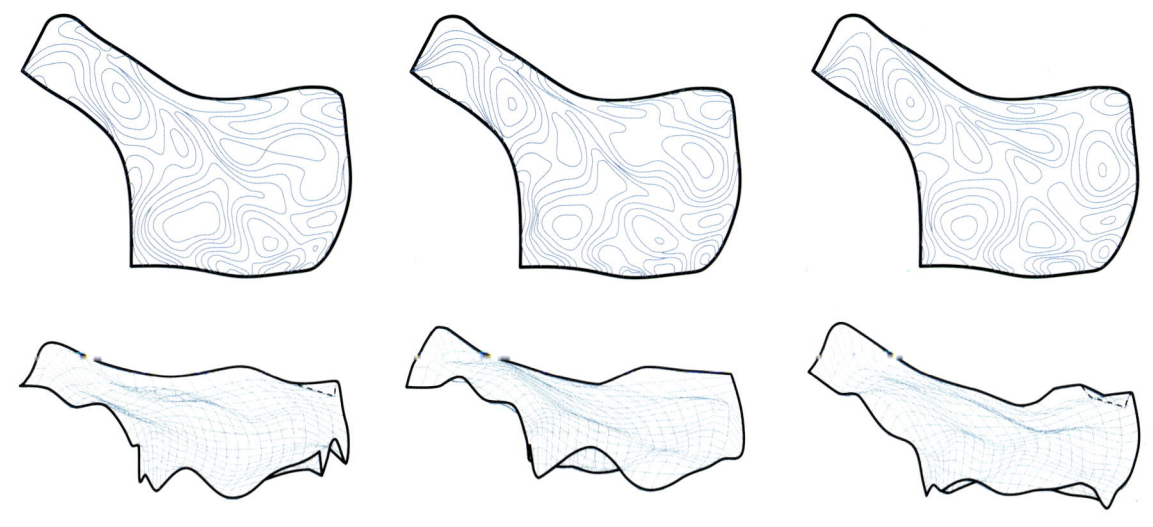

Figure 18.10: Three different shapes generated by the morphogenetic process

Exercises

- The 24m × 4m shell bridge was parameterized with just two variables. Generate a parametric (NURBS) model of the standard 10m × 10m footprint using more parameters, and in such a way that the forms generated can improve fitness function values. The parameters are z-coordinates of the control points of the NURBS surface.
- Generate a different parametric model that represents the same concept with different kinds of variables. See Chapters 5 and 16 for possible parameterizations.
- Computational morphogenesis can produce many near-optimal solutions. Using the design space developed in the previous exercises, can you extract and discuss very different-looking shapes?

Further reading

- *An Introduction to Genetic Algorithms*, Mitchell (1998). This book is a simple and complete introduction to genetic algorithms, implementing the main concepts of the work and research developed by Holland, Goldberg and Koza during the last four decades. It should be used to program a basic GA procedure, as well as for the development and tuning of the routines related to the three main GA operators.
- *Flux Structure*, Sasaki (2005). This book describes the philosophy of structures by Mutsuro Sasaki. The Evolutionary Structural Optimization (ESO) method is discussed, as well as Sensitivity Analysis (SA).

PART IV
Precedents

A NEW WORLD
ENGLAND'S FIRST VIEW OF AMERICA
UNTIL 17 JUNE 2007, ROOM 5
and let thy feet
millenniums hence

CHAPTER NINETEEN

The Multihalle and the British Museum

A comparison of two gridshells

Chris Williams

The terms 'gridshell', 'lattice shell' and 'reticulated shell' all mean essentially the same thing, a shell structure made of a grid, lattice or net of elements of any material, or possibly a continuous surface with lots of holes. The word 'reticulated' comes from the Latin 'reticulum', meaning a small net.

We are going to compare the Mannheim Multihalle and British Museum gridshells by classifying them according to a number of criteria. The reason for such a classification is to help the designer of a new gridshell understand the decisions to be made and how they may or may not interact. It is hoped that the lessons learnt on these two structures can be applied to other gridshells. It should be noted that the people who defined the geometry of these two structures were primarily interested in their structural action, following the edict that 'form ever follows function' (Sullivan, 1896), at least as far as the design constraints allowed.

In order to undertake a taxonomy of gridshell structures, that is, to classify them according to a logical system, we need to decide what are the important characteristics that define a particular structure.

Different people may emphasize different aspects. It would be perfectly reasonable for someone to include both bats and mice in the category 'small furry things', which would not include humans (not furry or small) or camels (not small). Many languages describe bats as flying mice – English has the words flitter-mouse and flutter-mouse. An aeronautical engineer may be interested in the aerodynamics of bats' flight in comparison with birds and categorize animals by if and how they fly. However, a biologist might classify using shared evolutionary ancestors and point out that bats are not rodents, indeed bats are in the group Laurasiatheria, along with pigs, camels, whales, dogs and cats, whereas rodents are in the group Euarchontoglires, as are humans.

No one system of classification is right or wrong, it just depends upon which characteristics are particularly important to an individual. The same individual may emphasize different characteristics at different times depending upon what is uppermost in their minds.

19.1 The two structures

We will only give brief details at this stage and consider things in more detail when we compare the structures.

Figure 19.1 The gridshells for the Mannheim Multihalle

Figure 19.2 Load test where dustbins full of water were hung from the roof by wires

The Multihalle gridshells (Fig. 19.1) in the Herzogriedenpark, Mannheim, were constructed for the 1975 Bundesgartenschau (Federal Garden Exhibition). The architects were Carlfried Mutschler + Partners with Professor Frei Otto, the engineers were Ted Happold's group at Ove Arup & Partners and the timber contractor was Wilhelm Poppensieker. The geometrical form was derived from Frei Otto's hanging model (Fig. 4.14). Büro Linkwitz replicated the physical model with a computer model which was used for fabrication and erection (see also Chapter 12). With an 80m span, 7,400m^2 of roofed area and a self-weight of 20kgm^{-2}, this structure remains the world's largest application of a timber gridshell, and in general, one of the largest and lightest compression structures ever built. The German authorities were understandably and rightly concerned about the safety of the structure and asked for a load test (Fig. 19.2). Much to the relief of all concerned, the results were almost identical to the predictions that we had made prior to the test. It was conceived as a temporary structure with a life of twenty years, but is still there.

The British Museum Great Court gridshell in London was completed in 2000, the architects were Foster + Partners, the engineers were Buro Happold and the steel and glass contractor was Waagner-Biro (see page 238). The surrounding building cannot provide any horizontal restraint and the roof therefore sits on sliding bearings. Thus form-finding techniques such as hanging models could not be used because the shell has to work in both tension and compression. Hence the form was determined geometrically (Williams, 2001). I was responsible for the structural analysis of the Multihalle when I worked for Arup. I was also responsible for the geometric definition of the British Museum, working with Filomena Russo at Foster + Partners, and the structural analysis, working with Andrew Chan at Buro Happold.

Category	Mannheim gridshell	British Museum gridshell
Inextensional deformation possible?	Yes	Yes
Probable collapse mechanism	Creep buckling	Elasto-plastic buckling
Dominant controlling collapse parameter	Bending stiffness (including effect of creep)	Bending stiffness and bending strength
Structural analysis	Physical model and nonlinear computer analysis using Newton–Raphson algorithm	Nonlinear computer analysis using dynamic relaxation. Program written specifically for this project.
Geometry definition	Hanging model, both physical and numerical	Mathematical function for surface. Grid relaxed numerically on surface.
Span	Approximately 60m	Approximately 35m
Material	Timber – hemlock, a straight grained conifer	Steel plate welded to form box sections and welded to nodes
Continuous members?	Yes, members cross at nodes in different planes (Fig. 19.4)	No, members all lie in same plane at a node and are welded to a machined node piece
Structural grid pattern	Quadrilateral, constant edge length of 0.5m	Triangles, varying sizes and angles
Cladding material	Flexible plastic sheet	Double glazed units
Cladding grid	Same as structural grid	Same as structural grid
Erection technique	Pushed up from below (Fig. 19.3)	Area fully scaffolded and structure fully propped. Partly prefabricated and partly welded in situ.

Table 19.1 Comparison of the Mannheim Multihalle and the British Museum Great Court roofs

19.2 Comparison

In Table 19.1 we have chosen twelve categories to compare and contrast these two shells. They are largely independent; for example, while timber suggests continuous members, that is not always the case – the timber gridshell over the courtyard in Portcullis House, London by Hopkins Architects has discontinuous timber members. To do this they had to overcome the problem of transferring force and moment from member to member across the joints, a not inconsiderable problem in timber.

Figure 19.3 Erection process

19.2.1 Inextensional deformation

We have chosen to put inextensional deformation first in the category table. What does this mean? In general when a structure deforms it involves changes in both length of elements and bending of elements. It is easier to bend than stretch or compress most structural elements and inextensional deformation is deformation that only involves bending. Thus structures which have inextensional modes tend to be more flexible and less efficient. The two structures have inextensional modes for different reasons. The surface of the British Museum gridshell is fully triangulated, but the structure is on sliding supports around the rectangular outer boundary which means that it can only exert a horizontal thrust into the corners where the force is resisted by the edge beam in tension.

The Mannheim gridshells are nicely supported around the boundary but the surface is not fully

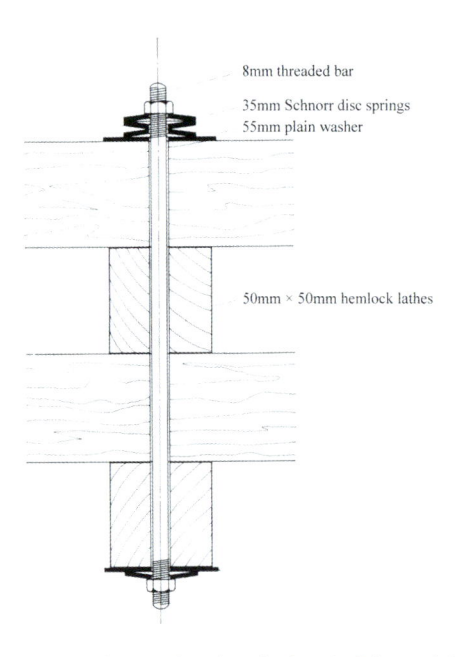

Figure 19.4 Timber grid and node detail of the Multihalle

triangulated and therefore there are inextensional modes, exactly as for a kitchen sieve made from a woven wire mesh that is formed into a hemisphere from a flat sheet by bending only. The Mannheim gridshells do have partial restraint of inextensional modes by light bracing cables. Two-way steel shells with moment connections between the two directions can use Vierendeel action to restrain this in-plane shear deformation due to relative rotation of the two sets of members. By Vierendeel action we mean carrying shear force around a quadrilateral by bending moments, as in a Vierendeel truss.

19.2.2 Collapse mechanisms

If a shell or gridshell structure can deform inextensionally, then bending stiffness and strength will probably be the dominant criteria controlling collapse. As a gridshell deflects under load, its shape becomes less efficient at carrying the load, at least if the load causes compressive forces in the structure. This might cause collapse even without material breaking or yielding – purely elastic buckling. If yielding occurs then we have elasto-plastic buckling, the probable collapse mode for the British Museum structure, were it to be overloaded. The nodes were designed and tested to ensure a plastic collapse, rather than brittle failure leading to progressive collapse.

Timber will creep and therefore the most likely collapse mechanism for Mannheim would be creep buckling in which the structure slowly moves and as it moves the moments and stresses increase leading to an increase in the rate of creep strain and so on in a vicious circle. This means that the hanging chain geometry is optimum, because the self-weight is the load that is there all the time.

If inextensional deformation is not possible, then Wright (1965) suggests that an estimate of the buckling of a shell can be found by considering an equivalent conventional shell of uniform thickness. The classical eigenvalue or linear buckling load for a spherical shell, radius a and thickness h, under uniform pressure is

$$p_{cr} = \frac{2Eh^2}{a^2\sqrt{3(1-\nu^2)}}, \qquad (19.1)$$

where E is the Young's modulus and ν is Poisson's ratio.

We can rewrite this as

$$p_{cr} = \frac{4}{a^2}\sqrt{Eh}\sqrt{\frac{Eh^3}{12(1-\nu^2)}} \qquad (19.2)$$

$$= \frac{4h^2}{a^2}\sqrt{\text{membrane stiffness}}\sqrt{\text{bending stiffness}}$$

in which bending stiffness assumes zero curvature in the orthogonal direction, while the membrane stiffness assumes zero stress in the orthogonal direction. But these distinctions are not relevant given that $(1-\nu)^2$ is close to 1.

Thus if we can estimate the membrane and bending stiffnesses per unit width, we can estimate the buckling load. However, this classical eigenvalue buckling load for shells is highly optimistic and no experiments have ever got near. The reason is that shells which cannot undergo inextensional deformation are extremely efficient and extremely stiff, but they are also very imperfection sensitive so that the slightest deviation from the ideal shape leads to sudden collapse.

19.2.3 Imperfection sensitivity

Thus the fact that the Mannheim and British Museum gridshells can undergo inextensional deformation makes them less efficient, but also less sensitive to imperfections. The Mannheim model tests showed that the collapse was definitely nonlinear since large deflections took place prior to collapse and some parts of the structure were actually hanging in tension of other, still stiff areas.

The membrane flexibility of the Mannheim timber grid due to relative rotation of the two sets of laths effectively reduces the membrane stiffness in comparison to the bending stiffness (Fig. 19.4). This is like making a conventional shell from a very thick but flexible material, such as sponge rubber, to give an increased bending stiffness to membrane stiffness ratio. Such a shell would not be particularly imperfection sensitive.

19.2.4 Structural analysis

We have seen that collapse loads predicted by classical linear eigenvalue buckling of shells can be wildly optimistic, and anyway not applicable to structures such as the Mannheim or British Museum gridshells. At the time that Arup was working on Mannheim, there were still people there such as Ronald Jenkins and John Blanchard, experts on shell theory, who had worked on the Sydney Opera House. Between us we were not able to come up with hand calculations which would predict the collapse loads of the Mannheim shells. Theories based on arch buckling were far too pessimistic. This led to only two possibilities: physical model tests and computer analysis. The analysis of the Mannheim Multihalle used both approaches, whereas the British Museum gridshell, designed a quarter century later, used computational models only.

Happold and Liddell (1975) give details of how the results from a physical model test can be scaled. Scaling laws are also discussed in Section 8.3. Physical models give an understanding of impending collapse as bits of structure lose their stiffness, which can be felt by gingerly touching the model. One might be concerned that suggesting a physical model test implies a lack of knowledge of gridshell structures, but the reverse is true; the more one knows, the more one realizes the complexity of their behaviour.

19.3 Conclusion

The discussion of membrane and bending stiffness tells us that a gridshell with pinned connections should never be built. Local snap-through buckling will occur unless the structure has a very coarse grid, in which case it is not really a gridshell, but instead just a three-dimensional arrangement of bars. Note that we are here talking about transmitting moments out of the plane of the structure, a triangulated shell does not need to transmit moments across the nodes in the plane of the structure.

The situation in engineering practice is much the same today as it was during the design of both projects, except for the increased power of computers and availability of software. Nonlinear buckling analysis, with or without material nonlinearity is still something which stretches software capability. Note that two pieces of software from different vendors may both give the same wrong answer because they both contain the same inappropriate assumptions. For example, rotation is *not* a vector unless the rotation is *small*. It is often not clear whether computer programs make the assumption of small rotations. For such reasons, one should be careful to ensure that any computer results are checked, either by other software or physical model tests.

Further reading

- 'Timber lattice roof for the Mannheim Bundesgartenschau', Happold and Liddell (1975). This journal paper provides a comprehensive overview of the design and construction of this project.

- 'The analytic and numerical definition of the geometry of the British Museum Great Court roof', Williams (2001). This book chapter describes the mathematical and computational methods in sufficient detail to be able to replicate the process.
- *IL 10 Gitterschalen (Grid Shells)*, Hennicke (1975). This book is from Frei Otto's Institut für Leichte Flächentragwerke (IL), renamed as the Institut für Leichtbau Entwerfen und Konstruieren (ILEK) since Professor Otto's retirement. The engineering content is similar to that in Happold & Liddell and was written by the same authors.
- 'Membrane forces and buckling in reticulated shells', Wright (1965). This paper relates the properties of a gridshell to an equivalent continuum, making the theoretical results for shell structures available for analysing gridshells by hand calculations. Computer calculations for gridshells have no need for this approach since modern computers can handle models containing every member of a gridshell.

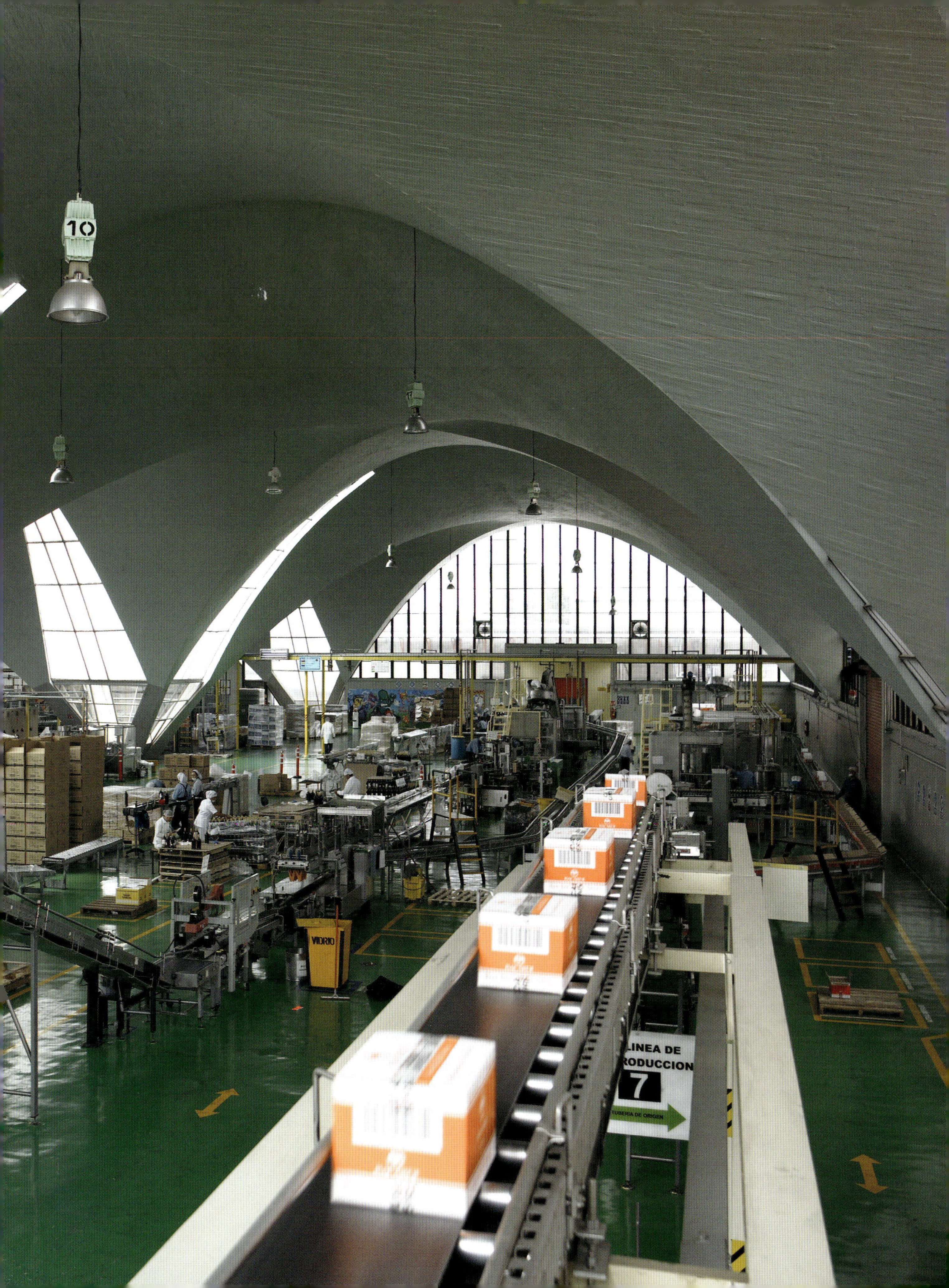
10
VIDRIO
LINEA DE
RODUCCION
7

CHAPTER TWENTY

Félix Candela and Heinz Isler

A comparison of two structural artists

Maria E. Moreyra Garlock and David P. Billington

Structural artists exhibit three characteristics that are fundamental to creating the best-engineered structures: the ethos of efficiency, the ethic of economy and the aesthetic motivation in design. Efficiency in this sense means the search for forms that use a minimum of materials consistent with sound performance and assured safety; economy signifies a minimum of construction costs consistent with low expense for maintenance. These two fundamentals imply a plan that pays attention to both design and construction.

One of the greatest myths in our structural engineering profession is that elegance in structural engineering is the province of architects and that while engineers ensure that it will stand, only architects can make it a work of art. This argument is contradicted by the most talented structural engineers over the last 200 years whose motivation included appearance along with efficiency and economy. Two such engineers were Félix Candela (1910–1997) and Heinz Isler (1926–2009) (Fig. 20.1).

20.1 Early development as shell designers

Both Candela and Isler were designers of thin-shell concrete structures who found the forms for their shells in different ways, but the outcome was the same: efficiency, economy and elegance. In this chapter, we examine the background of each that formed their development as shell designers, then we study the forms that each favoured, and finally we evaluate how each achieved economy of construction.

20.1.1 Candela

Félix Candela was born in Madrid, Spain, on 27 January 1910. He was trained as an architect at the Escuela Superior de Arquitectura in Madrid. He had chosen architecture as his discipline rather arbitrarily, but possessed a strong talent and interest in geometry and mathematics.

Due to the Spanish Civil War, Candela was exiled to Mexico in 1939, and it is here that he would become a shell builder. When Candela arrived in Mexico, he embarked on several projects and gained valuable experience in design and construction of traditional beam-column-type structures. It was not until 1949 that he built his first experimental shell (a funicular vault), but not without extensive study of the analysis of such shells that he learned from published papers. In 1950, excited about the success of this experimental shell, Candela formed the construction company Cubiertas Ala, with his friend Fernando

Figure 20.1 (top) Félix Candela and (bottom) Heinz Isler

Figure 20.2 Cosmic Rays Laboratory, 1951

Fernandez. The company specialized in shells, and also performed the engineering calculations.

Candela's first significant shell structure was the Cosmic Rays Laboratory (Fig. 20.2), built on Mexico City's university campus (UNAM) in 1951, which could be no more than 15mm thick at the top so that cosmic ray measurements can be made inside the laboratory. Candela – the engineer – gave the shell double curvature the required stiffness and stability.

Candela – the builder – used straight boards to create that double curvature and economy of construction by using the hyperbolic paraboloid form, which is a doubly curved surface generated with straight lines. And, finally, Candela – the artist – took a design given to him by an architect and modified it to create his own work of art. The laboratory was a huge success and elevated Candela to international fame.

20.1.2 Isler

Isler was born in Zurich, Switzerland, on 26 July 1926. He had a great talent for drawing and watercolours, where his art reflected a love of the natural world and Swiss landscape. He was attracted to a career in painting, but his father insisted on a professional degree. Therefore he entered the ETH in Zurich, in 1945, and graduated five years later with a degree in Civil Engineering.

While Candela was born physically in Madrid, Heinz Isler was born 'professionally' in Madrid. That is where a young Isler, coming directly from military service, would make his first appearance in front of many of the world's leading shell designers: in 1959, at the first congress of the International Association for Shell Structures (IASS). His presentation, the last of twenty-five, created the greatest impact and largest discussion of all. The paper was titled 'New shapes for shells' (1960) and it described three methods for shaping shells:

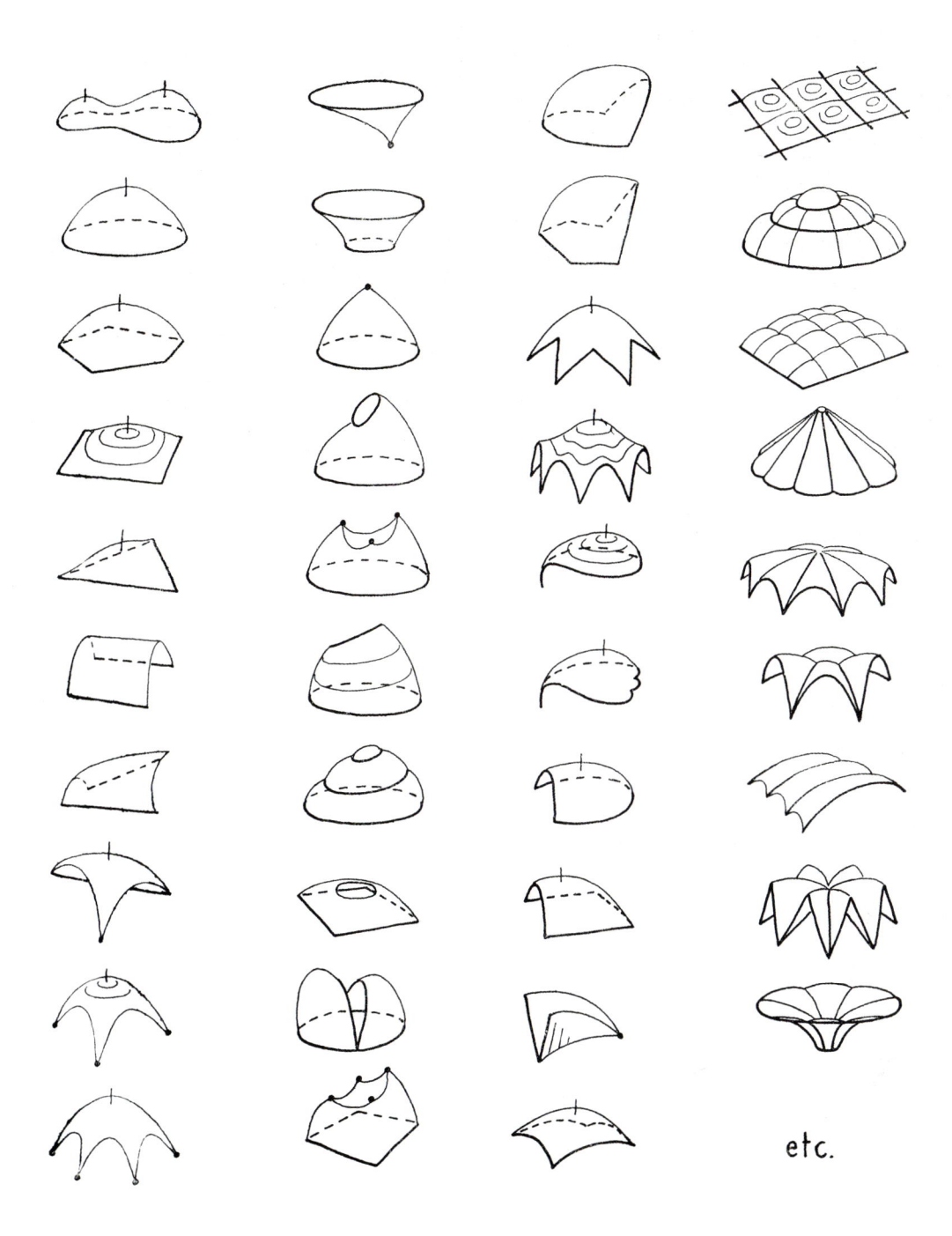

CHAPTER TWENTY: FÉLIX CANDELA AND HEINZ ISLER   **249**

Figure 20.3 Isler's examples of the endless forms possible for shells, from the 1959 IASS conference (Isler, 1960)

1. The freely shaped hill where the concrete is formed by the earth that is carved to its desired form, and after the concrete hardens the shell is lifted or the earth is excavated.
2. The membrane under pressure.
3. The hanging reversed membrane, which he refers to as 'the best method for design'.

The methodologies were unconventional and therefore brought out a rash of discussion from distinguished designers such as Eduardo Torroja, Nicolas Esquillan and Ove Arup.

The last figure of Isler's paper (Fig. 20.3) shows thirty-nine shapes for shells that are possible, with the fortieth spot showing 'etc', indicating that the variety is endless. Isler's enthusiasm and creativity stems from his artistic talents, and also from the encouragement of his engineering professor at the ETH Zurich, Pierre Lardy. Isler has written:

> I think that the most important contribution we students got from our teacher Lardy is this:
> He reminded us, the engineering students,
>
> - that we have in us a sense for esthetics
> - that we have the right to use it
> - that we are allowed to mention our opinion
> - and that we can find and express it in our projects
>
> This to my opinion was the invaluable, great, and unique contribution he gave to us. Not the statics, not the theories, not the investigations were his greatest and lasting influence, but encouraging us to find and apply esthetics from within us. And for this I am very grateful to him.
>
> (Isler, 2002)

Like Candela, Isler founded his own company. While it was not a construction company, as was Candela's, Isler worked closely with the builder and planned the construction – he had a 'builder's mentality' approach to his designs and nearly all his designs were the result of innovations developed by taking forms from the field.

20.2 Creativity of form

The success of the companies of both Candela and Isler arose from developing markets for their widely repeated designs: the hyperbolic paraboloid umbrella roofs for Candela and the pneumatic roofs for Isler. We examine these forms here.

20.2.1 Candela and the hyperbolic paraboloid

Candela has written that the simplest shape to give a shell and the easiest and most practical to build is the hyperbolic paraboloid. This shape is best understood as a saddle in which there are a set of arches in one direction and a set of cables, or inverted arches, in the other (Fig. 20.4). The shape also has the property of being defined by straight lines. The boundaries, or edges, of the hyperbolic paraboloid (also referred to as a hypar) can be straight or curved. The edges in the second case are developed by planes 'cutting through' the hypar surface.

All Candela's significant structures were of this form and with that discipline he could build them only 4cm thick. Chapel Lomas de Cuernavaca (Fig. 20.5) is an example of a curved-edge hypar, while Our Lady of the Miraculous Medal Church (Milagrosa) (Fig. 20.6) is an example of straight-edged hypars.

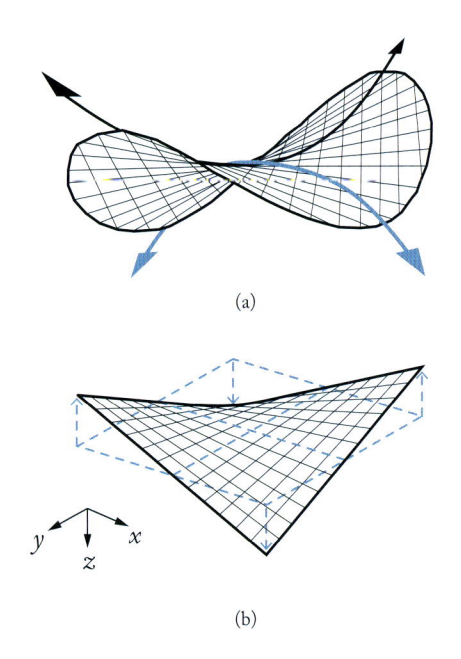

Figure 20.4 The hyperbolic paraboloid for (a) curved edges and (b) straight edges

CHAPTER TWENTY: FÉLIX CANDELA AND HEINZ ISLER 251

Figure 20.5 (top) Chapel Lomas de Cuernavaca, 1958, and (bottom) falsework used during construction showing straight wood formwork

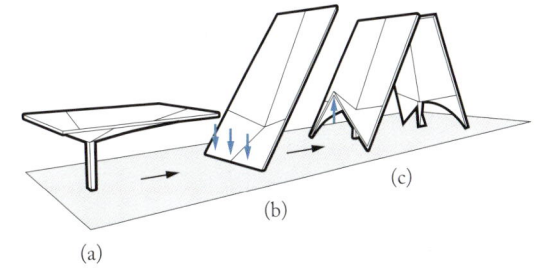

Figure 20.6 Our Lady of the Miraculous Medal Church, Navarte, Mexico City, 1953–1955, and its design concept in three stages (a–c)

The form for Milagrosa was derived from an *umbrella* form (Fig. 20.6). Candela devised the umbrella form by combining four straight-edge hypar surfaces. By placing umbrellas side by side, large roof spaces were generated, as shown in Figure 20.7, and by tilting the umbrellas, a saw-tooth profile was created, which allows light to enter the space.

20.2.2 Isler and the pneumatic form and the hanging reversed membrane form

While the climate in Mexico is moderate, Isler's shells were mostly built in harsh environmental conditions such as Switzerland and Germany. Isler's shells were typically 8cm thick, twice as thick as Candela's shells, but they still expressed an elegance of thinness. But not all of Isler's shells revealed the shell thickness. One of his first innovative forms was a pneumatic form, devised from a simple wood frame holding a rubber membrane inflated with a bicycle pump to achieve a form (Fig. 20.8). These pneumatic forms were used for roofs such as the one shown in Figure 20.9. The exterior shows the prestressed beams surrounding the shell for edge support, thus hiding the thinness.

In the early 1960s, Isler saw in a store window a book cover with a photograph of the Restaurant Los Manantiales designed and built by Candela (Fig. 20.10). Its form inspired Isler to rethink his forms for thin-shell concrete structures and it stimulated him to think about how he might express the same kind of thinness which he could not do with his traditional pneumatic forms. However, he succeeded in expressing the thinness with his hanging reversed membrane forms, as demonstrated, for example, by the BP service station built in 1968 (Fig. 20.11), and the Grötzingen outdoor theatre in 1977 (Fig. 20.12).

Isler describes the process of finding the hanging reversed membrane form (Figs. 4.8, 5.2 and 20.13):

Figure 20.7 A series of umbrellas comprising Rio's Warehouse, Mexico City, 1954

'The process consists of pouring a plastic material onto a cloth resting on a solid surface. Once the material is evenly spread on the cloth, the solid surface is lowered and the plastic-covered cloth, now in pure tension, is freely suspended from its corners. In that position, the plastic hardens and the solid shell model is turned upside down, giving a shell form in pure compression' (Isler, 1980b).

20.2.3 Comparison of shell stresses

Candela identified two types of shells: proper shells, which avoid bending stresses and are doubly curved (like the hyperbolic paraboloid), and improper shells, which carry the load through some bending action. In a shell structure, the thickness is significantly smaller than its width and length. Bending stresses should be avoided since they can lead to rupture or significant tension leading to cracking. An example of a thin

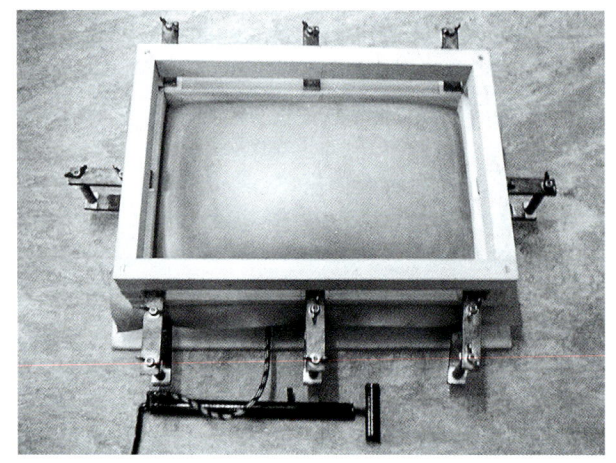

Figure 20.8 Wood frame holding a membrane inflated to a pneumatic form

Figure 20.9 Pneumatic form roof. (top) Interior of the Eschmann Company, 1958, and (bottom) exterior of one of Isler's designs with prestressed tension ties concealing the shell's thinness

concrete structure under pure bending is a horizontal slab. By giving the concrete curvature, the structure can span greater distances, and if properly designed,

CHAPTER TWENTY: FÉLIX CANDELA AND HEINZ ISLER   253

Figure 20.10 Restaurant Los Manantiales, Xochimilco, Mexico, 1958

Figure 20.11 BP Service Station, Deitingen, 1968

Figure 20.12 Grötzingen outdoor theatre near Stuttgart, Germany, 1977

Figure 20.13 The hanging membrane, once hardened, is inverted to create a shell form in pure compression

eliminate or significantly reduce bending. A proper shell carries the load through membrane stresses, which means that the stresses in the slab thickness are evenly distributed. Since concrete has relatively little strength in tension, the desired membrane stress for concrete thin shells is compression (see also Chapter 3).

Both Candela and Isler's forms were such that the shells were almost in a pure state of compressive membrane stress (i.e. negligible bending). Even if the concrete can carry the compression or tension developed in the shell, steel reinforcing bars are added to protect the shell against cracks that could be caused by creep, shrinkage and temperature effects.

Isler's shells, by virtue of his form-finding methods, were essentially in a state of pure compression. In the pneumatic roofs, the form found in the membrane (Fig. 20.8) was in pure tension under 'upwards' pressure from the inside. For a relatively flat shell, that pressure load is close to the exact opposite of gravity load, so that under the shell's self-weight, the concrete would be essentially under pure compression. For the reversed hanging membrane forms, the gravity by which the membrane hung upside-down (Fig. 20.13) produces pure tension. This tension is changed to pure compression once the membrane is reversed.

The double-curvature forms of Candela's hyperbolic paraboloid were also mostly in a state of pure compressive membrane stress, even though his form-finding approach did not necessarily make that an inherent characteristic as did Isler's method. For example, finite element studies of Our Lady of the Miraculous Medal Church (Milagrosa) (Fig. 20.6), designed by Candela, showed that very little bending stresses develop in the shell of this church (Fig. 20.14), which is consistent with finite element studies of Candela's other major works done by us.

Figure 20.14 shows the maximum principal stresses at the top and bottom of two bays of the church. The sign convention is such that positive represents tension, and negative compression. The maximum principal stress thus represents the maximum tension (or least compression). Although the minimum principal stresses (representing the largest compression) stresses were larger in absolute magnitude, we only show the positive (tension) stresses since these would indicate danger of cracking and can also give an indication of bending. It is seen that the stresses are generally quite low, and also close to the same value on top and on bottom, indicating that there is no significant bending in the shell. The same finite element study of Milagrosa also showed that the scalloped ridge (the thickening of the shell from 4cm to 14cm at the top of the roof) serves a structural function first (eliminates dangerous tension), and an aesthetic one second.

20.3 Economy of construction

The construction of concrete shells requires scaffolding (sometimes referred to as falsework), a temporary structure of wood or metal that supports the form boards and wet concrete. Form boards are placed on top of the scaffolding and mould the concrete, while it hardens, into its proper form. Designers and contractors have often shied away from curved concrete forms, believing that they would be expensive since they are custom, curved forms. Candela, however, understood that hyperbolic paraboloids, while doubly curved, can actually be formed from straight lines (Fig. 20.4). And Isler understood that by reusing the curved forms on several projects, economy of construction can be achieved. This section examines in more detail the different ways that each designer, Candela and Isler, planned the construction for their works.

20.3.1 Candela the builder

Candela's practice was that of a builder and contractor, and that is how he identified himself: 'I must say … that although an architect by training, in practice, I am a constructor and building contractor' (Candela, 1955). Candela found the construction process of paramount importance in a project. In his words: 'few people realize that the only way to be an artist in this difficult specialty of building is to be your own contractor. … it may be shocking to think of a contractor as an artist; but it is indeed the only way to have in your hands the whole set of tools or instruments to perform the forgotten art of building, to produce "works of art"' (Candela, 1973).

The doubly curved surface of hyperbolic paraboloids appealed to Candela for three major reasons: added stiffness, ease of construction and visually elegant. The ease of construction is due to the fact that the surface, while doubly curved, can be constructed

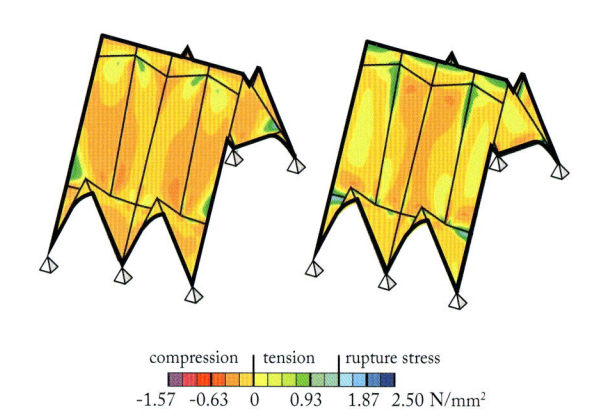

Figure 20.14 Shell stresses in two bays of the Miraculous Medal Church (see Fig. 20.6) (Thrall and Garlock, 2010)

compression | tension | rupture stress
-1.57 -0.63 0 0.93 1.87 2.50 N/mm²

with straight form boards of wood (Figs. 20.4–20.5). By not using curved form boards, and reusing the wood, Candela was able to bid projects at a competitive price. Once the forms are placed, steel reinforcing is added. Because thin shells typically are not vibrated (a process in concrete construction to move the wet concrete through the reinforcement and ensure its even distribution in the form), Candela placed a layer of cement grout over the form to obtain a smooth interior surface before the concrete (which contains gravel) was poured. Then, the concrete, typically 4cm thick, was placed by the labourers who carried it in buckets (Fig. 20.15).

Structures such as Cuernavaca, Milagrosa and Los Manantiales are examples of 'custom' structures because they were not repeated. However, Candela had a standardized, 'bread and butter' structure that he designed and built repeatedly, which kept his construction company in business: the umbrella (Fig. 20.7), used for markets, warehouses and factories. The steady income from umbrella constructions allowed Candela the time and resources needed to design and build his custom forms, including his favourites: the Miraculous Medal Church (Fig. 20.6), the Chapel at Cuernavaca (Fig. 20.5), Restaurant Los Manantiales at Xochimilco (Fig. 20.10) and the Bacardi Rum factory (Fig. 21.1 and page 246).

20.3.2 Isler the builder

Like Candela, Isler had a standardized form that kept his company in business. For Isler, this form was the pneumatic roof (Fig. 20.9). This relatively standardized business allowed Isler to build highly original structures, such as the BP Service Station of 1968 (Fig. 20.11), the Sicli Building of 1969 (Fig. 20.16 and page 44), the Grötzingen outdoor theatre of 1977 (Fig. 20.12), and the Heimberg tennis court shells of 1979 (Fig. 20.17). Isler based each design on his reversed hanging membrane method, and, as with Candela, each structure, apart from the Heimberg design, was unrepeated.

Isler's construction process begins by erecting metal falsework (scaffold) to support the curved laminated wooden arches (that will be reused a number of times), on top of which are placed wooden slats. Since Isler needed expensive curved wooden girders to create his forms, he gained economy by constant reuse of those forming members (Fig. 20.18). He could sometimes use them for his non-standard forms as well.

Figure 20.15 Laborers placing buckets of concrete on the shell for Los Manantiales Restaurant (see Fig. 20.10)

Figure 20.16 Sicli Company Building, Geneva, 1969

Figure 20.17 Heimberg tennis court shells, 1979

On these wooden slats, builders place flat fibreboards (Fig. 20.18), on which they place the reinforcing steel bars. The concrete goes over the steel but with no further cover. When concrete hardens, the falsework, arches and wooden slats are removed leaving the fibreboards to act as insulation which serves to prevent the concrete structure from having cracks due to differential strains from the cold exposed surface above and the warm internal spaces below. This construction plan also does away with the need for the waterproof covering normally required for concrete surfaces, keeping them crack-free, even in the harsh environment typical of Switzerland.

20.4 Conclusion

For both Candela and Isler, their central idea came from construction: the need to make forms in the field, rather than to make calculations first. Both engineers first built full-scale shells of traditional forms, which they found to be unsatisfactory, but useful to learn about construction. Both men sought to develop shapes that could, however, satisfy their desires for simplicity of form, yet did not require complexity of calculation. In the process of thinking about this problem, they hit upon different forms. Each engineer came to differing conclusions, but with the same basic results: first, the structures could be easily built; second, they could be formed in such a way as to require much thought but to eliminate complex calculations; and, third, those shapes could be made to be visually elegant.

Figure 20.18 Leuzlinger Sons Building, 1979. Isler standing beside glued laminated timber arches, and formwork of fibreboard insulation placed on wooden boards

In 1979, twenty years after Isler's stunning debut at the first IASS Congress, both Isler and Candela were invited to give plenary lectures to mark this special anniversary. This honour illustrates that both men were the leading shell designers of that time. Isler called his lecture 'New shapes for shells – twenty years after'. He concluded in this presentation that 'form finding is one of the most important factors in shell design. I would say the most important one', and that 'the method of hanging reversed membrane seems to be the most efficient one'.

In a lecture given in honour of Candela after his death, Isler said of Candela: 'His structures have a lightness and elegance that had never before been achieved.' Candela inspired Isler to achieve this lightness of form, which he did with his inverted membrane. Both men demonstrated a playful spirit in designing the forms for their structures. But, that play was disciplined by efficiency and economy. Candela used the following words to explain this ethos that was embraced by both designers: 'an efficient and economical structure has not necessarily to be ugly. Beauty has no price tag and there is never one single solution to an engineering problem. Therefore, it is always possible to modify the whole or the parts until the ugliness disappears' (Candela, 1973).

Further reading

- *Candela: The Shell Builder*, Faber (1963). This is an excellent compilation of Candela's designs that includes discussion, analysis, dimensions and drawings of Candela's works.
- *Seven Structural Engineers: The Félix Candela Lectures*, Nordenson (2008). This collection of essays, based on a lecture series in honour of Candela that started in 1997, discusses the importance and influence of Candela's works. Isler was one of the contributors.
- *Félix Candela: Engineer, Builder, Structural Artist*, Garlock and Billington (2008). This book examines in detail Candela's greatest works, the evolution of his forms, and his life, education and experiences. The book was a companion to a museum exhibition with the same name.
- In this chapter, some general discussion of the structural behaviour of Candela's most important shells is given. More detailed structural evaluations are found in 'Structural analysis of the cosmic rays laboratory', Kelly et al. (2010); 'Analysis of the design concept for the Iglesia de la Virgen de la Medalla Milagrosa', Thrall and Garlock (2010); 'Finite-element analysis of Félix Candela's Chapel of Lomas de Cuernavaca', Draper et al. (2008); 'A comparative analysis of the Bacardí Rum Factory and the Lambert-St. Louis airport terminal', Segal et al. (2008); and 'Félix Candela, elegance and endurance: an examination of the Xochimilco shell', Burger and Billington (2006).
- *Heinz Isler as Structural Artist*, Isler (1980a). This book went together with the first major art museum exhibition on Isler's work.
- *The Engineer's Contribution to Contemporary Architecture: Heinz Isler*, Chilton (2000). This book is a culmination of numerous articles by Isler and a good summary of Isler's approach to model making and form finding, both for the design of concrete shells and for teaching structural understanding.
- *The Art of Structural Design: A Swiss Legacy*, Billington (2003). This book examines four Swiss structural engineers (Robert Maillart, Othmar Ammann, Heinz Isler and Christian Menn) and their teachers.

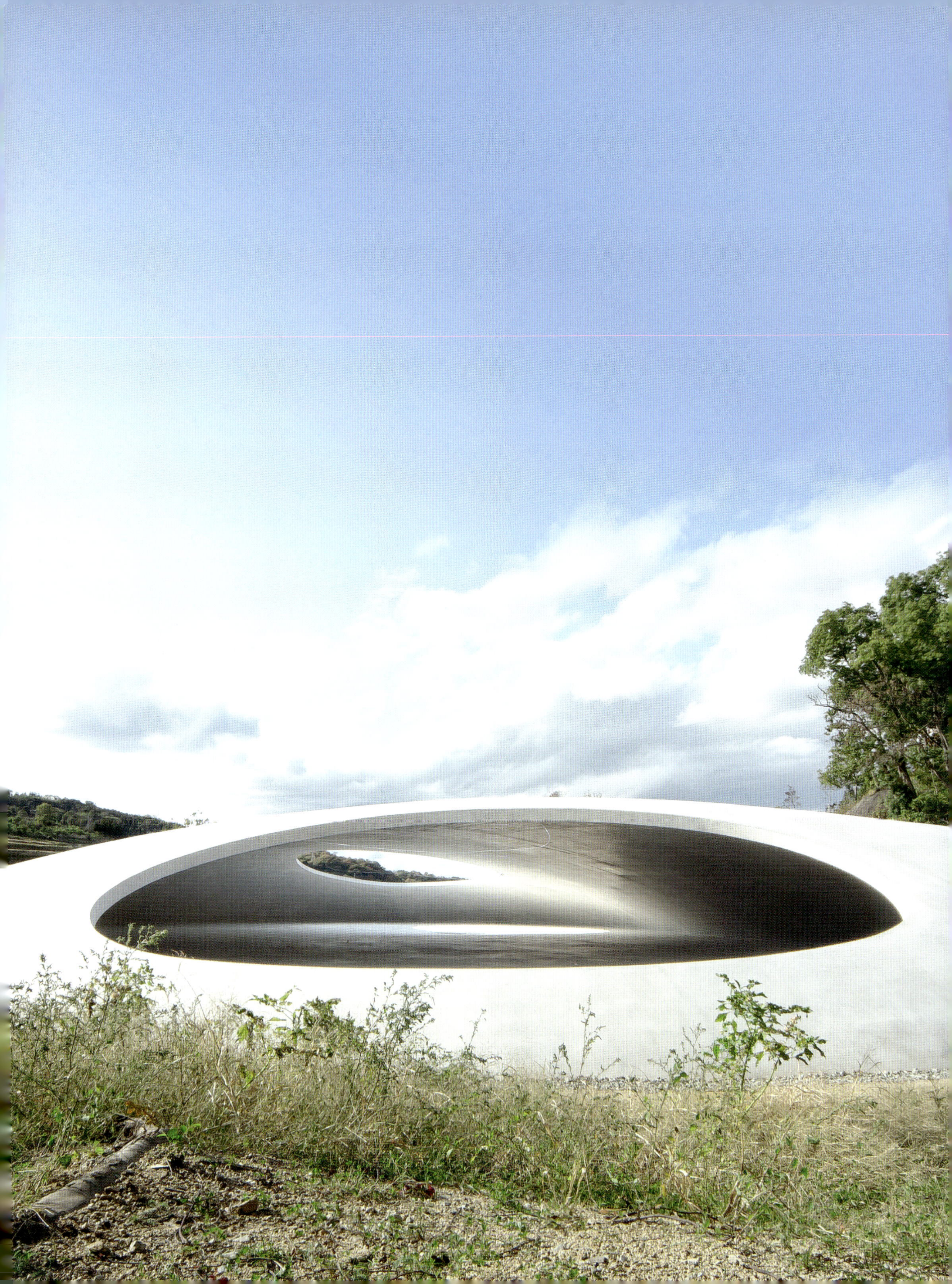

CHAPTER TWENTY-ONE

Structural design of free-curved RC shells

An overview of built works

Mutsuro Sasaki

After many twists and turns, shells developed into a main system for large-span structures. Almost five decades ago, during a significant point in their evolution, these systems became part of an architectural trend that sought to break the deadlock of functionalist architecture. As a result, Reinforced Concrete (RC) shell structures flourished throughout the world during the 1950s and 1960s. This time is often referred to as the era of Structural Expressionism – a period that generated new types of architectural spaces from structurally rational shells – and remembered as a peculiar page in the history of architecture. Notable architects and structural engineers such as Pier Luigi Nervi (1891–1979), Eduardo Torroja (1899–1961), Yoshikatsu Tsuboi (1907–1990), Félix Candela (1910–1997), Eero Saarinen (1910–1961) and Kenzo Tange (1913–2005) are key figures in this Structural Expressionist period.

With the exception of Heinz Isler's works, the construction of RC shells declined rapidly after the 1960s. These structures are rarely constructed today. Scarcity of skilled workers, rising prices for formwork, cost and schedule challenges, inefficiency of on-site fabrication, large deformations and deteriorations in the concrete, and a transition to steel gridshells to satisfy new demands are perhaps the primary causes of their decline. In the end, their potential for architectural expression faced its limits, and architects lost interest.

In the half century following the decline of shell construction, the circumstances in architecture drastically changed. Comparing only the capabilities of architectural expression between then and now, today is an entirely different age of architecture. Today's architectural expression is a child of its time – these characteristics are evident in architectural shapes. A particular trend in the international field of contemporary architectural design is the inclination to express informal three-dimensional shapes that possess free, complex, mutable, fluid and organic characteristics. The structural design of free-curved RC shells, the subject of this chapter, is one manifestation of this trend. In this chapter, the aforementioned trend is treated as a temporary variation in shell shapes from classical shells to contemporary shells, and a structural design method for free-curved RC shells is described as an example of contemporary shell design.

21.1 From classical to contemporary shells

Curved-surface shapes are at the core of shell structures. From a structural point of view, the principle of a form-resistant structure is inherent in the curvature of these shapes. The shells resist loads by in-plane stresses. In a state with no bending stress and only in-plane stress, the shell exhibits membrane behaviour. This state is an optimal and efficient stress state for structures: stresses are uniformly distributed through the shell's thickness, and the full cross section allows for effective resistance. Though the ideal membrane stress state is unrealizable, this state can nonetheless be obtained in nearly the entire shell surface, since bending stresses near the boundary sharply decline. Here, it goes without saying that the shell's boundaries are critical to achieving a state of membrane stress. Until now, only structures composed of various geometric forms have been studied as shapes of curved surfaces. Classical shells adopted curved surfaces that are geometrically traceable because this allowed easier analysis, simpler understanding of structural behaviour, and more convenient construction.

The RC shell shapes built before 1960 show distinctive changes throughout their history: starting from the enclosed and semi-spherical shell covering the 1932 Algeciras Market Hall by Manuel Sánchez Arcas and Eduardo Torroja, to the dynamic and open Hyperbolic Paraboloid (HP) surface of the 1959 Bacardi Rum Factory by Félix Candela (Fig. 21.1), to the formative and lively free-curved surface found in the 1961 TWA Flight Centre, by Eero Saarinen

Figure 21.1 The Bacardi Rum Factory with dynamic and open hyperbolic paraboloid surfaces, Mexico, 1959

Figure 21.2 TWA Flight Centre with a formative and lively free-curved surface, New York, 1961

and Associates (Fig. 21.2). Together with modern preferences, shell shapes have certainly evolved. These changes are interesting in light of the idea that architecture is an artistic expression of the spirit and the sense of the time, as there is an obvious correlation between the shapes of shells and the time period.

For contemporary RC shell developments, it is essential to reflect back on the works of pioneers such as Torroja and Candela, but it is not necessary to revert to the conventional shells in all aspects. Today's design environment – for example, the high-degree application of computers – has changed from that of half a century ago in many ways. In facing these changes in our time and environment, adhering to the past is not only against the fundamental human desire to create, but also shuts out the potential to open up the future. As a contemporary structural designer, I would like to challenge the norm of classical RC shell design.

First, do not always persist in geometric curved surfaces. Especially in light of the contemporary design aesthetics of structures, geometric shapes might cause the feeling of déjà vu to cutting-edge designers. I find enormous aesthetic potential in free-curved surfaces, which remain largely unexplored. Restraints imposed by structural analysis or construction techniques hardly exist. Highly figurative shells provide more than just an economically rational solution.

Second, do not always strive for the ideal membrane stress state. In order to design earthquake-resistant shells, it is inevitable to deal with bending stresses; shells designed for dynamic loads are fundamentally different from shells that consider only gravity loads. Depending on the scale of the structures, designs in which bending stresses are purposely allowed are necessary in some cases. With respect to preventing progressive collapses, redundant shells are also desired.

Incidentally, the TWA Flight Centre is a pseudo shell with a complex geometry of four elliptic paraboloid surfaces and boundary structures. Although not completely structurally rational, the structure is a formative free, RC shell. More specifically, the shell suggested a way to break through the barriers imposed by engineering and geometry restrictions, which tended to reduce the visual potential of shells. Compression-only is still the main concern for contemporary shells, but flexible designs that allow tensile and bending stresses to some degree are needed. Modern shells need to take full advantage of today's design environment and balance engineering knowledge with visual expression.

21.2 Shape design of free-curved RC shells

One clear example of an engineering shell design method approach that makes the best use of today's structural design environment is the shape design method of free-curved RC shells, which uses sensitivity analysis (Sasaki, 2005). The principle of the method is very simple: the rational shell shape is determined, with the help of computer technology, by minimizing strain energy (especially bending strain energy) in the shell (Ebata et al., 2003). In a sense, this shape-defining method uses the same processes as physical model experiments, but substitutes those processes with numerical algorithms.

The hanging membrane inverted to form a membrane shell by Heinz Isler (Fig. 20.13) is an expansion of Antoni Gaudí's inverted hanging arches (Fig. 4.6) to curved surfaces. The curved surfaces obtained by this method are free-curved surfaces, shaped only by the force of gravity. The well-known technique realizes optimal shell shapes with only compressive stresses acting under self-weight. The numerical shape analysis determines these shapes exactly and also does not require a craftsman's skill and time. The proposed method is effective as a modern design tool for the initial phase of a preliminary design.

Generally, shape generation is formulated as an optimization problem. These nonlinear problems can be divided into two problems: the local optimum solution problem and the global optimum solution problem. The sensitivity analysis method I adopted is a typical method suitable to search local optimum solutions. From a structural viewpoint, it is desirable to find a local optimum solution that is as close as possible to the initially conceived shape. This objective is quite intuitive, since in real architectural and structural design, the first thing that comes across one's mind is the desired image of a shape. However, imagined shapes are often not capable to carry loads to the foundation – these shapes will require theoretical modifications. By establishing an initial digital shape from the imagined curved surface in a study model, and then seeking an engineered shape close to the initial imagined shape, a structurally optimal form with minimum strain energy can be found with minimal modifications. In technical terms, for each node the gradient of the strain energy with respect to change in the vertical coordinate z is computed. The value of z is revised in the direction that will reduce the strain energy. The new strain energy is computed in FE analysis, and the procedure continues until there is no appreciable change in the strain energy after an iteration.

Using the method as a modification tool generally yields structurally complex curved forms that satisfy both architectural and structural constraints. The analysis starts from an initial shape in which architectural boundary conditions are accounted for. Therefore, the method is expected to work effectively as a practical tool in the preliminary design. Solutions of the sensitivity analysis depend heavily on the choice of the initial shapes. A comprehensive study is crucial to expand the designs by preparing multiple candidates for the initial forms. Additionally, to achieve a high design density, it is important to have a feedback loop until an architecturally and structurally satisfying shape is obtained. This loop allows for a comprehensive investigation of the acquired solution shapes, and if it becomes necessary, allows for re-establishing the initial shape and conducting the shape analysis literately. In either case, the ability to define in a short time desired free structural shapes, specifically at the preliminary design phase, is the advantage of this method, and it enables the design of modern RC shells.

Since 2000, several contemporary free-curved RC shells have been designed and realized worldwide,

Project, location	Design phase, construction phase (MM/YYYY)	Architect	Area (m²)	Span (m)	Thickness (cm)
Kitagata Community Centre, Gifu, Japan	05/2001–03/2004, 01/2004–10/2005	Arata Isozaki	4,495	25	15
The Island City Park 'Gringrin', Fukuoka, Japan	10/2002–11/2003, 03/2004–05/2005	Toyo Ito	5,040	70	40
Kakamigahara Crematorium, Gifu, Japan	05/2004–03/2005, 04/2005–05/2006	Toyo Ito	2,265	20	20
Rolex Learning Centre, Lausanne, Switzerland	01/2005–12/2009, 01/2007–11/2009	SANAA	39,000	80	40–80
Teshima Art Museum, Kagawa, Japan	09/2008–08/2009, 10/2009–07/2010	Ryue Nishizawa	2,040	43	25

Table 21.1 List of free-curved RC shells

using a theoretical shape design method called flux structure. While the geometry changed from an analytical surface to a free-curved surface, the design concept still strives for structural rationalism and expression of beauty. As of 2011, ten years after the original idea, five structural designs of free-curved RC shells have been realized through collaboration with architects including Arata Isozaki, Toyo Ito and SANAA (Kazuyo Sejima + Ryue Nishizawa). Table 21.1 and Figures 21.3–21.7 give more information about these works.

21.3 Teshima Art Museum

Using the most recent design, the Teshima Art Museum (see page 258), as a case study, this section details the structural design method of free-curved RC shells. Our focus lies on the structural planning (shape design), structural designing (stability study) and preliminary design planning (precision control) that characterize the structural design of the museum's shell roof. The main building of the Teshima Art Museum (Fig. 21.7) is a single-storey art gallery with an irregular elliptical footprint spanning 60m in its longitudinal direction and 43m in the other direction. The roof is a three-dimensional free-curved RC shell (with a maximum rise of 5.12m and a thickness 25cm). The roof has two openings, and the interior of the shell is a single, semi-covered exhibition hall. The shell's finish is architectural concrete. The main building itself is a part of the museum's art collection. Figure 21.8 show cross sections, plans and details of the reinforcement.

Figure 21.3 Kitagata Community Centre, Gifu, Japan, 2005

21.3.1 Structural planning

The concept of the roof is a 'water drop'. Based on this notion, the architect gave the roof its original curved-surface shape. Unfortunately, the original shape had little structural rationality. Without changing the image of the original shape, the curved surface was

Figure 21.4 Island City Park 'Gringrin', Fukuoka, 2005

Figure 21.5 Kakamigahara Crematorium, Gifu, 2006

modified, and then proper structural planning was initiated. The shape design method used sensitivity analysis to generate the structurally rational shape for the curved surface. This method captures the curved shape in the form of Non-Uniform Rational Basis Splines (NURBS), which reduces the number of unknown quantities whilst preserving a shape with a high level of freedom. Minimizing the strain energy across the whole shell surface under self-weight as an objective function and the coordinates of the NURBS control nodes as design variables, the surface shape is formulated as an optimization problem and solved by the gradient descent method to obtain a shell shape close to the architect's vision and with minimal strain energy. This approach generates a curved-surface shape with reasonable structural rationality while maintaining the architect's image and satisfying programmatic and visual requirements. The structure's layout is shown in Figure 21.9a.

The shell has a thickness of 25cm and a maximum rise of 5.12m. The results of the roof's shell shape derived from the shape design method are presented in Figure 21.9. Figure 21.9d shows the history of the total strain energy and the maximum vertical displacement under self-weight. Figure 21.9e is the change between the initial and the final shapes, and the distributions of vertical displacements throughout the evolution process are shown in Figure 21.10. Figures 21.9d and 21.10 show that the total strain energy and the vertical displacements get smaller throughout the process. Figure 21.9e shows the maximum displacement between the initial and the final shapes as approximately 400mm. According to these results, a rational shell shape is obtained without losing the original image of the curved surface.

21.3.2 Structural design

We skip the details about the general structural analysis (static stress and earthquake response analysis) of the

Figure 21.6 Rolex Learning Centre, Lausanne, 2010

Figure 21.7 Teshima Art Museum, Kagawa, 2010

shell because these analyses are conducted according to the Japanese code. Instead, we focus on the stability of the shell, which is specific to this structural design.

Specification of the safety factor

The stability evaluation formulae are defined as follows for static load and earthquake load respectively:

$$P_{cr} \geq \gamma P_L \qquad (21.1)$$

$$P_{cr} \geq P_L + \gamma P_E \qquad (21.2)$$

Herein, P_{cr} is the buckling load defined by the factors explained next, and γ, P_L and P_E are, respectively, a safety factor, the static load to the shell, and the earthquake load, which rarely occurs. The value of P_{cr} is defined by:

- concrete creep;
- cracks and rebar arrangements in concrete;
- initial imperfections;
- differences between the designed value and the construction of the shell thickness;
- eccentricity of the acting point of the resultant stress;
- plastification effect of concrete.

In a stability analysis, the computation of the buckling load of the shell, P_{cr}, and the specification method of the safety factor are of great importance. The safety factor γ is judged and defined comprehensively in two ways: the IASS recommendation for buckling of reinforced-concrete shells (Dulácska, 1981), and the limit state design method. We will not discuss the details closely here, but in this design, it has been confirmed that setting the safety factor to $\gamma = 3.0$ provides enough safety for both stationary and short-term load.

CHAPTER TWENTY-ONE: STRUCTURAL DESIGN OF FREE-CURVED RC SHELLS   265

Design load

Since the nonlinear buckling analysis by incremental analysis is used for the seismic stability analysis, the seismic force is applied as an equivalent static seismic force. Because of the shape in plan of the shell, calculating the static seismic force with natural mode analysis is too complicated. For that reason, the distribution of absolute acceleration response at the time when the focused response value shows the maximum in the time history response analysis is employed to calculate the static seismic force. Figure 21.11 shows the seismic intensity distributions of the static seismic force, which is the most important factor obtained for conducting a stability analysis.

Nonlinear stability analysis

Two numerical analysis methods, an abbreviated calculation method based on the IASS recommendation for buckling and an exact calculation method by complex nonlinear incremental analysis, are used to compute the buckling load considering plastic deformation P_{cr}. From the abbreviated method in

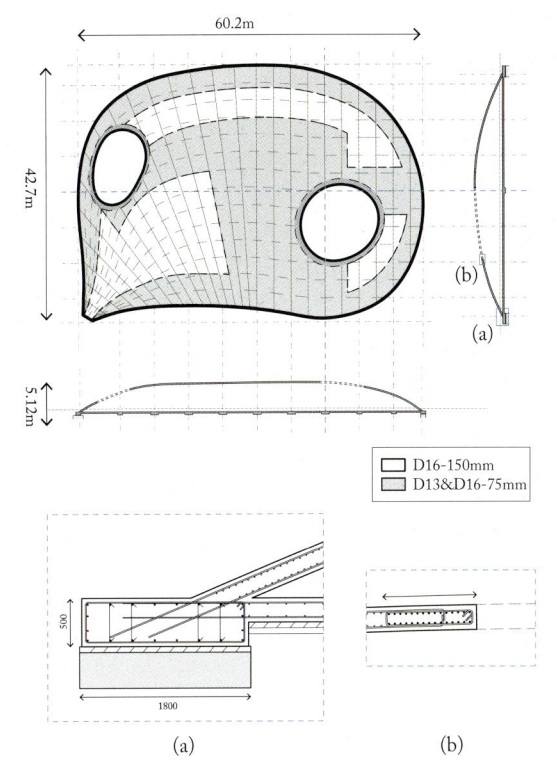

Figure 21.8 Arrangement of reinforcement, cross sections, and structural details of (a) the connection of the shell to the foundations, and (b) the edges of the openings in the shell

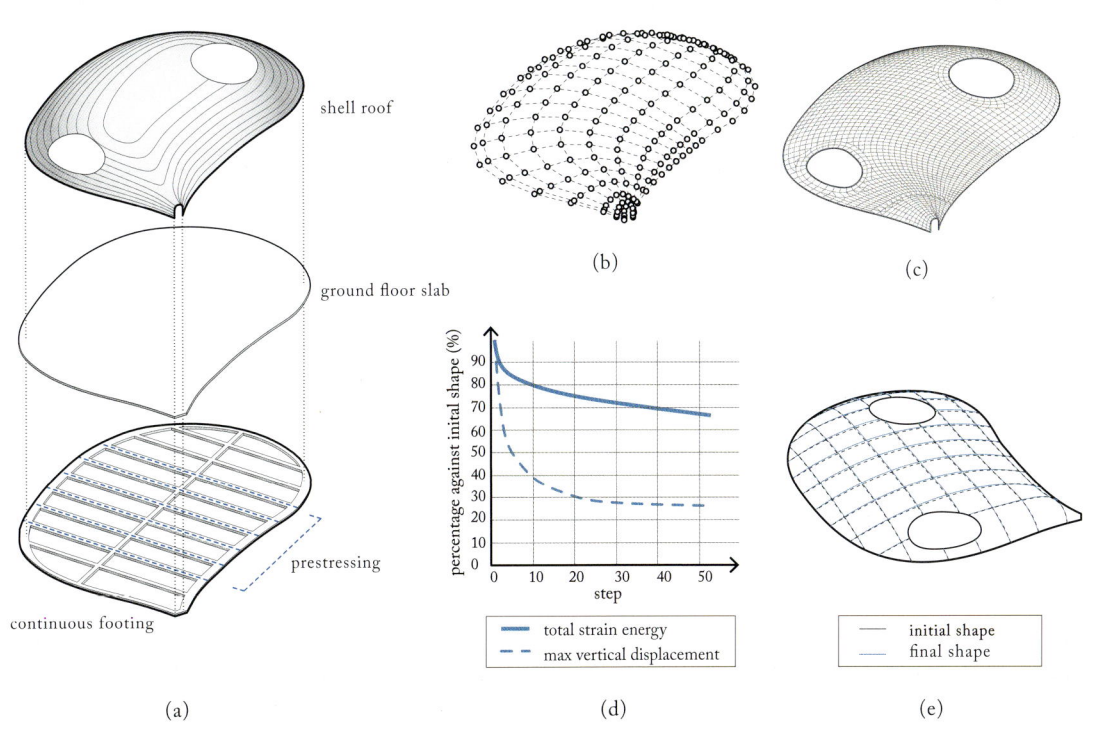

Figure 21.9 Teshima Art Museum, (a) Structural diagram, (b) control points of NURBS, (c) FE analytical model, (d) evolution histories and (e) changes in shape

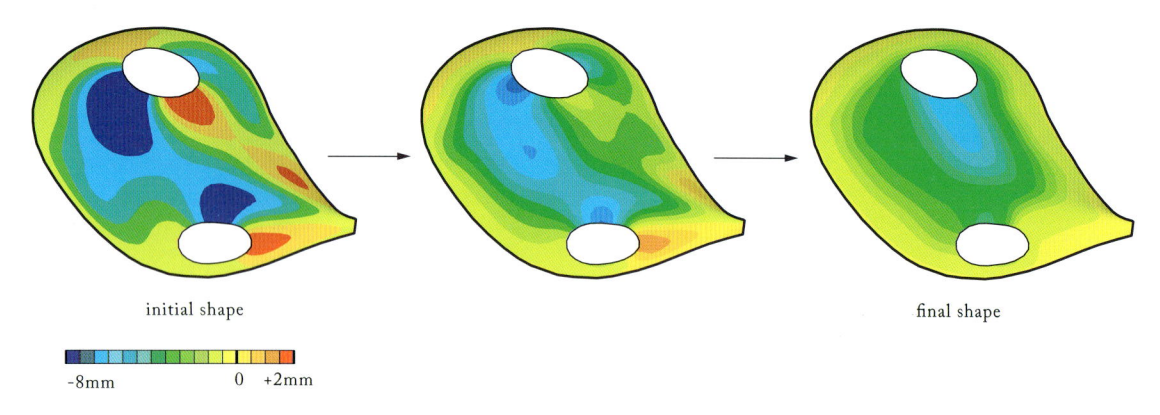

Figure 21.10 Distribution of vertical displacements through process

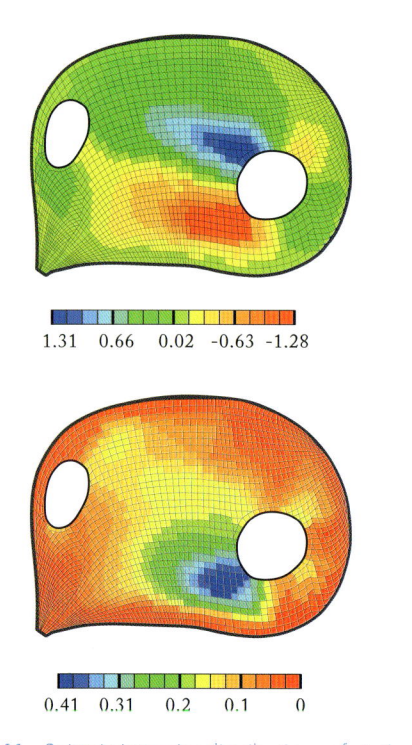

Figure 21.11 Seismic intensity distributions of static seismic force in horizontal and vertical direction

1. Assuming a construction tolerance of 5mm, define the thickness of the shell as 245mm.
2. Reinforced concrete is modelled as a monolithic material (reinforcements and concrete are not modelled separately).
3. The shell is divided into five layers over its thickness so that the plastic region progresses into the cross section.
4. 20mm of the maximum construction tolerance is assumed as the initial shape imperfection for stationary loading.
5. Imperfections by construction error and deformations under stationary load, including creep, are assumed as the initial shape imperfections for short-term loading.

General descriptions such as the material properties of reinforced concrete and yield curves are omitted here. Figure 21.12 shows the load-displacement curves at the maximally displaced points by static and short-term loading. In Figure 21.12a, a decrease in the load-bearing capacity related to geometric nonlinearities shows a major effect, and the shell is thought to collapse due to buckling phenomena and additional stress from large displacements. For the short-term loading, as plotted in Figure 21.12b, the rate of a decrease in the bearing capacity related to material nonlinearity is larger compared with geometric nonlinearity. The maximum compressive strain at the maximum bearing force gets up to 0.3%, which is almost equal to the ultimate strain value. As indicated in Figure 21.12, buckling loads in the static loading and short-term loading cases are $P_{cr} = 5.77 P_L$ and $P_{cr} = P_L + 3.20 P_E$, respectively.

the IASS recommendation (the rest of the details are omitted), the buckling load by stationary loading, $P_{cr} = 5.23 P_L$, and the buckling load by short-term loading, $P_{cr} = P_L + 3.14 P_E$, are derived. Hereafter, the exact calculation method by complex nonlinear analysis is described.

The buckling load is computed by the nonlinear incremental analysis, taking into account concrete creep, geometric nonlinearity and material nonlinearity. The overview of the modelling is as follows:

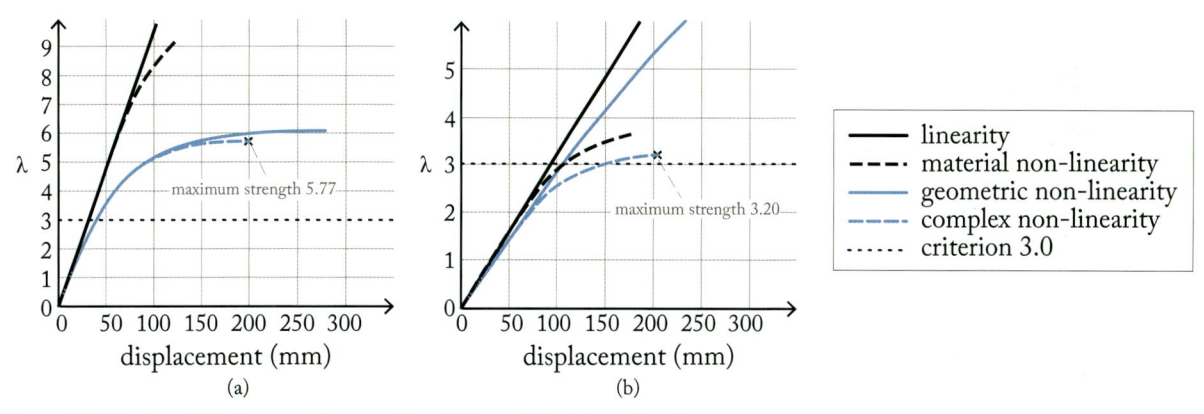

Figure 21.12 Load-displacement curves for static and short-term loading

Validation of stability

Both buckling loads, obtained by the IASS recommendation, denoted by (I), and the complex nonlinear analysis, denoted by (II), respectively, satisfy the stability evaluation formula for static load (21.1),

$$P_{cr} = 5.23 P_L \geq 3.0 P_L \quad \text{(I)} \quad (21.3)$$

$$P_{cr} = 5.77 P_L \geq 3.0 P_L \quad \text{(II)} \quad (21.4)$$

and earthquake load (short-term loading) according to equation (21.2),

$$P_{cr} = P_L + 3.14 P_E \geq P_L + 3.0 P_E \quad \text{(I)} \quad (21.5)$$

$$P_{cr} = P_L + 3.20 P_E \geq P_L + 3.0 P_E \quad \text{(II)} \quad (21.6)$$

Therefore, this shell has sufficient stability for static and earthquake loads. In addition, in both loading cases, buckling loads required by the IASS recommendation and the complex nonlinear analysis are in good agreement with each other.

21.3.3 Preliminary design planning

The shell's interior finish is architectural concrete, a smooth surface without traces of the formwork, seams or mould joints. Additionally, errors of concrete pouring were targeted to ±5mm for shell thickness and ±10mm for shell shape as objectives of construction quality control; hence, a precise construction control was required. Satisfying these two requirements was believed to be problematic with the use of conventional moulded slab formwork. Instead, an earth-fill concrete laying method was used. The shell was cast on an earth mound for support. The earth-fill was coated with mortar to produce a mould for a three-dimensional free-curved surface. All quality control targets were achieved with this method. The construction scheme is shown in Figure 21.13, and some construction scenes are shown in Figures 21.14 to 21.16.

21.4 Conclusion

In this chapter we focused on the holistic structural design of contemporary shells using design examples

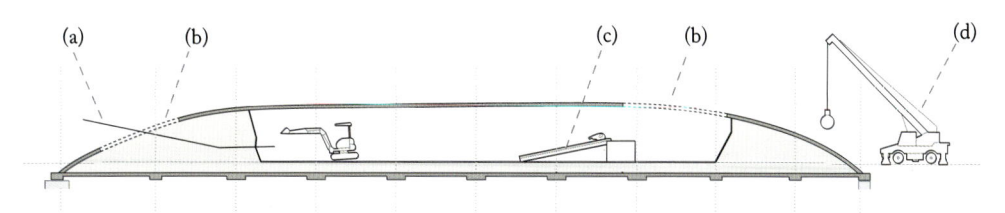

Figure 21.13 Construction scheme using an earth mound as formwork, with (a) a ventilation duct, (b) openings in the shell, (c) conveyor belt for transporting soil and (d) rough terrain crane for further soil removal

Figure 21.14 Formation of the earth-filled formwork

Figure 21.15 Shape measurement of the formwork

Figure 21.16 Concrete placement

a sensitivity analysis to obtain a free-curved form. By comparing their structural behaviour, interesting findings are obtained. The whole factory consists of six intersecting HP shells that have about 30m square planes. In this study, a single bay is analysed as the base unit of the building. Figure 21.17a shows the initial shape (HP surface). The shell has a global thickness of 40mm. The diagonal ribs vary in thickness from 140mm to 331mm. Only a quarter of the shell is analysed due to symmetry. Each corner has a pinned support. Self-weight and uniformly distributed load are applied as the load condition. Conditions such as volume, thickness (40–331mm), and height constraints, the final shape (free-curved surface), shown in Figure 21.17b, are obtained by applying the sensitivity analysis to minimize the evaluation function, strain energy, for two variables: shape and thickness.

From Figure 21.17a, the height has changed slightly (about 5cm) at the end of the free edge, but significant changes to the original HP shell are not

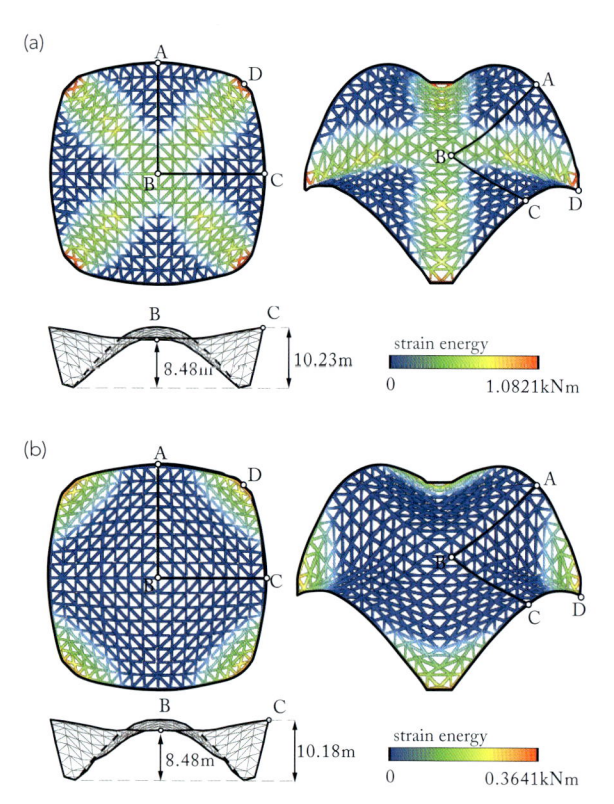

Figure 21.17 Distribution of shell thickness for the (a) initial shape (HP surface) with a range of 40–331mm, and (b) for the final shape (free-curved surface) with a range of 40–281mm

and methods. This included the shape design method for preliminary design planning, five examples of real works in which the method was applied, and discussions of the respective structural design methods.

To conclude, we turn to Candela's Bacardi Rum Factory. Taking its original hyperbolic parabolic shell (a geometric shape) as the initial form, we conduct

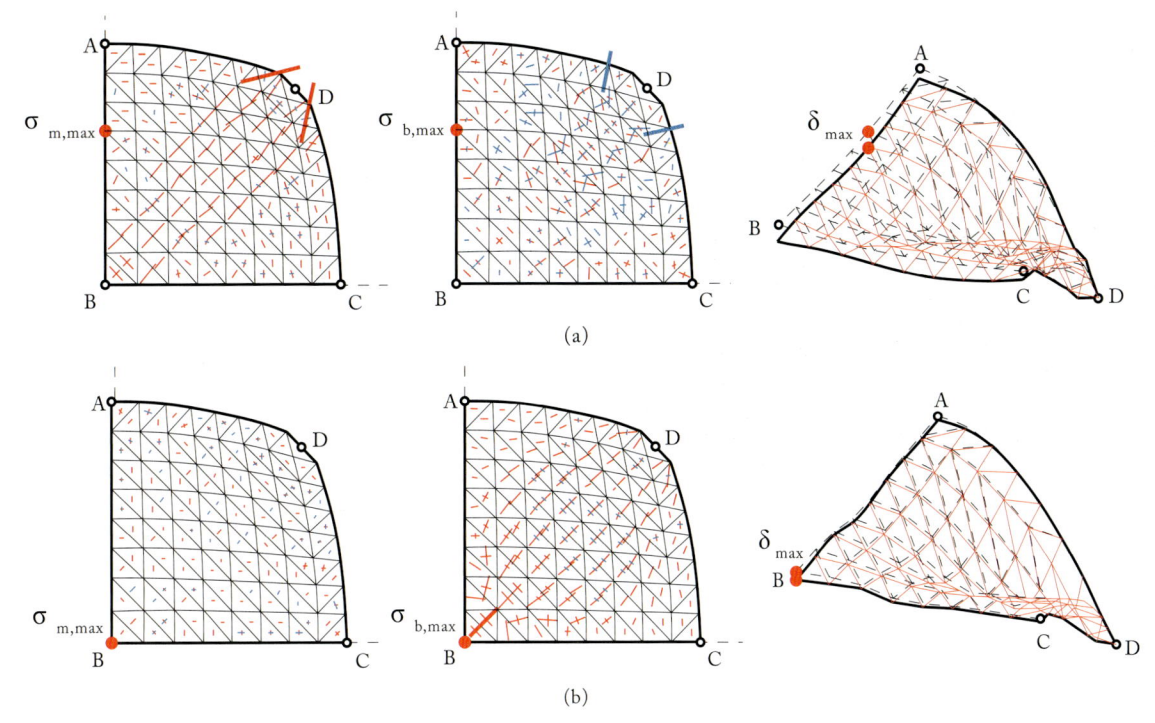

Figure 21.18 Membrane stress, bending stress, and deformation of the initial shape, for the (a) HP surface and (b) final free-curved shape

evident. These results show that the initial shape itself has strong structural rationality, and the sensitivity analysis method is a suitable method to search for the local optimal solution. However, as can be seen in Figure 21.17, the thicknesses of the two shapes show remarkable differences. In order to form plane and smooth membrane stress distributions over the entire shell, the diagonal ribs in the initial shape have disappeared, and the thickness of the shell has changed as if to reduce the self-weight of the top part. Only around the point supports it becomes thicker to carry concentrated stresses. The obtained shape is very natural when the load is limited to the gravitational forces and indicates that crossing arches seen in traditional intersecting shells are not necessarily needed.

In addition to the slight changes in the appearance, such as the curved surface shape and the distribution of thickness, as illustrated in Figure 21.18, considerable differences are observed in the distribution of principal stresses and displacements. Table 21.2 shows several structural response values of the two shapes.

These results show that by optimizing the shape and thickness of the initial shape (HP surface) to achieve minimum strain energy, the distribution of principal stresses and displacements of the final shape (free-curved surface) are improved. Needless to say, it does not mean the free-curved surface shape is structurally superior. Only gravity forces are considered to obtain the final form. The shape is also optimized for a plane membrane stress state. The initial form, which resists bending moment partly through the use of ribs, is more successful in resisting other types of load cases such as earthquakes. These also show that a change in shape or section can improve strain distributions.

	strain energy (kNm)	$\sigma_{m,max}$ (kNm^{-2})	$\sigma_{b,max}$ (kNm^{-2})	$\sigma_{m,mean}$ (kNm^{-2})	$\sigma_{b,mean}$ (kNm^{-2})	δ_{max} (mm)
Initial	1.0821	3,137.3	1,596.1	494.5	467.3	9.88
Final	0.3641	1,982.9	499.4	467.7	132.7	4.63

Table 21.2 Structural responses for initial (HP) and final (free-curved) surface, membrane stress σ_m, bending stress σ_b, vertical displacement δ

Conversely, a small construction error can cause correspondingly large defections.

The structural design of conventional free-curved RC shells relies not only on optimization techniques to arrive at a structurally rational solution, but also requires a holistic approach that uses modern design methods and encompasses all basic designs, detailed designs and construction works.

CONCLUSION

The congeniality of architecture and engineering

The future potential and relevance of shell structures in architecture

Patrik Schumacher

My interest in shell structures is part and parcel of a more general interest in advanced structural engineering and its capacity to handle, shape and exploit complex, differentiated geometries via relative optimization strategies. In structural engineering research – in the tradition pioneered by Frei Otto – I find an exciting coincidence of pursuits that is congenial to my architectural striving for a richly differentiated and clearly articulated architectural order. The emerging style that I try to contribute to and promote builds upon recent advances in engineering and needs the congenial contribution of sophisticated, creative engineers to achieve its global ambitions (Schumacher, 2009, 2012).

The premise of my contribution here is a double thesis that implies both the strictest demarcation and the closest collaboration between architecture and engineering as preconditions for the productive advancement of the built environment. The underlying division of labour might be posited as follows: architecture is responsible for the built environment's social performance. Engineering is responsible for the built environment's technical performance. Technical performance is a basic precondition of social performance. In this sense engineering might be argued to be primary. Social performance is the goal. In this sense architecture might be argued to be primary. Thus the relation cannot be brought into a hierarchy. Rather it is a relation of mutual dependency and dialectical advancement. Architectural goals must be defined within a technically delimited space of possibilities. Engineering research and development thus expands the universe of possibilities that constrains architectural invention. However, it cannot be taken for granted that engineering research and development expands the universe of possibilities in relevant, desired directions without being prompted and inspired by architectural goals. In turn, architectural goals and inventions might be prompted and inspired by recent engineering advances. The two disciplines co-evolve in mutual adaptation. Although there can be no doubt that architecture remains a discourse that is distinct from engineering, a close collaboration with the engineering discipline's as well as the architect's acquisition of reliable intuitions about their respective logics are increasingly important conditions for the design of contemporary high-performance built environments.

Shell structures are among my favourite devices to differentiate and articulate spatial compositions. The convex curvature of the shells makes spatial units easily recognizable and traceable, even if such spatial units proliferate and interpenetrate in complex arrangements. While intersecting rectangles soon produce an undecipherable cacophony of corners, intersecting shells remain perceptually tractable. Their size is indicated by their height, and locally by the degree of curvature. While straight walls are mute with respect to whether one is inside or outside of a certain territory, in the case of shells the distinction of concave and convex clarifies one's relation to the spatial unit in question. Thus shell configurations are in many ways conducive to the ordered, legible build-up of organizational complexity. Shape-optimized shell configurations are inherently information-rich artefacts that give clues allowing for local-to-global inferences as well as for

Figure 22.1 Field of Domes, exterior view of the commissioned design, 2012, by Zaha Hadid Architects

Figure 22.2 Banquette Hall, interior view of the commissioned design, 2012, by Zaha Hadid Architects

local-to-local inferences. This is a direct consequence of the calculative information processing that has generated their final form.

In my design work I am now trying more and more to move away from the freeform play with complex curvature towards the disciplined use of structural form-finding algorithms. The increased computational power to handle complex formations makes this possible. The lightness that can be achieved with structural optimization in general and with shells in particular is a much appreciated factor here. However, the fact that I prefer a structurally constrained search-space to an unconstrained one is not only motivated by technical efficiency considerations or by the achievement of relative lightness. What motivates me here as well is the morphological coherence that comes with rule-based or law-governed design logics. This allows me to give a unifying character to a morphological world I choose for a particular project or part of a project. These unifying rules give a character and identity that remains recognizable despite the rich differentiation that remains available within these logics. I am not only talking about the overall spatial forms here but also include sub-articulations such as grids (gridshells), ribs (ribbed vaults), perforations and tessellations. All these sub-articulations are driven by scripted rules that include structural logics, fabrication logics or environmental logics. Here, the collaboration between engineering and design (tectonic articulation) takes the form of feeding data that come from engineering simulations into geometric responses that serve as articulating patterns, that then deliver further character enhancements and orienting information. The rule-based form-production that computes results from multi-variable inputs in turn allows – at least in principle – for inferences to be drawn in the inverse direction: from the resultant output variables to the input variables. An intuitive grasp of the embodied logic of such rule-based spatial formations gives users a sense of orientation and successful, intuitive navigation.

It is this search for coincidences between technical and communicative morphologies where the key project of tectonic articulation resides.

Why is this important? The ability to navigate dense and complex urban environments is an important aspect of our overall productivity today. Post-Fordist network society demands that we keep continuously connected and informed. We cannot afford to beaver away in isolation when innovation accelerates all around. In order to remain relevant and productive we need to network all the time and coordinate our efforts with what everybody else is doing. Everything must communicate with everything. The speed and confidence with which one can make new experiences and meaningful connections is decisive. The design of environments that facilitate such hyper-connectivity must be very dense and complex and yet highly ordered and legible. As urban complexity and density increase, effective articulation becomes more important. Computational techniques and attendant formal–spatial repertoire are maturing, allowing us to build up and order unprecedented levels of spatial and morphological complexity.

APPENDIX A

The finite element method in a nutshell

Chris Williams

The finite element method *approximates* a continuum with a finite number of finite-sized elements. The larger the number of elements, the more accurate the result, unless there are so many elements that numerical errors build up. The elements can have any number of sides or faces: triangles, quadrilaterals and so on in two dimensions, and usually tetrahedra or hexahedra in three dimensions. A cube is an example of a hexahedron, a polyhedron with six faces.

Shells are represented by two-dimensional surface elements in three dimensions using curved triangles or quadrilaterals. A gridshell might be modelled with one-dimensional line or 'beam' elements.

Adjacent elements share nodes. For example, each triangle in Figure A.1 has six nodes and each pair of neighbouring triangles share three nodes. This is sufficient for the elements to fit together with no gaps along their curved edges. However, for the bending theory of plates and shells we not only want the elements to fit together, but we also want there to be *no kinks or folds between the elements.*

This is exactly the problem solved in computer-aided design by using techniques such as biquadratic or bicubic B-splines, NURBS or subdivision surfaces, and these methods are now being employed in the finite element method as an alternative to the more traditional plate and shell elements.

Triangular elements might just have three nodes, one at each corner, in which case the triangle is flat with straight sides. Alternatively one might have more than six nodes, ten would be the next obvious number, one more on each edge and one in the middle, away from all the edges. Sometimes nodes can 'merge', in

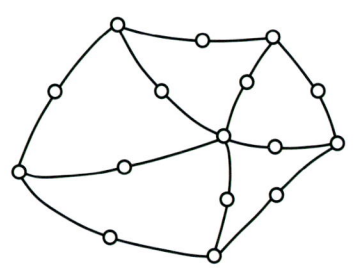

Figure A.1 Curvilinear triangular elements with six nodes per element. Adjacent elements share three nodes

which case they can specify not just a position, but also an orientation; for example, to define the tangent plane to a surface. There are advantages and disadvantages in including orientations, and therefore rotations, in a finite element analysis of a shell. Rotations have to be included to transfer moment between a shell and its supports, but buckling may involve large rotations which some computer programs do not handle well.

In three dimensions the continuum may be a solid or fluid and it may be stationary or moving. In fact there is no such thing as a continuum since all materials; for example, air, water, steel and sand are made from individual molecules or grains. Even though there might be the possibility of computer modelling every grain of sand in a small sample, this is never possible with molecules – one cubic millimetre of air contains 300×10^{15} molecules, way beyond any conceivable computer.

The finite element method started with the analysis of structures, but has been extended to all sorts of problems in engineering and science. The fundamental ideas behind the finite element method are very simple, but the details are quite complicated so it is easily possible to get needlessly confused.

A.1 The virtual work theorem

In structural analysis we have only three types of equation:

- **Equilibrium.** Every infinitesimal part of a continuum obeys Newton's second law, $F = ma$, resultant force equals mass times acceleration. In statics we have simply $F = 0$.
- **Compatibility.** The deformation or strain is determined by the relative displacement of different parts of the material.
- **Constitutive.** The stress within a material is determined by its strain and possibly strain rate, past history of strain, temperature and so on. If a ductile material has yielded the stress will depend upon the past plastic strain.

These three equations can be combined in various ways, often using work-like concepts such as strain energy, complementary strain energy, virtual work and so on. Of these virtual work is perhaps the most useful since it does not require the material to be elastic or the loads to be associated with a potential such as gravitational potential energy.

The virtual work theorem is always the same:

Sum of internal 'stresses' times increments of 'strain' = sum of external 'loads' times increments of 'displacements'.

The internal 'stresses' include stresses, tensions and moments and the corresponding increments of 'strain' are increments of strain, elongation and curvature. External 'loads' include applied loads, but also support reactions, both forces and moments. Increments of 'displacements' include increments of rotation.

The sums may be replaced by integrals, as appropriate.

In dynamics the external 'loads' include inertia forces acting in the *opposite direction to the acceleration*.

The virtual work equation is proved starting only from the equilibrium equations and the compatibility equations. It therefore always applies, regardless of the material properties. Corresponding 'stress or force-like' and 'increment of strain or displacement-like' quantities are chosen such that their product is a 'work-like' quantity. The proof is very short but rather mathematical and involves the use of Green's theorem – essentially multiplying stresses by strain increments and then using a sort of 'integrating by parts' in three dimensions. A version of the proof is given in Section A.9.

The internal 'stresses' are in equilibrium with the external 'loads' and they form the *equilibrium set*.

The increments of 'strain' are compatible with the increments of the 'displacements' and they form the *compatibility set*.

We have, perhaps rather pedantically, always talked of 'increments' of displacement and strain. This is because a structure could completely change its shape and the virtual work equation would then apply to each increment of displacement. In French the virtual work principal is called *le principe des puissances virtuelles* (virtual power) which is far more logical because an increment of displacement is caused by a velocity times an increment of time.

If displacements are small, then we no longer have to use increments of displacement, we can use the whole displacement.

The reason why virtual work is so useful is that there is absolutely no requirement that the equilibrium set and the compatibility set exist simultaneously. Hence the 'virtual'. The equilibrium set satisfy the equilibrium equations and the compatibility set satisfy the compatibility equations, but the equations are independent and both apply provided that the geometry of the structure is the same – hence the *increment* of displacement.

We have described the use of the equilibrium and compatibility equations to prove the virtual work theorem; however, *the proof can be reversed to use virtual work and the compatibility equations to prove the equilibrium equations*. This is done using the idea that the virtual work equation applies for *any* virtual increment of displacement field (or virtual velocity field).

Thus the application of the virtual work theorem to the finite element method involves:

- virtual work;
- compatibility equations;
- constitutive equations.

The equilibrium equations are not used at all.

Perhaps the most confusing aspect of the finite element method is that similar and sometimes identical formulations can be obtained using other techniques, in particular Galerkin's method (Boris Galerkin, 1871–1945), which does use a weighted form of the equilibrium equations, and also Lagrangian mechanics which avoid the equilibrium equations in a similar way to virtual work, but use 'real' kinetic energy, strain energy and gravitational potential energy.

Again rather confusingly, if one is applying virtual work to an elastic material (linear or nonlinear) then the stress–strain relationships should be written in terms of derivatives of the strain energy.

A.2 Shape functions, degrees of freedom and generalized coordinates

The finite element method is based upon interpolation to define the geometry of the structure at points other than the nodes. As we have already noted, this is exactly the same problem as in computer-aided design, and that is why techniques such as NURBS and subdivision surfaces can be used as finite elements for shells.

The term 'shape function' is used for the functions that are used for interpolation. The shape functions are controlled by the values of the degrees of freedom, displacements and rotations for solids. The degrees of freedom are associated with the nodes and hence adjacent elements share some of their degrees of freedom.

It is logical to use the same functions for interpolating the displacement of a structure as for interpolating its initial shape. Such an approach is described as *isoparametric*.

A.3 Finite element formulation

People usually first learn the finite element method as applied to linear elastic structures. They then have to 'unlearn' quite large parts of this to apply the method to nonlinear problems such as shell buckling. This is very confusing and so we will avoid making the linear assumption.

Real structures move with time and each bit of the structure will accelerate according to the resultant force upon it. However, we will not allow each bit of the structure that freedom, we will only allow the finite number of degrees of freedom to change with time, hence the approximation.

At any given time the degrees of freedom will have certain values and hence we can find the current strain at all points in the structure using the shape functions. We can then use the constitutive relations to find the current stress, which may also depend upon the current strain rate and the past history of strain.

We can also use the shape functions to find the virtual strain increment associated with a virtual increment of any one of the degrees of freedom.

We can integrate the real current stress multiplied by the virtual strain increment over the elements which share the degree of freedom undergoing the virtual increment. This gives us the nodal load or moment necessary to 'balance' the internal stresses. There are no real nodal loads or moments, but they are a useful concept in evaluating the different contributions to the virtual work equation.

We can also integrate the real load applied to the structure multiplied by the virtual displacements. This gives us the 'actual' nodal loads.

The difference between an actual nodal load and the load required to balance the internal stresses is a resultant nodal force or moment which causes the structure to accelerate.

If we had included the inertia forces due to acceleration in with the real loads, then the resultant would be zero; in other words, the virtual work equation would be satisfied.

A.4 The mass matrix

So the question now is 'how can we find the nodal accelerations and angular accelerations necessary to balance the virtual work equation?' The simplest answer is to assume that the mass of the structure is lumped at the nodes, in which case the nodal acceleration is just the resultant force divided by the mass. For moments and rotations we would have a lumped moment of inertia.

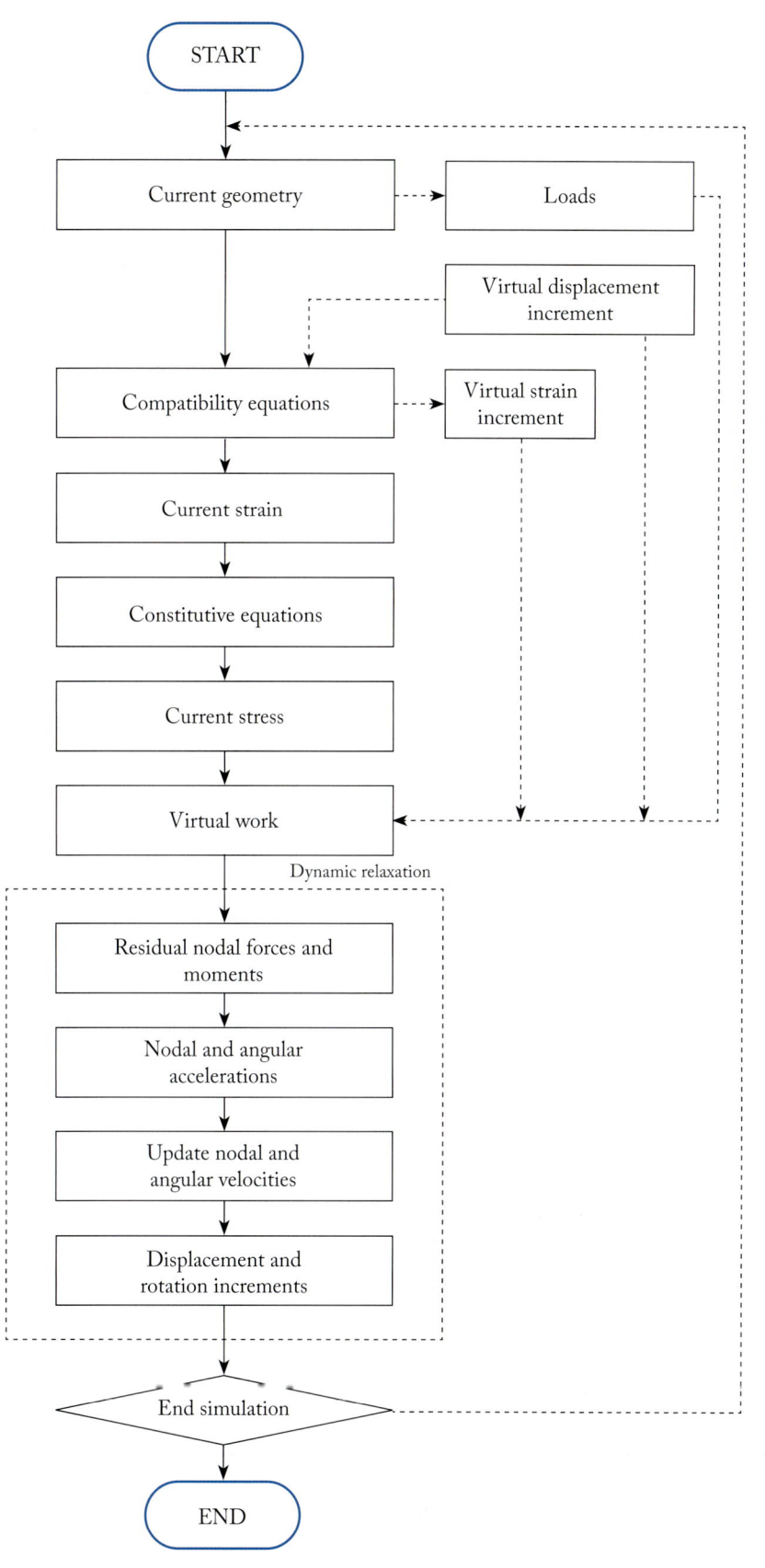

APPENDIX A: THE FINITE ELEMENT METHOD IN A NUTSHELL 277

Figure A.2 Flowchart for explicit dynamic analysis of an elastic structure with lumped masses using Verlet integration or dynamic relaxation

However, the mass of a structure is not lumped at the nodes and so acceleration at one node contributes to the resultant force and moment of its neighbours via the shape functions. This effect can be included via a 'consistent' mass matrix which would have to be inverted to find the nodal accelerations. We have used virtual work in our discussion; however, if the total real kinetic energy is T, then the elements of the mass matrix are

$$M_{ij} = \frac{\partial^2 T}{\partial \dot{q}_i \partial \dot{q}_j}, \quad (A.1)$$

where $\dot{q}_i$ is the rate of change of the degree of freedom q_i.

However, the approximation caused by assuming lumped masses is small and reduces if the size of the elements is reduced and therefore inversion of the consistent mass matrix is not necessary.

A.5 Verlet integration

Verlet integration (which is also known by a number of other names) is essentially the same as dynamic relaxation, the only difference being that Verlet is intended for dynamic problems whereas dynamic relaxation is more associated with the solution of a static problem by considering a damped dynamic problem. Thus one might use dynamic relaxation to solve the static problem of the buckling of a shell.

They are both explicit methods that step through time using the finite element method (or something equivalent) to calculate accelerations and hence the change in velocity in a certain time interval. The new values of the degrees of freedom are then found using the new velocities multiplied by the time interval.

A.6 The entire process

Figure A.2 is a flowchart of the entire process for explicit dynamic analysis of an elastic structure with lumped masses using Verlet integration or dynamic relaxation. It can be seen that the compatibility equations and the virtual work theorem lie at the core of the process and that the equilibrium equations do not appear at all. The compatibility equations are needed to calculate the current strain and also the virtual strain increments due to the virtual displacement increments.

The 'Stress at some initial strain state' is included because the structure may have some initial prestress or we might be modelling a soap film in which the stress state is independent of strain.

Strains include changes of curvature as well as membrane strains and they are both calculated from the coefficients of the first and second fundamental forms described in Appendix B. Strain (and curvature) increments are calculated from increments of the coefficients of the first and second fundamental forms. Usually we do not need the Christoffel symbols, unless we are interested in changes in the in-plane (geodesic) curvature of gridshell members. Also we do not use Gauss's Theorema Egregium and Peterson–Mainardi–Codazzi equations because we are calculating the coefficients of the first and second fundamental forms directly from the current nodal coordinates and rotations.

It is often not possible to do the integrations required for the application of the virtual work theorem analytically and instead they are done numerically using Gaussian quadrature in which the values at Gauss points are weighted and added.

Note that the stiffness matrix (Section A.7) does not appear at all in the formulation for nonlinear explicit analysis. We shall see in the next section that the elastic stiffness matrix involves the second derivative of the strain energy (or the equivalent in virtual work terms). Explicit analysis only requires the equivalent of the first derivative of the strain energy, making the whole process much simpler and also avoiding the storage and manipulation of large matrices.

The more complicated and nonlinear the analysis, the more it makes sense to use explicit methods rather than implicit methods involving the elastic stiffness matrix.

A.7 The elastic stiffness matrix

The stiffness matrix is at the heart of the finite element method for linear elastic problems. We are discussing it here for completeness, even though explicit methods are often preferable.

For a linear elastic structure

$$\mathbf{p} = \mathbf{Kq}, \quad (A.2)$$

where **q** is a column matrix of degrees of freedom, **p** is a column matrix of loads and **K** is the square stiffness matrix. For nonlinear structures the stiffness matrix gives the *change* in loads caused by a *change* in displacements.

The elements of the stiffness matrix are

$$K_{ij} = \frac{\partial^2 U}{\partial q_i \partial q_j}, \quad (A.3)$$

where U is the total strain energy. This equation applies for all elastic structures, but for a linear elastic structure the elements K_{ij} are constant, they do not change as the structure deflects.

It makes no difference whether the stiffness matrix is derived by considerations of virtual work or strain energy, the result is the same.

For nonlinear structures we have to establish out-of-balance forces using the first derivatives of the strain energy (or virtual work if one prefers) since the second derivatives in the stiffness matrix only tell us about the changes in the load. We can approach static equilibrium bit by bit using an implicit method such as Newton–Raphson which requires fewer steps than dynamic relaxation, but each step requires more computation.

A.8 The geometric stiffness matrix

Geometric stiffness is what gives a violin string its resistance to lateral movement. It is called geometric stiffness because it depends upon the geometry of a structure and the state of stress within it, but *not* the elastic properties of the material.

Compression causes negative geometric stiffness and reduction in stability, possibly leading to buckling.

If one makes assumptions about linear behaviour, buckling loads can be found using eigenvalue–eigenvector analysis. Shell buckling is usually nonlinear so eigenvalue techniques are of questionable value and may be very unsafe.

Our discussion of shape functions and virtual work makes no assumptions of linear behaviour and automatically includes the effect of geometric stiffness via the change to the virtual work equation as the structure changes shape. Thus one could derive the linear eigenvalue buckling equations from the fully nonlinear equations, but there does not seem much point. Similarly there does not seem much point in starting with linear equations and then struggling to add in nonlinear bits as best one can.

A.9 Proof of the virtual work theorem

The following proof is in two dimensions. The three-dimensional version is identical, it just has more terms. We start with the compatibility equations. u and v are *increments* of displacement and ε_x, ε_y and γ_{xy} are *increments* of strain. It would be better to follow the French, in which case u and v would be velocities and ε_x, ε_y and γ_{xy} would be strain rates. Either way the compatibility equations are

$$\varepsilon_x = \frac{\partial u}{\partial x} \quad (A.4)$$

$$\varepsilon_y = \frac{\partial v}{\partial y} \quad (A.5)$$

$$\gamma_{xy} = \gamma_{yx} = \frac{1}{2}\left(\frac{\partial v}{\partial x} + \frac{\partial u}{\partial y}\right) \quad (A.6)$$

$$\omega_{xy} = -\omega_{yx} = \frac{1}{2}\left(\frac{\partial v}{\partial x} - \frac{\partial u}{\partial y}\right). \quad (A.7)$$

Note the 1/2 in equations (A.6) in the definition of the increment of shear strain γ_{xy}. This is the 'mathematical' definition of increment of shear strain, the 'engineering' definition does not have the 1/2. Many formulae are more logical using the mathematical definition. The engineering equivalent of $(\tau_{xy}\gamma_{xy} + \tau_{yx}\gamma_{yx})$ is $\tau_{xy}\gamma_{xy}$. The variable ω_{xy} is the increment of average rotation about the z axis, or vorticity if one is using velocities.

The equilibrium equations are

$$\frac{\partial \sigma_x}{\partial x} + \frac{\partial \tau_{yx}}{\partial y} + p_x = 0 \quad (A.8)$$

$$\frac{\partial \tau_{xy}}{\partial x} + \frac{\partial \sigma_y}{\partial y} + p_y = 0 \quad (A.9)$$

$$\tau_{xy} = \tau_{yx} \quad (A.10)$$

in which p_x and p_y are the body forces per unit area, including self-weight and inertia forces.

We now need to do a bit of manipulation using both the compatibility equations and the equilibrium equations:

$$\sigma_x \varepsilon_x + \tau_{xy}\gamma_{xy} + \tau_{yx}\gamma_{yx} + \sigma_y \varepsilon_y$$

$$= \sigma_x \varepsilon_x + \tau_{xy}(\gamma_{xy} + \omega_{xy}) + \tau_{yx}(\gamma_{yx} + \omega_{yx}) + \sigma_y \varepsilon_y$$

$$= \sigma_x \frac{\partial u}{\partial x} + \tau_{xy}\frac{\partial v}{\partial x} + \tau_{yx}\frac{\partial u}{\partial y} + \sigma_y \frac{\partial v}{\partial y}$$

$$= \frac{\partial}{\partial x}(\sigma_x u + \tau_{xy} v) + \frac{\partial}{\partial y}(\sigma_y v + \tau_{yx} u)$$

$$- \left(\frac{\partial \sigma_x}{\partial x} + \frac{\partial \tau_{yx}}{\partial y}\right)u - \left(\frac{\partial \tau_{xy}}{\partial x} - \frac{\partial \sigma_y}{\partial y}\right)v$$

$$= \frac{\partial}{\partial x}(\sigma_x u + \tau_{xy} v) + \frac{\partial}{\partial y}(\sigma_y v + \tau_{yx} u) + p_x u + p_y v. \quad (A.11)$$

If we integrate over an area A, we have

$$\int_A (\sigma_x \varepsilon_x + \tau_{xy}\gamma_{xy} + \tau_{yx}\gamma_{yx} + \sigma_y \varepsilon_y)dA$$
$$= \oint_{\partial A} (\sigma_x u + \tau_{xy} v)dy + \oint_{\partial A} (\sigma_y v + \tau_{yx} u)dx$$
$$+ \int_A (p_x u + p_y v)dA, \quad (A.12)$$

which is the virtual work equation. We have used Green's theorem (much the same as integration by parts) to obtain the boundary integrals

$$\oint_{\partial A} (\sigma_x u + \tau_{xy} v)dy + \oint_{\partial A} (\sigma_y v + \tau_{yx} u)dx, \quad (A.13)$$

which is the increment of virtual work done by the stresses applied to the boundary. The increment of virtual work done by the body forces is

$$\int_A (p_x u + p_y v)dA. \quad (A.14)$$

Note that we could run the proof backwards to prove the equilibrium equations starting from virtual work and the compatibility equations. To do this we have to note that the equations apply for *any* virtual increment of displacement field.

Further reading

- *History of Strength of Materials*, Timoshenko (1953). This book is too early to include the finite element method, but it is of interest in showing how the theory developed up to the introduction of the finite element method, including virtual work and Castigliano's theorems. It does, however, include the finite difference and Rayleigh–Ritz methods which are precursors of the finite element method.
- 'Some historic comments on finite elements', Oden (1987). This paper gives a fascinating historical account of the development of the finite element method.

APPENDIX B

Differential geometry and shell theory

Chris Williams

Differential geometry is the study of curved things – lines, surfaces and space-time curved by mass and stress to produce gravity. The question is whether a person who is designing a curved object will benefit from knowing anything about the theory of their geometry. In truth, most shell structures are designed by people who know nothing about geometry and they are still perfectly good, safe structures.

However, there are times when geometric insight may add something to a design, perhaps to use a geodesic coordinate system for a pattern of stones in a vault or for the fabric's cutting pattern of a tent structure.

The following is a brief introduction to all the important aspects of differential geometry, as far as shell structures are concerned. It is relatively short so that the reader can see that learning the subject is not an insuperable task. There are many excellent books on the subject, but some of the more mathematical and abstract ones are difficult. Here, we have used tensor notation as in Green and Zerna (1968), mainly because the tensor notation is ideal for writing structural equations as well as geometric relationships. 'Elementary' books, such as those by Struik (1988) and Eisenhart (1909), are more accessible because they do not use tensors, but they do not make the connection to shell theory as Green and Zerna do. The tensor notation is indispensible for the general theory of relativity, and Dirac (1975) is only sixty-nine pages long. Shell theory and the general theory of relativity are closely linked because they are both concerned with curvature and stress. Mass curves space-time through the stress–energy–momentum tensor, which has four principal values, density and the three principal stresses. Force is non-dimensional in the general theory, which means that there is an absolute measure of force, not relying on arbitrary units such as kilogrammes, metres and seconds.

A general surface can be written in parametric form as

$$\mathbf{r}(u,v) = x(u,v)\mathbf{i} + y(u,v)\mathbf{j} + z(u,v)\mathbf{k}, \quad (B.1)$$

in which u and v are surface parameters or surface coordinates, $\mathbf{r}$ is a position vector $\mathbf{i}$, $\mathbf{j}$ and $\mathbf{k}$ are unit vectors in the directions of the Cartesian x, y, z axes. Thus, for example, the sphere $x^2 + y^2 + z^2 = R^2$ can be written as

$$x = R \cos u \cos v, \quad (B.2)$$

$$y = R \sin u \cos v, \quad (B.3)$$

$$z = R \sin v. \quad (B.4)$$

In the case of the Earth, u would be the longitude and v the latitude. Clearly it is easier to specify where we are using latitude and longitude plus height rather than Cartesian coordinates with the origin at the centre of the Earth.

We noted in Appendix A that the finite element method starts with the geometric problem of interpolation between the nodes. Each finite element would have its own set of equations relating x, y and z to u and v. The elements need continuity along their edges and providing continuity of slope is rather tricky. In the parlance of computer-aided design, shell elements would be described as 'surface patches'. One could argue that it is rather inelegant to represent a surface by a patchwork of elements, and if possible one would use one mathematical description for the whole surface.

The same surface (or element or patch) can be represented in different parametric forms. Thus

$$\mathbf{r} = a\cos u \cosh v \mathbf{i} + b\sin u \cosh v \mathbf{j} + c\sinh v \mathbf{k} \quad (B.5)$$

and

$$\mathbf{r} = a\cos(u-v)\sec(u+v)\mathbf{i}$$
$$+ b\sin(u-v)\sec(u+v)\mathbf{j} + c\tan(u+v)\mathbf{k} \quad (B.6)$$

are two different parametric forms for the hyperboloid of one sheet (cooling tower shape)

$$\frac{x^2}{a^2} + \frac{y^2}{b^2} - \frac{z^2}{c^2} = 1. \quad (B.7)$$

Many books use the parameters u and v, but there are compelling reasons (which will gradually become apparent) for using parameters with superscripts, say u^1 and u^2, or θ^1 and θ^2, or even x^1 and x^2. We will use θ^1 and θ^2, and must note that θ^1 and θ^2 are two separate parameters replacing u and v, and *not* θ to the power 1 and θ squared.

Thus the general surface is written

$$\mathbf{r}(\theta^1,\theta^2) = x(\theta^1,\theta^2)\mathbf{i} + y(\theta^1,\theta^2)\mathbf{j} + z(\theta^1,\theta^2)\mathbf{k} \quad (B.8)$$

The equation θ^2 = constant produces a curve on the surface as θ^1 is varied. Varying the value of the constant will produce a family of curves. Similarly, θ^1 will produce a second family, giving a net over the surface, as shown in Figure B.1.

In general, the curves forming the net will not cross at right angles and the distances between the intersections will not be constant. We could imagine drawing a net or grid of squares on a flat sheet of rubber and then stretching and bending the sheet to lie on our general surface. In so doing, the squares would become deformed so that the lengths and angles vary.

We will use the geometry of the net to investigate the geometry of the underlying surface. Some geometers would say that this is unsatisfactory because

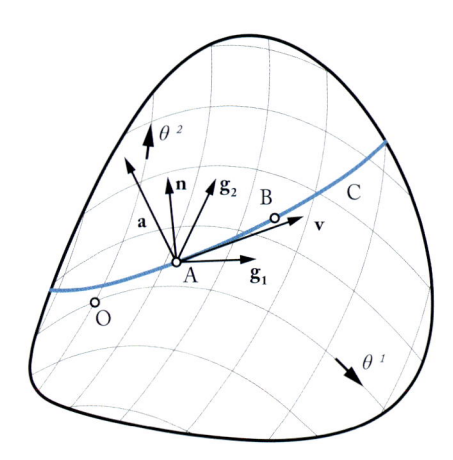

Figure B.1 Differential geometry notation

the surface exists independently of the net. But, the net is very useful for defining where we are and for specifying lengths and directions – and how directions change due to curvature.

What does seem odd is that we do not start by specifying a particular surface and the reason for this is that we want to find relationships that apply to all surfaces. If we want to investigate a particular surface, we do that right at the end. 'Elementary' books tend to start with particular special cases, which does sometimes make sense, if one is dealing with a simple shape such as a sphere.

B.1 Covariant base vectors and the first fundamental form

Figure B.1 also shows a number of vectors, $\mathbf{a}$, $\mathbf{g}_1$, $\mathbf{g}_2$, $\mathbf{v}$ and $\mathbf{n}$, all at the point A. All of these vectors lie in the plane tangent to the surface at A, except n which is the unit normal to this plane and therefore to the surface. The vector $\mathbf{v}$ is the unit vector tangent to the curve C and $\mathbf{a}$ is the unit vector perpendicular to $\mathbf{v}$ and $\mathbf{n}$.

Depending upon the application, C might be a real curve, perhaps a railway line on the Earth, while the lines of latitude and longitude are purely conceptual. The curve C can also be specified in parametric form as $\theta^1 = \theta^1(u)$ and $\theta^2 = \theta^2(u)$ so that as the parameter u varies, θ^1 and θ^2 change on the surface and so x, y and z vary in three-dimensional Cartesian space.

Let us suppose that u is the value of the parameter at A and $u + \delta u$ is the value at B. In Figure B.1, δu is finite, but we now have to imagine that $\delta u \to 0$ so that the straight line AB is tangent to the curve, that is in the direction v. Thus v is in the direction of the vector

$$\frac{d\mathbf{r}}{du} = \frac{\partial \mathbf{r}}{\partial \theta^1} \frac{d\theta^1}{du} + \frac{\partial \mathbf{r}}{\partial \theta^2} \frac{d\theta^2}{du}, \quad (B.9)$$

where the vectors

$$\frac{\partial \mathbf{r}}{\partial \theta^i} = \frac{\partial x}{\partial \theta^i} \mathbf{i} + \frac{\partial y}{\partial \theta^i} \mathbf{j} + \frac{\partial z}{\partial \theta^i} \mathbf{k}, \quad (B.10)$$

for $i=1$ and $i=2$. Note the use of both straight (d) and curly (∂) 'd's in equation (B.9). This because θ^1 and θ^2 are only functions of u along the curve C, whereas $\mathbf{r}$ is a function of *both* θ^1 and θ^2.

Because the vectors in equation (B.10) are used so much, they are given their own symbols with subscripts,

$$\mathbf{g}_i = \frac{\partial \mathbf{r}}{\partial \theta^i}, \quad (B.11)$$

again for $i=1$ and $i=2$. They are called the *covariant base vectors*. Green and Zerna (1968) use a_i and on a surface instead of $\mathbf{g}_i$.

The vector

$$d\mathbf{r}/du = \sum_{i=1}^{2} \left(\mathbf{g}_i (d\theta^i/du) \right) \quad (B.12)$$

is tangent to the curve C and $\mathbf{g}_1$ and $\mathbf{g}_2$ are tangent to the curves $\theta^2 =$ constant and $\theta^1 =$ constant respectively.

The Einstein summation convention states that an index is summed if it is repeated in the same term as a subscript and superscript. Thus, we can write

$$\frac{d\mathbf{r}}{du} = \mathbf{g}_i \frac{d\theta^i}{du} \quad (B.13)$$

and infer the $\sum_{i=1}^{2}$ from the repeated i.

In general, $\mathbf{g}_1$ and $\mathbf{g}_2$ will not be unit vectors, and neither will $\mathbf{g}_1$ and $\mathbf{g}_2$ be perpendicular to each other. If the length AB is equal to δs, then

$$\delta s^2 = \delta \mathbf{r} \cdot \delta \mathbf{r}$$

$$= \sum_{i=1}^{2} \sum_{j=1}^{2} \left[(\mathbf{g}_i \delta \theta^i) \cdot (\mathbf{g}_j \delta \theta^j) \right]$$

$$= \mathbf{g}_i \cdot \mathbf{g}_j \delta \theta^i \delta \theta^j = g_{ij} \delta \theta^i \delta \theta^j, \quad (B.14)$$

which is known as the *first fundamental form* (note the use of the summation convention). The scalar products

$$g_{ij} = g_{ji} = \mathbf{g}_i \cdot \mathbf{g}_j \quad (B.15)$$

are known as the *coefficients of the first fundamental form* or the *components of the metric tensor*. Equation (B.14) appears in exactly this form in the general theory of relativity, the only change is that the implied summations are from 0 to 3, corresponding to the four dimensions of space-time, and δs is now the time elapsed as experienced by observer at event A and subsequently B.

The unit tangent to the curve can now be written

$$\mathbf{v} = \frac{d\mathbf{r}}{ds} = \frac{\left(\frac{d\mathbf{r}}{du}\right)}{\left(\frac{ds}{du}\right)} = \frac{\frac{d\theta^i}{du}\mathbf{g}_i}{\sqrt{g_{jk}\frac{d\theta^j}{du}\frac{d\theta^k}{du}}}. \quad (B.16)$$

The unit vector

$$\mathbf{n} = \frac{\mathbf{g}_1 \times \mathbf{g}_2}{\sqrt{g}} \quad (B.17)$$

is normal to the surface and therefore also to the curve C. The quantity $g = g_{11}g_{22} - (g_{12})^2$ is equal to the square of the magnitude of the vector product $\mathbf{g}_1 \times \mathbf{g}_2$.

The unit vector $\mathbf{a} = \mathbf{n} \times \mathbf{v}$ and $\mathbf{a}$, $\mathbf{n}$ and $\mathbf{v}$ form a set of mutually perpendicular unit vectors.

B.2 The contravariant base vectors

It would seem that we have enough base vectors with the covariant base vectors, $\mathbf{g}_1$ and $\mathbf{g}_2$, tangent to the surface and the unit normal to the surface, $\mathbf{n}$. Nevertheless, it is very useful to have some more, the *contravariant base vectors* $\mathbf{g}^1$ and $\mathbf{g}^2$, defined by

$$\mathbf{g}^i \cdot \mathbf{g}_j = \delta^i_j, \quad (B.18)$$

$$\mathbf{g}^i \cdot \mathbf{n} = 0, \quad (B.19)$$

where the Kronecker delta, δ^i_j is equal to 1 if $i=j$ and 0 if $i \neq j$.

The product $\mathbf{g}^1 \cdot \mathbf{g}_2 = 0$ so that the direction of $\mathbf{g}^1$ is specified by the fact that it lies in the plane of $\mathbf{g}_1$ and $\mathbf{g}_2$, but is perpendicular to $\mathbf{g}_2$. The magnitude of $\mathbf{g}^1$ is then determined by $\mathbf{g}^1 \cdot \mathbf{g}_1 = 1$. The direction and magnitude of $\mathbf{g}^2$ are determined in a similar way.

The reason for using the two sets of base vectors can be seen by considering the vector,

$$\mathbf{p} = p_x \mathbf{i} + p_y \mathbf{j} + p_z \mathbf{k}$$
$$= \sum_{i=1}^{2}\left(p^i \mathbf{g}_i\right) + p\mathbf{n} = p^i \mathbf{g}_i + p\mathbf{n}$$
$$= \sum_{i=1}^{2}\left(p_i \mathbf{g}^i\right) + p\mathbf{n} = p_i \mathbf{g}^i + p\mathbf{n}, \quad \text{(B.20)}$$

whose components

$$p^i = \mathbf{p} \cdot \mathbf{g}^i, \quad \text{(B.21)}$$

$$p_i = \mathbf{p} \cdot \mathbf{g}_i, \quad \text{(B.22)}$$

$$p = \mathbf{p} \cdot \mathbf{n}. \quad \text{(B.23)}$$

Note that we can use either $\mathbf{p} = p^i \mathbf{g}_i + p\mathbf{n}$ or $\mathbf{p} = p_i \mathbf{g}^i + p\mathbf{n}$, or sometimes one and sometimes the other, at our convenience. But, we need $\mathbf{g}^i$ to find p^i and $\mathbf{g}_i$ to find p_i.

A vector is a first-order tensor and a scalar is a zero[th]-order tensor. In n dimensions an mth order tensor will have n^m components. A surface is a two-dimensional object (θ^1 and θ^2) embedded in three-dimensional space. This explains why p, the normal component of $\mathbf{p}$, does not have a superscript or subscript. Stress and strain are second-order tensors and the elastic stiffness of a material is a fourth-order tensor. Thus, in three dimensions, stress and strain each have $3^2 = 9$ components and elastic stiffness has $3^4 = 9 \times 9 = 81$ components. However, both stress and strain are symmetric tensors ($\tau_{xy} = \tau_{yx}$ and so on) and also stiffness has to be the second derivative of strain energy with respect to strain and therefore we end up with only $(6 \times 7)/2 = 21$ elastic constants. For an isotropic material many of these are zero and the remainder can be expressed in terms of Young's modulus and Poisson's ratio. In two dimensions, as in the surface of a shell, we have only $(3 \times 4)/2 = 6$ elastic constants, which again can be expressed using only Young's modulus and Poisson's ratio for an isotropic shell.

The scalar products $g^{ij} = \mathbf{g}^i \cdot \mathbf{g}^j$ are the contravariant components of the metric tensor and

$$\mathbf{g}^i = \sum_{j=1}^{2}\left(g^{ij}\mathbf{g}_j\right) = g^{ij}\mathbf{g}_j$$
$$\mathbf{g}_i = g_{ij}\mathbf{g}^j$$
$$p_i = g_{ij}p^j$$
$$p^i = g^{ij}p_j \quad \text{(B.24)}$$

If we have a second vector, $\mathbf{f}$, the scalar product

$$\mathbf{f} \cdot \mathbf{p} = \left(f^i \mathbf{g}_i + f\mathbf{n}\right) \cdot \left(p^j \mathbf{g}_j + p\mathbf{n}\right)$$
$$= f^i p^j g_{ij} + fp = f_i p_j g^{ij} + fp$$
$$= f_i p^i + fp = f^i p_i + fp \quad \text{(B.25)}$$

Finally, $g^{ij}g_{jk} = \delta^i_k$ which can be solved to give

$$g^{11} = \frac{g_{22}}{g},$$
$$g^{12} = -\frac{g_{12}}{g},$$
$$g^{22} = \frac{g_{11}}{g}. \quad \text{(B.26)}$$

We can use the contravariant base vectors to find an expression for the components of the unit vector $\mathbf{a}$ lying in the plane of the surface perpendicular to $\mathbf{v}$:

$$\mathbf{a} = \varepsilon_{ij} v^i \mathbf{g}^j \quad \text{(B.27)}$$

in which $\varepsilon_{11} = 0$, $\varepsilon_{22} = 0$ and $\varepsilon_{12} = -\varepsilon_{21} = \sqrt{g}$ are the components of the *Levi-Civita permutation pseudo tensor*.

B.3 The Christoffel symbols and the second fundamental form

We are now going to look at curvature. Returning to Figure B.1, $\mathbf{v}$ is a unit vector tangent to the curve C and

$$\kappa = \frac{d\mathbf{v}}{ds} \quad \text{(B.28)}$$

is defined to be the curvature vector of C.

The curvature vector κ is perpendicular to $\mathbf{v}$ so that $\mathbf{v} \cdot \kappa = 0$ and the magnitude of κ is equal to one over the radius of curvature of C.

The curvature vector can be resolved into two components, $\kappa = \kappa_{geodesic} + \kappa_{normal}$ in which the geodesic curvature is the curvature in the local tangent plane to the surface and the normal curvature is the curvature perpendicular to the surface. A geodesic is a line on the surface with zero geodesic curvature and the shortest distance between two points on a surface is a geodesic.

We can calculate the magnitude of the curvature by differentiating $\mathbf{r}$ a second time. A comma is often used to denote partial differentiation and thus

$$\mathbf{g}_{i,j} = \frac{\partial \mathbf{g}_i}{\partial \theta^j} = \frac{\partial^2 \mathbf{r}}{\partial \theta^i \partial \theta^j} = \mathbf{g}_{j,i} = \Gamma_{ij}^k \mathbf{g}_k + b_{ij}\mathbf{n}. \quad (B.29)$$

The magnitudes of the geodesic and normal curvatures are

$$|\kappa_{\text{geodesic}}| = \frac{\left(\frac{d^2\theta^i}{du^2} + \frac{d\theta^j}{du}\frac{d\theta^k}{du}\Gamma_{jk}^i\right)a_i}{g_{mn}\frac{d\theta^m}{du}\frac{d\theta^n}{du}} \quad (B.30)$$

and

$$|\kappa_{\text{normal}}| = \frac{b_{ij}\frac{d\theta^i}{du}\frac{d\theta^j}{du}}{g_{mn}\frac{d\theta^m}{du}\frac{d\theta^n}{du}}. \quad (B.31)$$

The *Christoffel symbols of the first and second kind*, Γ_{ijm} and Γ_{ij}^k,

$$\Gamma_{ijm} = \mathbf{g}_m \cdot \mathbf{g}_{i,j} = \frac{1}{2}(g_{im,j} + g_{mj,i} - g_{ij,m})$$
$$\Gamma_{ij}^k = \Gamma_{ji}^k = \mathbf{g}^k \cdot \mathbf{g}_{i,j} = -\mathbf{g}_{,j}^k \cdot \mathbf{g} = g^{km}\Gamma_{ijm} \quad (B.32)$$

look like they are components of a tensor, but are not because they do not obey the rules for what happens when the coordinate system is changed. They are essentially properties of the coordinate system, describing how it stretches and bends over the surface. Tensors and their components describe only real, physical things such as stress and the curvature of a surface itself, rather than just a coordinate system we have drawn upon it.

The quantities,

$$b_{ij} = b_{ji} = \mathbf{n} \cdot \mathbf{g}_{i,j} = -\mathbf{n}_{,j} \cdot \mathbf{g}_i, \quad (B.33)$$

are the components of a symmetric second-order tensor since they do represent something real, the curvature of the surface which exists independently of the particular coordinate system that we put on it.

$$\delta\mathbf{n} \cdot \delta\mathbf{r} = -b_{ij}\delta\theta^i\delta\theta^j \quad (B.34)$$

is known as *second fundamental form* and therefore b_{ij} are sometimes called the *coefficients of the second fundamental form*. As we move along the curve C the change in the normal, $\delta\mathbf{n} = -b_{ij}\mathbf{g}^i(d\theta^j/du)\delta u$. The component of $\delta\mathbf{n}$ parallel to the curve is due to the normal curvature, while the component perpendicular to the curve is due to the 'twist' of the surface in the direction of the tangent.

The second-order symmetric tensor

$$\mathbf{b} = b_{ij}\mathbf{g}^i\mathbf{g}^j \quad (B.35)$$

behaves like all second-order symmetric tensors, particularly stress, and thus we have principal curvatures in orthogonal directions in the same way that we have principal stresses. This is also closely connected with eigenvalues and eigenvectors, and the principal curvatures are the values of κ for which the determinate

$$\begin{vmatrix} (b_{11} - \kappa g_{11}) & (b_{12} - \kappa g_{12}) \\ (b_{21} - \kappa g_{21}) & (b_{22} - \kappa g_{22}) \end{vmatrix} = 0. \quad (B.36)$$

Thus

$$\left(g_{11}g_{22} - (g_{12})^2\right)\kappa^2 - \left(g_{11}b_{22} - 2g_{12}b_{12} + g_{22}b_{11}\right)\kappa$$
$$+ b_{11}b_{22} - (b_{12})^2 = 0, \quad (B.37)$$

and the two principal curvatures are

$$\kappa_{\text{I}} = H + \sqrt{H^2 - K} \quad (B.38)$$

and

$$\kappa_{\text{II}} = H - \sqrt{H^2 - K}, \quad (B.39)$$

where

$$H = \frac{\kappa_{\text{I}} + \kappa_{\text{II}}}{2} = \frac{g_{11}b_{22} - 2g_{12}b_{12} + g_{22}b_{11}}{2\left(g_{11}g_{22} - (g_{12})^2\right)} = \frac{1}{2}g^{ij}b_{ij} = \frac{1}{2}b_i^i \quad (B.40)$$

is the mean or Germain (after Marie-Sophie Germain) curvature and

$$K = \kappa_{\text{I}}\kappa_{\text{II}} = \frac{b_{11}b_{22} - (b_{12})^2}{g_{11}g_{22} - (g_{12})^2} = b_1^1 b_2^2 - b_1^2 b_2^1 \quad (B.41)$$

is the Gaussian curvature. The Gaussian curvature is positive for synclastic (dome-like) surfaces and negative for anticlastic (saddle-like) surfaces. Minimal surfaces (soap films) have $H = 0$.

B.4 The fundamental theorem of surface theory and the deformation of shells

The fundamental theorem of surface theory states that a surface is uniquely determined up to a rigid body movement and rotation if we know the coefficients of the first and second fundamental forms, g_{ij} and b_{ij}, as functions of the surface coordinates.

This is relevant to shell theory in that *deformation of a shell* is related to *changes* in g_{ij} and b_{ij}. If we differentiate the first fundamental form (B.14) with respect to time, we obtain the rate of strain

$$\text{rate of increase of } \delta s / \delta s = \frac{\gamma_{ij}\delta\theta^i\delta\theta^j}{g_{ij}\delta\theta^i\delta\theta^j}, \tag{B.42}$$

where

$$\gamma_{ij} = \frac{1}{2}\frac{\partial g_{ij}}{\partial t} \tag{B.43}$$

are the components of the rate of membrane strain tensor.

Things are a bit more complicated for bending since changes in both g_{ij} and b_{ij} are usually involved. We also have to be careful not to define rate of bending as rate of change of curvature. Imagine a spherical shell expanding uniformly, its curvature is reducing, but it is not bending since bending must involve some relative rotation.

However, the values of g_{ij} and b_{ij} are not independent. We have compatibility conditions that ensure a shell fits together and continues to do so as it deforms. These conditions are Gauss's Theorema Egregium and Peterson–Mainardi–Codazzi equations.

B.5 Gauss's Theorema Egregium and Peterson–Mainardi–Codazzi equations

So far we have found out nothing particularly startling. However, Gauss's Theorema Egregium (Latin for 'excellent theorem') is one of the most surprising and elegant results in all mathematics, as well as having all sorts of geometric and structural implications for shell structures. It is often just called Gauss's theorem, but there are other Gauss's theorems.

The basic idea is very simple, we have three functions of θ^1 and θ^2, namely x, y and z which define the surface. We then have eight quantities, g_{11}, $g_{12} = g_{21}$, b_{11}, $b_{12} = b_{21}$, $b_{12} = b_{21}$ – but as $g_{12} = g_{21}$ and $b_{12} = b_{21}$ there are really only six – which define lengths on the surface and its curvature. But since the six quantities depend on only three, there must be three equations relating to the six quantities. These equations ensure that the surface 'fits together'.

We have already differentiated twice to obtain

$$\mathbf{g}_{i,j} = \frac{\partial^2 \mathbf{r}}{\partial\theta^i\partial\theta^j} = \mathbf{g}_{j,i} = \Gamma^k_{ij}\mathbf{g}_k + b_{ij}\mathbf{n}, \tag{B.44}$$

and we are now going to differentiate for a third and final time,

$$\mathbf{g}_{k,ij} = \frac{1}{2}(g_{nk,ij} + g_{in,kj} - g_{ki,nj})\mathbf{g}^n$$
$$+ \Gamma^n_{kin}\mathbf{g}^n_{,j} + b_{ki,j}\mathbf{n} + b_{ki}\mathbf{n}_{,j}. \tag{B.45}$$

Order of partial differentiation does not matter, and therefore writing $\mathbf{g}_{k,ij} - \mathbf{g}_{k,ji} = 0$ and substituting for $\mathbf{g}_{k,ij}$ and $\mathbf{n}_{,j}$,

$$\left(R_{ijkm} - (b_{ki}b_{jm} - b_{kj}b_{im})\right)\mathbf{g}^m$$
$$+ \left(b_{ki,j} + b_{jp}\Gamma^p_{ki} - (b_{kj,i} + b_{ip}\Gamma^p_{kj})\right)\mathbf{n} = 0, \tag{B.46}$$

where

$$R_{ijkm} = \frac{1}{2}(g_{im,kj} + g_{kj,mi} - g_{ki,mj} - g_{jm,ki})$$
$$+ g^{np}\left(\Gamma_{kjn}\Gamma_{mip} - \Gamma_{kin}\Gamma_{mjp}\right) \tag{B.47}$$

and R_{ijkm} are the components of the *Riemann–Christoffel tensor*, which appears in exactly this form in the general theory of relativity.

Thus, we have

$$\nabla_j b_{ki} = \nabla_i b_{kj}, \tag{B.48}$$

which are the Peterson–Mainardi–Codazzi equations and

$$R_{ijkm} = b_{ki}b_{jm} - b_{kj}b_{im}, \tag{B.49}$$

which is Gauss's Theorem. The symbol ∇ denotes the covariant derivative, Green and Zerna (1968) use a vertical line, rather than a nabla or del symbol (∇),

$$\nabla_i b_{jk} = b_{jk,i} - b_{pj}\Gamma^p_{ik} - b_{pk}\Gamma^p_{ij}. \qquad (B.50)$$

The Riemann–Christoffel tensor has various symmetries which means that the only relevant term in two dimensions is

$$R_{1212} = \tfrac{1}{2}(2g_{12,12} - g_{11,22} - g_{22,11})$$
$$+ g^{np}(\Gamma_{12n}\Gamma_{12p} - \Gamma_{11n}\Gamma_{22p}) = b_{11}b_{22} - (b_{12})^2. \qquad (B.51)$$

Thus, Gauss's theorem is often written as

$$K = \frac{R_{1212}}{g} \qquad (B.52)$$

and it tells us that we can find the Gaussian curvature simply by measuring lengths on a surface.

It follows also that we can only change the Gaussian curvature by changing lengths on a surface.

B.6 Monge form

In the Monge form a surface is expressed as a heightfield, that is, $z = z(x,y)$, which is equivalent to writing $x = \theta^1$ and $y = \theta^2$. Then we have:

$$g_{11} = 1 + z_{,x}^2 = gg^{22},$$
$$g_{12} = z_{,x}z_{,y} = -gg^{12},$$
$$g_{22} = 1 + z_{,y}^2 = gg^{11},$$
$$\mathbf{n} = \frac{-z_{,x}\mathbf{i} - z_{,y}\mathbf{j} + \mathbf{k}}{\sqrt{g}},$$
$$g = 1 + z_{,x}^2 + z_{,y}^2,$$
$$b_{11} = \frac{z_{,xx}}{\sqrt{g}},$$
$$b_{12} = \frac{z_{,xy}}{\sqrt{g}},$$
$$b_{22} = \frac{z_{,yy}}{\sqrt{g}}, \qquad (B.53)$$

in which $z_{,x} = \partial z/\partial x$ and so on. The mean and Gaussian curvature are

$$H = \frac{(1 + z_{,y}^2)z_{,xx} - 2z_{,x}z_{,y}z_{,xy} + (1 + z_{,x}^2)z_{,yy}}{2(1 + z_{,x}^2 + z_{,y}^2)^{\frac{3}{2}}} \qquad (B.54)$$

and

$$K = \frac{z_{,xx}z_{,yy} - z_{,xy}^2}{(1 + z_{,x}^2 + z_{,y}^2)^2} \qquad (B.55)$$

respectively.

B.7 Cylindrical polar coordinates

In cylindrical polar coordinates

$$x = r\cos\theta,$$
$$y = r\sin\theta,$$
$$z = z(r,\theta),$$
$$r = \theta^1,$$
$$\theta = \theta^2, \qquad (B.56)$$

and

$$H = \frac{(r^2 + z_{,\theta}^2)rz_{,rr} + 2z_{,r}z_{,\theta}(z_{,\theta} - rz_{,r\theta}) + (1 + z_{,r}^2)r(rz_{,r} + z_{,\theta\theta})}{2[(1 + z_{,r}^2)r^2 + z_{,\theta}^2]^{\frac{3}{2}}}$$

$$K = \frac{r^2 z_{,rr}(rz_{,r} + z_{,\theta\theta}) - (z_{,\theta} - rz_{,r\theta})^2}{[(1 + z_{,r}^2)r^2 + z_{,\theta}^2]^2} \qquad (B.57)$$

In the case of radial symmetry, derivatives with respect to θ are zero and so

$$H = \frac{z_{,rr}}{2(1 + z_{,r}^2)^{\frac{3}{2}}} + \frac{z_{,r}}{2r\sqrt{1 + z_{,r}^2}},$$
$$K = \frac{z_{,rr}z_{,r}}{r(1 + z_{,r}^2)^2}. \qquad (B.58)$$

B.8 Equal mesh net

An equal mesh net might be a fishing net, a string bag, a cable net or a gridshell.

We can write

$$g_{11} = g_{22} = L^2 = \text{constant},$$
$$g_{12} = L^2\cos\alpha, \qquad (B.59)$$

where α is the angle between the cables, gridshell laths etc. The Christoffel symbols of the first kind now become

$$\Gamma_{111} = 0,$$
$$\Gamma_{112} = g_{12,1} = -L^2 \sin\alpha\, \alpha_{,1},$$
$$\Gamma_{121} = \Gamma_{211} = 0,$$
$$\Gamma_{122} = \Gamma_{212} = 0,$$
$$\Gamma_{221} = g_{12,2} = -L^2 \sin\alpha\, \alpha_{,2},$$
$$\Gamma_{222} = 0, \tag{B.60}$$

so that Gauss's theorem reduces to

$$K = -\frac{1}{L^2 \sin\alpha} \frac{\partial^2 \alpha}{\partial \theta^1 \partial \theta^2}. \tag{B.61}$$

This result is attributed to Chebyshev as a result of studying the problem of cutting cloth to make clothes.

B.9 Geometry and structural action

A pin-jointed truss in two dimensions or a space truss in three dimensions is a structure which resists loads by tensions and compressions in its members which try to prevent the distance between the nodes changing. A beam resists loads by using bending moments to try to prevent its curvature changing.

In the case of shell structures we have lengths on the surface defined by the first fundamental form, equation (B.14) and we have curvature of the surface defined by the second fundamental form, equation (B.34). Because shells are thin it is easier to change their curvature than lengths on the surface and therefore we want shells to work primarily by *membrane action* rather than bending – although we know that we must rely on bending action to help resist buckling.

We will now derive the membrane equilibrium equations from what we know about geometry. The same technique also works for bending action, it is just a bit more complicated. We will use virtual work to be consistent with our discussion of the finite element method, and also because it allows us to define membrane stress in terms of strain rate and rate of work.

Imagine a surface which is moving with a *virtual* velocity **u**. The rate of virtual membrane strain is then

$$\gamma_{ij} = \frac{1}{2} \frac{\partial g_{ij}}{\partial t}$$
$$= \frac{1}{2}\left(\frac{\partial \mathbf{r}}{\partial \theta^i} \cdot \frac{\partial \mathbf{u}}{\partial \theta^j} + \frac{\partial \mathbf{u}}{\partial \theta^i} \cdot \frac{\partial \mathbf{r}}{\partial \theta^j} \right)$$
$$= \frac{1}{2}\left(\mathbf{g}_i \cdot \frac{\partial \mathbf{u}}{\partial \theta^j} + \frac{\partial \mathbf{u}}{\partial \theta^i} \cdot \mathbf{g}_j \right). \tag{B.62}$$

We can now *define* the membrane stress components $\sigma^{ij} = \sigma^{ji}$ such that $\sigma^{ij}\gamma_{ij}$ is the rate of virtual work being done on the surface per unit surface area. Then the total rate of virtual work being done on the shell S is

$$\int_S \sigma^{ij}\gamma_{ij}\sqrt{g}\,d\theta^1 d\theta^2 = \int_S \sigma^{ij}\frac{1}{2}\left(\mathbf{g}_i \cdot \frac{\partial \mathbf{u}}{\partial \theta^j} + \frac{\partial \mathbf{u}}{\partial \theta^i} \cdot \mathbf{g}_j \right)\sqrt{g}\,d\theta^1 d\theta^2$$
$$= \int_S \sigma^{ij}\frac{\partial \mathbf{u}}{\partial \theta^i} \cdot \mathbf{g}_j \sqrt{g}\,d\theta^1 d\theta^2$$
$$= \int_S \frac{\partial}{\partial \theta^i}\left(\sigma^{ij}\mathbf{u} \cdot \mathbf{g}_j \sqrt{g} \right)d\theta^1 d\theta^2$$
$$- \int_S \mathbf{u} \cdot \frac{\partial}{\partial \theta^i}\left(\sigma^{ij}\mathbf{g}_j \sqrt{g} \right)d\theta^1 d\theta^2. \tag{B.63}$$

The first part of the last line can be integrated to produce a boundary integral which is the rate of virtual work being done on the shell by membrane stresses at the boundary,

$$\int_S \frac{\partial}{\partial \theta^i}\left(\sigma^{ij}\mathbf{u} \cdot \mathbf{g}_j \sqrt{g} \right)d\theta^1 d\theta^2 = \oint_{\partial S} \mathbf{u} \cdot \left(\sigma^{ij}\mathbf{g}_j \varepsilon_{ik}d\theta^k \right). \tag{B.64}$$

We therefore have the virtual work equation,

$$\int_S \sigma^{ij}\gamma_{ij}\sqrt{g}\,d\theta^1 d\theta^2$$
$$= \oint_{\partial S} \mathbf{u} \cdot \left(\sigma^{ij}\mathbf{g}_j \varepsilon_{ik}d\theta^k \right) + \int_S (\mathbf{u} \cdot \mathbf{p})d\theta^1 d\theta^2, \tag{B.65}$$

in which **p** is the load applied to the shell per unit area. Thus

$$\int_S \mathbf{u} \cdot \left(\mathbf{p} + \frac{\partial}{\partial \theta^i}\left(\sigma^{ij}\mathbf{g}_j \sqrt{g} \right) \right)d\theta^1 d\theta^2 = 0. \tag{B.66}$$

However the virtual velocity **u** can be chosen arbitrarily so that

$$\mathbf{p} + \frac{\partial}{\partial \theta^i}\left(\sigma^{ij}\mathbf{g}_j \sqrt{g} \right) = 0, \tag{B.67}$$

which reduces to

$$\mathbf{p} + \nabla_i \sigma^{ij}\mathbf{g}_j + b_{ij}\sigma^{ij}\mathbf{n} = 0.$$

The covariant derivative

$$\nabla_i \sigma^{ij} \mathbf{g}_j = \left(\sigma^{ij}_{,i} + \sigma^{kj}\Gamma^i_{ki} + \sigma^{ik}\Gamma^j_{ki} \right) \mathbf{g}_j$$

$$= \left(\frac{\partial \sigma^{ij}}{\partial \theta^i} + \sigma^{kj}\Gamma^i_{ki} + \sigma^{ik}\Gamma^j_{ki} \right) \mathbf{g}_j \quad \text{(B.68)}$$

balances the in-plane components of load. The Christoffel symbols look after the fact that the base vectors change magnitude and direction in the plane of the surface.

The curvature multiplied by the membrane stress $b_{ij}\sigma^{ij}$ balances the normal component of load. This result is particularly elegant and the membrane equilibrium equations can be made even nicer by writing them in symbolic notation without the indices:

$$\mathbf{p} + \nabla \cdot \sigma + (\mathbf{b}{:}\sigma)\mathbf{n} = 0. \quad \text{(B.69)}$$

The membrane equilibrium equations can be derived and written in many different forms. In Chapter 3 they are written in plane form in which the equilibrium equations are resolved in the horizontal and vertical directions. Horizontal equilibrium can be ensured using the Airy stress function leaving just one equation for vertical equilibrium. This has practical advantages for architectural shells where the vertical and horizontal have special significance, but it does blur the distinction between what happens in the plane of a shell and in the normal direction.

Further reading

- *Lectures on Classical Differential Geometry*, Struik (1988). This classic work is probably the most accessible book on differential geometry. It uses u and v coordinates instead of θ^1 and θ^2. It also uses E, F, G, e, f and g instead of $g_{11}, g_{12}, g_{22}, b_{11}, b_{12}$ and b_{22} respectively.
- *Theoretical Elasticity*, Green and Zerna (1968). In Chapter 3 it said that if you only own one book on shell theory, it has to be this one. The same applies to the differential geometry of surfaces which is contained in the first thirty-nine pages. Green and Zerna (1968) use the letter 'g' for quantities in three dimensions – base vectors, coefficients of the metric tensor and so on – but the letter 'a' for the corresponding quantities on a surface. We have used 'g' on a surface. In addition they use Greek letters for indices in the range 1, 2 whereas we have used Latin.

APPENDIX C

Genetic algorithms for structural design

Rajan Filomeno Coelho, Tomás Méndez Echenagucia, Alberto Pugnale and James N. Richardson

Genetic algorithms are a subclass of evolutionary algorithms. A Genetic Algorithm (GA) is defined by Goldberg (1989) as a 'search algorithm based on the mechanics of natural selection and natural genetics', but another definition, more focused on its way of functioning, is provided by John R. Koza in his 1992 book, *Genetic Programming: On the Programming of Computers by Means of Natural Selection*:

> The GA is a highly parallel mathematical algorithm that transforms a set (population) of individual mathematical objects […], each with an associated fitness value, into a new population (i.e. the next generation) using operations patterned after the Darwinian principle of reproduction and survival of the fittest and after naturally occurring genetic operations (notably sexual recombination).

In other words, in a random population of potential solutions, the best individuals are favoured and combined in order to create better individuals at the next generation. In the 1980s, genetic algorithms received an increasing recognition by scientists, and studies in fields ranging from biology, artificial intelligence, engineering and business to social sciences began to appear. At present, GAs provide a robust and flexible tool to solve complex problems, including air-traffic programming, weather forecasts, share portfolios balance and electronic circuits design, in which a consolidated analytic way of resolution is unknown. Moreover, as far as the world of construction is concerned, GAs are increasingly being used to deal with the optimization of bridges and large-span structures, the morphogenesis of shells and membranes, and the spatial configuration of reciprocal structures.

C.1 Main characteristics

With respect to other traditional optimization and search procedures, GAs differ in four fundamental aspects, described by Goldberg (1989):

- GAs search using a 'population' of candidate solutions, and not a unique solution;
- GAs work with a coded version of the parameter set, and not the parameters themselves;
- GAs are based on stochastic transition rules (they use randomized operators);
- GAs are blind to auxiliary information; they only need an objective function (fitness function).

All these characteristics contribute to the typical GA's robustness.

C.2 Terminology

The specific terminology for GAs derives from natural systems as well as from computer science technical vocabulary. For this reason, it is possible to find in technical literature the same concept expressed with two different, but equivalent, terms (see Table C.1). Many of these terms specifically refer to the world of natural systems and cannot be found in other optimization procedures. Examples are terms such

Meaning	Natural systems	Computer science
genetic codes	chromosome	string
genetic constitution of an individual	genotype	structure
observable characteristics of an individual	phenotype	parameter set, solution alternative, point
basic unit of a genetic code	gene	feature, character, detector
possible settings of a gene	allele	feature value
the position of a gene in a genetic code	locus	string position

Table C.1 Comparison of natural and artificial GA terminology

as 'individual', which represents a candidate solution to the evaluated problem, 'population', indicating a set of individuals considered at the same iterative step of the evolutionary process, and 'generation', synonymous to iteration, to refer to a specific step of the algorithm procedure. This is also the case with the three main operators of GAs – selection, reproduction and mutation – which are described in Section C.3.2.

C.3 Elements of a genetic algorithm

As with any systematic approach to search problems, GAs require two main elements in order for them to provide an effective and reliable result:

1. A representation scheme, describing each possible solution (individual) with a set of variables (chromosome), as well as the limits in which these variables can operate. This is done by means of parameterization and the definition of the parameters domain.
2. A fitness function, measuring the generated solutions (individuals) on the basis of a well-defined performance parameter, and the respective evaluation criterion.

The definition of a problem for a GA implementation requires also several parameters and termination criteria to control the algorithm. The parameters include the number of generations, the population size, the number of parents and various coefficients that are applied on the GAs operators (selection, crossover, mutation, elitism). A termination criterion can be related to the fitness function (the algorithm can stop once a minimum required fitness value has been reached), it can be related to the number of solutions considered, number of generations, or even calculation time.

C.3.1 Procedure

The conventional procedure of a GA can be summarized by the following steps:

1. Generate an initial, random population of individuals, or candidate solutions to the problem.
2. Evaluate the performance (fitness) of each individual.
3. Generate a new population of candidate solutions applying the following three genetic operators (or at least the first one):
   a. Selection: select best individuals for the reproduction to the new population.
   b. Crossover or reproduction: recombine genetic codes of selected individuals, creating new candidate solutions.
   c. Mutation: apply random mutations to the genetic codes of new individuals.
4. Repeat steps two and three to evolve the population over a number of generations, until a satisfactory result is achieved.

C.3.2 Operators

Successive populations are generated from the previous population by way of three genetic operators: the 'selection' of best individuals, their 'reproduction' or 'crossover', and 'mutation'. In order to further improve the general efficiency of the algorithm, some other secondary operators could be added to the main procedure, most commonly the 'elitism' operator. It is worth noting that the various user-defined parameters controlling these operators will be of great importance to the efficacy of the algorithm. As yet no universal method exists to optimally choose these parameters and the experience of the user will play a dominant role in the choice of values for the parameters.

Selection

The selection operator retains individuals from the previous population. Different methods have been developed which all employ the fitness values in their selection, while some methods also incorporate a degree of randomness. Individuals with higher fitness generally have a greater probability of making a contribution to the new generation of individuals. The end result of the selection operator is the 'mating pool'; it is a list of pairs of individuals which are to be used in reproduction (and crossover).

Crossover

The reproduction or crossover operator acts on the selected individuals, creating a new population of individuals. Crossover involves the 'mating' of individuals to produce offspring with characteristics of its predecessors. The offspring individual is created by copying part of the genetic code of one parent, and the rest from the other parent (Fig. C.1). Different crossover operators perform this operation in diverse ways, introducing some level of randomness. The probability of a crossover operation taking place is set by the user. This operator is traditionally considered as the 'core' of the GA, because it is the main cause of variation and innovation of candidate solutions.

Mutation

The mutation operator performs genetic variations in chromosomes of the individuals of a population. The mutation operator is also implemented with a probability chosen by the user. Similarly, the scale of the mutation can also be set. Mutation consists of replacing entries in the chromosome by random values within the permitted range of the variable. The mutation rate is usually relatively small compared to the crossover rate, as is the mutation scale, the relative number of mutations in the chromosome. While crossover is generally considered as a constructor of new candidate solutions, mutation works as a disruptor of existing configurations; for this reason, it plays a secondary role with as main purpose avoiding genetic drift (when all individuals become identical, evolution is no longer possible).

Elitism

The 'elitism' operator, introduced by Kenneth Alan De Jong in his 1975 doctoral thesis 'An analysis of the behavior of a class of genetic adaptive systems', is the most common secondary operator. It works in addition to the selection method, forcing the GA to retain a fixed number of best individuals at each generation in order to save their chromosomes from destruction due to crossover and/or mutation, therefore avoiding a maximum performance decrease during the evolutionary process. Generally, it significantly improves the algorithm's efficiency.

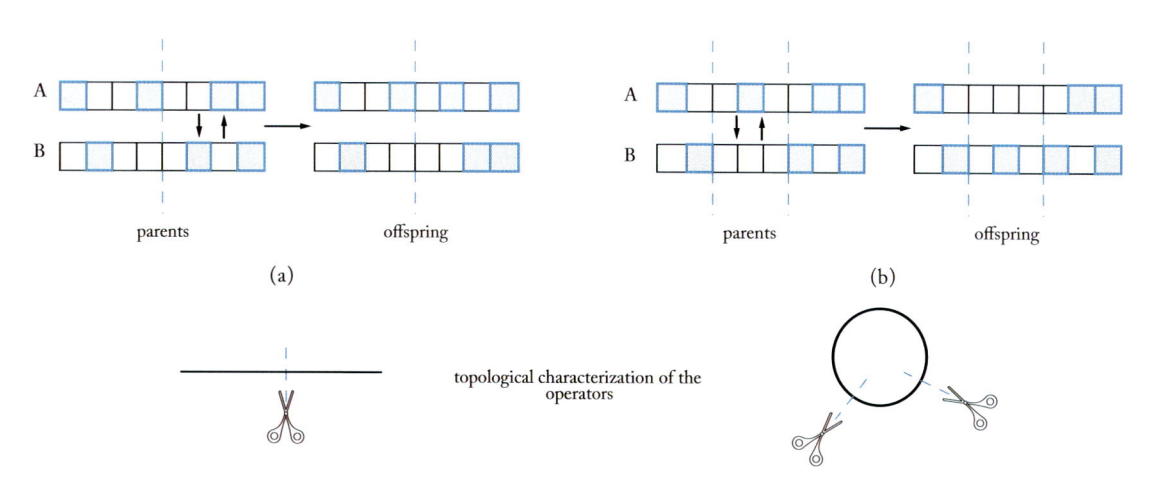

Figure C.1 Simple (a) single-point and (b) two-point crossover operators

C.4 Multi-objective genetic algorithms

When we study multiple and contrasting objective functions, with any search method, we have to consider the fact that the solutions will not be optimal for all functions. Therefore, the final result of a multi-objective search is inherently a set of solutions, not a single individual. This group of solutions is called 'trade-off set', 'Pareto front' or 'non-dominated set'. It comprises solutions that are said to be not dominated. The concept of dominance and non-dominance is defined as follows:

> In order for solution A to dominate solution B, solution A has to outperform, or equal B in all functions, as well as outperform B in at least one function. If solution A outperforms or equals solution B in all objective functions except in one in which solution B outperforms A, then A and B are non-dominated solutions.

GAs can easily be adapted to multi-objective search and optimization thanks to their inherent handling of multiple potential solutions, thereby leading to various trade-off designs. Different ways of transforming a GA into a Multi-Objective Genetic Algorithm (MOGA) by modifying some of the operators have been proposed (Deb, 2001). Various MOGAs employ different genetic operators in order to introduce the characteristics of multi-objective search and optimization, but one of the most important modifications in most MOGAs is done in the selection operator. The main differences in the selection operator (see Section C.3.1) are:

- Pareto-optimality: solutions are not ranked by their performance in the objective functions directly, but are ranked by their level of domination in the population.
- Clustering: how similar is this design to other designs in the population? Promoting diversity in the population leads to better exploration of the design space.

For a full explanation on MOGAs, the reader is referred to Deb (2001).

C.5 Application to structural design and optimization

For structural design and optimization, the use of GAs is very attractive, for the following reasons:

- The nature of the variables: structural optimization problems may be characterized by mixed variables (continuous, discrete, integer and/or categorical). GAs handle these variations naturally, whereas gradient-based algorithms, for instance, are mainly devoted to problems with continuous variables and differentiable functions.
- The nature of the functions: as the functions involved in structural optimization (e.g. the maximum stress among all elements of a truss) may be non-differentiable, and sometimes discontinuous, gradient-based techniques are excluded. Only algorithms requiring only the values of the functions, and not their derivatives, are applicable.
- Exploration of the search space: as they work on a population of solutions instead of a single point (at each iteration) – even while blind to any specific knowledge about the problem – GAs are less likely to be trapped in a local minimum, and effectively find the optimal global value. The schemata theorem has shown that the way recombination of individuals is performed allows the algorithm to explore widely the whole design space. GAs are thus very well suited for noisy and multi-modal functions.

C.5.1 Chromosomes

When considering structural problems, the variables describing each individual, that is, its chromosome, depend on the type of optimization considered. For instance, the chromosome can consist of member sizing, nodal position (shape) or topology variables, or of some combination of these three types. It can consist of bit string representation or real-valued representation, though mutation and crossover operators are typically a bit more involved for real-valued representation. The capability to handle variables of different types at once is one of the great strengths of GAs as an optimization procedure. GAs allow, for example, for combining continuous shape variables with discrete topology variables in a single chromosome representation.

Figure C.2 shows a chromosome representation of the shape and topology variables for a two-dimensional, four-node structure. Each entry in the chromosome refers to one of the variables considered. In the example, the first four entries correspond to the shape variables x_1, y_1, x_2 and y_2, the coordinates of nodes 1 and 2. The remaining entries in the chromosome correspond to the elements defined by the nodes they connect. The binary 0/1 topology variables refer to the existence or non-existence of the element.

C.5.2 Fitness function

Besides the abstract definition of the set of potential solutions, a measure of the structural performance (used in step two of the procedure described in Section C.3.1) must be chosen in order to drive the optimization process. In single objective optimization, basic choices could be measures of mass or stiffness. The maximum vertical displacement of a structure is a suitable structural parameter to have a raw evaluation of its structural behaviour. However, analogous results could be obtained by calculating the fitness function from other integral parameters of the structure, such as total strain energy or buckling multiplier.

Furthermore, there may be structural properties associated with the chromosome that we wish to know in order to ensure that the structure adheres to certain structural requirements. For example, the maximal stresses in the structure should not exceed a certain limit, ensuring satisfaction of the stress constraint. In general, violation of the constraints will either be used to augment or penalize the fitness function, so as to influence the selection of unfeasible solutions, or eliminate them as candidate designs. The evaluation of the individual's fitness often occurs by way of structural analysis such as the force method or displacement method (e.g. finite element analysis). Whatever method of structural analysis, the evaluation will be called at least once per iteration of the genetic algorithm. It is noted though, that for very complex problems, alternative methods to evaluate the objective functions exist; for example, response surface methods, surrogate modelling. Since the GA itself is a black-box type procedure, the operators (selection, crossover and so on) are not specific to structural design and optimization.

Further reading

- *An Introduction to Genetic Algorithms*, Mitchell (1998). A simple and complete introductory book to genetic algorithms, implementing the main concepts of the work and research developed by Holland, Goldberg and Koza during the last four decades. It should be used to program a basic GA procedure, as well as for the development and tuning of the routines related to the three main GA operators.
- *Genetic Algorithms in Search, Optimization, and Machine Learning*, Goldberg (1989). The initial reference book on genetic algorithms.
- *Multi-Objective Optimization using Evolutionary Algorithms*, Deb (2001). Widely regarded as the definitive book in its field.

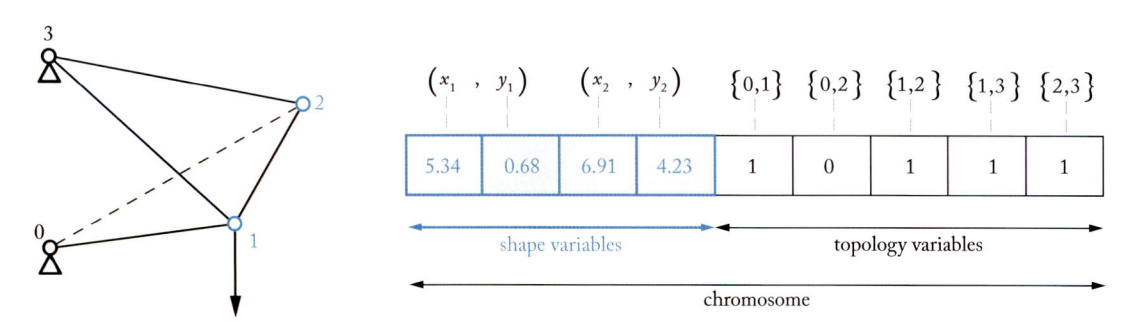

Figure C.2 Chromosome representation of a four-node, pin-jointed structure. Entries in the chromosome correspond to shape variables (coordinates of nodes) and topology variables (existence or non-existence of elements)

APPENDIX D
Subdivision surfaces

Paul Shepherd

This appendix introduces the subdivision surface as an efficient method of representing shell geometry in an optimization framework.

Many shell structures are designed using smooth continuous surface modelling techniques such as NURBS surfaces. However, whilst these models are easy to manipulate in terms of geometry, they are extremely difficult to manufacture accurately and inevitably need to be split into subfields in order to provide supporting structure for façade panels. Even building shells, which might be single monocoque structures when completed, usually require formwork built by traditional methods using discrete elements. Therefore, such projects are often modelled using a discrete mesh representation of the surface, rather than a parameterized continuous NURBS surface, at least in the engineering and fabrication phases.

A mesh representation uses polygons (usually triangles or quadrilaterals) to describe the continuous underlying surface. Each polygon is known as a face, made up of straight lines (edges) which go between the corners (vertices) of each polygon. Meshes are particularly suited to engineering disciplines since the calculation software usually requires a finite element mesh for the analysis of structural behaviour. This is driven by the fact that the underlying equations are solved only at discrete points on the surface (the vertices) and mathematical assumptions are made as to how the quantities being modelled vary in between. The surface mesh may be extended into three dimensions for a finite element or finite volume analysis. The disadvantage of a mesh representation is that through splitting the surface up into triangular or quadrilateral elements the smoothness of the underlying surface is lost. There is therefore always a compromise between an increasingly accurate representation of the smooth surface and keeping the number of faces which need to be stored and manipulated down to a manageable size.

D.1 Description

Subdivision surfaces offer a solution to this dichotomy. At their heart is a carefully calculated subdivision algorithm (known as a scheme) for taking an input base mesh and refining it, replacing each face with a number of smaller faces. The resulting finer mesh can then have the same algorithm applied to it to produce an even finer mesh, and so on. This results in a sequence of finer and finer meshes, each being a more accurate representation of an underlying smooth surface.

Subdivision surfaces have been used widely by the computer gaming and animation industries as an efficient way of representing the surface geometry of animated objects. Only a very coarse base mesh representation of the object, using perhaps only hundreds of polygons, is required to be stored, leaving precious computer memory available for other objects, sounds and so on. If such an object then needs to be rendered very close to the camera, a large number of subdivision steps can be performed on the base mesh to generate the smooth surface detail required. If, however, the object is placed very far from the camera, a representation of the object with very few subdivision steps is sufficient, since the extra detail provided by creating more polygons will not be noticeable.

D.2 Implementation

There are many different subdivision schemes, some require triangular base meshes, others quadrilateral meshes and others can cope with higher order polygons or even a mixture. The basic process of a subdivision scheme starts by taking each face in the base mesh and refining it topologically in some way. For example, the Loop subdivision scheme (Loop, 1992), which is used for the examples here, is applied to triangular meshes such as the one shown in Figure D.1a, which represents a single face of a spherical mesh. Each edge of every triangle is split into two by the introduction of a new child vertex, and each triangle is then replaced by four new smaller triangles, as shown in Figure D.1b.

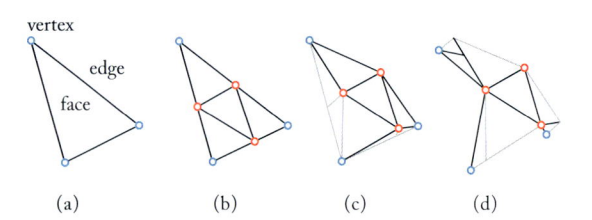

Figure D.1 Successive stages of a subdivision scheme with (a) base mesh, (b) topological split, (c) child vertices smoothed and (d) parent vertices smoothed

Clearly, whilst this topological step does indeed create a representation of the surface with more faces (in this case four times as many), nothing is gained in terms of surface representation, since the geometry of the original planar faces remains. Subdivision schemes therefore always involve a second step, which moves the newly created child vertices to a new position, based on a weighted average of the parent vertices around it, as shown in Figure D.1c. This new position allows the refined mesh to more closely represent the underlying smooth surface. For some subdivision schemes, known as interpolating schemes, this second step is sufficient, and the subdivision interpolates a smooth surface through the vertices of the base mesh. However, approximating schemes also exist, such as the Loop scheme, whereby a third step is performed which also moves the parent vertices to a new weighted-average position, as shown in Figure D.1d. Approximating schemes do not respect the original base mesh completely, but approximate it with a very smooth surface.

The new positions of the vertices are calculated using very specific weightings of the surrounding vertices to ensure that the surface is smooth. For example, the Loop scheme places child vertices at an average position by weighting the vertices at the ends of its edge (shown orange in Figure D.2a) by 3/8 and the other vertex of the faces touching the edge (shown in green in Figure D.2a) by 1/8. The approximating Loop scheme also moves the parent vertices to a new position, in this case weighted at 5/8 of its old position and 3/8 of the average position of the surrounding vertices, as shown in Figure D.2b.

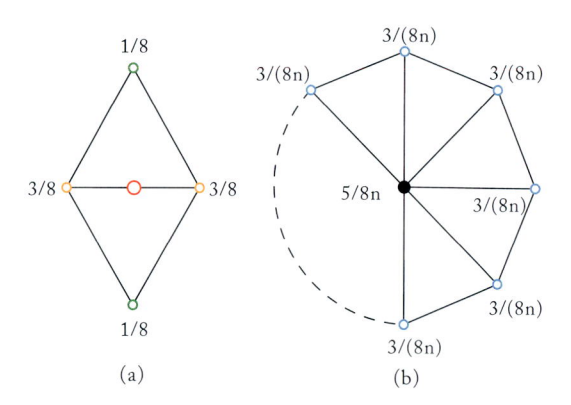

Figure D.2 Weighting of internal vertices for Loop subdivision scheme with (a) child vertices and (b) parent vertices

The edges of a surface usually require special cases of the subdivision scheme to be applied since, for example, there will only be one face touching an edge. For the Loop scheme, child vertices created along a boundary edge are placed at the midpoint of the edge, as shown in Figure D.3a. Parent vertices on an edge are positioned with a 3/4 weighting of their original position and 1/8 of the position of each of the two neighbouring boundary vertices, irrespective of the locations of their internal neighbours, as shown in Figure D.3b.

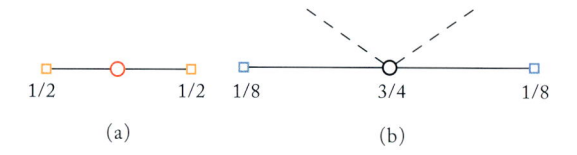

Figure D.3 Weighting of boundary vertices for Loop subdivision scheme

If a base mesh is repeatedly subdivided using such schemes, the geometry converges onto what is known as the limit surface, and it is this limit surface which it is proposed should be used to represent the geometry of a shell.

D.3 Benefits for building design

Apart from the general advantages that a mesh representation of a surface has in terms of buildability, subdivision surfaces offer two main benefits over traditional tools for modelling the geometry of building shells, benefits for analysis and for optimization.

The hierarchic nature of the mesh topology means that a single subdivision surface can simultaneously be represented at many different levels of detail. Therefore, independent of the actual shape of the underlying surface, a range of mesh topologies can be extracted, which can represent façade cladding panels or supporting structure with a required size or spacing. For example, if a given surface is to be clad with glass panels, and each panel is to be cut from a 2m wide sheet of glass, then the mesh can be subdivided down until each polygon (triangle or quadrilateral) is less than 2m across. This allows the designer easy control of the mesh size, and therefore to find a balance between the size of panels and the number of connections, without changing the geometry of the underlying surface.

Whilst useful in itself, this multiple level-of-detail representation is especially beneficial when multi-objective optimization is to be performed on the surface.

As part of the design and engineering process, a given building surface is often optimized in some way to improve its performance. Structural optimization will attempt to identify small changes in the shape which will lead to disproportionately large savings in material. This can be achieved either by placing structural members only where they are needed, and thereby minimizing the number of members required, and/or by aligning structural members so that they can effectively transmit loads to the supports, thereby reducing the size of each member. Wind and snow loads can be reduced by modifying a building's shape. Similarly the environmental performance of a building can be improved by considering the thermal, ventilation and daylighting implications of its orientation on the site, shape and self-shading.

For each of these objectives, an analysis of the performance of a given shape will need to be carried out, using a particular representation of the building envelope. As mentioned above, the majority of these analyses will require a discrete mesh-like representation, and the granularity of the mesh required will vary depending on the type of analysis being performed. For example, a finite element structural analysis might require the mesh to be discretized in a way that reflects the proposed supporting structure, whereas a computational wind-flow analysis might need a much finer analysis mesh and a solar energy-generation calculation could be sufficiently accurate even with a very coarse mesh representation. Finite element structural analysis of a continuous shell requires shape functions to interpolate the geometry between nodes (see Appendix A) and the subdivision surface itself could be used for this.

The hierarchical nature of a subdivision surface is therefore ideally suited to such multi-objective optimization. Given the same underlying proposal for a particular building's shape, many different meshes, representing many different levels of detail, can be quickly extracted and used for performance analysis.

Subdivision can be seen as a smoothing process, whereby the coordinates of the divided mesh vertices are a smoothed average of their surrounding parent vertex neighbours. In this way the coordinates of a mesh are smoothed to form successively close approximations to the underlying limit surface. In fact it can be shown (Zorin et al., 2000) that in general the limit surface is G2-continuous, at least away from the boundaries or vertices with unusual numbers of neighbours. This means that there is no sudden change in shape (the surface does not have gaps), in surface tangent (no creases), or in rate-of-change of tangent (no distortions in visual reflections). This can make subdivision surfaces particularly desirable for aesthetic reasons, since the resulting building surface would appear smooth.

Another benefit of the smoothing process is that it is not only the coordinates of the vertices which can be smoothed. Any set of numerical values can be associated with each vertex, and these values can be smoothed in exactly the same way as the coordinates, by taking the same weighted average of the values of the neighbouring vertices. This means that, for example, a colour could be applied to each vertex in the base (coarse) mesh, and all successive subdivided meshes would automatically smooth out the colours across the surface. This could similarly be applied to more practical parameters such as façade permeability or transparency values, louvre angles or cladding offset distances. Such quantities can be defined on the base mesh just where required and the resulting subdivided surface would distribute them evenly over the entire surface in a G2-continuous manner in exactly the same way as the vertex positions.

D.4 Subdivision mesh

Using a subdivision surface framework involves an inherent triangular or quadrilateral mesh, and this is an obvious candidate for a structural mesh too. Since a subdivision surface can be sampled at various levels of subdivision, many possible grids are available at many different densities (Fig. D.4a–c). This means that sufficiently many levels of subdivision can be applied such that the individual triangular panels are small enough to be manufactured in a cost-effective manner, but the mesh is not so dense as to lead to prohibitively many members and connecting nodes.

Of course, not all the potential members need be used as a structural element. Subsets of the subdivision mesh edges could be used for the structure to simplify its construction and change its aesthetic (Fig. D.4d, e). Similarly not all potential members need be the same size, and a system of primary and secondary structure could be introduced (Fig. D.4f).

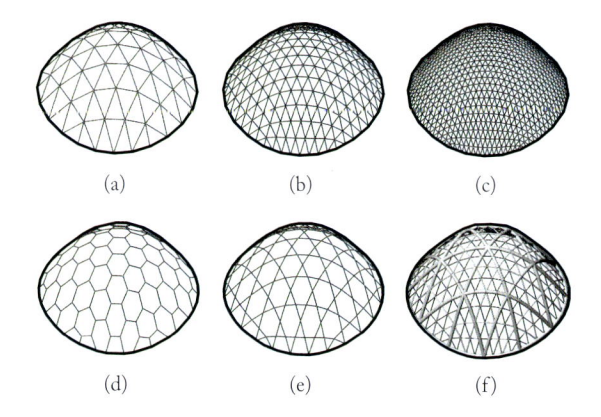

Figure D.4 Possible structural grids (a–c) resulting directly from subdivision, (d–e) alternatives subsets of the mesh edges and (f) applying member hierarchy

D.5 Advantages

A subdivision surface framework allows a very fast parametric study to be performed on an underlying smooth surface. It offers meshes with multiple levels of detail to be extracted with no extra effort, which is particularly useful if many different types of analysis are to be performed, each requiring a mesh with a different density. This efficiency can be especially advantageous if automated multi-objective optimization routines are to be applied, since each analysis can be performed on a suitable mesh and the results directly compared.

Despite these advantages, there remain a number of issues which will need to be addressed if subdivision surfaces are to be adopted for shell design by the mainstream architectural design community.

D.6 Disadvantages

The main challenge in using subdivision surfaces as part of a standard building design process is the lack of Boolean operations such as intersection, union and difference.

The more general problem of calculating the intersection between two subdivision surfaces is yet to be solved satisfactorily, not least because there is no obvious way of representing the resulting curve. Until more robust methods of representing subdivision surfaces, if found, and of performing Boolean operations on them, there will inevitably be a point in the design of every building where the subdivision framework must be left behind and more traditional methods of representing the geometry (such as NURBS) will have to be adopted.

Bibliography

Addis, B. (2007), *Building: 3000 Years of Design Engineering and Construction*, Phaidon, New York and London.

Addis, B. (2013), '"Toys that save millions" – a history of using physical models in structural design', *The Structural Engineer*, 91(4), 12–27.

Addis, W. (1990), *Structural Engineering – the Nature of Theory and Design*, Ellis Horwood, Chichester, UK.

Adriaenssens, S. M. L. and Barnes, M. R. (2001), 'Tensegrity spline beam and grid shell structures', *Engineering Structures*, 23, 29–39.

Allen, E. and Zalewski, W. (2009), *Form and Forces: Designing Efficient, Expressive Structures*, Wiley, New York.

Asghar Bhatti, M. (2006), *Advanced Topics in Finite Element Analysis of Structures: With Mathematica and MATLAB Computations*, Wiley, New York.

Baraff, D. and Witkin, A. (1998), 'Large steps in cloth animation', in *SIGGRAPH 98 Computer Graphics Proceedings*, Orlando, FL, USA.

Barnes, M. and Wakefield, D. (1984), 'Dynamic relaxation applied to interactive form finding and analysis of air-supported structures', in *Proceedings of Conference on the Design of Airsupported Structures*, pp. 147–161.

Barnes, M. R., Adriaenssens, S. and Krupka, M. (2013), 'A novel torsion/bending element for dynamic relaxation modeling', *Computers and Structures*, 119, 60–67.

Barrett, R., Berry, M., Chan, T. F., Demmel, J., Donato, J., Dongarra, J., Eijkhout, V., Pozo, R., Romine, C. and der Vorst, H. V. (1994), *Templates for the Solution of Linear Systems: Building Blocks for Iterative Methods*, 2nd edn, SIAM, Philadelphia, PA.

Bendsøe, M. and Sigmund, O. (2003), *Topology Optimization: Theory, Methods and Applications*, Springer Verlag, Berlin and Heidelberg, Germany.

Billington, D. P. (1982), *Thin Shell Concrete Structures*, 2nd edn, McGraw Hill, New York.

Billington, D. P. (2003), *The Art of Structural Design: A Swiss Legacy*, Princeton University Art Museum, New Jersey.

Bletzinger, K.-U. (2011), 'Form finding and morphogenesis', in I. Munga and J. F. Abel, eds., *Fifty Years of Progress for Shell and Spatial Structures*, Multi-Science, Essex, pp. 459–482.

Bletzinger, K.-U. and Ramm, E. (2001), 'Structural optimization and form finding of light weight structures', *Computers and Structures*, 79, 2053–2062.

Bleuler, S., Laumanns, M., Thiele, L. and Zitzler, E. (2003), 'Pisa – a platform and programming language independent interface for search algorithms', in *Evolutionary Multi-Criterion Optimization Lecture Notes in Computer Science*, vol. 2632, Springer, Berlin and Heidelberg, pp. 494–508.

Block, P. (2009), 'Thrust network analysis: exploring three-dimensional analysis', PhD thesis, Massachusetts Institute of Technology, Cambridge, MA.

Block, P. and Lachauer, L. (2013), 'Advanced funicular analysis of masonry vaults', *Computers and Structures*.

Borgart, A. (2005), 'The relationship of form and force in (irregular) curved surfaces', in *Proceedings of the 5th International Conference on Computation of Shell and Spatial Structures*, Salzburg, Germany.

Burger, N. and Billington, D. P. (2006), 'Félix Candela, elegance and endurance: An examination of the Xochimilco shell', *Journal of the International Association of Shell and Spatial Structures*, 27(3), 271–278.

Calladine, C. R. (1983), *Theory of Shell Structures*, Cambridge University Press, Cambridge.

Candela, F. (1955), 'Encuestra espacios (an interview with Félix Candela)', *Espacios: Revista integral de arquitectura, plani cation, artes plasticos* (28).

Candela, F. (1973), 'New architecture', in D. P. Billington, J. F. Abel and R. Mark, eds., *The Maillart Papers*, Princeton, New Jersey, pp. 119–126.

Chilton, J. (2000), *The Engineer's Contribution to*

Contemporary Architecture: Heinz Isler, Thomas Telford, London.

Choi, K.-J. and Ko, H.-S. (2002), 'Stable but responsive cloth', *ACM Transactions on Graphics (TOG) – Proceedings of ACM SIGGRAPH 2002*, 21, pp. 604–611.

Cottrell, J. A., Hughes, T. J. R. and Bazilevs, Y. (2009), *Isogeometric Analysis: Toward Integration of CAD and FEA*, John Wiley and Sons, New York.

Cowan, H., Gero, J., Ding, G. and Muncey, R. (1968), *Models in Architecture*, Elsevier, Amsterdam and London.

Deb, K. (2001), *Multi-Objective Optimization using Evolutionary Algorithms*, Wiley, New York.

De Jong, K. A. (1975), Analysis of the behavior of a class of genetic adaptive systems', PhD thesis, Technical Report No. 185, University of Michigan.

Dems, K. (1991), 'First- and second-order shape sensitivity analysis of structures', *Structural Optimization*, 3(2), 79–88.

Dirac, P. (1975), *General Theory of Relativity*, Wiley, New York.

Draper, P., Garlock, M. E. and Billington, D. P. (2008), 'Finite-element analysis of Félix Candela's Chapel of Lomas de Cuernavaca', *Journal of Architectural Engineering*, 14(2), 47–52.

Dulácska, E. (1981), 'Explanation of the chapter on stability of the "recommendations for reinforced-concrete shells and folded plates", and a proposal to its improvement', *Bulletin of the IASS* (77).

Ebata, K., Cui, C. and Sasaki, M. (2003), 'Creation of free-formed shell by sensitivity analysis (application to structural design)' (in Japanese), in *Proceedings of Architectural Institute of Japan 2003*, pp. 269–270.

Eisenhart, L. P. (1909), *A Treatise on the Differential Geometry of Curves and Surfaces*, Ginn and Company, Boston.

Faber, C. (1963), *Candela: The Shell Builder*, Reinhold Publishing Corporation, New York.

Firl, M., Wüchner, R. and Bletzinger, K.-U. (2012), 'Regularization of shape optimization problems using FE-based parametrization, *Structural and Multidisciplinary Optimization*, 47(4), 507–521.

Fisher, M., Schröder, P., Desbrun, M., and Hoppe, H. (2007), 'Design of tangent vector fields', *ACM Transactions on Graphics (Proc. SIGGRAPH 2007)*, 26(3), 79–88.

Flügge, W. (1960), *Stresses in Shells*, Springer-Verlag, Berlin and Heidelburg.

Garlock, M. E. M. and Billington, D. P. (2008), *Félix Candela: Engineer, Builder, Structural Artist*, Princeton University Art Museum; Yale University Press, New Jersey; New Haven and New York.

Goldberg, D. (1989), *Genetic Algorithms in Search, Optimization, and Machine Learning*, Addison Wesley, Upper Saddle River, New Jersey.

Green, A. E. and Zerna, W. (1968), *Theoretical Elasticity*, 2nd edn, Oxford University Press, Oxford.

Gründig, L. (1976), 'Die Berechnung von vorgespannten Seil- und Hängenetzen unter Berücksichtigung ihrer topologischen und physikalischen Eigenschaften und der Ausgleichungsrechnung', PhD thesis, IAGB, University of Stuttgart, Munich.

Gründig, L. and Schek, H.-J. (1974), 'Analytical form finding and analysis of prestressed cable networks', in *International Conference on Tension Roof Structures*, London, UK.

Gründig, L., Linkwitz, K., Bahndorf, J. and Ströbel, D. (1988), 'Formfinding and Computer Aided Generation of the working drawings for the timber shell roof at Bad Dürrheim', *Structural Engineering Review*, 1, 83–90.

Haeckel, E. (1904), *Kunstformen der Natur – Kunstformen aus dem Meer*, Prestel Verlag.

Hajela, P. and Lee, E. (1995), 'Genetic algorithms in truss topological optimization', *International Journal of Solids and Structures* 32, 3341–3357.

Happold, E. and Liddell, W. I. (1975), 'Timber lattice roof for the Mannheim Bundesgartenschau', *The Structural Engineer*, 53(3), 99–135.

Harris, R., Romer, J., Kelly, O. and Johnson, S. (2003), 'Design and construction of the Downland gridshell', *Building Research and Information*, 31(6), 427–454.

Hennicke, J. (1975), *IL 10 Gitterschalen (Grid Shells)*, Institut für Leichte Flächentragwerke, Universität Stuttgart, Germany.

Heyman, J. (1977), *Equilibrium of Shell Structures*, Oxford Engineering Science Series, Clarendon Press, Oxford.

Heyman, J. (1995), *The Stone Skeleton: Structural Engineering of Masonry Architecture*, Cambridge University Press, Cambridge.

Heyman, J. (1996), *Arches, Vaults and Buttresses:*

Masonry Structures and their Engineering, Variorum, Aldershot, UK.

Heyman, J. (1998), 'Hooke's cubico-parabolical conoid', *Notes and Records of the Royal Society of London*, 52(1), 39–50.

Heyman, J. (2007), *Structural Analysis: A Historical Approach*, Cambridge University Press, Cambridge.

Hooke, R. (1676), *A Description of Helioscopes, and Some Other Instruments*, John Martyn, London.

Hormann, K., Lévy, B. and Sheer, A. (2007), 'Mesh parameterization: theory and practice', in *ACM SIGGRAPH 2007 Courses*, SIGGRAPH '07, ACM, New York.

Hossdorf, H. (1974), *Model Analysis of Structures*, Van Nostrand, New York.

Isler, H. (1960), 'New shapes for shells', *Bulletin of the International Association for Shell Structures* (8).

Isler, H. (1980a), *Heinz Isler as Structural Artist*, Princeton University Art Museum, New Jersey.

Isler, H. (1980b), 'New shapes for shells – twenty years after', *Bulletin of the International Association for Shell Structures* (71/72).

Isler, H. (1993), 'Generating shell shapes by physical experiments', *Bulletin of the International Association for Shell Structures*, 34, 53–63.

Isler, H. (2002), 'Letter from Heinz and Maria Isler to Phyllis and David Billington'.

Kawamura, H., Ohmori, H. and Kito, N. (2002), 'Truss topology optimization by a modified genetic algorithm', *Structural and Multidisciplinary Optimization*, 23, 467–472.

Kelly, K., Garlock, M. E. M. and Billington, D. P. (2010), 'Structural analysis of the cosmic rays laboratory', *Journal of the International Association of Shell and Spatial Structures*, 51(1), 17–24.

Kelvin, W. T. and Tait, P. G. (1867), *Treatise on Natural philosophy*, vol. 1, Clarendon Press, Oxford.

Kiendl, J. (2011), 'Isogeometric analysis and shape optimal design of shell structures', PhD thesis, Technische Universität München, Germany.

Kilian, A. and Ochsendorf, J. (2005), 'Particle-spring systems for structural form finding', *Journal of the International Association for Shell and Spatial Structures*, 46(148), 77–84.

Koza, J. R. (1992) *Genetic Programming: On the Programming of Computers by Means of Natural Selection*, MIT Press, Cambridge, MA.

Kueh, A. and Pellegrino, S. (2007), 'ABD matrix of single-ply triaxial weave fabric composites', in *48th AIAA Structures, Structural Dynamics, and Materials Conference*, Honolulu, Hawaii.

Lancaster, P. and Salkauskas, K. (1981), 'Surfaces generated by moving least squares methods', *Mathematics of Computation*, 37(155), 141–158.

Linkwitz, K. (1972), 'New methods for the determination of cutting pattern of prestressed cable nets and their application to the Olympic Roofs Münich', in Y. Yokoo, ed., *Proc. 1971 IASS Pacific Symposium Part II on Tension Structures and Space Frames*, Architectural Institute of Japan, Tokyo and Kyoto, pp. 13–26.

Linkwitz, K. (1999), 'Formfinding by the "direct approach" and pertinent strategies for the conceptual design of prestressed and hanging structures', *International Journal of Space Structures*, 14(2), 73–87.

Linkwitz, K. and Schek, H. J. (1971), 'Einige Bemerkungen zur Berechnung von vorgespannten Seilnetzkonstruktionen', *Ingenieur Archiv*, 40, 145–158.

Linkwitz, K., Schek, H.-J. and Gründig, L. (1974), 'Die Gleichgewichtsberechnung von Seilnetzen unter Zusatzbedingungen', *Ingenieur Archiv*, 43, 183–191.

Loop, C. T. (1992), 'Generalized B-spline surfaces of arbitrary topological type', PhD thesis, University of Washington.

Maurin, B. and Motro, R. (2004), 'Concrete shells form-finding with surface stress density method', *Journal of Structural Engineering*, 130(6), 961–968.

Meek, J. L. and Xia, X. (1999), 'Computer shape finding of form structures', *International Journal of Space Structures*, 14(1), 35–55.

Michalewicz, Z. (1996), *Genetic Algorithms + Data Structures = Evolution Programs*, Springer Verlag, Berlin and Heidelberg.

Mitchell, M. (1998), *An Introduction to Genetic Algorithms*, MIT Press, Boston.

Nervi, P. L. (1956), *Structures*, F. W. Dodge Corporation, New York.

Nocedal, J. and Wright, S. J. (2000), *Numerical Optimization*, Springer Verlag, Berlin and Heidelberg.

Nordenson, G. (2008), *Seven Structural Engineers: The Félix Candela Lectures*, The Museum of Modern Art, New York.

Novozhilov, V. V. (1959), *The Theory of Thin Shells*, Noordhoff, Groningen.

Ochsendorf, J. (2010), *Guastavino Vaulting: The Art of Structural Tile*, Princeton Architectural Press, Princeton.

Oden, T. (1987), 'Some historic comments on finite elements', in *Proceedings of the ACM Conference on History of Scientific and Numeric Computation*, HSNC '87, ACM, New York, pp. 125–130.

Pauletti, R. M. O. and Pimenta, P. M. (2008), 'The natural force density method for the shape finding of taut structures', *Computer Methods in Applied Mechanics and Engineering*, 197, 4419–4428.

Pedersen, P. (1989), 'On optimal orientation of orthotropic materials', *Structural Optimization*, 1(2), 101–106.

Petersen, K. B. and Pedersen, M. S. (2008), 'The matrix cookbook', version 20081110, www2.imm.dtu.dk/pubdb/p.php?3274.

Piga, C. (1996), *Storia dei modelli dal tempio di Salomone alla realtà virtuale*, Istituto Sperimentale Modelli e Strutture (ISMES), Italy.

Poleni, G. (1748), *Memorie storiche della gran cupola del Tempio Vaticano*, Padua.

Press, W., Teukolsky, S., Vetterling, W. and Flannery, B. (2007), *Numerical Recipes 3rd Edition: The Art of Scientific Computing*, Cambridge University Press, Cambridge.

Ramm, E. (2004), 'Shape finding of concrete shell roofs', *Journal of the International Association for Shell and Spatial Structures*, 45(144), 29–39.

Ramm, E. (2011), 'Heinz Isler shells – the priority of form', *Journal of the International Association for Shell and Spatial Structures*, 52(3), 143–154.

Ray, N., Li, W. C., Lévy, B., Sheffer, A. and Alliez, P. (2006), 'Periodic global parameterization', *ACM Transactions on Graphics*, 25(4), 1460–1485.

Rayleigh, L. (1890), 'I. on bells', *Philosophical Magazine Series 5*, 29(176), 1–17.

Reutera, M., Biasotti, S., Giorgi, D., Patane, G. and Spagnuolo, M. (2009), 'Discrete Laplace-Beltrami operators for shape analysis and segmentation', *Computers and Graphics*, 33(3), 381–390.

Richardson, J. N., Adriaenssens, S., Bouillard, P. and Filomeno Coelho, R. (2012), 'Multiobjective topology optimization of truss structures with kinematic stability repair', *Structural and Multidisciplinary Optimization*, 46, 513–532.

Rippmann, M. and Block, P. (2013), 'Rethinking structural masonry: Unreinforced, stone-cut shells', *Proceedings of the ICE – Construction Materials*.

Rippmann, M., Lachauer, L. and Block, P. (2012), 'Interactive vault design', *International Journal of Space Structures*, 27(4), 219–230.

Roland, C. (1965), *Frei Otto – Spannweiten. Ideen und Versuche zum Leichtbau. Ein Werkstattbericht*, Ullstein, Berlin.

Saad, Y. (2003), *Iterative Methods for Sparse Linear Systems*, 2nd edn, Society for Industrial and Applied Mathematics.

Sasaki, M. (2005), *Flux Structure*, Toto Publishers, Tokyo.

Schek, H.-J. (1974), 'The force density method for form finding and computation of general networks', *Computer Methods in Applied Mechanics and Engineering*, 3(1), 115–134.

Schumacher, A. (2013), *Optimierung mechanischer Strukturen. Grundlagen und industrielle Anwendungen*, Springer Verlag, Berlin and Heidelberg.

Schumacher, P. (2009), 'Parametricism – a new global style for architecture and urban design', *AD Architectural Design – Digital Cities* 79(4). Reprinted in: Mario Carpo, ed., *The Digital Turn in Architecture 1992-2010: AD Reader*, John Wiley and Sons, 2012.

Schumacher, P. (2012), *The Autopoiesis of Architecture, Volume 2, A New Agenda for Architecture,* John Wiley and Sons, New York.

Segal, E., Garlock, M. and Billington, D. (2008), 'A comparative analysis of the Bacardí Rum factory and the Lambert-St. Louis airport terminal', in *Proceedings of the International Symposium IASS-SLTE 2008*, Acapulco, Mexico.

Singer, P. (1995), 'Die Berechnung von Minimalflächen, Seifenblasen, Membrane und Pneus aus geodätischer Sicht', PhD thesis, University of Stuttgart.

Stam, J. (2009), 'Nucleus: Towards a unified dynamics solver for computer graphics', in *2009 Conference Proceedings: IEEE International Conference on*

Stevin, S. (1586), *De Beghinselen der Weeghconst*, Druckerye van Christoffel Plantijn, Leyden.

Struik, D. J. (1988), *Lectures on Classical Differential Geometry*, 2nd edn, Dover Publications, New York.

Sullivan, L. H. (1896), 'The tall office building artistically considered', *Lippincott's Magazine*, 57.

Thrall, A. P. and Garlock, M. E. M. (2010), 'Analysis of the design concept for the Iglesia de la Virgen de la Medalla Milagrosa', *Journal of the International Association of Shell and Spatial Structures*, 51(1), 27–34.

Timoshenko, S. (1953), *History of Strength of Materials*, McGraw Hill (Reprint: Dover Publications, 1983), New York.

Timoshenko, S. P. and Goodier, J. N. (1970), *Theory of Elasticity*, 3rd edn, McGraw-Hill, New York.

Tomlow, J. (2011), 'Gaudí's reluctant attitude towards the inverted catenary', *Proceedings of the Institution of Civil Engineers, Engineering History and Heritage*, 164(4), 219–233.

Varignon, P. (1725), *Nouvelle mécanique, ou Statique, dont le projt fut donné en M. DC. LXXXVII. Tome 1*, Jombert, Paris.

Veenendaal, D. and Block, P. (2012), 'An overview and comparison of structural form finding methods for general networks', *International Journal of Solids and Structures*, 49, 3741–3753.

Waller, R. (1705), *The Posthumous Works of Robert Hooke Containing his Cutlerian Lectures and Other Discourses*, Robert Waller, London.

Williams, C. J. K. (2001), 'The analytic and numerical definition of the geometry of the British Museum Great Court roof', in M. Burry, S. Datta, A. Dawson and A. J. Rollo, eds, *Mathematics and Design*, Deakin University, Geelong, Victoria, Australia, pp. 434–440.

Winslow, P., Pellegrino, S. and Sharma, S. (2010), 'Multi-objective optimization of free-form grid structures', *Structural and Multidisciplinary Optimization*, 40(1), 257–269.

Wright, D. T. (1965), 'Membrane forces and buckling in reticulated shells', *Journal of the Structural Division, ASCE*, 91(ST 1).

Zorin, D., Schroder, P., DeRose, A., Kobbelt, L., Levin, A. and Sweldens, W. (2000), 'Subdivision for modeling and animation', *SIGGRAPH 2000 Course Notes*.

List of contributors

Editors

Sigrid Adriaenssens is a structural engineer and Assistant Professor at the Department of Civil and Environmental Engineering at Princeton University, USA, where she directs the Form Finding Lab. She has held a PhD in lightweight structures from the University of Bath since 2000, adapting the method of dynamic relaxation to strained gridshells. She worked as a project engineer for Jane Wernick Associates, London, and Ney + Partners, Brussels, Belgium, where she found the form for the 2011 courtyard roof of the Dutch National Maritime Museum in Amsterdam. At Princeton, she co-curated the exhibition 'German Shells: Efficiency in Form' which examined a number of landmark German shell projects.

Philippe Block is a structural engineer and architect and Assistant Professor at the Institute of Technology in Architecture, ETH Zurich, Switzerland, where he directs the BLOCK Research Group. He studied at the Vrije Universiteit Brussel (VUB), Belgium and MIT, USA, where he earned his PhD in 2009. He is a founding partner of Ochsendorf, DeJong & Block LLC, where he consults on the stability of historic masonry structures and develops new, unreinforced stone and brick structures. He has received the Hangai Prize (2007) and Tsuboi Award (2010) from the International Association of Shells and Spatial Structures (IASS) as well as the Edoardo Benvenuto Prize (2012). He developed thrust network analysis for the analysis of historic vaulted masonry and design of new funicular shells.

Diederik Veenendaal is a civil engineer and a research assistant at the BLOCK Research Group, ETH Zurich, Switzerland. He received his Masters from TU Delft, Netherlands, on the form finding and evolutionary optimization of fabric formed beams. He started his career at Witteveen+Bos engineering consultants, Netherlands, working on groundfreezing calculations and safety analysis for the downtown subway stations of the North/South subway line in Amsterdam and the structural design for the largest tensioned membrane roof in the Netherlands, the ice skating arena De Scheg. In 2010, he started his doctoral research at ETH, comparing existing form-finding methods and developing new ones for flexibly formed shells and other structural systems.

Chris Williams is a structural engineer and a Senior Lecturer at the University of Bath, UK. He has been Visiting Professor at the Royal Academy of Fine Arts, Copenhagen, Denmark. He specializes in computational geometry and structural mechanics, in particular for lightweight structures and tall buildings, and his work has been applied by architects and engineers, including Foster + Partners, Rogers Stirk Harbour + Partners and Buro Happold. He worked at Ove Arup and Partners, where he was responsible for structural analysis of the Mannheim Multihalle. Since then, he has defined the geometry and performed nonlinear structural analysis for such projects as the British Museum Great Court roof, Weald & Downland Museum gridshell, and the Savill Gardens gridshell.

From left to right: Sigrid Adriaenssens, Philippe Block, Diederik Veenendaal and Chris Williams

Co-authors

Bill Addis works as a consulting engineer and is an Affiliated Lecturer at the School of Architecture, University of Cambridge, where he had studied engineering. He worked as a design engineer with Rolls-Royce Aeroengines and later studied at the University of Reading where he completed his PhD in the history and philosophy of engineering. He has authored many publications including *Building: 3,000 Years of Design, Engineering and Construction*, and is co-editor of *Construction History*.

Shigeru Ban is an architect, head of Shigeru Ban Architects and Professor at Kyoto University of Art and Design. He studied at the Southern California Institute of Architecture and the Cooper Union, graduating in 1984. He worked for Arata Isozaki, before opening his own studio in Tokyo in 1985. He has won many honours and awards, and worked on seminal projects such as the Japan Pavilion at Expo 2000 in Hannover with Frei Otto and the 2010 Centre Pompidou in Metz.

Mike Barnes is a structural engineer, and Professor Emeritus of Building Engineering at Bath University, where he originally succeeded Ted Happold. He is winner of the IStructE Lewis Kent Award (2005) and was Visiting Professor at the Bauhaus. He is involved in the assessment of the Montreal Olympic Stadium roof, and performed checking analysis of the Millennium Dome. He is famous for applying and developing dynamic relaxation for the form finding and structural design of lightweight and widespan structures.

Shajay Bhooshan is an architect and Lead Designer and head of the computational design group ZHA|CODE at Zaha Hadid Architects, having previously worked at Populous on projects such as the Millennium Dome. He completed his Masters at the Architectural Association Design Research Laboratory, London in 2006. He is studio tutor at the AA-DRL and director of the AA Visiting School India. He has applied his expertise in computation and form finding to ZHA's design practice and several shell pavilions.

David P. Billington is a structural engineer and the Gordon Y. S. Wu Professor of Engineering Emeritus in the Department of Civil and Environmental Engineering at Princeton University. He is a member of the National Academy of Engineering and a Fellow of the American Academy of Arts and Sciences. He is an eminent authority on thin-shell concrete structures, and has authored numerous detailed books on structural engineers such as Robert Maillart, Christian Menn, Heinz Isler and Félix Candela.

Kai-Uwe Bletzinger is a civil engineer and Professor at the Chair of Structural Analysis at TUM, Germany. He obtained his doctorate in 1990 at the University of Stuttgart and in 1996 accepted the Associate Professorship of Numerical Methods in Statics at the University of Karlsruhe. He is on the editorial board of *Computers & Structures* and review editor of *Structural and Multidisciplinary Optimization*. He has developed methods for form finding, shape optimization and patterning of lightweight shell and membrane structures.

Philippe Bouillard is a civil engineer and became Associate Professor in 1999 and Full Professor in 2005 at Université Libre de Bruxelles (ULB), Belgium. He started his career as a project manager for French contractors Dumez EPS and SAE. His main research activity is computational dynamics and vibro-acoustics with a special emphasis on verification and validations of numerical models and on the development of generalized finite element formulations.

Rajan Filomeno Coelho is a civil engineer and Associate Professor at the Building, Architecture & Town Planning Department (BATir) at ULB, Belgium. He completed his PhD in 2004 at the ULB, followed by postdoctoral studies at Cenaero, Belgium, and at the Université de Technologie de Compiègne, France. His current research interests include multidisciplinary design optimization, multicriteria evolutionary algorithm, metamodelling techniques and optimization under uncertainty.

Maria E. Moreyra Garlock is a civil engineer and Assistant Professor of Civil Engineering at Princeton University. Before becoming a professor, she spent several years as a consulting structural engineer in New York City designing concrete and steel structures with Leslie E. Robertson Associates. She has published numerous articles on the works of Félix Candela and co-authored the 2008 book *Félix Candela: Engineer, Builder, Structural Artist*.

Richard Harris is a structural engineer and Professor of Timber Engineering at the University of Bath, after working as Technical Director with Buro Happold. He worked with large international contractors, on projects in the UK and overseas, before joining Buro Happold in 1984. He has published several award-winning papers and was responsible for a wide range of timber engineering projects, including the Globe Theatre, the Downland Gridshell and the Savill Building.

Sawako Kaijima is an architect and Assistant Professor in Architecture and Sustainable Design at the Singapore University of Technology and Design. She obtained her Masters in 2005 from MIT after her Bachelors in Environmental Information from Keio University, Japan. While working at AKT, she provided consultancy for high-profile projects by architecture practices such as ZHA, Thomas Heatherwick, Fosters + Partners, etc. She co-developed software for the intuitive use of structural engineering methods in design.

Axel Kilian is an architect and Assistant Professor for Computational Design at Princeton University. He previously taught Computational Design at TU Delft and at the Department of Architecture at MIT, and is co-author of the book *Architectural Geometry*. He has lectured and published widely on his research on the role of computational design in design exploration. He is known for pioneering the application of particle-spring systems to structural form finding and design.

Lorenz Lachauer is an architect and a research assistant at the Chair of Structural Design and the BLOCK Research Group, both ETH Zurich. He started his career at Herzog & de Meuron, first as project architect and later as member of the Digital Technology Group. His research is focused on equilibrium-based design methods for structural design in architecture, and concepts which allow for the integration of structural constraints into design processes.

Klaus Linkwitz is a retired geodetic and structural engineer. He received his doctorate in Munich in 1960 and was appointed professor at the University of Stuttgart in 1960, receiving honorary degrees from ETH Zurich (1993) and Donetsk National Technical University (1995). He worked on the Munich Olympic Roofs in 1972 and is famous for developing the force density method which is widely used in the design of membrane, but also shell structures.

Irmgard Lochner-Aldinger is a structural engineer and Professor at Biberach University of Applied Sciences, as well as associate at Peter & Lochner Consulting Engineers, Stuttgart. Following work as a project engineer for Professor Polónyi and Arup, both in Berlin, she obtained her doctorate from the Institute for Lightweight Structures and Conceptual Design (ILEK) in 2005. She teaches and publishes on the application of optimization methods in structural and architectural design.

Tomás Méndez Echenagucia is an architect and a PhD candidate at the Polytechnic University of Turin. He holds a double degree in Architecture from Central University of Venezuela (UCV) and Politecnico di Torino. He was awarded the IASS Hangai Prize in 2008. His research focuses on multi-disciplinary search and optimization tools for the early stages of architectural design, including acoustic, structural and environmental design.

Panagiotis Michalatos is an architect and Assistant Professor of Architectural Technology at the Harvard Graduate School of Design. He holds a Master of Science in Applied IT from Chalmers Technical University, Sweden. While working as a computational design researcher for structural engineering firm AKT, he provided consultancy and computational solutions for a range of high-profile projects. He co-developed a range of software applications for the intuitive and creative use of structural engineering methods in design.

Laurent Ney is a structural engineer and founder of Ney & Partners in Brussels, Luxembourg and Japan. He previously worked from 1989 to 1996 at Bureau d'études Greisch in Liège, and has lectured on construction stability since 2005 in several Belgian universities. His practice is characterized on design by research: optimization and form finding. Constructability and sustainability are integral parts of his designs.

John Ochsendorf is a structural engineer and the Class of 1942 Professor of Architecture and Civil and Environmental Engineering at the MIT, where he directs the Structural Design Lab and the Masonry Research Group. He is a founding partner of Ochsendorf, DeJong & Block LLC. He is the author of the 2010 book *Guastavino Vaulting: The Art of Structural Tile*. He won the Rome Prize and was named a MacArthur Fellow in 2008.

Daniele Panozzo is a computer scientist and a post-doctoral researcher at the Interactive Geometry Lab, ETH Zurich. He studied at the University of Maryland, was a visiting researcher at the Courant Institute of Mathematical Sciences, New York University and earned his PhD in Computer Science from the University of Genova in 2012. His research interests are digital geometry processing, shape modelling, computational photography and architectural geometry.

Alberto Pugnale is an architect and a Lecturer in Architectural Design at the University of Melbourne. Previously, he was Assistant Professor at Aalborg University in Denmark. He has won the IASS Hangai Prize (2007) and a grant from the ISI Foundation in Turin (2008), related to the computational morphogenesis of freeform structures and complex architectural–structural bodies. His research interests are also in the fields of reciprocal structures and construction history.

Ekkehard Ramm is a civil engineer and Professor Emeritus of the Department of Civil and Environmental Engineering at University of Stuttgart. After postdoctoral positions in Berkeley and Stuttgart, he became an Associate Professor in 1976, and Full Professor (Chair) in 1983 and was Head of the Institute of Structural Mechanics until 2006. He received the IASS Tsuboi Prize and Torroja Medal, and the IACM Gauss–Newton Medal. The central theme of his research on structural mechanics is shell structures.

James N. Richardson is a structural engineer and architect and an FNRS fellow and PhD candidate at the Universite Libre de Bruxelles. He studied applied mathematics at the University of Cape Town, civil engineering and architecture at the Vrije Universiteit Brussel and the Technical University of Eindhoven, and worked as a structural engineer at Guy Nordenson & Associates. His research interests include discrete topology optimization, optimization under uncertainty and topology optimization of large-scale structures.

Matthias Rippmann is an architect, a research assistant at the BLOCK Research Group, ETH Zurich, and founding partner of design and consulting firm ROK. He graduated from the University of Stuttgart in 2007. He worked for LAVA and Werner Sobek Engineers as an architect and programmer on projects such as Stuttgart 21 and the Heydar Aliyev Centre, and studied at the Institute for Lightweight Structures (ILEK). His research is focused on structural form finding linked to construction-aware design strategies.

Mutsuro Sasaki is a structural engineer, head of the structural consulting office Sasaki and Partners (SAPS) and Professor at Hosei University in Tokyo. He established SAPS in 1980, after working with Kimura Structural Engineers for ten years. Since, he has pushed the envelope on advanced structural design techniques and collaborated with renowned architects such as Toyo Ito and SANAA, engineering several recent reinforced concrete shells, such as the EPFL Rolex Learning Centre in Lausanne, Switzerland.

Mario Sassone is a structural engineer and Assistant Professor at the Polytechnic University of Turin. His research concerns the computational analysis of time-dependent effects on steel, concrete and composite structures, and the structural and architectural optimization of reinforced concrete shells by means of artificial intelligence techniques. He is currently working on bridging the gap between the architectural issues and the computational approaches to structural and physical problems.

Jörg Schlaich is a structural engineer, founding partner of Schlaich Bergermann und Partner (SBP), and Professor Emeritus of the University of Stuttgart. He studied architecture and civil engineering in Stuttgart and Berlin, before obtaining his Masters from Case Western Reserve University in Cleveland in 1960. He joined Leonhardt und Andrä as design engineer and became partner in 1970, before founding SBP in 1980, focusing on lightweight structures. He is the developer of the Solar Updraft Tower and is largely credited for promoting strut-and-tie modelling for reinforced concrete.

Axel Schumacher is a mechanical engineer and Professor of Numerical Methods and Optimization in Product development at the University of Wuppertal (BUW). He studied at the Universities of Duisburg and Aachen, and performed research at the University of Siegen. Between 1999 and 2002 he was project leader for structural optimization at Adam Opel AG, and previously was professor of automotive engineering at the University of Applied Sciences in Hamburg.

Patrik Schumacher is an architect and partner and managing director at Zaha Hadid Architects in London. He has been the co-author of many key projects such as the National Italian Museum for Art and Architecture of the 21st Century (MAXXI) in Rome. His interest in shell structures is part and parcel of a more general interest in advanced structural engineering and its capacity to handle, shape and exploit complex, differentiated geometries via relative optimization strategies.

Paul Shepherd is a structural engineer and mathematician and Lecturer in Digital Architectonics at the University of Bath. He studied at Cambridge University and obtained a PhD in Structural Engineering at Sheffield University. At Buro Happold, he developed modelling and analysis software, and worked on projects such as Stuttgart 21 and the Japan Pavilion for the Expo 2000. He now applies this practical knowledge to his research into the design and optimization of complex geometry buildings.

Olga Sorkine-Hornung is a computer scientist and Assistant Professor of Computer Science at ETH Zurich, where she leads the Interactive Geometry Lab at the Institute of Visual Computing. She earned her degrees at Tel Aviv University, and became Assistant Professor at the Courant Institute of Mathematical Sciences, New York University in 2008. She has received many honour and awards such as the EUROGRAPHICS Young Researcher Award (2008) and the ACM SIGGRAPH Significant New Researcher Award (2011).

Tom Van Mele is a structural engineer and architect and a postdoctoral researcher at the BLOCK Research Group, ETH Zurich. He studied at the Vrije Universiteit Brussel, Belgium, where he earned his PhD in 2009 on scissor-hinged membrane structures. He won the Tsuboi Award (2010) from the IASS. His current research involves three-dimensional collapse behaviour of masonry vaults and graphic-statics-based form-finding methods. He is also the project lead and lead developer of eQuilibrium, an online graphic statics platform.

Peter Winslow is a structural engineer at Expedition Engineering in London. His PhD, awarded in 2009 by the University of Cambridge, involved developing new tools for the design of freeform structures with the Buro Happold SMART group. Awarded the IASS Tsuboi Prize in 2007, he has worked on several high-profile projects including the London 2012 Olympic Velodrome, Stockton Infinity Footbridge and the Stavros Niarchos Foundation Cultural Centre.

List of credits

Unless otherwise noted, all copyrights and credits of the figures belong to the authors of the corresponding chapter. The following images are in the public domain: Figs. 1.1, 1.2, 14a, 3.8, 4.1–5, 17.16 (left, middle). Several images have been licensed under Creative Commons. The corresponding legal codes can be found here:

- CC BY 2.0: http://creativecommons.org/licenses/by/2.0/legalcode
- CC BY 3.0: http://creativecommons.org/licenses/by/3.0/legalcode
- CC BY-SA 1.0: http://creativecommons.org/licenses/by-sa/1.0/legalcode
- CC BY-SA 3.0: http://creativecommons.org/licenses/by-sa/3.0/legalcode

Foreword by Jörg Schlaich: © Schlaich, Bergermann und Partner, Figs. 0.1, 0.4, 0.5, 0.6b, 0.6c, 0.7, courtesy Pier Luigi Nervi Project Association, Fig. 0.2, courtesy Avery Architectural and Fine Arts Library, Columbia University, from Faber (1963), Fig. 0.3, © Prof. Volkwin Marg, gmp Architekten von Gerkan, Marg und Partner, Fig. 0.6a.

Foreword by Shigeru Ban: © Shigeru Ban Architects, Hiroyuki Hirai, Fig. 0.8. CC BY 2.0 Jean-Pierre Dalbéra, Fig. 0.9. © Blumer-Lehmann AG, Fig. 0.10.

Chapter 1: © Iwan Baan, cover, courtesy Avery Architectural and Fine Arts Library, Columbia University, Figs. 1.4a, 1.6a, © Michael Freeman, The Bay Trust, Fig. 1.4b © Helionix Designs, Kevin Francis, Fig. 1.4c, © Philippe Block, Fig. 1.6b.

Chapter 2: © Ney and Partners, Jean-Luc Deru, photo-daylight, cover, Figs. 2.1, 2.6, © Robert Hyrum Hirschi, Fig. 2.3, © Yiannis Kouzoumis, Architect N.T.U.A., Greece, Fig. 2.4, © Ney and Partners, Fig. 2.5.

Chapter 3: © Roberto dello Noce, cover, CC BY 2.0, Horia Varlan, Fig. 3.3, CC BY-SA 1.0, Hennessy, Fig. 3.10, © Heinz Isler Archive, GTA Archive, ETH Zurich, Fig. 3.14.

Chapter 4: CC BY-SA 3.0 Peter Lebato, cover, © Institute for Lightweight Structures and Conceptual Design (ILEK), University of Stuttgart, Figs. 4.9, 4.12–14, © Heinz Isler Archive, GTA Archive, ETH Zurich, Fig. 4.8, © Arup, Figs. 4.10–11, © Ian Liddell, Fig. 4.15.

Chapter 5: © Heinz Isler Archive, GTA Archive, ETH Zurich, cover, Fig. 5.2a, © Institute for Lightweight Structures and Conceptual Design (ILEK), University of Stuttgart, Fig. 5.3, © Ekkehard Ramm, Fig. 5.6.

Chapter 6: CC BY 3.0 Hubert Berberich, cover, © Diederik Veenendaal, Figs. 6.1, 6.7, 6.9–11, © Klaus Linkwitz and Diederik Veenendaal, Fig. 6.8, CC BY-SA 3.0 Sigrid Adriaenssens, Philippe Block, Diederik Veenendaal and Chris Williams, Fig. 6.12.

Chapter 7: © BLOCK Research Group, ETH Zurich, Klemen Breitfuss, cover, Fig. 7.1, © Philippe Block, Figs. 7.5–9, 7.11, © Diederik Veenendaal and Philippe Block, Fig. 7.13, © Matthias Rippmann, Fig. 7.15.

Chapter 8: CC BY-SA 3.0 Oosoom, cover, Fig. 8.1, © Robert Greshoff, Buro Happold, Fig. 8.4, © Maryanne Wachter, Figs. 8.5–6, © Tony Jones, Fig. 8.12 (top), © Richard Harris, Fig. 8.12 (bottom), © Green Oak Carpentry Company, Fig. 8.13.

Chapter 9: © Shajay Bhooshan, Zaha Hadid Architects, cover, Figs. 9.1–2, 9.4–11.

Chapter 10: CC BY-SA 3.0 Daniel Schwen, cover.

Chapter 11: CC BY-SA 2.0, Cecilia Roussel, cover.

Chapter 12: © Kur- und Bäder GmbH, Bad Dürrheim, cover, Fig. 12.1 (top), CC BY 3.0 Hubert Berberich, Fig. 12.1 (bottom), © Herbert Schötz Fig. 12.4 (top), © Fritz Wenzel and Klaus Linkwitz, Figs. 12.4 (bottom), 12.7–8, 12.9 (top), © Elsevier Science and Technology Journals, from Gründig, Linkwitz, Bahndorf and Ströbel (1988), Figs. 12.5–6, © Lothar Gründig, Fig. 12.9 (bottom), Deutsche Geodätische Kommission (DGK), from Gründig (1976), Figs. 12.12–13.

Chapter 13: © Bert Kaufmann, cover.

Chapter 14: ⓒ ⓘ 2.0 Simon Whitehead, cover.

Chapter 15: © Evan Chakroff, evanchakroff.com, © Studio Fuksas, photo by Ramon Prat, Fig. 15.1, © Peter Winslow, Buro Happold, Fig. 15.2.

Chapter 16: courtesy Pier Luigi Nervi Project Association, cover.

Chapter 17: © Nicholas Worley, cover, © Gianni Birindelli, Fig. 17.1, © Diederik Veenendaal, Fig. 17.7.

Chapter 18: © Ken Lee, cover.

Chapter 19: ⓒ ⓘ-ⓢ 2.0, Ian Muttoo, © Institute for Lightweight Structures and Conceptual Design (ILEK), University of Stuttgart, Figs. 19.1, 19.2 (top), 19.3, 19.4 (top), © Ian Liddell, Fig. 19.2 (bottom), 19.4 (bottom).

Chapter 20: © Princeton University Library, cover, Figs. 20.1a, 20.5, 20.7, 20.10, 20.15, © Frank Döring, Figs. 20.1b, 20.13, © Maria E. Moreyra Garlock, Fig. 20.2, © Heinz Isler Archive, GTA Archives, ETH Zurich, Figs. 20.3, from Isler (1960), 20.17, courtesy Avery Architectural and Fine Arts Library, Columbia University, Figs. 20.4, 20.6 (bottom) from Faber (1963) © Princeton University Library, Bruce Whitehead, Fig. 20.6 (top), © David P. Billington, Figs. 20.8–9, 20.11, 20.13, 20.18.

Chapter 21: © Office of Ryue Nishizawa, cover, © Princeton University Library, Fig. 21.1, © Sasaki and Partners, SAPS, Fig. 21.2–3. 21.4 (top), 21.5 (bottom), 21.6 (bottom), 21.8–18, © Takenaka Corporation Co. Ltd., Fig. 21.4 (bottom), © Toyo Ito and Associates, Fig. 21.5 (top), © SANAA, Fig. 21.6 (top), © Teshima Art Museum, photo by Noboru Morikawa, Fig. 21.7 (top), © Teshima Art Museum, photo by Ken'ichi Suzuki Fig. 21.7 (bottom).

Conclusion by Patrik Schumacher: © Zaha Hadid Architects, Fig. 21.1–2.

Appendix C: © Alberto Pugnale and Tomás Méndez Echenagucia, Fig. C.1, © James N. Richardson and Rajan Filomeno Coelho, Fig. C.2.

List of Contributors: © Sigrid Adriaenssens, Philippe Block, Diederik Veenendaal and Chris Williams, photo by Regine van Limmeren.

List of projects

Plan or projected areas are denoted with an asterisk (*). Estimated numbers are denoted with a plus-minus sign (±). The thickness of gridshells is given as width times (×) height of the lattice elements. Demolished structures are denoted by a dagger symbol (†) behind their design and construction period. If ranges (–) are not given, then for spans the value is a maximum, and for thicknesses it is a minimum.

Project; Location Coordinates	Year	Architect; Engineer	Area (m²)	Span (m)	Thickness (mm)	Page
Aichtal Outdoor Theatre, or Naturtheater Grötzingen; Grötzingen, Germany 48°37′20.23″N 9°15′58.15″E	1977	Michael Balz; Heinz Isler	600	42	90–120	31, 251, 253, 255
Algeciras Market Hall; Algeciras, Spain 36°7′44.37″N 5°26′44.05″W	1932	Manuel Sánchez Arcas; Eduardo Torroja	±1,775*	47.5	89–457	260
Alster-Schwimmhalle; Hamburg, Germany 53°33′36.44″N 10°1′17.49″E	1967	Niessen und Störmer; Schlaich, Bergermann und Partner		96		vii, x
Bacardi Rum Bottling Factory; Cuautitlan, Mexico 19°37′41.08″N 99°11′28.57″W	1959	Saenz-Cancio-Martin and Gutierrez; Félix Candela	±4,056	36.8	40	246, 255, 257, 260, 268
British Museum, Queen Elizabeth II Great Court roof; London, UK 50°31′9.07″N 0°7′45.99″W	2000	Foster + Partners; Buro Happold	6,700	13–39	92×73	238–44
Bronx Zoo Elephant House dome; New York, USA 40°51′5.32″N 73°52′45.74″W	1909	Heins & LaFarge; Guastavino Company				9
Bundesgartenschau Pavilion; Stuttgart, Germany 48°47′55.63″N 9°12′18.29″E	1977†	Hans Luz und Partner, Landschaftsarchitekten; Schlaich, Bergermann und Partner	±3,000*	10–26	12–15	viii–ix
Centre Pompidou Metz; Metz, France 49°6′29.54″N 6°10′53.80″E	2006–10	Shigeru Ban; Blumer-Lehmann AG, sjb.kempter.fitze AG	±8,000*	20–45, 22 cantilever	140×440	xii–xiii
Chapel Lomas de Cuernavaca; Morelos, Mexico 18°52′33.83″N 99°12′9.89″W	1958–59	Guillermo Rossell and Manuel Larossa; Félix Candela	±360*	18–31	40	250–1, 255, 257
Chiddingstone Orangery; Kent, UK 51°11′5.61″N 0°8′28.20″E	2007	Peter Hulbert Arquitectos; Buro Happold	±95*	5–12	30×45	91–2

LIST OF PROJECTS

Project; Location Coordinates	Year	Architect; Engineer	Area (m²)	Span (m)	Thickness (mm)	Page
Church of Our Lady of the Miraculous Medal (Milagrosa); Naruate, Mexico 19°22'46.14"N 99°9'26.37"W	1954–55	Félix Candela	1,530*	21–11	40	250–1, 254–5, 257
Church of San José Obrero (St Joseph the Laborer); Monterrey, Neuvo León, Mexico 25°43'48.54"N 100°17'47.79"W	1959	Enrique de la Mora and Fernando Lopez Carmona; Félix Candela	±830*	30	40	viii–ix
Colònia Güell; Barcelona, Spain 41°21'49.90"N 2°1'40.25"E	1908–17	Antoni Gaudí	±866*	±7		28, 37–8, 130, 132
Cosmic Rays Laboratory; Mexico City, Mexico 19°20'2.19"N 99°10'52.47"W	1951	Jorge Gonzales Reyna; Félix Candela	±129	12	15–50	248, 257
Deitingen motorway BP service station; Deitingen, Switzerland 47°13'33.34"N 7°36'59.38"W	1968	Heinz Isler	±316*	31.6	90	253, 255
Deubau; Essen, Germany 51°25'38.45"N 6°59'38.22"E	1962†	Frei Otto	225	15	40×60	41
Downland gridshell; Chichester, UK 50°54'24.10"N 0°45'23.94"W	2002	Edward Cullinan Architects; Buro Happold	50×14.25	12.5–16	23.5×50	98–101
Duomo, Basilica di Santa Maria del Fiore (Basilica of Saint Mary of the Flower); Florence, Italy 43°46'23.09"N 11°15'25.37"E	1296–1436	Arnolfo di Cambio; Filippo Brunelleschi	±2,500*	42	2,200–2,400 (inner dome) 450–900 (outer dome)	36
Gaoliang, or Jade Belt Bridge, Summer Palace; Beijing, China 39°58'48.68"N 116°16'25.53"E	1751–64	...		11.38		28–9
Garden Centre Florélite; Plaisir, France 48°47'50.38"N 1°56'59.20"E	1966	Heinz Isler	1,400*	41	80	38
Gateway Arch; St Louis, USA 38°37'28.27"N 90°11'4.96"W	1947–65	Eero Saarinen, Saarinen and Associates; Hannskarl Bandel, Severud Associated	N/A	192 (height)	16,000–5,200 (triangle side)	114, 117
Gatti Wool Mill; Rome, Italy 41°53'29.48"N 12°35'44.48"E	1951–53	Aldo Arcangeli, Pier Luigi Nervi, Nervi & Bartoli	±2,400	5	300 (ribs)	194, 196
Gießhauss; Kassel, Germany 51°19'14.11"N 9°30'26.16"E	1837	Carl-Anton Henschel	±200	16	320–175	36–7

LIST OF PROJECTS

Project; Location Coordinates	Year	Architect; Engineer	Area (m²)	Span (m)	Thickness (mm)	Page
Gringrin Island City Park; Fukuoka, Japan 33°39'49.86"N 130°25'9.12"E	2002–05	Toyo Ito; Mutsuro Sasaki	5,040	70	400	262–3
Haesley Nine Bridges Golf Clubhouse; Yeoju, South Korea 33°21'2.96"N 126°23'14.48"E	2006–09	Shigeru Ban, Kyeongsik Yoon; sjb.kempter.fitze AG, Blumer-Lehmann AG	2,592*			xiii
Hagia Sophia; Istanbul, Turkey 41°0'29.95"N 28°58'48.32"E	532–37	Isidore of Miletus and Anthemius of Tralles	±5,986*	33		18
Heimberg tennis court; Heimberg, Switzerland 46°47'33.64"N 7°35'43.26"E	1979	Heinz Isler	±900*	48.5	90–100	255–6
Hippo House, Berlin Zoo; Berlin, Germany 52°30'32.77"N 13°20'11.61"E	1996	J. Gribl; Schlaich, Bergermann und Partner	±520*	29	60–40	92–3
Hyperthreads, AA Visiting School; Bangalore, India 12°56'30.29"N 77°33'57.43"E	2011	Zaha Hadid Architects, ZHA|CODE, Abhishek Bij, Design plus India; CS Yadunandan and Deepak	48	6	80	102–6, 110–13
Hyperthreads, AA Visiting School; Mexico City 19°21'35.21"N 99°15'29.61"W	2011	Shajay Bhooshan, Mostafa El Sayed (Zaha Hadid Architects, AADRL), Joshua Zabel (Kreysler and Associates), Alicia Nahmad and Knut Brunier (AADRL); Nathaniel M. Stanton and CRAFT	62.5	4.95; 2.75 cantilever	80	112
Japan Pavilion, 2000 Expo; Hannover, Germany 52°19'2.43"N 9°49'9.03"E	2000	Shigeru Ban, Frei Otto; Buro Happold	±3,090*	25.1	50–250	xii
Jerónimos monastery, Church of Santa Maria of Bélem Lisbon, Portugal 38°41'51.69"N 9°12'20.25"W	c. 1519–22 (vault)	Joã de Castilho, Diogo de Boitaca	±1,138* (nave)	±10 (nave)	70–100	156, 166–7
Kakamigahara Crematorium; Gifu, Japan 35°25'14.25"N 136°50'11.31"E	2004–06	Toyo Ito; Mutsuro Sasaki	2,265	20	200	224–5, 229, 262–3
Kitagata Community Centre; Gifu, Japan 35°26'8.78"N 136°40'57.62"E	2001–05	Arata Isozaki; Mutsuro Sasaki	4,495	25	150	262
Kresge Auditorium; Boston, USA 42°21'29.21"N 71°5'42.14"W	1955	Eero Saarinen; Anderson, Beckwith & Haible Ammann & Whitney	±1,030*	48.77	75–455	51–2

LIST OF PROJECTS

Project; Location Coordinates	Year	Architect; Engineer	Area (m²)	Span (m)	Thickness (mm)	Page
Los Manantiales Restaurant; Xochimilco, Mexico 19°14′54.81″N 99°5′34.56″W	1957–58	Félix Candela; Joaquin and Fernando Alvarez Ordoñez	±1,600*	42.4	40	251, 253, 255
Mapungubwe National Park Interpretive Centre; South Africa 22°14′29.86″S 29°24′17.30″E	2006–09	Peter Rich Architects; John Ochsendorf and Michael Ramage,	1,130*	5–14.5 (10 vaults)	300	6–7
Multihalle; Mannheim, Germany 49°30′16.73″N 8°28′46.14″E	1973–76	Atelier Frei Otto Warmbronn, Mutschler & Partners; Ove Arup & Partners, Bräuer Spaeh, Büro Linkwitz und Preuß with Lothar Gründig; Prof. Fritz Wenzel (checking engineer)	9,500	80	50×50	28, 40–2, 58–9, 99, 143–4, 146, 151–5, 238–44
Munich Olympic Roofs; Munich, Germany 48°10′23.09″N 11°32′47.79″E	1972	Behnisch und Partner, freie Architeckten, Munich/Stuttgart, Prof. Günther Behnisch, Fritz Auer, Winfried Büxel, Erhard Tränkner, Karlheinz Weber (design, concept and planning); Prof. Frei Otto, Warmbronn (development); Prof. Dr-Ing. Fritz Leonhardt, Dipl.-Ing. Wolfhard Andrä, Dipl.-Ing. Dr Jörg Schlaich (engineering); Prof. Dr-Ing. Klaus Linkwitz, Institut für Anwendungen der Geodäsie, University of Stuttgart, Prof. John H. Argyris, Institut für Statik und Dynamik der Luft- und Raumfahrtkonstruktionen, University of Stuttgart (cutting patterns)	74,800	450	750 × 750 mesh, 2· (Ø 11.7/16.5)	ix–x, 143, 147, 155
Murinsel; Graz, Austria 47°4′23.46″N 15°26′4.47″E	2003	Vito Acconci; Zenckner & Handl; Kurt Kratzer	958	20–50		92–3
Museum of Hamburg History; Hamburg, Germany 53°33′3.98″N 9°58′23.10″E	1989	Gerkan, Marg und Partner; Schlaich, Bergermann und Partner	±1,000*	17	Ø60–40	x–xi
New Milan Trade Fair; Milan, Italy 45°31′12.36″N 9°4′44.06″E	2002–05	Massimiliano and Doriana Fuksas architects; Schlaich Bergermann and Partners, Mero	50,600		60–160×60–350; 10–20 (thickness)	180–1

LIST OF PROJECTS

Project; Location Coordinates	Year	Architect; Engineer	Area (m²)	Span (m)	Thickness (mm)	Page
Palazzetto dello Sport; Rome, Italy 41°55'46.61"N 12°28'14.54"E	1956–58	Annibale Vitellozzi; Pier Luigi Nervi	±2,820*	60	25	viii–ix, 210–12
Pines Calyx, The; St Margaret's Bay, UK 51°8'56.36"N 1°22'54.69"E	2007	Phil Cooper; John Ochsendorf, Michael Ramage, Wanda Lau	±220	12	±70	9
Portcullis courtyard; London, UK 51°30'4.79"N 0°7'29.75"W	2001	Hopkins Architects; Ove Arup & Partners	1,250*	25	100×200	241
Rolex Learning Centre; Lausanne, Switzerland 46°31'5.86"N 6°34'6.11"E	2005–09	SANAA; Mutsuro Sasaki	39,000	80	400–800	262, 264
Rio's Warehouse; Lindavista, Mexico 19°29'7.17"N 99°9'57.51"W	1954	Félix Candela	150	15.25	40	251–2
Savill Building; Windsor Great Park, UK 51°25'35.45"N 0°35'50.22"W	2006	Glenn Howells Architects; Buro Happold	±2,250*	±25	2· (80×50)	88, 90, 99
Scheepvaartmuseum, or Dutch National Maritime Museum; Amsterdam, Netherlands 52°22'17.82"N 4°54'53.25"E	2011	Ney & Partners	±1,156	34	40/60×100–180	cover, 14, 16, 18–19
Shukhov tower, or Shabolovka tower; Moscow, Russia 55°43'2.50"N 37°36'41.64"E	1922	Vladimir Shukhov	±1,250*	40, 60 (height)	100×10	92
Sicli factory; Geneva, Switzerland 46°11'20.43"N 6°7'47.01"E	1969	C. Hiberer; Heinz Isler	±1,500	58	100	44, 48, 255
Solemar Therme; Bad Dürrheim, Germany 48°0'54.83"N 8°32'5.88"E	1987	Geier + Geier; Wenzel, Frese, Pörtner, Haller, Büro für Baukonstruktionen with Dipl.-Ing. R. Barthel (structural design); Linkwitz, Preuß, Büro für geodätische Meß- und Rechentechnik, with Dr-Ing. L. Gründig and Dipl.-Ing. J. Bahndorf	2,500	17	200×250	59, 142–4, 146–51, 154–5
St Paul's Cathedral, London, UK 51°30'49.08"N 0°5'54.05"W	1708	Robert Hooke, Christopher Wren	±900*	34	450	9, 32, 35–6

LIST OF PROJECTS

Project; Location Coordinates	Year	Architect; Engineer	Area (m²)	Span (m)	Thickness (mm)	Page
St Peter's Cathedral; Vatican City 41°54'7.55"N 12°27'11.65"E	1506–1626	Donato Bramante, Michelangelo, Carlo Maderno and Gian Lorenzo Bernini	±1,350*	41.9	300	36–7, 42
Sydney Opera House; Sydney, Australia 33°51'24.65"S 151°12'54.66"E	1973	Jørn Utzon; Ove Arup & Partners	16,000	50		11–12, 39–40, 42, 243
Temple of Mercury; Baiae, Italy 40°49'4.20"N 14°4'11.66"E	1st c. BC		±360*	21.5	600–1,600	20, 22, 87
Teshima Art Museum; Kagawa, Japan 34°29'22.49"N 134°5'28.56"E	2008–10	Ryue Nishizawa; Mutsuro Sasaki	2,040	43–60	250	258, 262–70
Thin tile vault at ETH Zurich; Zurich, Switzerland 47°24'31.22"N 8°30'21.59"E	2011†	BLOCK Research Group: Matthias Rippmann, Lara Davis, Philippe Block; Lara Davis, Tom Pawlofsky	28.6		90–140	70–2
TWA Flight Center; New York, USA 40°38'44.91"N 73°46'39.48"W	1962	Eero Saarinen and Associates; Ammann & Whitney			177–1,016	260–1
Wolfgang Meyer Sport Centre Hamburg-Stellingen; Hamburg, Germany 53°35'25.07"N 9°56'36.99"E	1994	Silcher, Werner und Partner; Schlaich, Bergermann und Partner	6,700	20	membrane	xi

Index

For an index of built shells and other structures mentioned throughout the book, please refer to the List of Projects. Pages in bold refer to the "Key concepts and terms" sections at the end of the chapters.

3D-print, 85

acceleration, 94, 107, 124, 276
active constraint, 177
Airy stress function, 24, 289
allele, 291
analytical shape, *see* mathematical shape
angle-preserving parameterization, 185
anisotropy, 45, 49, 184
anticlastic curvature, viii, 25, 60, 67, 104, 110, 286
Arcangeli, Aldo, 196
Arnoldi algorithm, 197, 205
Arup, Ove, 260
Auer, Fritz, x

backward Euler method, 107, 113, 126
Balz, Michael, 31
bending moment, 23, 125, 196, 288
bending stiffness, 26, 93, 242
Bergermann, Rudolf, x
BFGS, 187
BiCGSTAB, 109
biconjugate gradient stabilized method, *see* BiCGSTAB
biharmonic equation, 24
binary, 173, 212, 229, 293
Blanchard, John, 243
Blumer, Hermann, xiii
branch-node matrix, 63, **68**, 78, 117, 205
Brunelleschi, Filippo, 36
buckling, 22, 25, 31, 35, 91, 175, 242, 243, 279
load, 26, 41, 50, 93, 191, 242, 266–9

CAGD, *see* computer-aided geometric design
Candela, Félix, viii, 104, 247, 261–2
Catalan vault, 9
catenary, 2, 8, 28, 30, 35–6, 117, 122, 132, 135

Catmull-Clark subdivision, *see* subdivision surface
central difference form, 107
centroidal dual, 160, **167**, 204
CG, *see* conjugate gradient method
Cholesky decomposition, 109, 164, **167**
Christoffel symbol, 284
chromosome, 172, 229, 291, 295–6
circular arch, 29
classic Runge-Kutta method, *see* RK4
closed form expression, 162, **167**
clustering, 189, **192**, 293
Cohen-Vossen theorem, 25
colander, 22
collapse, 41, 52, 85, 91, 242
compatibility equation, 24, 62, 275, 279
compliance design, 212
computational morphogenesis, 3, 226, 231, **235**
computer-aided geometric design, 50–2
cone, 8
conformal parameterization, 185, **192**
conjugate gradient method, 109
    on the normal equations, 146
    with Jacobi preconditioning, 187
constant length, 122, 152
constant stress arch, 29
constitutive equation, 24, 62, 213, 275
constraint, 3, 50, 144, 148, 173, 175, **178**, 213, 294
construction tolerance, 266
contact face, 84
continuous variable, 213, 293
contravariant base vector, 283
convergence, 95, 97, 116, 122, 127
convex, parallel dual, 158, **167**
cooling tower, *see* hyperboloid of one sheet
coordinate difference, 64, 117
covariant base vector, 282
creep, 242

cross-bracing, 90
crossover, 172, 189, 292
curvature, 23
    anticlastic, *see* anticlastic curvature
    Gaussian, *see* Gaussian curvature
    Germain, *see* Germain curvature
    mean, *see* mean curvature
    synclastic, *see* synclastic curvature
cutting pattern, 48, 111, 143

damping, 107, 108, 125, 127
    kinetic, 95, 97
    viscous, 95, 97, 124
Day, Alistair, 93
deformation energy, 213
Delaunay triangulation, 175, **178**
design loading, 2, 48, 117
design noise, 46, 51
design patch, 51, 54
design space, 3, 46, 173–4, 212
design variable, 3, 226
developable surface, 25, 186
diagonal bracing, 91, 135
diagonal matrix, 63, 125, 205
diagram
    force, *see* force diagram
    form, *see* form diagram
    reciprocal, *see* reciprocal diagram
    Voronoi, *see* Voronoi diagram
diatom, 222
difference form, 94
    central, 107
    forward, 107
differential geometry, 281
dimensional analysis, 41, 93
direct approach, *see* force density method
directed graph, 79
direction cosine, 62, 64, 118
discrete Laplacian, 201, 205, **208**
discrete topology optimization, **178**
discrete variable, 212, 293
displacement, 275
domain integral based function, 213
dome, 8–9, 11, 17, 22, 25, 35–7, 221
dominance, 293
DR, *see* dynamic relaxation
drag, 108, 126

dual
    centroidal, *see* centroidal dual
    convex, parallel, *see* convex, parallel dual
    graph, 75, 167
dynamic equilibrium methods, 115, 128
dynamic relaxation, 89, **101**, 107, 123, 278

eigenfunction, 197, 201, **208**
eigenshell, 196
eigenvalue, 197, 242, 285
eigenvalue buckling load, *see* buckling load
eigenvector, 205, **208**, 285
Einstein summation convention, 283
elastic stiffness, 284
elastic stiffness matrix, 213, 278
elitism, 292
elliptic paraboloid, 261
equilibrium equation, 23–4, 30, 62, 116, 128, 275, 279, 289
error, 42, 86, 145, 268–9
Esquillan, Nicolas, 250
Euler method, 107, 109, 113
    backward, *see* backward Euler method
    semi-explicit, 107
evaluation, 189, 231
evolutionary algorithm, 4, 172
explicit integration, 26, 107, **113**, 125, 278

fabric guidework, 104, 111
falsework, 85–6, 91, 111, 251, 256–8, *see also* scaffolding
FDM, *see* force density method
FE, *see* finite element
filter, 46–7, 52
    radius, 52, 216
finite element, 53, 185, 187, 215
    analysis, 50–3, 189, 206, 209, 254, 257
    isoparametric, 53, 276
    method, 26, 274
first fundamental form, 283
fitness
    function, 172, 231, 235, 293
    landscape, **235**
flexibility, *see* mean compliance
flow of forces, 73, 83–4, 138–9, 166, 220
flux structure, 262
force density, 62, 65, **68**, 80, 106–7, 110, 127, 158

method, 59, 118–23, 128
    nonlinear, 128, 143, 162
force
    diagram, 72–3, **86**, 158, 160, 164
    pattern, 166
    polygon, 120
form
    diagram, 72–3, **86**, 158–9
    finding, 2, 48, 115, 131
form-active, 1, 22
form-found shape, 2, 34
form-passive, 1, 22
forward difference form, 107
forward substitution, 164
free material optimization, 185
free-curved shape, *see* freeform shape
freeform shape, 2, 86, 92, 181, 259, 271–2
funicular, 8, 15, 26–8, 34–8, 71–2, **86**, 157
    polygon, 27, 72, 120

GA, *see* genetic algorithm
Gabriel, Knut, x
Galerkin's method, 276
Galilei, Galileo, 35
Gaudí, Antoni, vii, 53–4, 132
Gauss's Theorema Egregium, 25, 286
Gauss–Jordan elimination, 119
Gauss–Newton's method, 145
Gaussian curvature, 25, 110, 285, 287
Gaussian elimination, *see* Gauss-Jordan elimination
general theory of relativity, 283
genetic algorithm, 171–2, **178**, 225, 227, 290
    multi-objective, *see* MOGA
    non-dominated sorting genetic algorithm II, 189
geometric stiffness, 94
    matrix, 279
    methods, 115–16
geometrical anisotropy, 45
geometrical shape, *see* mathematical shape, 63
Germain curvature, *see* mean curvature
Giambattista della Porta, 36
global optimum, 4
glulam, 150
Godzilla, 166
gradient, 64, 68, 108, 162
    descent, 162, **167**, 263

graph, **167**
    dual, 75, **167**, 204
    planar, 158, **167**
graph theory, 63
graphic statics, 72, **86**
Green's theorem, 280
grid
    layout, 90, 182
    quadrangulated, x, 93, 151, 197
    triangulated, 135, 184, 243–5
gridshell, 2, 89, **101**, 171, 181, 239, 259
    construction, 90, 99–100, 150, 153, 241
    multi-layer, *see* multi-layer gridshell
    strained, *see* strained gridshell
    unstrained, *see* unstrained gridshell
Gründig, Lothar, 143
Guastavino method of construction, 9
Gösling, Friedrich, 37

hanging
    chain, 7–8, 12, 34, 117
    model, 35–41, 48–50, 72, 152, 240, 253–6, 261
Happold, Ted, 240
Heyman, Jacques, 73
homogenization method, 185, 211, **223**
Hooke's law of
    elasticity, 7, 61–2, 66, 106
    inversion, 8, 34, 48
Hooke, Robert, 7, 9, 35
hypar, *see* hyperbolic paraboloid
hyperbolic paraboloid, ix–x, 250–7, 260, 270–1
hyperboloid of one sheet, 25, 282
Hübsch, Heinrich, 36

imperfection, 26, 54, 91, 266
    sensitivity, 63, 243
implicit integration, 26, 107–9, **113**, 126
incidence matrix, *see* branch-node matrix
indeterminacy, *see* static indeterminacy
inextensional deformation, 25, 45, 241
initial length, 60, 66, 94, 106, 116, 123–5, 128, *see also* rest length
instability, 45, 72
integration, 109, 116, *see also* Euler method, midpoint method, Runge-Kutta methods, Verlet method
    explicit, *see* explicit integration

implicit, *see* implicit integration
inversion, *see* Hooke's law of inversion
Isler, Heinz, x, 16, 31, 37–9, 48–9, 249–59
isogeometric analysis, 53
isoline, 204
isoparametric, *see* finite element
isosurface, 217
isotropy, 184, 213, 284
Ito, Toyo, 229, 262

Jacobian, 64, **68**, 108–9
Jenkins, Ronald, 243

Kelvin, Lord William Thomson, 25
kern, 74
kinematic equation, *see* compatibility equation
kinematic stability, 173, **178**
kinetic damping, 95–6
kinetic energy peak, 95–6

Lagrange multiplier, 144, **154**, 161
Lagrangian, *see* Lagrange multiplier
Laplace-Beltrami operator, 205, **208**
Laplacian, **208**
    discrete, 201, 205, 208
    graph, 205
    matrix, 201, 205
lattice shell, *see* gridshell
leapfrog method, 93–4, 109, 124, 128
least squares, 144–5, **154**, 161
    moving, *see* MLS
length 61, 64, 118
length-preserving parameterization, 185, **192**
Leonhardt, Fritz, x
line search, 162, **167**
linear buckling load, *see* buckling load
linear system, 60, 63, 108, 117–18, 145, *see also* BiCGSTAB, Cholesky decomposition, conjugate gradient method, Gauss–Jordan elimination
linearization, 62, 108, 148
local optimum, 3, 52
Loop subdivision, 296, *see also* subdivision surface
lower-bound theorem, 73

macroscopic stiffness, 231
Marg, Volkwin, x
masonry, 9, 71–4, 165

mass, 94, 107, 124–6
mass matrix, 276
mathematical shape, 3, 33
mean compliance, 212–13
mean curvature, 285
membrane,
    action, 25, 31, 45, 260, 288
    stiffness, 242
    stress, 1, 23–4, 90, 253, 288
method of Lagrange multipliers, *see* Lagrange multiplier
method of least squares, *see* least squares
Michelangelo, 36
micro-cell, 213–14
middle third rule, 74, **86**
midpoint method, 107, 109
minimal squared length net, 110
minimal surface, 110, 285
mixed variables, 293
MLS, 174, **179**
MOGA, 189, 293
Mohrmann, Karl, 37
Monge form, 287
Moore-Penrose pseudoinverse, 145
moving least squares, *see* MLS multi-layer gridshell, 90–1, 98–9, 151
multi-objective genetic algorithm, *see* MOGA
multi-objective optimization, 3, 50, 183, 193
mutation, 175, 177, 189–90, 292
Mutschler, Carlfried, 40

Nervi, Pier Luigi, viii, 195, 212, 259
Newton's second law, 94, 106–7, 275
Newton–Rhapson's method, 108, 113, 116, 145, 148
noise, 197 201
    design, *see* design noise
non-dominance, 293
non-dominated set, *see* Pareto front
non-uniform rational basis spline, *see* NURBS
nonlinear programming, 50
normal equations, 144–5, **154**
normal matrix, 145
normal stress, 23
NP, *see* nonlinear programming
NURBS, 2, 51, 53, 81, 104, 111, 181, 185, **192**, 197, 202, 230, 263, 274, 276, 295

objective, 3, 45–6, 50, 181, 183, 189–90, 212–13
    function, 3, 167, 171–3, **178**, 201, 227
oculus, 8–9, 11, 23, 25
optimality condition, 144
optimization, 3–4, 45, 48, 50
    problem, 3, 80, 144–5, 154, 160–1, 173
ordinary differential equation, 113
orthotropy, 48, 184, 213
Otto, Frei, ix–x, xii, 38–41, 49, 151, 242
overdetermined system, 144

parabola, 8, 28, 119, 122–3
parabolic arch, 23
parallel dual graph, 75, 158, 167
parameterization, 50, 197
    angle-preserving, 185, **192**
    CAGD-based, 50
    conformal, 186, **192**
    FE-based, 52
    length-preserving, 186, **192**
    surface, 185, **192**
parametric definition, 226, 230, **235**
Pareto
    front, 3, 16, 183, 293
    optimal, 182–3, 189, 191–**2**, 293
    optimal set, *see* Pareto front
particle-spring method, 104, **113**, 116, 125, 131, 135
pattern optimization, 195–7
pattern scaling, 197, 200, 204
PCGM, *see* conjugate gradient method with Jacobi
    preconditioning
penalty power, 215
performance objective, *see* objective
periodic global reparameterization algorithm, 186,
    197, 203, 207
Peterson-Mainardi-Codazzi equations, 278, 286
physical
    form finding, 33, *see also* hanging model
    model, 33, 85, 93, 240, 243, 261
pixel, 212–13, **223**
planar graph, 158, 167
plane stress, 23
plate bending, 23
pneumatic form, 251
Poleni, Giovanni, 36–7
Poppensieker, Wilhelm, 240
population, 172, 189, 290

porosity, 215
principal curvature, 25, 30, 186, 202, 208, 285
principal stress, 186, 197, 202, 206, 208, 281, 285
proper cell decomposition of the plane, 158
PS, *see* particle-spring method

quadrangulation, *see* grid
quadratic program, 144, **154**
quasi-Newton Broyden-Fletcher-Goldfarb-Shanno
    method, *see* BFGS

rainflow analogy, 83, 138
Rayleigh, Lord John William Strutt, 25
RC, *see* reinforced concrete
reciprocal diagram, 11, 72–6, 79, **86**, 120, 158, 164,
    167
regularization, 47
reinforced concrete, 259
reinforcement, 112, 255, 265
reparameterization, 52, 186, 203, 207
reproduction, 293–4
residual, 145
residual force, 94, 106–7, 122, 124, 144, 154
rest length, 106, 110, 116, 125–7, 132, 134–5,
    see also initial length
reticulated shell, *see* gridshell
Riemann-Christoffel tensor, 287
RK4, 109, 117, 126, 128
rod spacing, 186
Runge-Kutta methods, 127, 203, *see also* RK4

SA, *see* sensitivity analysis, *see* simulated
    annealing
Saarinen, Eero, 51, 117, 261–2
safe theorem, 73 , 165
safety factor, 264
Sasaki, Mutsuro, 226, 229, 259
scaffolding, 52, 99, 241, 256–8, *see also* falsework
scalar potential function, 186
Schlaich, Jörg, iix
scripting, **235**
sculptural shape, *see* freeform shape
search space, 173–4, 226, 230
Segelschalen, 53
seismic force, 265
selection, 293–4
semi-explicit Euler method, 107

sensitivity analysis, 230, **235**, 261
shape dependent loading, 122
shape function, 50, 53, 276
shape modifier, 197
shape optimization, 4, 48, 50, 200
shear,
    force, viii, 96
    resistance, 48
    stiffness, 90–1
    stress, 23, 202, 213
    strain, 279
shell, 1, 21
    action, 23, 26, 92
    construction, 99
    decline of, 259
sieve, 22, 93
simulated annealing, 197
singularity, 204
sizing optimization, 4
smoothing, 47, 297
spectral method, **208**
spherical dome, 8, 25, 221, *see also* dome
spline, 89, 97–9, **101**
    element, 96, 125
spring damping, 125
spring stiffness, 106, 110, 127, 132
square-cube law, 35
stability analysis, 266–7
static indeterminacy, 11, 24, 75–7, 133
steel, x, 92, 171, 175, 242
Stevin, Simon, 35
stiffness
    bending, 22, 25–6, 46, 93–4, 127, 244–5
    membrane, 23–4, 244–5
    matrix, 125, 173, 213–5, 280–1
    matrix methods, 115
stone cutting, 84
strain, 24, 62, 123, 213, 215, 227, 275, 279, 286
    energy, 50, 227, 231, 261, 279
strained gridshell, 89–90, 97–8, **101**, 125
stress, 23–4, 275
    membrane, *see* membrane stress
    normal, *see* normal stress
    plane, *see* plane stress
    shear, *see* shear stress
strong Wolfe conditions, 162, **167**
Structural Expressionism, 259

structural
    feedback, 136
    optimization, 3–4, 50, 293
    pattern, 195, 202
Stuttgart direct approach, *see* force density method
subdivision surface, 52, 104, **113**, 295
support condition, 133, 211, 218
surface
    coordinate system, 185
    developable, 25, 186
    of revolution, 25, 30, 92
    parameterization, 185, **192**, 281
synclastic curvature, 25, 110, 285
system of linear equations, *see* linear system

Tange, Kenzo, 259
Taylor series, 108, 148
tension coefficient, *see* force density
tensor, 284
tessellation, 83
Theorema Egregium, *see* Gauss's Theorema Egregium
thrust line, 72–4
thrust network, 74–6, **86**, 158
thrust network analysis, 11, 71, 116, 120, 157
timber, 2, 98, 243–5
TNA, *see* thrust network analysis
topology, 2, 4, 63, 78, 135, 173
topology optimization, 4, 211, *see also* discrete topology optimization
Torroja, Eduardo, 250, 260
trade-off set, *see* Pareto front
translational surface, 92
triangulation, 26, *see also* Delaunay triangulation
tributary load, 80, 174–5
Tsuboi, Yoshikatsu, 259

umbrella form, 251
underdetermined system, 144, 146
uniform stress shell, 30
unit cell, 184, **192**
unstrained gridshell, 90–1, **101**,
Utzon, Jørn, 39

valency, 76
velocity, 94–5, 106–7, 124, 126, 279
    relative, 125
    virtual, 288

Verlet method, 26, 101, 125, 278
Vierendeel action, 242
virtual work, 275, 279, 288
viscous damping, 95, 124
volume fraction, 215, 217, **223**
Voronoi diagram, 80, 174, **179**
voussoir, 36, 72–3, 83–6

voxel, *see* pixel

weighting, 3, 145, 149
Wren, Christopher, 9, 35

zero-length spring, 110, 126